Special Publication No. 10
of the Society for Geology
Applied to Mineral Deposits

Sediment-Hosted Zn-Pb Ores

Edited by

L. Fontboté M. Boni

With 144 Figures, Some in Colour

Springer-Verlag Berlin Heidelberg GmbH

Prof. Lluís Fontboté
Département de Minéralogie
Université de Genève
13, rue des Maraîchers
1211 Genève
Switzerland

Prof. Maria Boni
Dipartimento di Scienze della Terra
Università di Napoli
Largo S. Marcellino 10
80138 Napoli
Italy

ISBN 978-3-662-03056-1

Library of Congress Cataloging-in-Publication Data. Sediment-hosted Zn-Pb ores / edited by L. Fontboté, M. Boni. p. cm. – (Special publication no. 10 of the Society for Geology Applied to Mineral Deposits) Includes bibliographical references and index.
ISBN 978-3-662-03056-1 ISBN 978-3-662-03054-7 (eBook)
DOI 10.1007/978-3-662-03054-7
1. Zinc ores.
2. Lead ores. 3. Metallogeny. 4. Sedimentation and deposition. I. Fontboté, L. (Lluís) II. Boni, M. (Maria), 1948– . III. Series: Special publication ... of the Society for Geology Applied to Mineral Deposits; no. 10. QE390.2.Z54S43 1994 553.4′52—dc20 93-43690

Originally published by Springer-Verlag Berlin Heidelberg New York in 1994
Softcover reprint of the hardcover 1st edition 1994

Typesetting: Best-set Typesetter Ltd., Hong Kong
32/3130/SPS – 543210 – Printed on acid-free paper

Preface

Sediment-hosted zinc-lead ores, among numerous other subjects, have always been one of the main interest of Professor Dr. Dr. h. c. G. Christian Amstutz. The classical books, *Sedimentology and Ore Genesis*, edited by G.C. Amstutz in 1964, and *Ores in Sediments*, edited together with A.J. Bernard in 1973, as well as the chapter of Amstutz and Bubenicek entitled, "Diagenesis in sedimentary mineral deposits", contained in the book, *Diagenesis in Sediments*, edited by Larsen and Chillingar in 1967, illustrate the strong impulse given by Amstutz to the study of ores in sedimentary rocks; certainly one of the most innovative aspects in his still very active career. Concepts such as facies analysis, paleogeography, diagenesis, and basin evolution had to be slowly introduced into the metallogenetic world, which at that time was still mainly interested in "hard rocks". Christian Amstutz, as teacher, author and editor, as well as cofounder of the "Society for Geology Applied to Mineralium Deposits" and managing editor for the first 6 years of its journal *Mineralium Deposita* (1966–1971), contributed decisively to the change of perspective in the study of ores hosted in sedimentary rocks, particularly of those consisting of base metals.

As Professor G. Christian Amstutz retired at the end of March 1991 from his position as Director of the Mineralogisch-Petrographisches Institut of the University of Heidelberg, which he had held with remarkable success since 1964, he expressed the wish that no celebration be organized. During his career, Christian Amstutz provided intellectual stimulation, an open scientific environment and friendship to a large number of colleagues and students throughout the world. The occasion of his 70th birthday was seen by former students, coworkers and colleagues as an opportunity to demonstrate their appreciation and gratitude through a volume which should update, in the perspective of 1992, the knowledge of some aspects of sediment-hosted zinc-lead deposits. The present volume was presented to Professor Amstutz in form of a manuscript during a colloquium held in Heidelberg on November 27, 1992, the date of his 70th birthday.

We wish to thank all the contributors for their papers and the following reviewers for their thoughtful comments and suggestions: J. Barbier, T. Bechstädt, M. Götzinger, I. Hedley, B.

Henry, S.E. Kesler, D. Large, P. Lattanzi, D. Leach, R. Lehne, P. MacArdle, R. Moritz, J. Parnell, W. Püttmann, K. Shelton, H.-J. Schneider, J. Spangenberg, H. Sperling, E. Stumpfl, J.-F. Sureau, and E. Walcher. The cooperation of the staffs of the Departments of Mineralogy of the Universities of Geneva and Naples, in particular Ms. Jacqueline Berthoud and Mr. Michael Doppler, is warmly acknowledged. Finally, we would like to express our special thanks to Dr. W. Engel, Springer-Verlag, Heidelberg, for the excellent cooperation during the preparation of this book.

Geneva, Switzerland
Napoli, Italy
December 1993

L. Fontboté
M. Boni

Contents

Australian, Chinese, and North African Deposits

Exploration and Economics

List of Contributors

You will find the addresses at the beginning of the respective contribution

Introduction and General Aspects

Sediment-Hosted Zinc-Lead Ores – An Introduction

L. Fontboté[1] and M. Boni[2]

1 Introduction

Sediment-hosted ore deposits are the main source of lead and zinc, representing more than 50% of the world production and more than 60% of the reserves of these metals (see Wellmer et al., this Vol.). They include two main types of ore deposits, the definitions of which in part overlap (Fig. 1).

1. Sediment-hosted massive sulfide deposits of Zn and Pb (Large 1980, 1988; Sangster 1983). These massive to semimassive deposits occur in extensional basins characterized by a strong thermal subsidence and are hosted by clastic sedimentary rocks, mainly shales (hence, also the term "shale-hosted") but also by other sedimentary lithologies, including carbonate rocks. They are typically fine-grained and stratiform down to the hand-specimen scale. They are also commonly referred to as "sediment-hosted stratiform deposits" or "sedimentary-exhalative lead-zinc", or "sedex" deposits because they are considered to mainly have formed through exhalation of basinal brines at the sea floor and/or "inhalation" into poorly consolidated sediments below the surface.
2. Mississippi Valley-type ore deposits (e.g. Anderson and Macqueen 1988). These stratabound zinc-lead (F-Ba) deposits are mainly, but not exclusively, hosted by carbonate rocks hosted by carbonate rocks deposited at the margins of tectonically stable platforms. Even if at the deposit scale they may present in places tabular morphologies parallel to bedding, closer examination often reveals cross-cutting relationships at outcrop and hand-specimen scale. In comparison with the former type, they are generally coarser-grained. Vein-type morphologies with similar mineral assemblages are considered to be the analogous product of mineralization processes. Mississippi Valley-type deposits are usually interpreted as having been formed by precipitation from saline hot basinal brines during burial diagenesis and later evolutionary stages of the host rocks.

A detailed comparison between these two types of ore deposits has been published by Sangster (1990) which gives abundant references. Additional recent reviews can be found in Eidel (1991) and Russell and Skauli (1991).

[1] Départment de Minéralogie, rue des Maraîchers 13, 1211 Géneve, Switzerland
[2] Dipartimento di Scienze della Terra, Universitá di Napoli, Largo S. Marcelliao 10, 80138 Napoli, Italy

Spec. Publ. No. 10 Soc. Geol. Applied to Mineral Deposits
Fontboté/Boni (Eds.), Sediment-Hosted Zn-Pb Ores

This volume dedicated to Prof. Dr. G.C. Amstutz on the occasion of his 70th birthday presents recent achievements in the research on zinc-lead deposits. In fact, a significant part of G.C. Amstutz's scientific research, as well as teaching activity/PhD tutorship, has been dedicated to Zn-Pb deposits (e.g. Amstutz 1958, 1959, 1982; Amstutz et al. 1964; Amstutz and Bubenicek 1967; Amstutz and Park 1967, 1971; Amstutz and Bernard 1973; Amstutz et al. 1982).

2 Classification Problems

Sediment-hosted massive sulfide deposits of zinc and lead, in particular if they display a clear stratiform geometry, are often referred to as "sedimentary-exhalative" or "sedex" deposits. Quoting Brown (1989, p. 41) we also think that "most geologists would agree that a deposit-type name should normally be based on descriptive rather than genetic terms". We prefer, for this reason, to avoid as much as possible the use of the genetic term "sedex" as a deposit type. The example of the Irish deposit of Navan, which bears many of the characteristics of sediment-hosted stratiform deposits, but which does not appear to be completely "sedimentary-exhalative" but rather largely a product of replacement (Ashton et al. 1986), should prevent excessive eagerness in attributing genetic names to deposit types.

On the other hand, it could be argued that since the term Mississippi Valley type implies generally an epigenetic origin, the genetic neutral term "carbonate-hosted lead-zinc" deposits should be preferred. This would present two problems. Firstly, there are several significant examples of MVT districts hosted in carbonate units developed as transgressive sequences overlying a detrital base which in places is also ore-bearing. This is the case, for instance, in southwest Missouri concerning the Lamotte sandstone-hosted Old Lead Belt and the carbonate-hosted Viburnum Trend. A similar situation can be recognized in southern France where the Triassic sandstone-hosted lead-zinc deposit of Largentière (Foglierini et al. 1980) and the nearby carbonate-hosted Les Malines district (Charef and Sheppard 1988) occur. Both these clastic- and carbonate-hosted deposits share many geometrical and geochemical characteristics and their genesis is normally regarded to involve different aspects of an essentially similar process. Thus, current use considers sandstone-hosted ore deposits of this kind as a variant of the Mississippi Valley type (Bjørlykke and Sangster 1981). Another obvious reason for not using the descriptive "carbonate-hosted zinc-lead" as a synonym of Mississippi Valley type is that several examples of sediment-hosted massive sulfide deposits are also hosted in carbonate rocks. For example, we should mention again the Irish deposits zinc-lead province, which bear "prototypes" of sedimentary-exhalative types as Silvermines (Andrew 1986), which is mainly carbonate-hosted.

In Europe, an additional ambiguity exists regarding the Mississippi Valley-type definition. In occasions, the term "Alpine-type" or "Bleiberg-type" is used for lead-zinc deposits in the Eastern Alps, e.g., Bleiberg-Kreuth in Austria, Mežica in Slovenia and Raibl and Salafossa in Italy, which are hosted by Middle-Upper Triassic carbonate successions. A problem with this is that the term "Alpine-type" has often been accompanied by a genetic connotation.

Several authors (e.g. Schneider 1964; Brigo et al. 1977; Klau and Mostler 1983) interpreted the Alpine lead-zinc carbonate-hosted deposits at least in part syngenetically in opposition to the American Mississippi Valley-type deposits which were admittedly epigenetic. However, recent investigations on Bleiberg and on other deposits in the Eastern Alps (e.g. Zeeh and Bechstädt, this Vol. and references therein) show that the main characteristics of these ore deposits are similar to the typical MVT and that also the Alpine deposits appear to have formed long after sedimentation of the carbonate host rock.

Thus, we prefer to include these deposits under the general term "Mississippi Valley-type" without referring specifically to an "Alpine-type". In this sense, already Sawkins (1990, p. 322) writes that the "Alpine-type" deposits "... appear to be of fairly typical Mississippi Valley-type affiliation, but their tectonic setting can be related to widespread rifting events in the area".

The main difficulties in the classification derive from the fact that the two types mentioned above are two end members within the broad group of sediment-hosted zinc-lead deposits. Sangster (1990) in his "comparative examination of Mississippi Valley-type and sedex lead-zinc deposits" concluded that the two types "... are most dissimilar in their morphological characteristics. In all other features examined, however, large overlaps exist, probably reflecting the common ultimate derivation of both deposit types from fluids that emanated from sedimentary basins". The example of the Irish zinc-lead province with its coexistence of exhalative and diagenetic replacement ores

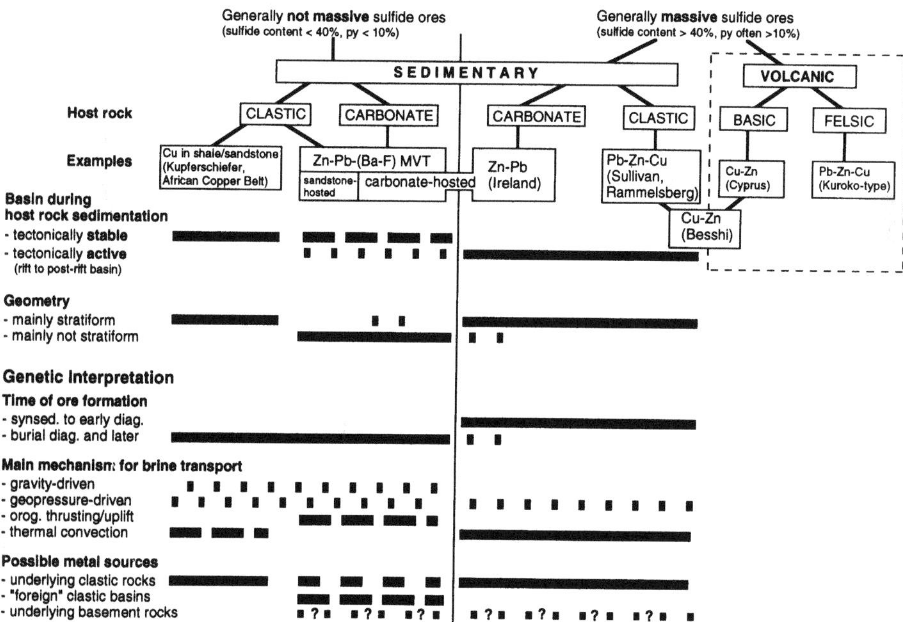

Fig. 1. Selected characteristics and current genetic hypotheses of sediment-hosted base metal deposits. Stratabound volcanic-associated ore deposits are shown for comparison. Ore deposit classification modified from D. Large (pers. comm., 1985)

supports Sangster's conclusion. In the present volume two contributions may provide additional examples of possible coexistence of exhalative-sedimentary processes and epigenetic late diagenetic replacement. They concern deposits in the Lower Cretaceous Basque-Cantabrian basin (Velasco et al., this Vol.) and the carbonate sequences of the same age in the Tunesian Atlas (Orgeval, this Vol.).

A problem occurring when only descriptive terms are considered, however, is that obvious similarities with deposits containing other commodities not included in the description may be overlooked. It appears that processes essentially equivalent to those controlling lead-zinc deposits may not only form fluorite and barite ores, but also iron deposits as in the Basque-Cantabrian basin (Velasco et al., this Vol.). In this sense, the iron ores could be regarded as "MVT Fe" deposits, similar to the term "MVT F-Ba" proposed by Sangster (1990, p. B23) for the F-(Ba) dominant districts in Illinois, Kentucky, and the English Pennines (see also Jones et al., this Vol.).

The existence of transitions to and similarities with other types of deposits, in particular sediment-hosted stratiform copper deposits (Kirkham 1989), should be underlined. Some of the mechanisms of brine migration proposed for the Kupferschiefer and red-bed deposits may be not very different than those forming sediment-hosted zinc-lead deposits.

That the classification into types, although necessary, is particularly difficult when discussing sediment-hosted lead-zinc deposits, can be seen in the regional syntheses included in the present book. Legge and Lambert (this Vol.), in their comprehensive critical review of Australian world class sediment-hosted Zn-Pb deposits, like Broken Hill, Mount Isa, Hilton, McArthur River, and Century, build the conceptual bridge to other types of ore deposits, in particular volcanic-associated massive sulfides. A review of the less well known sediment-hosted Pb-Zn deposits in China is presented by Song (this Vol.).

3 Fluid Transport, Precipitation Mechanisms, and Other Genetic Considerations

Consensus exists that most sediment-hosted zinc-lead deposits were formed by hot metalliferous saline basinal brines (Sverjensky 1986; Hanor 1987; Anderson and Macqueen 1988; Sangster 1990). The different morphologies would mainly result from different times of hydrothermal discharge, the stratiform deposits being typically a product of early diagenetic and exhalative-sedimentary processes, the MVT deposits having formed later in the evolution of the basin, under considerable burial.

A main point are the transport mechanisms of the ore-forming brines. This subject is dealt with in several contributions of this volume. The main accepted mechanisms of transporting brines in sedimentary basins are geopressure drive, thermal convection, and gravity drive.

Compaction-driven migration firstly proposed in Pine Point by Jackson and Beales (1966) has been subsequently shown (e.g. Cathles and Smith 1983) as not explaining the temperature gradients between fluid and host rock recognized in many ore deposits, because of possible thermal reequilibration between fluids and host rock. Other authors suggest episodic dewatering from geopressure zones enabling rapid fluid transport and thus preventing thermal reequilibration, e.g. through seismic pumping (Sibson et al. 1975). Fowler and Anderson (1991) suggest that geopressure zones can act as proximal sources of hot mineralizing fluids in shale-dominated basins. In these models the fluids could be rapidly injected from depth, thus conserving their heat.

It should be noted that compaction-driven fluid migration was a mechanism already proposed by different authors in the 1950s and 1960s. For instance, Amstutz (1964, Fig. 5) explains "late diagenetic galena fillings" in Missouri through fluids moving as the result of differential compaction between shales and carbonate rocks. Another example of geological phenomena possibly controlled by overpressure-driven base metal-bearing fluids is the Decaturville sulfide breccia in Missouri (Zimmerman and Schidlowski, this Vol.).

In sediment-hosted massive sulfide ores, which form mainly in tectonically active rift to post-rift extensional basins (Large, 1988), thermal convection is possibly the main fluid transport mechanism. Russell (1988) and Russell and Skauli (1991) propose that the Irish lead-zinc deposits and the F-dominated district in the English Pennines were formed by brines circulating in convective cells at great depths, affecting the basement. This assumption, which in the case of the ore deposits in Ireland is mainly based on the need of temperatures higher than those reached at the bottom of the basin, is discussed and partly rejected by Jones et al. (this Vol.) on the basis of the existence of high geothermal gradients during the formation of the Irish deposits. According to Jones et al., the Irish deposits formed during early Carboniferous crustal extension and were associated with high geothermal gradients. In contrast, the Pennine deposits would have formed during a period of a declining geothermal gradient. As illustrated by the example of the Cobar basin deposits in Australia, discussed in the contribution of Legge and Lambert (this Vol.), the question whether the basement is permeable enough to enable effective mass transport and can serve as an ore source, is one of the main issues presently under debate.

For a long time it was considered that MVT deposits did not have any relationship with orogenic events. A different view emerged from the studies of Leach and Rowan (1986) and Oliver (1986, 1992), who suggested that the formation of foldbelts is an effective mechanism to initiate the migration of large amounts of fluids. Gravity-driven migration (Garven 1985; Bethke 1986) may be caused by hinterland recharge at a tectonic uplifted hydraulic head. Kesler (this Vol.) evaluates the possible importance of regional thrusting as a mechanism controlling brine transport. Taking the example of the Appalachian-Caledonian orogen, he supports the hypothesis that large-scale thrusting is the dominant factor for expulsion of basinal brines and the development of MVT mineralization in the East Tennessee area of the southern Appalachians.

Further evidence in this respect is given again by Leach (this Vol.) in his contribution on the Ozark region in the United States which includes the Old Lead Belt, the Viburnum Trend, and the Tri-State deposits and other smaller districts. He presents multidisciplinary data supporting a large-scale regional fluid migration in Late Paleozoic in response to convergent plate tectonics in the Ouachita foldbelt.

In contrast to "sedex" deposits where fluid transport is essentially vertical along synsedimentary faults, Sangster (1990) notes that large-scale lateral fluid transport as great as several hundred kilometers would help to explain, for example, anomalously radiogenic leads determined in a number of MVT deposits. Without denying the existence of large-scale lateral brine transport, it should be pointed out that many MVT deposits are locally controlled by high-angle fractures (e.g. Leach, this Vol.; Jones et al., this Vol.; Pelissonier 1967; Rowe et al. 1993). The recent discovery of the carbonate-hosted Lisheen Zn-Pb-Ag deposit in Ireland (Hitzman 1992) illustrates the essential role played by fractures for fluid transport at the district scale. In addition, it should be noted that vertical transfer along faults is one of the few mechanisms to produce significant thermal anomalies in sedimentary basins (e.g. Vasseur and Demongodin 1993).

Diapiric salt structures may also vertically channel hydrothermal fluids. The relationship between zinc-lead ores and salt diapirs is the topic dealt with in the contributions by Posey et al. (this Vol.) and Orgeval (this Vol.). Although the spatial relationship between salt domes and base metal deposits has been recognized for a long time, in particular in southern Europe and North Africa (see references in Rouvier et al. 1985 and in Nicolini, 1990), only recently have they been studied in detail. Posey et al. (this Vol.) note that although known salt dome-hosted zinc-lead metal deposits are of modest grade and tonnage, they provide an excellent laboratory for the study of other classes of sediment-hosted base metal deposits. This is mainly because the evolution of the salt domes gives more precise physicochemical and time constraints for the ore formation than is usually the case in other environments. Orgeval (this Vol.) presents the first extensive description of the peridiapiric metal concentrations of the Bou Grine Zn-Pb deposit (Tunesian Atlas) which will enter in production shortly.

The different hypotheses suggested to explain fluid transport are difficult to test because of the problems faced when trying to date directly the mineralization process. Symons and Sangster (this Vol.) give an overview of the different analytical dating methods applied to MVT deposits and present a summary of paleomagnetic methods that have been proven successful in selected North American examples. From their results the age of mineralization can be correlated with an adjacent orogenic event, thus supporting the gravity-driven flow from an adjacent orogenic uplift.

Two contributions in this volume address the role played by organic matter in the precipitation mechanisms of zinc-lead carbonate-hosted ore deposits. Gize and Barnes (this Vol.) note that MVT deposits appear to be associated with thermally altered Type I kerogen of phytoplankton origin but not to Type III kerogen derived from terrestrial plant organics. Although Type I kerogens are inadequate to cause ore deposition, they may play an important role by selectively complexing metals. Spirakis and Heyl (this Vol.) point out the importance of thiosulfates, as they can be transported together with metals without sulfide precipitation. According to the model presented in their contribution, where the thiosulfate-bearing hot solutions encounter organic matter, thiosulfates may be reduced to provide reduced sulfur for precipitation of sulfides. It can be added that Kucha and Viaene (1993) report peak shifts in microprobe analyses, indicating the presence of thiosulfates in different MVT deposits.

4 Textural Aspects, Geochemistry, and Exploration

One of the main scientific interests of G.C. Amstutz, is the objective description of geometric relationships in ores, in particular the study of ore textures. In the present book this field is less developed compared to other aspects. However, a detailed textural study is found in the contribution by Sass-Gustkiewicz and Mochnacka (this Vol.) on certain rhythmic ore fabrics in the Upper Silesian zinc-lead deposits.

It is often disregarded that minor occurrences of zinc and lead sulfides occur, without an apparent association to ore deposits, as a "normal" result of basin evolution in different geological situations. Two contributions illustrate this. Friedman (this Vol.) reports on dolostone-hosted sulfide occurrences in Silurian strata of the Appalachian basin of New York. Swaine (this Vol.) presents a review of galena and sphalerite occurrences hosted by coal seams.

Two contributions of this volume try to characterize geochemically ore and host rock of some ore deposits in the Triassic of the Eastern Alps. Brigo and Cerrato (this Vol.) present a case study on the fracture-controlled lead-zinc deposit of Raibl, northern Italy. They are able to recognize a distribution zoning for Ge, Cd, Ga, Tl, As, and Sb which can be correlated with major structures controlling the ore deposit. Schroll et al. (this Vol.), on the basis of abundant trace element and isotope data available for the MVT deposit of Bleiberg (Austria), introduce the concept of "geochemometry" in an attempt to characterize geochemically ore and host rock with the help of statistical tools.

One of the main difficulties in defining exploration programs for carbonate-hosted ore deposits is the frequent absence of easily interpretable alteration patterns. Lavery et al. (this Vol.) present an extensive summary of lithogeochemical and other geochemical investigations applied to the exploration for sediment-hosted Zn-Pb deposits. They present a broad survey of geochemical signatures commonly considered to be associated with the formation of ore and observe that designing an exploration program based on the mechanical superposition of a large number of *necessary* anomalies may lead to failure. Rather, the recognition of the sufficient anomaly related to the ore-forming processe(s) may be the key issue.

In conclusion, research on sediment-hosted zinc-lead deposits increasingly shows that we are confronted with different aspects of the global process of ore precipitation from fluids at different stages of basin evolution. Therefore, integrative approaches interfacing research on the origin and nature of fluids, fluid transport, and interaction with host rock, both in the hydrocarbon and ore mineral fields, as attempted in the Geofluids Conference in Torquay (Parnell et al. 1993), may be a step in the right direction. This research direction was decisively influenced by G.C. Amstutz, who 30 years earlier, in 1963, first brought earth scientists of different backgrounds together with economic geologists at a meeting of the International Sedimentological Congress in Delft (Amstutz 1964).

References

Amstutz GC (1958) Syngenetic zoning in ore deposits. Proc Geol Assoc Can 11:95–113

Amstutz GC (1959) Syngenese und Epigenese in Petrographie und Lagerstättenkunde. Schweiz Miner Petrogr Mitt 39:1–84

Amstutz GC (ed) (1964) Sedimentology and ore genesis. Elsevier, Amsterdam, 184 pp

Amstutz GC (1982) Preface and outlook. In: Amstutz GC, El Goresy A, Frenzel G, Kluth C, Moh G, Wauschkuhn A, Zimmermann R (eds) Ore genesis, the state of the art. Springer, Berlin Heidelberg New York, pp v–x

Amstutz GC, Bernard AJ (eds) (1976) Ores in sediments. Springer, Berlin Heidelberg New York, 350 pp

Amstutz GC, Bubenicek L (1967) Diagenesis in sedimentary mineral deposits. In: Larsen G, Chillingar GV (eds) Diagenesis in sediments. Elsevier, Amsterdam, pp 417–475

Amstutz GC, Park WC (1967) Stylolites of diagenetic age and their role in the interpretation of the Southern Illinois Fluorspar District. Miner Deposit 2:44–53

Amstutz GC, Park WC (1971) The paragenetic position of sulfides in the diagenetic crystallization sequence. Soc Mining Geol Jpn (Spec Issue 3):280–282

Amstutz GC, Ramdohr P, El Baz F, Park WC (1964) Diagenetic behaviour of sulphides. In: Amstutz GC (ed) Sedimentology and ore genesis. Elsevier, Amsterdam, pp 65–90

Amstutz GC, El Goresy A, Frenzel G, Kluth C, Moh G, Wauschkuhn A, Zimmermann R (eds) (1982) Ore genesis, the state of the art. Springer, Berlin Heidelberg New York, 804 pp

Anderson GM (1991) Organic maturation and ore precipitation in southeast Missouri. Econ Geol 86:909–926

Anderson GM, Macqueen RW (1988) Mississippi Valley-type lead-zinc deposits. In: Roberts RG, Sheahan PA (eds) Ore deposit models. Geoscience Canada, Reprint Series 3, pp 79–90

Andrew CJ (1986) The tectono-stratigraphic controls to mineralization in the Silvermines area, County Tipperary, Ireland. In: Andrew CJ, Crowe RWA, Finlay S, Pennell WM, Pyne JF (eds) Geology and genesis of mineral deposits in Ireland. Irish Assoc Econ Geol, Dublin, pp 377–417

Ashton JH, Downing DT, Finlay S (1986) The geology of the Navan Zn-Pb orebody. In: Andrew CJ, Crowe RWA, Finlay S, Pennell WM, Pyne JF (eds) Geology and genesis of mineral deposits in Ireland. Irish Assoc Econ Geol, Dublin, pp 243–280

Beales FW, Jackson SA (1966) Precipitation of lead-zinc ores in carbonate rocks as illustrated by Pine Point ore field. Can Inst Mining Metallur Transa 79:B278–B285

Bethke CM (1986) Hydrologic constraints on the genesis of the Upper Mississippi Valley mineral district from Illinois basin brines. Econ Geol 81:233–249

Bjørlykke A, Sangster DF (1981) An overview of sandstone lead deposits and their relation to red-bed copper and carbonate-hosted lead-zinc deposits. Econ Geol, 75th Anniv Vol: 179–213

Bogacz K, Dzulynski S, Haranczyk C, Sobczynsky P (1973) Sphalerite ores reflecting the pattern of primary stratification in the Triassic of the Cracow-Silesian region. Ann Soc Geol Pol (Krakow) 43, 3:285–300

Brigo L, Kostelka L, Omenetto P, Schneider HJ, Schroll E, Schulz O (1977) Comparative reflections on four Alpine Pb-Zn deposits. In: Klemm DD, Schneider HJ (eds) Time and strata-bound ore deposits. Springer, Berlin Heidelberg New York, pp 273–293

Brown AC (1989) Sediment-hosted stratiform copper deposits: deposit-type name and related terminology. In: Boyle RW, Brown AC, Jefferson CW, Jowett EC, Kirkham RV (eds) Sediment-hosted stratiform copper deposits. Geol Assoc Can, Spec Pap 26:39–51

Cathles LM, Smith AT (1983) Thermal constraints on the formation of Mississippi Valley-type lead zinc deposits and their implications for episodic basin dewatering and deposit genesis. Econ Geol 78: 983–1002

Charef A, Sheppard SMF (1988) The Malines Cambrian carbonate-shale-hosted Pb-Zn deposit, France: thermometric and isotopic (H, O) evidence for pulsating hydrothermal mineralization. Miner Deposit 25:86–95

Eidel JJ (1991) Basin analysis for the mineral industry. In: Force ER, Eidel J, Maynard JB (eds) Sedimentary and diagenetic mineral deposits: a basin analysis approach to exploration. Rev Econ Geol, Society of Economic Geologists (El Paso) 5:1–15

Foglierini F, Samama JC, Rey M (1980) Le gisement stratiforme de Largentière (Ardèche) Pb (Ag, Zn, Sb). 26th Int Geol Congr, Paris, Gisements Français, Fasc E-4, 55 pp

Fowler AD, Anderson MT (1991) Geopressure zones as proximal sources of hydrothermal fluids in sedimentary basins and the origin of Mississippi Valley-type deposits in shale rich sediments. Trans Inst Mining Metallur B 100:B14–B18

Garven G (1985) The role of regional fluid flow in the genesis of the Pine Point deposit, Western Canada Sedimentary Basin. Econ Geol 80:307–324

Gayer R (1993) The effect of fluid over-pressuring on deformation, mineralisation and gas migration in coal-bearing strata. In: Parnell J, Ruffell AH, Moles NR (eds) Geofluids '93. 4th–7th May 1993, Torquay, England, pp 186–189

Hanor JS (1987) Origin and migration of subsurface sedimentary brines. Society of Economic Geologists and Paleontologists. Lecture notes for short course No 21, 247 pp

Hitzmann MV (1992) Discovery of the Lisheen Zn-Pb-Ag deposit, Ireland. SEG Newsletter, April 1992, No 9, pp 1–15

Kirkham RV (1989) Distribution, settings, and genesis of sediment-hosted stratiform copper deposits. In: Boyle RW, Brown AC, Jefferson CW, Jowett EC, Kirkham RV (eds) Sediment-hosted stratiform copper deposits. Geol Assoc Can Spec Pap 26:3–38

Klau W, Mostler H (1983) Alpine Middle and Upper Triassic Pb-Zn deposits. In: Kisvarsanyi G, Grant SK, Pratt WP, Koenig JW (eds) International Conference on Mississippi Valley type lead-zinc deposits. Proc Vol, Rolla, University of Missouri Rolla, pp 113–128

Kucha H, Viaene W (1993) Compounds with mixed and intermediate sulfur valences as precursors of banded sulfides in carbonate-hosted Zn-Pb deposits in Belgium and Poland. Miner Deposit 28:13–21

Large DE (1980) Geological parameters associated with sediment-hosted, submarine exhalative Pb-Zn deposits: an empirical model for mineral exploration. Geol Jahrb (Hannover) D 40:59–129

Large DE (1988) The evaluation of sedimentary basins for massive sulfide mineralization. In: Friedrich GH, Herzig PM (eds) Base metal sulfide deposits. Springer, Berlin Heidelberg New York, pp 3–12

Leach DL, Rowan EL (1986) Genetic link between Ouachita foldbelt tectonism and the Mississippi Valley-type deposits of the Ozarks. Geology 19:190–191

Maucher A, Schneider H-J (1967) The Alpine lead-zinc ores. In: Brown JS (ed) Genesis of stratiform lead-zinc-barite-fluorite deposits in carbonate rocks (the so-called Mississippi Valley type deposits). Econ Geol Monogr 3:71–89

Nicolini P (1990) Gîtologie et exploration minière. Lavoisier, Paris, 589 pp

Oliver J (1986) Fluids expelled tectonically from orogenic belts: their role in hydrocarbon migration and other geologic phenomena. Geology 14:99–102

Oliver J (1992) The spots and stains of plate tectonics. Earth Sci Rev 32:77–106

Parnell J, Ruffell AH, Moles NR (eds) (1993) Geofluids '93. Torquay, England, 471 pp

Pelissonnier H (1967) Analyse paléohydrologique des gisements stratiformes de plomb, zinc, baryte, fluorite du type "Mississippi Valley". In: Brown JS (ed) Genesis of stratiform lead-zinc-barite-fluorite deposits in carbonate rocks (the so-called Mississippi Valley type deposits). Econ Geol Monogr 3:234–252

Rouvier H, Perthuisot V, Mansouri A (1985) Pb-Zn deposits and salt bearing diapirs in southern Europe and North Africa. Econ Geol 80:666–687

Rowe J, Burley S, Gawthorpe R, Cowan C, Hardman M (1993) Palaeo-fluid flow in the East Irish sea basin and its margins. In: Parnell J, Ruffell AH, Moles NR (eds) Geofluids '93, 4th–7th May, 1993, Torquay, England, pp 358–362

Russell MJ (1988) A model for the genesis of sediment-hosted exhalative (SEDEX) ore deposits. In: Zachrisson E (ed) Proceedings of the seventh quadrennial IAGOD symposium held in Luleå, Sweden. Schweizerbart, Stuttgart, pp 59–66

Russell MJ, Skauli H (1991) A history of theoretical developments in carbonate-hosted base metal deposits and a new tri-level enthalpy classification. Econ Geol Monogr 8:96–116

Sangster DF (1983) Mississippi Valley-type deposits: a geological mélange. In: Kisvarsanyi G, Grant SK, Pratt WP, Koenig JW (eds) International Conference on Mississippi Valley type lead-zinc deposits. Proc Vol, Rolla, University of Missouri Rolla, pp 7–19

Sangster DF (1990) Mississippi Valley-type and sedex lead-zinc deposits: a comparative examination. Trans Inst Mining Metallur B 99:B21–B42

Sawkins FJ (1990) Metal deposits in relation to plate tectonics, 2nd edn. Springer, Berlin Heidelberg New York, 461 pp

Schneider HJ (1964) Facies differentiation and controlling factors for the depositional lead-zinc concentration in the Ladinian geosyncline of the Eastern Alps. In: Amstutz GC (ed) Sedimentology and ore genesis. Elsevier, Amsterdam, pp 29–45

Sibson RH, Moore JMcM, Rankin AH (1975) Seismic pumping – a hydrothermal fluid transport mechanism. J Geol Soc Lond 131:653–659

Sverjensky DA (1986) Genesis of Mississippi Valley-type lead-zinc deposits. Annu Rev Earth Planet Sci 14:177–199

Vasseur G, Demongodin L (1993) Convective and conductive heat trasnfer in sedimentary basins. In: Parnell J, Ruffell AH, Moles NR (eds) Geofluids '93. 4th–7th May 1993, Torquay, England, pp 84–87

Organic Contributions to Mississippi Valley-Type Lead-Zinc Genesis – A Critical Assessment

A.P. Giże[1] and H.L. Barnes[2]

Abstract

Mississippi Valley-Type deposits are associated in time and space with marine basins which contain thermally altered, type I kerogen of phytoplankton origins, but not type III derived from terrestrial plant organics. The thermal maturity of the organic matter at the time of mineralization is at the end of organic diagenesis and the beginning of the "oil window". Type I kerogens are H-rich, O-poor, and in these environments, sulphide-rich also. They are inadequate, both kinetically and in abundance, to cause ore deposition by reduction of dissolved sulphates or metal sulphate complexes. However, they are chemically potentially capable of selectively complexing both the divalent metals of these deposits and the specific Zn/Pb ratios characteristic of individual districts, while under the required strongly reducing conditions at ore-forming temperatures for millions of years. Possible complexing species which can account for the divalent ore assemblages include short-chain aliphatic anions and their sulphur analogues. The necessary concentrations of organic anions, above about 10 ppm, are geologically common.

1 Introduction

Mississippi Valley-Type (MVT) lead-zinc deposits are, at least superficially, amongst the simplest of metalliferous ore types. The metallic mineralogy is dominantly galena, sphalerite, and pyrite, with subordinate amounts of marcasite, and other minor sulphides. The host rocks are predominantly dolomitized carbonates or, less often, sandstones. The sulphur isotope ratios tend to be $\delta^{34}S$ high. Fluid inclusions indicate precipitation from brines (200-000 ppm total dissolved solids) at temperatures below about 200 °C (Roedder 1984). MVT deposits have formed in the Phanerozoic with a few possible earlier exceptions. Apparently, the mineralization of MVT deposits formed in basins located during sediment deposition in palaeolatitudes between the Tropics (Dunsmore 1975; Symons and Sangster 1991). These general observations, even when coupled with intensive experimental geochemical research

[1] Department of Geology, University of Manchester, Manchester, M13 9PL, UK
[2] Ore Deposits Research Section, The Pennsylvania State University, University Park, PA 16802, USA

Spec. Publ. No. 10 Soc. Geol. Applied to Mineral Deposits
Fontboté/Boni (Eds.), Sediment-Hosted Zn-Pb Ores

over the past few decades, have failed to provide a satisfying model of the genesis of these simple deposits. Not yet understood are the processes of ore transport and deposition, the most fundamental of genetic problems.

2 Proposed Organic Contributions

Organic matter is ubiquitous as a component of MVT deposits, and consequently has been proposed as a key genetic reactant (e.g. Anderson 1991). Possible roles include preconcentration of metals and sulphur in source rocks, complexing of metals in the ore solution, or in precipitating the ores by sulphate reduction. There are very few detailed studies of organic geochemical factors that permit evaluation of these possibilities. The conclusions from these limited studies are varied.

3 Previous Organic Studies

Commonly, the organic source has been deduced. In the Derbyshire (UK) fluorite lead-zinc deposit, Pering (1973) and Xuemin et al. (1987) have shown that the probable source of the Windy Knoll bitumens was the local Namurian (Carboniferous) Edale Shales. At the Saint Privat deposit (France), Connan and Orgeval (1973) showed local Autunian shales to be the bitumen source rock. Similarly, at Pine Point (Canada), Macqueen and Powell (1983) argued for local bitumen derivation.

Petroleum-associated brines have been invoked as the mineralizing fluids for MVT deposits (Jackson and Beales 1967). A consequence of this concept is that a MVT deposit should be a site of mixing of organic matter indigenous to the host rocks with a second, epigenetic generation introduced with the oil-field brines. The transport of the Laisvall (Sweden) ores by such fluids was proposed by Rickard et al. (1975, 1979), but the evidence remains equivocal using the published data. In a study of the Gays River (Canada) deposit, Giże and Barnes (1987) detected a hydrocarbon composition in the probable ore solution aquifer markedly different from the indigenous organics, and suggested that the variations associated with the ores could result from hydrocarbon mixing.

4 Organic Alteration

Where heated brines are involved in MVT deposits, there must be thermal alteration of local, indigenous organic matter. This thermal alteration has been recorded from several deposits. For example, Macqueen and Powell (1983) reported local heating at the Pine Point deposit. In the Elmwood deposit (USA), Giże (1990) found petrographic evidence for multiple generations of bitumens, reflected in differing thermal maturities. Recently, Henry et al. (1992) reported local bitumen generation and increased reflectance associated with the Viburnum Trend mineralization in southeastern Missouri.

Biodegradation of the ore-associated organic matter has been frequently reported, including the examples at Windy Knoll (Pering 1973; Xuemin et al. 1987), Saint Privat (Connan and Orgeval 1973), and Pine Point (Macqueen and Powell 1983). Evidence of biodegradation has been used to imply the presence of sulphate-reducing bacteria during ore deposition. This argument is permissible, but not conclusive. These three deposits occur near the surface in open-pit mines, where bacteria are abundant. At the Gays River deposit, Giże and Barnes (1987) showed that the extent of biodegradation decreased with depth, independently of ore grade. Consequently, reported biodegradation may simply reflect surface alteration, rather than the presence of sulphate-reducing bacteria during mineralization. The role of sulphate-reducing bacteria has been discussed with respect to carbonate-hosted lead-zinc deposits from an isotopic and microbial viewpoint by Trudinger (1982), with the conclusion that sulphate-reducing bacteria could have only played a minor role in the bulk of the sulphur geochemistry of major orebodies, with the possible exception of pyrite precipitation.

There are reactions to be expected between hydrothermal fluids and either epigenetic or indigenous organic compounds, especially hydrolysis and cleavage of larger organic molecules. The general types of reactions, together with examples, have recently been summarized by Siskin and Katritzky (1991). These reactions are analogous to those in the common experimental technique of organic hydrous pyrolysis. Indeed, the detailed organic geochemical studies of the Upper Mississippi Valley District (USA) by Giże (1984), Hatch et al. (1986), and Giże and Barnes (1987), showing increasing concentrations of low molecular weight hydrocarbons associated with mineralization, but no changes in biomarker distributions, reflect well the expected hydrothermal processes.

In summary, studies of the organic matter associated with ores to date have provided few insights beyond those predictable from the more extensive inorganic geochemical studies. General source-rock studies have shown, not surprisingly, local sources. The thermal effects of ore fluid migration at temperatures of up to 200 °C for somewhat longer than 0.25 million years (Lavery and Barnes 1971) should cause the observed alteration of the organic matter. Definitive evidence on the presence during ore formation of sulphate-reducing bacteria has not been presented. Consequently, genetic arguments based on the organic-MVT association, although promising, remain only theoretical possibilities.

5 Organic Composition

There are two major factors which will influence the composition of organic matter and their potential roles in MVT genesis. These two parameters are their biological source and their thermal maturity.

The original biological source is reflected in the "type" of organic matter, as derived from higher plants, algae, micro-organisms, etc. (van Krevelen 1961, 1984; Durand and Monin 1980). The classification is based on the atomic ratios of hydrogen, oxygen, and carbon of the insoluble sedimentary organic polymer, termed kerogen. Marine phytoplankton produce hydrogen-rich and

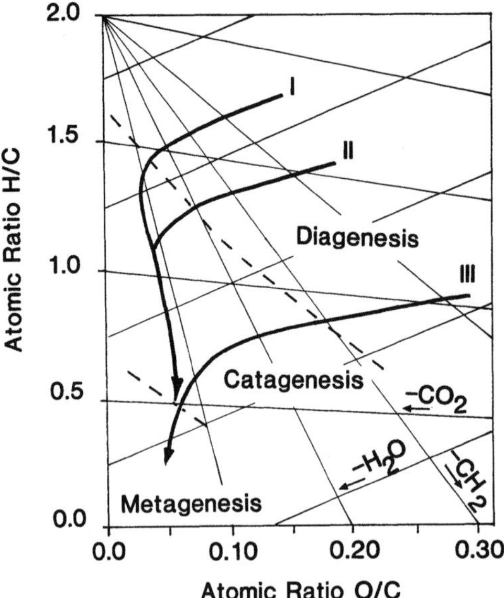

Fig. 1. The van Krevelen diagram, showing the elemental evolution of the kerogen types (I–III) with maturity. The evolution pathways can be simplistically described by the progressive loss of H_2O, CO_2 and CH_2. Note that diagenesis involves the loss of oxygen, and catagenesis (the "oil window") the loss of hydrogen. (After van Krevelen 1961, 1984; Tissot and Welte 1978)

oxygen-poor type I kerogens. Terrestrial plants are oxygen-rich, but relatively depleted in hydrogen, producing type III kerogens. The intermediate type II kerogen can be formed, for example, by mixing of marine and terrestrial organic matter in an estuary.

The second major parameter is the thermal maturity, or degree of maturation. The extent of maturation, which is a function of both time and temperature, can be approximated by assuming first-order kinetics (e.g. Connan 1974; Waples 1982). Simplistically, organic matter passes through three stages of progressive maturation. The first stage, below approximately 90 °C, referred to in organic geochemistry as diagenesis, is succeeded at about 90–120 °C by the main stage of petroleum generation, catagenesis (Fig. 1). The last stage of thermal alteration is metagenesis, during which organic matter approaches graphite. The extent to which organic matter will change as a result of increasing thermal maturity will depend upon the initial type.

The changes in the elemental composition of the three kerogen types with maturation, the van Krevelen diagram, are shown in Fig. 1. Fundamentally, diagenesis results in the loss of the bulk of oxygen and other hetero-atoms (e.g. sulphur), whereas catagenesis is primarily a loss of hydrogen.

6 Ore Composition

The sulphide mineralogy of MVT ores is simple, predominantly galena, sphalerite, pyrite, and marcasite. The Pb/Zn ratios, although variable between districts, tend to be diagnostic of individual districts. For example, Appalachian deposits, such as Friedensville, Elmwood and the Young Mine are sphalerite-rich deposits, with very little galena. Examples of galena-rich deposits include the Viburnum Trend (Missouri), Pine Point and Laisvall. Apparently, the metal-concentrating processes in an individual district are specifically selective for *divalent metals* (Zn, Pb, Fe, Mn, Cd, Ni and Hg) instead of monovalent metals (Cu, Ag) and, especially constraining chemically, for characteristic Pb/Zn ratios. It is interesting to note that inorganic complexing of metals depends more directly on the hardness of the electron shells (ionic being harder, covalent softer) than on valence, and discriminates little between divalent and monovalent metals. In contrast, organic complexes (or chelates) are quite effective in separating metals of different valence, and between specific metals such as lead and zinc. This observation is used to imply that organic complexes provide the high sulphide solubilities necessary to form MVT deposits.

7 Organic Type

The type of organic matter present in sediments must reflect plant evolution. Terrestrial plants (and hence type III organic matter) only appeared in the mid-Palaeozoic. If MVT deposits are genetically linked to type III organic matter, then MVT deposits should only appear in the geological record from the mid-Palaeozoic. There are many Early Palaeozoic MVT deposits (such as those in the Appalachians), and possible Proterozic deposits as well. Recently, Roberts (1992) has reported coked hydrocarbon residues in the late paragenetic stages of lead-zinc deposits in South Africa which bear similarities to Mississippi Valley-type deposits. Based on geological evidence, the mineralization and associated hydrocarbons were migrating between 2224–2060 Ma. The organic matter type present at the same time as these earlier deposits would be type I or II. In the South African example, the petroleum was suggested to be derived from local stromatolitic algae. Consequently, the MVT-organic association appears to be controlled by basinal type I (or II) organic matter, but not terrestrial type III.

The potential MVT type-I correlation is supported by two additional observations. Firstly, to the best of our knowledge, no MVT deposit shows a clear link with a coalfield (type III organic matter), but regional associations with oilfields (type I or II) are common. Secondly, the association with marine rather than terrestrial sediments is emphasized by plotting kerogen elemental analyses on a van Krevelen diagram (Fig. 2). It should be noted that the kerogens are from the ore host rocks, and are not those of implied source rocks. It can be seen from Fig. 2 that MVT deposits are associated with types I and II kerogen, but not type III.

8 Timing

Next we shall consider evidence on the stage of thermal maturation of the epigenetic organics during mineralization. Note that there need not be a simple

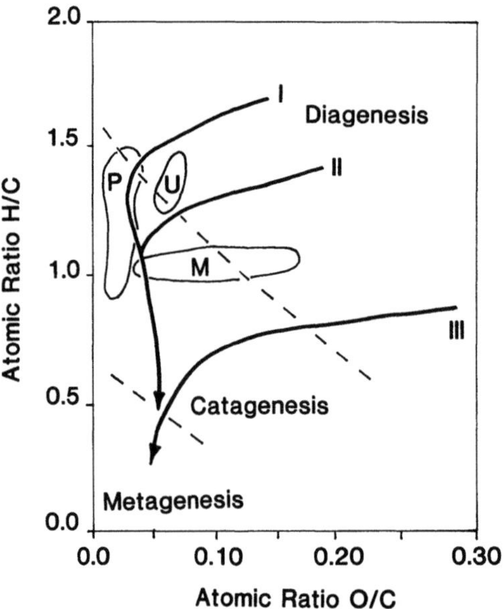

Fig. 2. Composition of MVT host-rock kerogens plotted on the van Krevelen diagram. *M* Southeast Missouri (Anderson 1991); *P* Pine Point (Macqueen and Powell 1983); *U* Upper Mississippi Valley District (Giże and Barnes 1987)

correlation between the temperatures of ore deposition and the maturity of organics in the source rocks of components of the ore solution.

There are two pertinent observations on the abundance and state of organics in MVT deposits. First, not all descriptions of MVT deposits mention organic matter. This implies that there is considerable variation in the organic content of these deposits. Secondly, when organic matter is reported, it usually occurs in the later stages of mineral parageneses, as bitumens (see, for example, Giże 1990) or petroleum inclusions (for example, Baines et al. 1991). Consequently, organic maturity during mineralization may be either at a stage when *liquid hydrocarbons* are not mobile, or at a stage during which liquid hydrocarbon migration is occurring. In other words, the thermal maturity has to be between the end of diagenesis (dissolved organics, but little or no liquid hydrocarbons) and start of catagenesis (mobile liquid hydrocarbons), or at the end of catagenesis and start of metagenesis (liquid hydrocarbons become immobile and gas prevalent). If the thermal maturity was at the end of catagenesis, then epigenetic organic occurrences should be less abundant and early in the paragenetic sequence, as the organic matter will have been through the oil window. More commonly, liquid hydrocarbons and bitumens are observed later in the paragenetic sequence, indicating that MVT genesis occurs at the end of organic diagenesis and early into the "oil window".

Where MVT deposits form simultaneously with organics entering the oil window, then there should be a similar areal distribution between hydrocarbon migration and MVT deposition. In the North Sea oilfields, sulphide mineraliza-

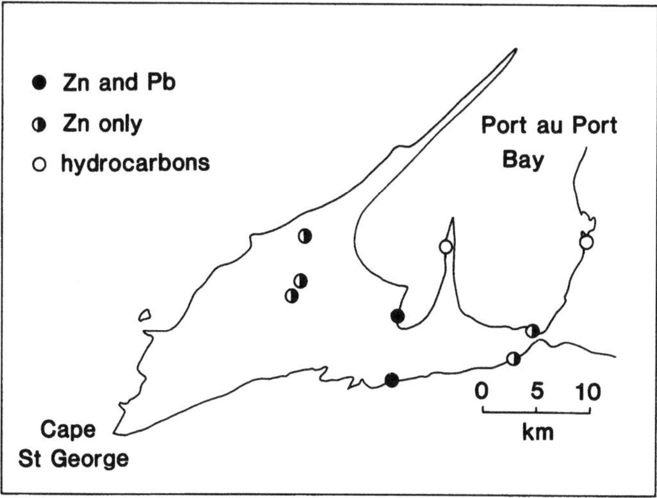

Fig. 3. Sketch map of the Port au Port Peninsula (Newfoundland) showing locations of hydrocarbons and lead-zinc showings in Palaeozoic carbonates (Douglas 1976). Although there is a close spatial relationship between the showings on a 10–100 km scale, on a 1-km scale, the hydrocarbons and ores do not overlap, implying local differences in the fluid timing

tion occurs within the oil reservoir and in local faults, suggesting co-migration of both mineralizing fluids and hydrocarbons (Baines et al. 1991). In other cases, such as the Illinois Basin (Bethke et al. 1991), the hydrocarbon occurrences are in the same area as mineralization, but slightly displaced. A second, simple example is the Port-au-Port Peninsula in western Newfoundland (Fig. 3), where the close hydrocarbon-ore spatial relationships suggest very similar timing, but some separation of the fluids into differing aquifers. The aquifer variation may reflect tectonic evolution of the basin, by evolving growth faults, for example. A thermal maturity at the end of diagenesis, and the onset of the oil window, best fit the field and paragenetic observations.

9 Solubility

In order to provide sufficient solubility for the transport of the metals in MVT deposits (a minimum of approximately 10 ppm of lead or zinc), complexing of the metals is required (e.g. Barnes 1979). Among the inorganic anions, only chloride and bisulphide are thermodynamically potential complexing agents. Central to the transport argument are estimates of how long it takes to form an orebody, and the rate of fluid flow during mineralization. Current estimates of the time taken to form a major orebody in the Upper Mississippi Valley District are of the order of 0.25 million years, based on zinc diffusion (Lavery and Barnes 1971), corroborated by organic thermal maturity arguments (Giże and Barnes 1987). Using these time estimates, coupled with fluid flow rates in sediments, it is possible to estimate that approximately 10 ppm total metals are required in the ore solutions. Thermodynamic calculations of lead and zinc solubility, as either chloride or bisulphide complexes, probably correct to ½-2x at the temperatures and pHs deduced during mineralization, show concentrations in equilibrium with the ore minerals which are two orders of magnitude lower than the nominal 10 ppm required for ore metal transport (Giordano and Barnes 1981; Bourcier 1983). The implication is that inorganic complexes are incapable of providing the necessary solubilities of lead and zinc sulphides to form MVT deposits. Other alternatives must exist.

10 Organo-Metallic Interactions

There is a possibility that organic complexes might increase the solubility of the metals in the ore solution. Consequently, the types of organo-metallic interactions to be considered as a basis for evaluating this possibility are limited to pi complexes, "sandwich" compounds, covalent linkages, or ionic bonds. Their nature and possible contributions to MVT genesis are as follows.

Pi complexes are formed by the interaction of a vacant pi orbital in an unsaturated carbon compound with a transition metal d-shell (for a review, see Cotton and Wilkinson 1980). Although the complexes meet the need for transition metal selectivity, the metal is typically in a low valence state (+1, 0, −1), rather than the divalent state required for MVT deposits. In addition, the thermal stability of such complexes tends to be low. In view of fluid inclusion data, indicating temperatures up to 200 °C during MVT genesis, the thermal weakness of pi complexes indicates that they are probably unimportant in MVT genesis.

Sandwich compounds are similar to pi complexes, in that they also arise by the interaction of arene (aromatic) pi electrons with a transition metal d-shell. Two aromatic ring compounds typically "sandwich" the cation, as in ferrocene, where iron is sandwiched between two cyclo-pentadiene rings. In contrast to pi complexes, sandwich compounds are more stable thermally, and have been recently reported geologically (Novgorodova and Buglaeva 1989). If sandwich compounds were important in ore genesis, then aromatic compounds should be present at high concentrations in orebodies. Although aromatic compounds are present in anomalously high concentrations in some ore deposits (especially mercury epithermal deposits; e.g. Blumer 1975), they are not prevalent in MVT deposits. Consequently, sandwich compounds will not be considered further.

Of the more traditional organo-metallic compounds, the two types are either those in which the metal is sigma-bonded directly to the carbon, or readily dissociable. In the former case, organo-lead compounds are both well known, and occur naturally, as in the example of methyl-lead compounds. The covalent bonding lends thermal stability. In the second situation, the interaction is essentially ionic, similar to an inorganic salt. Such interactions are well known geologically, as in the metallo-porphyrins which can complex a variety of metals.

Amongst the four types of organo-metallic interactions considered, pi complexes and sandwich compounds can be eliminated as geologically unimportant. Furthermore, sigma-bonded compounds are less probable because of MVT mineral textures. A characteristic of some MVT deposits is the remarkable delicacy of the banding in the sphalerite (e.g. McLimans et al. 1980; Baines et al. 1991). There are some deposits which do not show this, notable amongst which is Pine Point, where skeletal metal sulphides suggest very rapid precipitation. However, where found, the fine banding suggests a metal precipitation mechanism finely tuned to the fluid flow, chemical environment, and a relatively low activation energy to precipitation. The energy required to break sigma bonds is considered to be too high to account for the delicate textures observed, and we would argue that if an organo-metallic compound was involved, then only an ionic bonded compound would respond quickly enough. Consequently, we consider that the only organo-metallic compounds which may be feasible for MVT transport are ionic in nature. Ideally, such compounds incorporate stoichiometrically the necessary divalent metals and sulphide ions.

11 Precipitation

Perhaps one of the commonest hypotheses for MVT genesis is for the organic matter to act as a reductant to produce sulphide ions to precipitate the ores. The reductant proposed may be either sulphur or carbon species. Considering sulphur first, Skinner (1967) suggested that host-rock organics would be thermally decomposed by the hydrothermal solutions to free sulphide ions (or sulphur) to form ore sulphides. Although the organo-sulphur content of organic matter can be as exceptionally high as 10 wt% (Tissot and Welte 1978), mass balance calculations indicate that this mechanism cannot provide sufficient sulphide for an MVT deposit. For example, Giże (1984) estimated for the Gays River deposit that local organic sulphur could contribute at best, 0.1% of the sulphur in the ore deposit.

The potential role of carbon as a reductant can be through either biological or abiogenic mechanisms. In support of a biological reduction during ore deposition, biodegradation of the associated organic matter in an MVT deposit has been argued as evidence for active sulphate-reducing bacteria. However, deposition temperatures are above those in which such bacteria remain active. In addition, there is a question of their viability at depth as biodegradation has not been shown in any buried orebody.

The abiogenic reduction of sulphate to sulphide with concomitant ore precipitation was proposed by Barton (1967), and more recently by Anderson (1991). There are several arguments with respect to MVT deposits which weigh against such a mechanism being significant. First, an abundant organic reductant, optimally methane, is required. If methane and associated species were effective, then MVT deposits should be associated with gas fields. There is no such association, even when evaporites are in close association with gas fields, as in the southern North Sea and northwest Europe. Second, we have found above that the timing of ore precipitation coincides with early oil window maturity. At this thermal maturity, methane is derived microbially predominantly, and is isotopically very light; there is no evidence for such isotopically light carbon being associated with the ores. Thirdly, many experimental investigations have demonstrated that sulphate is not reducible at 100–200 °C, in weakly acidic solutions, by reductants as weak as relatively inert methane (Ohmoto and Lasaga 1982).

In addition to the above precipitation arguments which involve organic matter directly as a reductant, there may be indirect organic controls on the site of MVT precipitation. For example, in the Upper Mississippi Valley, the orebodies are located in solution breccias (Y-breccias) where minor fractures cross various carbonate horizons under the Ordovician Maquoketa shale. Near its base, this shale is generally organic-rich, and throughout has a very low permeability, both now and at the time of mineralization. This impermeability could limit MVT precipitation if caused by contact with an oxidizing groundwater (Howd and Barnes 1975; Barnes 1979). The following reactions can occur,

$$HS^- + 2O_2 \leftrightharpoons H^+ + SO_4^{2-} \quad (1)$$

$$2H^+ + CaCO_3 \leftrightharpoons H_2O + CO_2 + Ca^{2+} \quad (2)$$

$$Ba^{2+} + SO_4^{2-} \leftrightharpoons BaSO_4 \quad (3)$$

thereby accounting for both solution brecciation (reaction 2), and often, barite precipitation (reactions 1 and 3). In addition, diffusion of oxygen from oxidizing groundwaters into the reduced brine results in oxidation that could cause ore precipitation by either oxidizing the complexing ligand, or by lowering the pH to displace the metal sulphide by protons,

$$MeSL^- + H^+ \leftrightharpoons MeS + HL, \quad (4)$$

where L^- may be an organic complexing ligand, such as a thiol. Because the carbonate host will be dissolved to create cavities for the ore (reaction 2), this model implies that the solution breccias are an integral part of ore deposition in carbonates.

The indirect role of the organic-rich shale would therefore simply be to act as a redox boundary between underlying reduced basinal brines and the overlying oxidized groundwaters. The role is indirect as any impermeable horizon should be equally effective, such as lava flows, sills or tuffs.

12 Discussion

The above geological and geochemical observations have multiple implications for possible organic-MVT interactions. Firstly, the organic matter associated with the ores is predominantly types I and II, but not type III. Because the palaeolatitudes for the source rocks were probably tropical, the organics are characteristic of marine, tropical, highly reduced sediments. Of the several possible roles for organic matter in MVT genesis, including sulphate reduction, only the potential role of contributing to the metal solubility in the ore solutions remains therefore to be examined in detail.

The hypothesis of an organic complex or chelate in the ore solutions is based on the default argument that relatively weak stability constants of lead and zinc chloride and bisulphide complexes, at the pHs and temperatures estimated during ore transport, could not provide adequate solubilities inorganically. In addition, the selectivity of organic complexes for divalent metals, and for zinc or lead, favours transport by organic complexes. There are also requirements that such complexes persist under the requisite thermal conditions and strongly reduced state of the ores, shown by the FeS content of sphalerite with pyrite (Giordano and Barnes 1981).

Field evidence indicates that prior to ore precipitation, the hydrothermal solutions are slightly alkaline to neutral, as shown by the mineral assemblages along the solution path. For example, along the fractures under the ores of the Upper Mississippi Valley District, the alteration shows pH buffering between microcline and illite (Barnes 1983), about pH 7.1 at 100 °C. Any effective organic ligands must therefore have pKa's of 7 or less in order to be sufficiently ionized in order to be complexing agents. This requirement eliminates alcohols, and promotes short-chain (C_{1-4}) or aromatic carboxylic acids as potential complexing agents.

In order for an organic anion to be an effective ligand for lead and zinc transport, it must be stable and not break down significantly in the ore solution. Hydrodynamic evidence for the distances of ore solution migration extend up to 800 km, as in the case of the Illinois Basin, requiring stability over a time period of the order 10^6 years (Bethke 1986). Potential complexing agents can be stabilized thermally by one of two mechanisms. In the first, the molecule will be dipole-stabilized if it has a short carbon chain; the shorter, the more stable. Consequently, acetate will be the most dipole-stabilized. The second route to stability is for the carboxyl group to be attached to an aromatic, where the anion becomes resonance-stabilized.

Dipole-stabilized acetic acid is expected to be thermally the most stable aliphatic acid. This is confirmed by concentrations in petroleum brines and deep formation waters which can reach levels of 0.17 mol kg^{-1}. Bell (1991) has experimentally evaluated the stability of acetic acid and acetate under the pressure-temperature regimes of sedimentary basins and concluded that acetate is stable at temperatures of 100 °C for periods of 10^6 years in the presence of common aquifer minerals. Acetate therefore remains a viable, potential organic ligand.

The preceding discussion has focussed on oxygen-based functional groups. However, the chemistry of organic sulphur compounds also has parallels with oxygen-based functional groups. For example, the sulphur analogues of alcohols (ROH) are the thiols (RSH). The observation that the palaeolatitudes at the time of host-rock deposition appear to have been between the Tropics also suggests that sulphur-based functional groups may be important. Sediments typical of tropical latitudes are not only characterized by high organic concentrations, but also by thick accumulations of carbonates and evaporites. The organic matter in these lithologies tends to be characteristically enriched in sulphur. For example, the typical organic sulphur content in black shales is of the order of 0.1 wt%, but in carbonates and evaporites, where iron is not available to fix the sulphur as pyrite, the organic sulphur content can increase by an order of magnitude (Tissot and Welte 1978). An additional advantage in considering organo-sulphur compounds in the ore solutions is that the total hydrothermal sulphide concentration is increased, along with that of the ore metals.

The minimum concentrations necessary for organic ligands to be effective ore carriers can be readily estimated. If a minimum metal concentration of an ore solution is taken to be 10 ppm (Barnes 1979), then an approximately 10^{-4} molal ligand concentration is required, based on the following equation:

$$MeS + xL^{2-} \leftrightharpoons MeSL_x^{2x-}. \tag{5}$$

Assuming that the complex is strong [i.e. $(MeSL_x^{2x})/(L^{2-}) >$ approximately 2], perhaps a chelate, and that L^{2-} has a molecular weight of about 100, then 10 ppm of the ligand is required. Not only is this concentration easily achieved in basinal brines, but it also comes within the natural range of surface waters (Stumm and Morgan 1970). Therefore we see no stability or concentration problems to lead and zinc complexing by aliphatic ligands in MVT-forming hydrothermal fluids.

13 Conclusions

In spite of their deceptive simplicity, several aspects of MVT genesis remain unresolved, including the role of the apparently ubiquitous organic matter. The one probable role for organic matter is in metal complexing, as chloride and bisulphide complexes have been shown to be too weak, by two orders of magnitude, to transport sufficient combined metals in the hydrothermal solutions.

Consideration of the geological environment in which MVT deposits were formed indicates an association with marine type I kerogens, in contrast to terrigenous type III kerogens. The thermal maturity of the potential source rocks is probably at the end of diagenesis, and early into the "oil window", at which stage the kerogens have released hetero-atoms (especially oxygen and sulphur). Published data on acetate concentrations in oil-field brines indicate that these anions are present in sufficient quantities to contribute significantly to metal complexing. In addition, the tropical environments associated with the ores suggests an, as yet, unconsidered role for organo-sulphur compounds.

Organic complexes remain viable candidates for transporting the divalent metals under the conditions of MVT formation. Their gross chemical characteristics, including thermal stabilities, ionization constants, metal selectivities, possible concentrations, tolerance for strongly reduced conditions, and the absence of inorganic candidates, suggest that they deserve serious experimental investigation. The correlation with type-I kerogens shows that such investigations should begin with those organics of marine, tropical, strongly reduced environments as first candidates.

Acknowledgements. We wish to thank the editors for their kind invitation to publish this paper in honour of Prof. Amstutz's 70th birthday. Their comments helped clarify several aspects significantly.

References

Anderson GM (1991) Organic maturation and ore precipitation in southeast Missouri. Econ Geol 86:909–926

Baines SJ, Burley SD, Giże AP (1991) Sulphide mineralisation and hydrocarbon migration in North Sea oilfields. In: Pagel M, Leroy JL (eds) Source, transport and deposition of metals, Proc 25 years SGA Anniversary Meet, Nancy. Balkema, Rotterdam, pp 507–510

Barnes HL (1979) Solubilities of ore minerals. In: Barnes HL (ed) Geochemistry of hydrothermal ore deposits, 2nd edn. Wiley, New York, pp 404–460

Barnes HL (1983) Ore-depositing reactions in Mississippi Valley-type deposits. In: Kisvarsanyi G, Brant SK, Pratt WP, Koenig JW (eds) Proc Int Conf Mississippi Valley-type lead-zinc deposits. Rolla, Missouri, pp 77–85

Barton PB (1967) Possible role of organic matter in the precipitation of the Mississippi Valley ores. In: Brown JS (ed) Genesis of stratiform lead-zinc-barite-fluorite deposits. Econ Geol Monogr 3:371–378

Bell JLS (1991) Acetate decomposition in hydrothermal solutions. PhD Thesis, Dept Geosciences, The Pennsylvania State University, 229 pp

Bethke CM (1986) Hydrologic constraints on the genesis of the Upper Mississippi Valley Mineral District from Illinois basin brines. Econ Geol 81:233–249

Bethke CM, Reed JD, Oltz DF (1991) Long range petroleum migration in the Illinois Basin. Bull Am Assoc Petrol Geol 75:925–945

Blumer M (1975) Curtisite, idrialite and pendletonite, polycyclic aromatic hydrocarbon minerals: their composition and origin. Chem Geol 16:245–256

Bourcier WL (1983) Stabilities of chloride and bisulfide complexes of zinc in hydrothermal solutions. PhD Thesis, Dept Geosciences, The Pennsylvania State University, 179 pp

Connan J (1974) Time-temperature relation in oil genesis. Bull Am Assoc Petrol Geol 58:2516–2521

Connan J, Orgeval J-J (1973) Les bitumes des minéralisations barytiques et sulfurées de St Privat (Bassin de Lodève, France). Bull Cent Rech Pau-SNPA 6:195–214

Cotton FA, Wilkinson G (1980) Advanced inorganic chemistry, 4th edn. Wiley, New York, 1396 pp

Douglas C (1976) Mineral occurrences tables Newfoundland. Ministry Dev Div Dept Mines and Energy, Open File, St John's, Newfoundland

Dunsmore HE (1975) Origin of lead-zinc ores in carbonate rocks: a sedimentary-diagenetic model. PhD Thesis, University of London, 227 pp

Durand B, Monin JC (1980) Elemental analysis of kerogens. In: Durand B (ed) Kerogen. Editions Technip, Paris, pp 113–141

Giordano TH, Barnes HL (1981) Lead transport in Mississippi Valley-type ore solutions. Econ Geol 76:2200–2211

Giże AP (1984) The organic geochemistry of three Mississippi Valley-type ore deposits. PhD Thesis, Dept Geosciences, The Pennsylvania State University, 350 pp

Giże AP (1990) Petroleum derived cokes in sedimentary basins. In: Fermont WJJ, Weegink JW (eds) Proc Int Symp Organic petrology, Zeist 7–9, 1990. Meded Rijks Geol Dienst 45:65–70

Giże AP, Barnes HL (1987) The organic geochemistry of two Mississippi Valley-type lead-zinc deposits. Econ Geol 82:457–470

Hatch JR, Heyl AV, King JD (1986) Organic geochemistry of wall-rock alteration, Thompson-Timperly Pb-Zn deposits, Wisconsin. In: Dean WE (ed) Organics and ore deposits. Proc Denver Region Exploration Geologists Society Symp, Denver Region Exploration Geologists Society, pp 93–104

Henry AL, Anderson GM, Héroux Y (1992) Alteration of organic matter in the Viburnum Trend lead-zinc district of southeastern Missouri. Econ Geol 87:288–309

Howd FH, Barnes HL (1975) Ore solution chemistry IV. Replacement of marble by sulphides at 450 °C. Econ Geol 70:968–981

Jackson SA, Beales FW (1967) An aspect of sedimentary basin evolution: the concentration of MVT ores during late stages of diagenesis. Bull Can Petrol Geol 15:383–433

Lavery NG, Barnes HL (1971) Zinc dispersion in the Wisconsin zinc-lead district. Econ Geol 66:226–242

Macqueen RW, Powell TG (1983) Organic geochemistry of the Pine Point lead-zinc field and region, Northwest Territories, Canada. Econ Geol 78:1–25

McLimans RK, Barnes HL, Ohmoto H (1980) Sphalerite stratigraphy of the Upper Mississippi Valley lead-zinc district, southwest Wisconsin. Econ Geol 75:351–361

Novgorodova MI, Buglaeva EY (1989) Carbon in ore-forming process: metallo-organic compounds as migrating form, graphite and native carbides as members of mineral associations. Abstr 28th Int Geol Congr, Washington DC, pp 2–524

Ohmoto H, Lasaga AC (1982) Kinetics of reactions between aqueous sulfates and sulfides in hydrothermal systems. Geochim Cosmochim Acta 46:1727–1745

Pering KL (1973) Bitumens associated with lead, zinc and fluorite ore minerals in North Derbyshire, England. Geochim Cosmochim Acta 37:401–417

Rickard DT, Willdén MY, Marde Y, Ryhage R (1975) Hydrocarbons associated with lead-zinc ores at Laisvall, Sweden. Nature 255:131–133

Rickard DT, Willdén MY, Marinder N-E, Donelly TH (1979) Studies on the genesis of the Laisvall sandstone lead-zinc deposit, Sweden. Econ Geol 74:1255–1285

Roberts PJ (1992) The geology and geochemistry of selected Pb-Zn deposits in the Chuniespoort/-Ghaap Group of the Transvaal Sequence, South Africa. MSc Thesis, University of Witswatersrand

Roedder E (1984) Fluid inclusions. Reviews in mineralogy, vol 12. Mineralogical Society of America, Washington DC, 644 pp

Siskin M, Katritzky AR (1991) Reactivity of organic compounds in hot water: geochemical and technological implications. Science 254:231–237

Skinner BJ (1967) Precipitation in Mississippi Valley type ores: a possible mechanism. In: Brown JS (ed) Genesis of stratiform lead-zinc-barite-fluorite deposits. Econ Geol Monogr 3:363–370

Stumm W, Morgan JJ (1970) Aquatic chemistry. Wiley, New York, 583 pp

Symons DTA, Sangster DF (1991) Paleomagnetic age of the central Missouri barite deposits and its genetic implications. Econ Geol 86:1–12

Tissot BP, Welte DH (1978) Petroleum formation and occurrence, 2nd edn. Springer, Berlin Heidelberg New York, 538 pp

Trudinger PA (1982) Geological significance of sulphur oxidoreduction by bacteria. Philos Trans R Soc Lond B 298:563–581

van Krevelen W (1961) Coal, typology, chemistry, physics, and constitution. Elsevier, Amsterdam, 514 pp

van Krevelen W (1984) Organic geochemistry – old and new. Org Geochem 6:1–10

Waples D (1982) Organic geochemistry for exploration geologists. International Human Resources Development Corporation, Boston, 151 pp

Xuemin G, Fowler MG, Comet PA, Manning DAC, Douglas AG, McEvoy J, Giger W (1987) Investigation of three natural bitumens from central England by hydrous pyrolysis and gas chromatography-mass spectrometry. Chem Geol 64:181–195

Precipitation of Mississippi Valley-Type Ores: The Importance of Organic Matter and Thiosulphate

C.S. Spirakis[1] and A.V. Heyl[2]

Abstract

Certain constraints that apply to many Mississippi Valley-type deposits may be used to develop a model for the genesis of this group of deposits. These constraints include: (1) the occurrence of a wide variety of mineral types together at specific sites in a huge volume of potential host rock; (2) the minus-one oxidation state of sulphur in disulphides in the ores; (3) the ubiquitous occurrence of organic matter in the ores; and (4) the evidence of episodes of dissolution of the ore and gangue minerals. The most likely model for the genesis of Mississippi Valley-type ores involves the transport of metals and sulphur together in one solution in which sulphur is in the form of thiosulphate ($S_2O_3^{2-}$). Where this hot mineralizing solution encounters organic matter in the host rock, thiosulphate is reduced to provide the minus-one valent sulphur for disulphide (pyrite and marcasite) precipitation and the minus-two valent sulphur for sulphide (galena, sphalerite, etc.) precipitation. Reversal of some reactions among partly oxidized sulphur species may account for the oscillations between precipitation and dissolution of the sulphide minerals in the ores. Heating of the organic matter at the sites of mineralization produces organic acids, which initially dissolve carbonate host rocks. At somewhat higher temperatures, some organic acids slowly degrade to produce carbon dioxide, whereas others act as pH buffers. This addition of carbon dioxide to a solution with an organic acid pH buffer causes carbonates to precipitate. At higher temperatures still, organic acids quickly degrade so that the pH buffer is destroyed. In the absence of a buffer, the addition of carbon dioxide lowers the solution's pH and causes carbonates to dissolve. Carbon dioxide from organic acid degradation and from carbonate dissolution can trigger fluorite precipitation by forming the complex $MgHCO_3^+$ at the expense of the complex MgF^+. Because MgF^+ is an important species in the transport of fluorine, breaking the complex causes fluorite precipitation. In the cool late stage of mineralization, bacterial metabolism of thiosulphate in the presence of organic matter produces carbon dioxide and isotopically heavy sulphate. The former precipitates as late-stage calcite and the latter as late-stage baryte. Thus, the entire paragenesis can be linked to the presence of organic matter at the sites of mineralization, and the interaction of a hot thiosulphate-bearing solution with organic matter

[1] U.S. Geological Survey, P.O. Box 25046, MS 939, DFC, Denver, CO 80225, USA
[2] P.O. Box 1052, Evergreen, CO 80439, USA

Spec. Publ. No. 10 Soc. Geol. Applied to Mineral Deposits
Fontboté/Boni (Eds.), Sediment-Hosted Zn-Pb Ores

in the host rock may accommodate the widely applicable constraints on the genesis of the deposits.

1 Introduction

Mississippi Valley-type deposits derived their name from descriptions of the deposits in the Upper Mississippi Valley district. This use of a type locality in the characterization of the deposit type has led to some discussion as to which deposits should be included as members of the group. Although individual Mississippi Valley-type and related deposits have distinctive characteristics, they also share many common attributes, which suggests that they also share some common aspects of their genesis (Ohle 1959). Many previous studies of the genesis of Mississippi Valley-type deposits have focused on developing models for individual deposits rather than on models applicable to the entire group, and other previous studies have concentrated on the precipitation mechanisms of specific minerals in the Mississippi Valley-type association with little regard for the other minerals of the paragenesis or for what links all of the minerals of the association (including sulphides, disulphides, sulphates, carbonates, and fluorite) to the same sites. The occurrence of such a variety of mineral types together in very specific sites in a huge volume of potential host rock that was traversed by mineralizing solutions (Hoagland 1971; Erickson et al. 1978) is an important but often overlooked constraint on the genesis of the deposits—a constraint suggesting that some peculiar characteristic of the sites (hydrology, geology, chemistry) caused each of them to become a focus of mineralization.

Another generally applicable but overlooked constraint is the oxidation state of sulphur in iron disulphides (pyrite and marcasite), which are present in all of these ores. As discussed in Spirakis (1986a), the oxidation state of sulphur in pyrite is minus 1, whereas the oxidation states of sulphur in the thermodynamically stable forms in hydrothermal solutions of H_2S and SO_4^{2-} are minus 2 and plus 6, respectively. Disulphides cannot form unless minus one-valent sulphur is available. Oxidants to form minus one-valent sulphur from H_2S are scarce in the subsurface and do not appear to have been available at the sites of mineralization; although reductants are available, reduction of sulphate under the conditions of Mississippi Valley-type ore formation is inhibited (discussed below). Any viable model for the genesis of these deposits must accommodate the minus-one oxidation state of sulphur in disulphides. A possible solution involves the presence of metastable but reactive sulphur species such as thiosulphate in the mineralizing solution.

Like pyrite, organic matter is a "ubiquitous" component of Mississippi Valley-type and related deposits, a relationship which caused Skinner (1967) to pose the question of how could there not be a genetic relationship? Indeed, the presence of organic matter in Mississippi Valley-type and related deposits (see summary in Spirakis 1986b) and the evidence that organic matter reacted with the mineralizing solution in the Pine Point deposits (Macqueen and Powell 1983) and in the southeast Missouri district (Leventhal 1990) provide another constraint on ore genesis.

Detailed studies in some districts reveal delicate (microscopic) colour banding in sphalerite, referred to as sphalerite stratigraphy, which can be followed for tens of kilometres in the Upper Mississippi Valley district (McLimans 1977) and for kilometres in the Appalachian districts (Craig et al. 1983). In the southeast Missouri district (Horrall et al. 1983), central Kentucky district (Jolly and Heyl 1964), and Illinois-Kentucky district (Richardson and Pinckney 1984), traceable aspects of the mineral paragenesis or traceable microlayers in ore minerals have been identified. Although the detailed studies needed to detect traceable mineral or isotope banding have not been done in all districts, the number of districts in which district-wide banding or other traceable characteristics has been detected indicates that it is a widely applicable (if not universally applicable) constraint on genesis of these deposits—a constraint which indicates that the ores formed under uniform conditions affecting large areas simultaneously.

Another constraint derived from studies of several districts is the evidence of multiple episodes of precipitation and dissolution of ore minerals (southeast Missouri, Hagni and Trancynger 1977; Upper Mississippi Valley, Heyl et al. 1959; Illinois-Kentucky, Hall and Freidman 1963 and Cunningham and Heyl 1980; Tristate, Siebenthal 1915 and McKnight and Fischer 1970; Pine Point, Kesler et al. 1972; Canadian Arctic district, our observations of samples at the 1990 IAGOD meeting). As Barnes (1983) and Sverjensky (1981) pointed out, the repeated oscillations between dissolution and precipitation require that the precipitation mechanism be easily reversible.

These widely applicable constraints on the genesis of these deposits form the basis of our model and cast doubt on some of the previous models for the genesis of Mississippi Valley-type deposits. Specifically, they suggest that sulphate reduction and brine mixing models are unlikely.

2 Previous Models

Mixing of a sulphide-bearing solution with a metal-bearing solution has often been suggested as a means of precipitating sulphide minerals in specific Mississippi Valley-type deposits. There are some advantages to a mixing model. A mixing model could account for the mineralization of specific sites in huge volumes of potential host rock by the limited number of sites at which permeable zones intersect so that mixing could occur. Also, by carrying sulphide and metals in separate solutions, mixing accommodates the low solubility of metals with sulphide sulphur under conditions of Mississippi Valley-type mineralization (Barnes 1979), and there is little doubt that mixing of sulphide- and metal-bearing solutions will precipitate sulphide minerals. Mixing models, however, do not address some generally applicable constraints on the genesis of the ores. In particular, the traceability of microscopic banding in sphalerite, fluorite, or calcite over large distances in several districts (cited above) is incompatible with mixing of solutions because it is impossible to imagine that solutions could mix so uniformly over entire districts as to produce this banding. In a similar argument, Ohle (1980) pointed out that the continuation of the stages of the paragenesis throughout the complicated three-dimensional geometry of the ores was not compatible with solution mixing. The district-wide uniformity of temperature-salinity relationships in sphalerite is also an argument against mixing as the sulphide precipitation mechanism (Hanor 1979). Furthermore, mixing requires an improbable series of solutions (or pairs of solutions), all mixing at the same site in order to precipitate the various minerals of the paragenesis. One mineral in the Mississippi Valley association, fluorite, will not precipitate by mixing solutions containing its components. According to Richardson and Holland (1979), adding calcium to a fluorine-bearing solution by either mixing or dissolution of calcite will not cause fluorite to precipitate because under Mississippi Valley-type conditions the complex CaF^+ forms and enhances fluorite solubility. The dissolution of ore and gangue minerals in several important Mississippi Valley-type districts provides further reason to doubt that mixing of solutions was the precipitation mechanism. The low solubility of metals in sulphide-bearing solutions, which mixing models resolve by carrying each in a separate solution, prevents sulphide and metals from being carried away from the dissolving minerals in the same solution. However, it is impossible to unmix components of the minerals and have each carried away in a separate solution. Therefore, mineral dissolution is an argument against mixing as a precipitation mechanism. The minus-one oxidation state of sulphur in pyrite is not addressed in mixing models nor is the ubiquitous presence of organic matter in Mississippi Valley-type ores. As Crocetti and Holland (1989) pointed out, the correlation between sulphur and lead isotopes that they detected in galena in the southeast Missouri district could not be produced by solution mixing.

Another way to accommodate the low solubilities of metals with sulphide sulphur is to transport metals with sulphate sulphur and then reduce sulphate to sulphide at the sites of mineralization. According to this sulphate-reduction model, organic matter at the site of mineralization provides the reductant. Thus, the sulphate-reduction model is consistent with the observation that organic matter is ubiquitous in Mississippi Valley-type ores, and also like a one-solution (non-mixing) model, it is compatible with the sphalerite stratigraphy and uniform temperature-salinity relationships. Furthermore, the concentration of organic matter at specific sites in the host rock may account for the localization of the ores, and organic matter may be involved in other types of reactions required to form the other ore minerals.

Two problems arise for sulphate-reduction models. One is the dissolution textures in sulphides in the ores. The repeated oscillations between precipitation and dissolution indicated by the ore textures require that the precipitation mechanism be easily reversible. Thermodynamic calculations, however, clearly demonstrate that the reduction of sulphate by organic matter so strongly favours reduced sulphur in the presence of organic matter that the reaction is for all practical purposes irreversible (Sverjensky 1981; Barnes 1983). The other major problem with sulphate reduction is kinetics of the reaction. Several decades ago, Bastin (1926) concluded that reduction of sulphate was too slow at temperatures of Mississippi Valley-type mineralization (sulphide minerals typically form at 100 to 160 °C, Roedder 1967, and in some districts as high as 220 °C, McLimans 1977) to cause the precipitation of the ores. Since then, numerous experiments (summarized in Ohmoto and Lasaga 1982; Trudinger et al. 1985; Spirakis 1986b) using organic matter and other reducing agents have all shown that sulphate reduction is an extremely slow process at temperatures less than about 230 °C in solutions with amounts of sulphur and pH values believed to be typical of Mississippi Valley-type mineralizing solutions.

Geochemical evidence corroborates the experimental evidence. Studies of the distribution of sulphur isotopes between sulphate and sulphide minerals (summarized in Ohmoto and Rye 1979) indicate that sulphate reduction does not occur in hydrothermal systems at less than about 200 °C. From detailed investigations of the hydrothermal chemistry of the system at Creede, Colorado, Barton et al. (1977) concluded that equilibrium between sulphide and sulphate ceased (an indication that sulphate reduction ceased) at less than about 230 °C. Advocates of sulphate-reduction models often cite Toland (1960) as experimental evidence that the process occurs, but none of Toland's experiments at less than 300 °C showed evidence of sulphate reduction and Toland notes that below about 300 °C sulphate reduction becomes extremely slow. An experiment by Orr (1982) showed sulphate reduction at 175 °C, but the concentration of sulphur was unrealistically high. Drean (1978) showed that under conditions of high total sulphur, sulphate and hydrogen sulphide react to form native sulphur; native sulphur may then be reduced to sulphide. This mechanism for sulphate reduction does not apply to the genesis of Mississippi Valley-type deposits because chemical conditions at the sites of mineralization (Barnes 1979) are not such that native sulphur is stable. Heydari and Moore (1989) found native sulphur in wells in the Smackover Formation (Gulf Coast of the USA) only at temperatures greater than 150 °C. They concluded that sulphate reduction occurs at some temperature in excess of 150 °C. Although these studies indicate that sulphate reduction does occur above 150 °C, their studies actually reinforce the conclusion that sulphate reduction was not involved in the precipitation of Mississippi Valley-type ores. Much of the sulphide ore in Mississippi Valley-type deposits precipitated at temperatures of less than 150 °C, that is, at temperatures less than the minimum temperature suggested by their study. Also, the time available for sulphate reduction in Mississippi Valley-type ore zones was constrained by the residence time of any particular sulphate ion in the ore zone. Considering that the solution flowed at a few metres (Barnes 1983) or tens of metres in a year (Roedder 1967), any sulphate ion would flush through the ore zones in a few years at most. (Actually the abrupt change from unmineralized to mineralized rock requires a precipitation mechanism that forms sulphide minerals in much less time than the residence time for a sulphate ion in the ore zone.) In contrast, the time available for sulphate reduction in the Smackover Formation (Heydari and Moore's study) is several tens of millions of years. This difference in time available is important because it indicates that only a very small fraction (about one-millionth) of the sulphate in the solution could be reduced in the ore zone. Barnes (1979) estimated that the total sulphur in the mineralizing solution is about 10^{-2} mol/l. If only one-millionth of the sulphate can be reduced during its residence in the ore zone, then about 10^{-8} mol/l of sulphide would be produced. This is far less than the 10^{-4} mol/l of sulphide that is believed to be the minimum needed to form Mississippi Valley-type ores (Barnes 1979). Thus, all of the geologic, experimental, and isotopic evidence agrees that the sluggish kinetics of sulphate reduction precludes this process from precipitating sulphide minerals in Mississippi Valley-type ores—a conclusion corroborated by the non-reversibility of sulphate reduction, which is required to account for the oscillations between precipitation and dissolution of minerals in the ores.

The strong points of both the mixing model and the sulphate reduction model may be retained while their weak points are accommodated by a hybrid model in which sulphur as a partly oxidized species such as thiosulphate (not sulphate) and metals are carried together in one solution until the solution encounters organic matter (possibly by mixing with an oil reservoir). Reduction of thiosulphate by the organic matter then triggers the precipitation of sulphides, and other reactions with organic matter account for other aspects of the paragenesis.

3 Sulphide and Disulphide Precipitation

Generally, thiosulphate ($S_2O_3^{2-}$) and other partly oxidized sulphur species are not considered to be important to the chemistry of high-temperature ore-forming solutions. Indeed, at temperatures in excess of 230 °C or so, thiosulphate exists only in very small concentrations as a transient species representing the intermediate steps in redox reactions between sulphide and sulphate. However, kinetic data (Pryor 1960) indicate that, in the temperature range of the precipitation of Mississippi Valley-type deposits, thiosulphate may be very long-lived. The longevity of thiosulphate can be estimated by using the Arrhenius equation (rate = $Ae^{-Ea/RT}$) to re-evaluate Pryor's result at 270 °C in a solution with a pH buffer at a temperature typical of Mississippi Valley-type mineralization. Such a re-evaluation at 150 °C indicates that thiosulphate will endure 10^7 times longer at 150 °C than at 270 °C. In Pryor's 270 °C experiment, thiosulphate endured for a few tens of hours; so that at 150 °C it is predicted to endure for tens of thousands of years. In that amount of time, a Mississippi Valley-type mineralizing solution flowing a few metres or tens of metres per year could have carried thiosulphate for several tens of kilometres from its source to the site of mineralization. Processes to form thiosulphate or other partly oxidized sulphur species in sedimentary basins include oxidation of pyrite by atmospheric oxygen (Granger and Warren 1969); oxidation of hydrogen sulphide by ferric iron (Roberts et al. 1969); reaction between native sulphur and hydrogen sulphide (Hyne 1968; Muller and Hyne 1969); or possibly the oxidation of pyrite or hydrogen sulphide by ferric iron released as smectite converts to illite (Spirakis and Heyl 1987). Whatever the mechanism of their formation, the presence of thiosulphate and other partly oxidized sulphur species in groundwaters and thermal springs (Wilson 1941; Gundlach 1965; Boulegue 1977, 1981; Sulzhiyeva and Volkov 1982; Veldeman et al. 1991) and the longevity of thiosulphate, indicated by kinetic data, suggest that thiosulphate may have been an important component of Mississippi Valley-type mineralizing solutions. A genetic model involving thiosulphate instead of sulphate has some distinct advantages.

Thiosulphate differs from sulphate in that one of the minus two-valent oxygen atoms of sulphate has been replaced with a minus two-valent sulphur

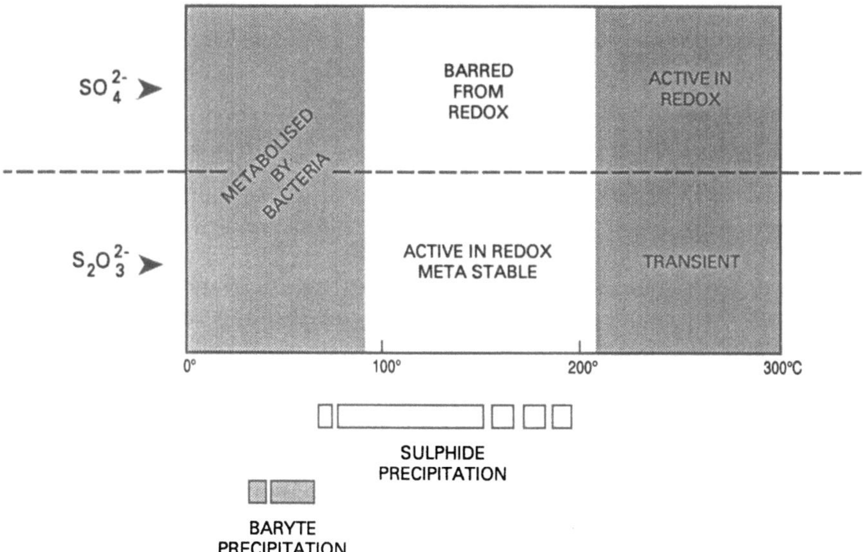

Fig. 1. Comparison of the reactivity of sulphate to thiosulphate. Note that sulphate does not participate in redox reactions in the temperature range of sulphide precipitation and that both sulphate and thiosulphate may be metabolized by bacteria at temperatures of the late baryte stage

atom. This seemingly small difference greatly alters the reactivity of the molecule; a particularly important difference is the ability of thiosulphate to participate in redox reactions at such low temperature that sulphate is barred from redox reactions. As Fig. 1 illustrates, both sulphate and thiosulphate are metabolized by bacteria at temperatures less than about 80 °C; throughout the temperature range of the sulphide stage of mineralization, thiosulphate is active in redox reactions, whereas sulphate is not. Because of this difference, it is reasonable to expect a mineralizing solution carrying thiosulphate, upon encountering organic matter, to react to form minus two-valent and minus one-valent sulphur, which precipitate as sulphide and disulphide minerals. It is not, however, reasonable to expect the same reaction from a sulphate-bearing solution. The restriction of Mississippi Valley-type sulphide precipitation to, and only to, the temperature range in which thiosulphate is long-lived and in which sulphate does not enter into redox reactions is consistent with the suggestion that thiosulphate is the sulphur species in the mineralizing solution.

In contrast to sulphate reduction, some reactions involving thiosulphate and other partly oxidized sulphur species may be reversed both to produce and to consume sulphide. One such example is:

$$9S_2O_3^{2-} + 2Pb^{2+} + 6H^+ \rightarrow 3H_2O + 2PbS + 4S_4O_6^{2-}. \quad (1)$$

Thermodynamic calculations using the free energy data for the partly oxidized sulphur species in Cobble et al. (1972) indicate that this reaction is reversible under the temperature and pH conditions of Mississippi Valley-type mineralizing solutions. This reaction is quite sensitive to pH. Thus, changes in pH that may have been caused by the reaction of the mineralizing solution with organic matter at the sites of mineralization (discussed in Sect. 5) might account for the multiple oscillations between precipitation and dissolution of the sulphide minerals in the ores.

Another advantage of a mineralizing solution containing thiosulphate is that a variety of metals may be transported in the same solution along with sulphur in the form of thiosulphate. At low temperatures, the solubility of lead is several orders of magnitude higher in thiosulphate solutions than in sulphide solutions and the solubility of barium is much higher in thiosulphate solutions than in sulphate solutions (Weast and Lide 1989). Experiments on the solubility of metals with thiosulphate, conducted at elevated temperatures (Oberste-Padtberg 1982), confirm the enhanced solubility of lead and barium in high-temperature thiosulphate-bearing solutions. Veldemen et al. (1991) found that in thermal spring water, thiosulphate enhanced copper solubility by forming the complex $Cu(S_2O_3)^-$. Lead and barium do not appear to form complexes with thiosulphate. Instead, the increased solubility of lead in thiosulphate solutions compared to sulphide solutions and of barium in thiosulphate solutions compared to sulphate solutions is more likely to be due to the fact that, as long as sulphur remains in the form of thiosulphate, neither sulphide nor sulphate minerals can precipitate. Because of the high solubility of metals with thiosulphate, only one metal- and thiosulphate-bearing solution is required to transport the elements needed for the precipitation of sulphide, disulphide, and sulphate minerals.

Baryte

Baryte is a common mineral in the late stage of Mississippi Valley-type deposits and in a few districts there is evidence of an early baryte or anhydrite stage that largely dissolved during the sulphide stage of mineralization (Upper Mississippi Valley, Heyl et al. 1959; Illinois-Kentucky, Heyl 1983; Viburnum trend, Marikos 1989). The small amounts of the early baryte and early anhydrite that remain leave few clues to their mechanism of precipitation; however, Marikos (1989) suggested that the precipitation of early sulphate minerals was due to the disproportionation of thiosulphate in the mineralizing solution catalyzed by hydrogen sulphide degassed from organic matter at the site of mineralization. Both thiosulphate and organic matter may also be involved in the precipitation of the late baryte.

Although fluid inclusions in baryte are prone to leakage and hence yield questionable temperatures, in the Upper Mississippi Valley district the fluid-inclusion temperatures of as low as 75 °C for sphalerite preceding the late baryte (McLimans 1977) and of 46 to 78 °C for calcite precipitated subsequent to baryte (Bailey and Cameron 1951; Erickson 1965) constrain the temperatures of baryte precipitation. In the Illinois-Kentucky district, fluid inclusion data from late-stage calcite (Spry et al. 1990) suggest that this baryte also formed at temperatures low enough for bacteria to survive. In the Illinois-Kentucky district, the abrupt shift to light carbon isotopes in the late-stage calcite associated with the baryte (Richardson et al. 1988) may be an indication of bacterial

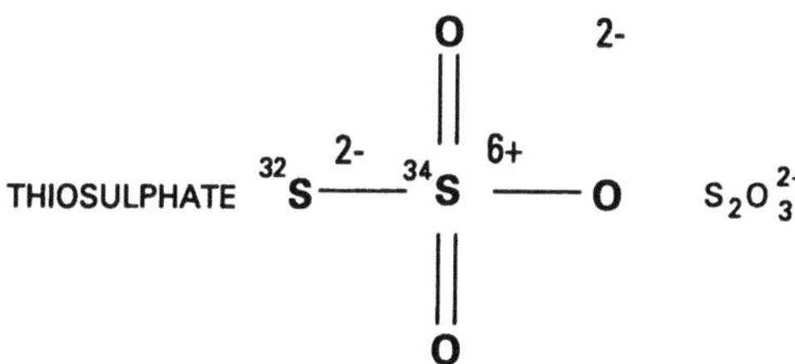

Fig. 2. Geometry of the thiosulphate molecule. Note that isotopically heavy sulphur is enriched in the inner site in which sulphur is in the +6 oxidation state

metabolism. In the presence of organic matter, bacteria may metabolize thiosulphate by reduction to hydrogen sulphide, dissimilatory reduction to sulphide and sulphite (SO_3^{2-}), disproportionation to sulphide and sulphate, or oxidation to sulphate (see references in Spirakis 1991 and in Fossing and Jorgensen 1990). Some of the strains of bacteria that metabolize thiosulphate have been found in oilfield brines (Semple et al. 1987). Bacterial metabolism of thiosulphate in the presence of organic matter may link baryte precipitation to the same sites as other aspects of the paragenesis.

The baryte-precipitating mechanism must account for the occurrence of disulphide minerals along with the late baryte, for the low solubility of barium in sulphate solutions, and for the very unusual sulphur isotopes in the late baryte of 25.8 to 35.9 per mil in the Upper Mississippi Valley district (McLimans 1977) and as high as 102 per mil in the Illinois-Kentucky district (Richardson et al. 1988). Under hydrothermal conditions, isotopically heavy sulphur tends to concentrate in the inner site in thiosulphate (see Fig. 2) and isotopically light sulphur in the outer site (Uyama et al. 1985) with a fractionation of up to 40 per mil. In some of the possible metabolic pathways, bacteria first reduce sulphur from the inner site to sulphite (SO_3^{2-}) and then disproportionate or oxidize a portion of the sulphite to sulphate (SO_4^{2-}), while the rest is recombined with sulphide to form more thiosulphate. The newly formed thiosulphate may then be metabolized. Fossing and Jorgensen (1990) found that several metabolic pathways occur simultaneously and that recycling of sulphur species through thiosulphate was indeed an important aspect of the metabolism of various sulphur species. The steps proposed above to metabolize thiosulphate are likely to produce sulphate that is enriched in isotopically heavy sulphur because bacteria more readily reduce isotopically light sulphite, thus leaving an isotopically heavy sulphite to form sulphate or more thiosulphate (Spirakis 1991). Note that the bacterial metabolism of thiosulphate simultaneously produces reduced and oxidized sulphur species. The increase in concentration of reduced sulphur may trigger the formation of disulphides, whereas the increase in concentration of sulphate may trigger baryte precipitation. The enrichment in heavy sulphur by bacteria along with the inorganic processes that concentrate heavy sulphur in the inner site in thiosulphate (Uyama et al. 1985) provides an explanation for the extremely S^{34}-enriched baryte. Organic matter is required for some of the metabolic pathways involving thiosulphate (Nriagu et al. 1979). Thus, it is the presence of organic matter that causes baryte to precipitate at the same sites at which sulphides had precipitated earlier in the paragenesis.

5 Carbonate Paragenesis

One of the earliest alteration effects related to mineralization in most Mississippi Valley-type districts is a dissolution of the carbonate rock hosting the deposits. In many districts, after the initial dissolution and prior to sulphide precipitation, dolomite or in some cases calcite precipitates in the ores and in halos surrounding the ores. Carbonate minerals, both early calcite and dolomite, then dissolve concomitant with sulphide formation. Finally, calcite or dolomite precipitates as one of the last minerals in the paragenesis. Work aimed at understanding the development of secondary porosity in carbonate rocks and the relation of this porosity to oil migration (Surdam and Crossey 1985) may also be used to explain the carbonate dissolution, precipitation and renewed dissolution related to Mississippi Valley-type mineralization (Spirakis and Heyl 1988). According to Surdam and Crossey (1985), heating of organic matter during burial diagenesis to temperatures in excess of about 80 °C produces organic acids that are quite effective at dissolving carbonate minerals (Meshri 1986). These two steps (organic acid production and carbonate dissolution) may be described by the equations:

$$nC \cdot H_2O \xrightarrow{T > 80°C} CH_3COOH^0 + \text{bitumen} \quad \text{and} \tag{2}$$

$$CH_3COOH^0 + CaCO_3 \rightarrow CaCH_3COO^+ + HCO_3^- . \tag{3}$$

With continued heating to approximately 120 °C, organic acids begin to slowly degrade:

$$CH_3COOH^0 \xrightarrow{120°C} CH_4 + CO_2 \quad \text{and} \tag{4}$$

$$CaCH_3COO^+ + H_2O \xrightarrow{120°C} CH_4 + HCO_3^- + Ca^{2+} . \tag{5}$$

Because the degradation is slow at temperatures around 120 °C, some organic acids are available to act as pH buffers while their neighbors thermally decompose:

$$CaCH_3COO^+ + H^+ \rightarrow Ca^{2+} + CH_3COOH^0 . \tag{6}$$

This buffer is important because in the presence of a pH buffer, the addition of carbonate to a solution decreases the solubility of carbonate minerals and causes carbonate precipitation:

$$Ca^{2+} + CO_2 + H_2O \rightarrow CaCO_3 + 2H^+ \quad \text{(with } H^+ \text{ removed by the buffer).} \tag{7}$$

As temperature increases, the rate of degradation of organic acids increases. Eventually, a temperature is reached (probably around 140 °C) at which organic acids quickly degrade. Consequently, the pH buffer is lost. Now in the absence of a pH buffer, the addition of CO_2 has the opposite effect as before; it lowers the solution's pH, and carbonate minerals dissolve:

$$CaCO_3 + CO_2 + H_2O \rightarrow 2HCO_3^- + Ca^{2+} . \tag{8}$$

Thus, as temperatures increase during diagenesis, the relative importance of organic acid and of carbonate pH buffers changes in such a manner that carbonate minerals dissolve, precipitate, and then dissolve again. In the case of Mississippi Valley-type deposits, the gradual heating of concentrations of organic matter in an initially cool host rock by the invasion of a hot mineralizing solution may have caused the ore-stage carbonate paragenesis typical of the ores. Small changes in temperature during mineralization might account for the oscillations between precipitation and dissolution of ore-stage carbonates. The very latest stage calcite, which contains isotopically light carbon (Richardson et al. 1988), may have formed as a by-product of bacterial metabolism of organic matter.

6 Fluorite Precipitation

Carbon dioxide from both carbonate dissolution and from degradation of organic matter at the sites of mineralization may be crucial to the precipitation of fluorite in Mississippi Valley-type ores. Richardson and Holland (1979) concluded that the complexes MgF^+ and CaF^+ were the most important complexes for transporting fluorine in Mississippi Valley-type mineralizing solutions. As pointed out in Spirakis and Heyl (1988), the addition of CO_2 to a solution transporting fluorine as MgF^+ or as CaF^+ forms the complexes $MgHCO_3{}^+$ and $CaHCO_3{}^+$ at the expense of MgF^+ and CaF^+. Breaking these fluorine-transporting complexes frees fluorine, lowers the solubility of fluorite, and provides a mechanism for fluorite precipitation. The data in Chiba (1986) show that the magnesium and calcium complexes with carbonate ($CO_3{}^{2-}$) are much stronger than those with bicarbonate ($HCO_3{}^-$). Consequently, the same effect may be produced without changing the total carbonate content of the system but by changing the relative amounts of carbonate ($CO_3{}^{2-}$), bicarbonate ($HCO_3{}^-$), and carbonic acid (H_2CO_3). (Carbonic acid does not complex with calcium and magnesium.) The relative amounts of the carbonate species is strongly dependent on the pH of the system. Therefore, an increase in pH of a solution in which MgF^+, $MgHCO_3{}^+$, and $MgCO_3{}^0$ are in equilibrium will produce more $MgCO_3{}^0$ at the expense of the other two complexes ($MgHCO_3{}^+$ and MgF^+). Because $CO_3{}^{2-}$ is even more effective at tying up magnesium than is $HCO_3{}^-$, a simple rise in pH and hence $CO_3{}^{2-}$ concentration will cause fluorite to precipitate.

Because these reactions are reversible, any shift in pH or total carbonate in the solution during mineralization may cause the oscillations between precipitation and dissolution of fluorite as observed in the ores. Also, because organic matter at the site of mineralization is the source of carbon dioxide (either directly through degradation or through carbonate dissolution), it is the presence of organic matter that causes certain sites to become mineralized with fluorite.

Many other types of fluorite deposits occur in environments where an increase in total carbonate due to carbonate dissolution or where an increase in pH of an initially acid-mineralizing solution are likely to have increased the formation of magnesium carbonate and calcium carbonate complexes (or carbonate and bicarbonate complexes with magnesium and calcium) at the expense of the fluoride complexes. Thus, the process may be widely applicable.

7 Conclusion

Organic matter is critical in each of the preceding precipitation mechanisms. In the case of sulphide and disulphide precipitation, organic matter acts as a reductant. Probably, the species being reduced is thiosulphate or one of the other partly oxidized sulphur species and not sulphate. There are theoretical advantages to a proposed mineralizing solution containing thiosulphate. Thiosulphate can be reduced under ore-forming conditions at a rate fast enough to cause sulphide and disulphide precipitation; sulphate cannot. An involvement of thiosulphate provides a reason for the restriction of Mississippi Valley-type ores to the temperature range in which thiosulphate exists as a long-lived species and the reversibility of reactions involving partly oxidized sulphur species is consistent with the dissolution of ore-stage sulphide minerals. The proposed thiosulphate solution also resolves the transport problem in that lead and zinc are more soluble in thiosulphate solutions than in sulphide solutions and barium is more soluble in thiosulphate solutions than in sulphate solutions. Thus, one thiosulphate- and metal-bearing solution may transport the ore constituents; additional solutions carrying ore components are not required. Organic matter is essential to the proposed baryte precipitation mechanism because certain pathways in the bacterial metabolism of thiosulphate can only occur in the presence of organic matter. The thermal generation and destruction of organic acids from organic matter links organic matter to the carbonate paragenesis. Carbon dioxide either directly from thermal decarboxylation of organic matter or from the carbonate dissolution caused by organic acids may trigger fluorite precipitation and tie fluorite precipitation to the presence of organic matter. The presence of organic matter at the sites of mineralization links the precipitation of all of the ore-related minerals to the same sites and a role for organic matter in ore formation is consistent with the ubiquitous occurrence of organics in Mississippi Valley-type ore deposits. The model presented here accommodates the minus-one oxidation state of sulphur in pyrite without the kinetic problems associated with sulphate reduction models, and, because it does not require mixing of solutions, the model is consistent with the regional traceability of banding and other features in the ores. Thus, the widely applicable constraints on ore genesis may all be explained by one thiosulphate-bearing solution reacting in various ways with organic matter at the sites of Mississippi Valley-type mineralization.

References

Bailey SW, Cameron EN (1951) Temperatures of mineral formation in bottom-run lead-zinc deposits of the Upper Mississippi Valley, as indicated by liquid inclusions. Econ Geol 46:626–651

Barnes HL (1979) Solubilities of ore metals. In: Barnes HL (ed) Geochemistry of hydrothermal ore deposits, 2nd edn. Wiley, New York, pp 404–460

Barnes HL (1983) Ore depositing reactions in Mississippi Valley-type deposits. In: Kisvarsanyi G, Grant SK, Pratt WP, Koenig JW (eds) International conference on Mississippi Valley-type lead-zinc deposits. University of Missouri, Rolla, pp 77–85

Barton PB Jr, Bethke PM, Roedder E (1977) Environment of ore deposition in the Creede Mining district, San Juan Mountains, Colorado: Part III. Progress toward interpretation of the chemistry of the OH vein. Econ Geol 72:1–25

Bastin ES (1926) The problem of the natural reduction of sulfates. Am Assoc Petrol Geol Bull 10:1270–1299

Boulegue J (1977) Equilibrium in a sulfide rich water from Enghien-les-Bains, France. Geochim Cosmochim Acta 41:1751–1758

Boulegue J (1981) Simultaneous determination of sulfide, polysulfide, and thiosulfate as an aid to óre exploration. J Geochem Explor 15:21–36

Chiba H (1986) Compilations of dissociation constants of major aqueous species present in geothermal fluids. Comparison of thermodynamic data base of aqueous speciation codes. Institute for the Study of the Earth's Interior, Tech Rep Ser B, 3, Okayma University, Misasa, Japan, 129 pp

Cobble JW, Stephens HP, McKinnon IR, Westrum EF Jr (1972) Thermodynamic properties of oxygenated sulfur complex ions. Heat capacity from 5 to 300 ·K for $K_2S_4O_6$(c) and from 273 to 373 ·K for $S_4O_6^{2-}$(aq). Revised thermodynamic functions for HSO_3^-(aq), $S_2O_3^{2-}$(aq) and $S_4O_6^{2-}$(aq) at 298 · K. Revised potential of the thiosulfate-tetrathionate electrode. Inorg Chem 11:1669–1674

Craig JR, Solberg TN, Vaughan DJ (1983) Growth characteristics in sphalerite in Appalachian zinc deposits. In: Kisvarsanyi G, Grant SK, Pratt WP, Koenig JW (eds) International conference on Mississippi Valley-type lead-zinc deposits. University of Missouri, Rolla, pp 317–326

Crocetti CA, Holland HD (1989) Sulfur-lead isotope systematics and the composition of fluid inclusions in galena from the Viburnum trend, Missouri. Econ Geol 84:2196–2116

Cunningham CG, Heyl AL (1980) Fluid inclusion homogenization temperatures throughout the sequence of mineral deposition in the Cave-in-Rock area, southern Illinois. Econ Geol 75:1226–1231

Drean TA (1978) Reduction of sulfate by methane, xylene, and iron at temperatures of 175 to 350 °C. Thesis, Pennsylvania State University, State College

Erickson AJ Jr (1965) Temperatures of calcite deposition in the Upper Mississippi Valley deposits. Econ Geol 60:506–528

Erickson RL, Mosier EL, Viets JG (1978) Generalized geologic and summary geochemical maps of the Rolla 1· × 2· quadrangle, Missouri. US Geol Surv Misc Field Studies Map MF-1004A

Fossing H, Jorgensen BB (1990) Oxidation and reduction of radiolabeled inorganic sulfur compounds in an estuarine sediment, Kysing Fiord, Denmark. Geochim Cosmochim Acta 54:2731–2742

Granger HC, Warren CG (1969) Unstable sulfur compounds and the origin of roll-type sulfur deposits. Econ Geol 64:160–171

Gundlach H (1965) Untersuchungen an einigen Schwefelquellen in Griechenland. Geol Jahrb 83:411–430

Hagni RD, Trancynger TC (1977) Sequence of deposition of ore minerals at the Magmont Mine, Viburnum trend, southeast Missouri. Econ Geol 72:451–464

Hall WE, Freidman I (1963) Composition of fluid inclusions Cave-in-Rock fluorite district, Illinois and Upper Mississippi Valley lead-zinc district. Econ Geol 58:886–911

Hanor JS (1979) The sedimentary genesis of hydrothermal fluids. In: Barnes HL (ed) Geochemistry of hydrothermal ore deposits, 2nd edn. Wiley, New York, pp 137–172

Heydari E, Moore CH (1989) Burial diagenesis and thermochemical sulfate reduction, Smackover Formation, southeastern Mississippi salt basin. Geology 17:1080–1084

Heyl AV (1983) Geologic characteristics of three major Mississippi Valley districts. In: Kisvarsanyi G, Grant SK, Pratt WP, Koenig JW (eds) International conference on Mississippi Valley-type lead-zinc deposits. University of Missouri, Rolla, pp 27–60

Heyl AV, Agnew AF, Lyons EJ, Behre CH Jr (1959) The geology of the Upper Mississippi Valley lead-zinc district. US Geol Surv Prof Pap 309, 310 pp

Hoagland AD (1971) Appalachian stratabound deposits: their essential features, genesis and the exploration problem. Econ Geol 66:805–810

Horrall KB, Hagni RD, Kisvarsanyi G (1983) Mineralogical, textural, and paragenetic studies of selected ore deposits of the southeast Missouri lead-zinc district and their genetic implications. In: Kisvarsanyi G, Grant SK, Pratt WP, Koenig JW (eds) International conference on Mississippi Valley-type lead-zinc deposits. University of Missouri, Rolla, pp 289–316

Hyne JB (1968) Sulfur deposition in sour gas wells. Alberta Sulfur Res Ltd Q Bull 5:2–18

Jolly JL, Heyl AV (1964) Mineral paragenesis and zoning in the central Kentucky mineral district. Econ Geol 59:596–624

Kesler SE, Stoiber RE, Billings GK (1972) Direction of flow of mineralizing solutions at Pine Point N.W.T. Econ Geol 67:19–24

Leventhal JS (1990) Organic matter and thermochemical sulfate reduction in the Viburnum trend, southeast Missouri. Econ Geol 85:622–632

Macqueen RW, Powell TG (1983) Organic geochemistry of the Pine Point lead-zinc ore field and region, Northwest Territories, Canada. Econ Geol 78:1–25

Marikos MA (1989) Gangue anhydrite from the Viburnum trend, southeast Missouri. Econ Geol 84:158–161

McKnight ET, Fischer RP (1970) Geology and ore deposits of the Picher Field Oklahoma and Kansas. US Geol Surv Prof Pap 588, 165 pp

McLimans RK (1977) Geological, fluid inclusion, and stable isotope studies of the Upper Mississippi Valley zinc-lead district, southwest Wisconsin. Thesis, Pennsylvania State University, State College

Meshri ID (1986) On the reactivity of carbonic and organic acids and the generation of secondary porosity. In: Gautier DL (ed) Roles of organic matter in sediment diagenesis. Soc Econ Paleontol Mineral Spec Publ 38:123–128

Muller E, Hyne JB (1969) Radiochemical S^{35} exchange in the $H_2S + S^{35}$ reaction. Alberta Sulfur Res Ltd Q Bull 6:16–18

Nriagu JO, Coker RD, Kemp ALW (1979) Thiosulfate, polythionates, and rhodanese activity in Lakes Erie and Ontario sediments. Limnol Oceanogr 24:383–389

Oberste-Padtberg R (1982) Die Bedeutung des intermediar oxidierten Schwefels für die Genese hydrothermaler Sulfat-Sulfid-Lagerstatten. Ber Bunsenges Phys Chem 86:1038–1041

Ohle EL (1959) Some considerations in determining the origin of ore deposits of the Mississippi Valley type. Econ Geol 54:769–789

Ohle EL (1980) Some considerations in determining the origin of ore deposits of the Mississippi Valley type – Part II. Econ Geol 75:161–172

Ohmoto H, Lasaga AC (1982) Kinetics of reactions between aqueous sulfates and sulfides in hydrothermal systems. Geochim Cosmochim Acta 46:1725–1745

Ohmoto H, Rye RO (1979) Isotopes of sulfur and carbon. In: Barnes HL (ed) Geochemistry of hydrothermal ore deposits, 2nd edn. Wiley, New York, pp 509–567

Orr WL (1982) Rate and mechanism of non-microbial sulfate reduction. GSA Abstr Prog 14:580

Pryor WA (1960) The kinetics of the disproportionation of sodium thiosulfate to sodium sulfide and sulfate. J Am Chem Soc 82:4794–4797

Richardson CK, Holland HD (1979) Fluorite deposition in hydrothermal systems. Geochim Cosmochim Acta 43:1327–1336

Richardson CK, Pinckney DM (1984) The chemical and thermal evolution of the fluids in the Cave-in-Rock fluorspar district, Illinois: mineralogy, paragenesis and fluid inclusions. Econ Geol 79:1833–1856

Richardson CK, Rye RO, Wasserman MD (1988) The chemical and thermal evolution of the fluids in the Cave-in-Rock district, Illinois: stable isotope systematics at the Deardorff Mine. Econ Geol 83:765–783

Roberts WMB, Walker AL, Buchanan A (1969) The chemistry of pyrite formation in aqueous solution and its relation to the depositional environment. Mineral Depos 4:18–29

Roedder E (1967) Fluid inclusion evidence on the genesis of ores in sedimentary and volcanic rocks. In: Wolf KH (ed) Handbook of stratabound and stratiform ore deposits. Elsevier, New York, pp 67–110

Semple KM, Westlake DWS, Krouse HR (1987) Sulfur isotope fractionation by strains of *Alteromonas putrefaciens* isolated from oil field fluids. Can J Microbiol 33:372–376

Siebenthal CE (1915) Origin of the zinc and lead deposits of the Joplin region, Missouri, Kansas, and Oklahoma. US Geol Surv Bull 606, 283 pp

Skinner BJ (1967) Precipitation of Mississippi Valley-type ore: a possible mechanism. In: Brown JS (ed) Genesis of stratiform lead-zinc-barite-fluorite deposits in carbonate rocks. Econ Geol Monograph 3:363–369

Spirakis CS (1986a) The valence of sulfur in disulfides—an overlooked clue to the genesis of Mississippi Valley-type lead-zinc deposits. Econ Geol 81:1544–1545

Spirakis CS (1986b) Occurrence of organic carbon in Mississippi Valley deposits and an evaluation of processes involving organic carbon in the genesis of these deposits. In: Dean WE (ed) Organics in ore deposits. Proc Denver Regional Exploration Geologists Society, Denver, Colorado, pp 85–92

Spirakis CS (1991) The possible role of thiosulfate in the precipitation of ^{34}S-enriched barite in some Mississippi Valley-type deposits. Mineral Depos 26:60–65

Spirakis CS, Heyl AV (1987) A link among ^{34}Sulfur-rich pyrite in greenish-gray shales, ferric iron from smectite as it converts to illite, kinetics of sulfur redox reactions, and the genesis of Mississippi Valley-type deposits. In: Sachs J (ed) USGS research on mineral deposits-1987. US Geol Survey Circ 995:68–69

Spirakis CS, Heyl AV (1988) Possible effects of thermal degradation of organic matter on the carbonate paragenesis and fluorite precipitation in Mississippi Valley-type deposits. Geology 16:1117–1120

Spry PG, Koellner MS, Richardson CK, Jones HD (1990) Thermochemical changes in ore fluid during deposition at the Denton mine, Cave-in-Rock, fluorspar district, Illinois. Econ Geol 85:172–181

Sulzhiyeva TM, Volkov II (1982) Thiosulfates and sulfites in thermal and hydrothermal water. Geochem Int 19:94–98

Surdam RC, Crossey LJ (1985) Mechanism of organic/inorganic reactions in sandstone/shale sequences. In: Relationship of organic matter and mineral diagenesis. Soc Econ Paleontol Mineral Short Course 17:177–272

Sverjensky DA (1981) The origin of a Mississippi Valley-type deposit in the Viburnum trend, southeast Missouri. Econ Geol 76:1848–1872

Toland WG (1960) Oxidation of organic compounds with aqueous sulfate. J Am Chem Soc 82:1911–1918

Trudinger PA, Chamber LA, Smith JW (1985) Low-temperature sulfate reduction: biological versus abiological. Can J Earth Sci 22:1910–1918

Uyama F, Chiba H, Kusakabe M, Sakai H (1985) Sulfur isotope exchange reactions in the aqueous system: thiosulfate-sulfide-sulfate at hydrothermal temperature. Geochem J 19:301–315

Veldeman E, Van't Dack L, Gijbels R, Pentcheva EN (1991) Sulfur species and associated trace elements in south-west Bulgarian thermal waters. Appl Geochem 6:49–62

Weast RC, Lide DR (eds) (1989) CRC handbook of chemistry and physics. 70th edn. CRC Press, Boca Raton, p B101

Wilson SH (1941) Natural occurrence of polythionic acids. Nature 148:502–503

Palaeomagnetic Methods for Dating the Genesis of Mississippi Valley-Type Lead-Zinc Deposits

D.T.A. Symons[1] and D.F. Sangster[2]

Abstract

Mississippi Valley-type (MVT) Pb-Zn(-Ba-F) ore deposits typically occur in undisturbed platform carbonates that preclude geologic dating, are comprised of minerals that are not amenable to radiometric age dating, and therefore have an uncertain age and genesis. This chapter summarizes palaeomagnetic methods that have been successful in dating several MVT deposits, paying particular attention to various tests that show that the measured magnetization event is coeval with the MVT mineralization event. All results to date are consistent with gravity-driven fluid flow from an adjacent orogenic uplift as the correct model for the genesis of MVT deposits.

1 Introduction

1.1 Purpose

The purpose of this chapter is to outline palaeomagnetic methods for dating Mississippi Valley-type (MVT) Pb-Zn(-Ba-F) mineralization, present results determined up to now, and discuss the genetic consequences of these results.

1.2 Characteristics of MVT Deposits

Lead-zinc deposits of the Mississippi Valley type typically occur in flat-lying or mildly deformed platformal carbonate sequences peripheral to clastic sedimentary basins (Anderson and Macqueen 1982); deposits in rift-related carbonates are known but these are relatively few in number (Leach and Sangster, in press). Fluid inclusion studies of ore-stage minerals reveal that these ores commonly formed in the 70–140 °C range and precipitated from highly saline fluids. Mineralogy is dominated by various proportions of galena, sphalerite, pyrite, and dolomite; lesser amounts of marcasite, barite, calcite, and fluorite occur in some deposits. Deposits usually result from massive or partial replacement of carbonates, open-space filling of collapse breccia zones or fractures, or combinations of these processes (Leach and Sangster, in press).

[1] Department of Geology, University of Windsor, Windsor, Ontario, Canada N9B 3P4
[2] Geological Survey of Canada, 601 Booth St., Ottawa, Canada K1A 0E8

Spec. Publ. No. 10 Soc. Geol. Applied to Mineral Deposits
Fontboté/Boni (Eds.), Sediment-Hosted Zn-Pb Ores

Several genetic models have been proposed for MVT deposits over the years and may be considered as variants of two main types: (1) compaction-driven fluid flow, and (2) gravity-driven fluid flow. Elaborating on the source-bed concept of Noble (1963), Beales and Jackson (1966) proposed that MVT ore fluids originated from steady-state continuous compaction of clastic sediments during sedimentation and that such fluids were driven out of the basin into the surrounding carbonate platform to form MVT deposits. Cathles and Smith (1983), suggesting that such a process would not sustain the temperatures found in MVT deposits, proposed an episodic fluid flow triggered by fluid overpressures in the clastic basin. Their calculations showed that maximum fluid flow by this process should take place a few tens of millions of years after initiation of the sedimentary basin. Finally, McGinnis (1968) suggested that outward fluid flow from sedimentary basins, resulting from the weight of ice during periods of maximum continental glaciation, would result in the formation of MVT deposits. Although suspect, critical evidence opposing this latter hypothesis is generally lacking.

The gravity-driven fluid flow model suggested by Garven and Freeze (1984a,b), Bethke (1985), and Oliver (1986) proposes to develop fluid flow by the development of a hydrologic gradient produced by tectonic uplift of one side of a sedimentary basin. The uplifted area serves as a fluid recharge zone in which fluid enters the sedimentary strata, penetrates to depth where it is heated, and is driven through the subjacent foreland shelf by the hydraulic gradient. As a variant on this model, Rowan and Leach (1989) proposed that MVT deposits formed preferentially in the area of the foreland bulge produced by tectonic uplift of the basin edge.

1.3 Need for Dating

The validity of each of the above models may be tested by determining the depositional age of the MVT deposits. The Beales-Jackson proposal would result in MVT mineralization early in the compactional history of a basin because maximum dewatering of sedimentary strata takes place soon after burial and deposits might be expected to form more or less continuously after that as long as the basin is being compacted. The Cathles-Smith episodic compaction process might be expected to produce MVT deposits about 40 Ma after the start of basin filling and the McGinnis model would yield MVT deposits with ages considerably less than 1 Ma. The gravity-driven fluid flow model would produce MVT deposits with ages correlatable with time(s) of tectonic uplift of the source basin.

1.4 Difficulties of Radiometric Dating Methods

MVT deposits typically occur in relatively undeformed shelf carbonate sequences. Post-ore dykes or veins are uncommon so that dating by geologic events is precluded. Radiometric dating has also proven difficult because galenas in the deposits contain isotopically anomalous leads, and because K-, Rb-, and U-bearing minerals are normally lacking.

The difficulties experienced in dating MVT deposits can be illustrated by recent attempts in the southeast Missouri lead district, USA (Fig. 1). The ores occur in flat-lying Cambrian carbonates and sandstones but smaller showings and occurrences are found in strata as young as Pennsylvanian. Direct dating of fluid inclusions in galena yielded a Rb-Sr isochron of 392 ± 21 Ma (Lange et al. 1983) and dating of pyrite gave a $^{40}Ar-^{39}Ar$ age of 549 ± 20 Ma (York et al. 1981). Indirect dating of the ore using Rb-Sr on glauconite from the host strata has given ages ranging from 350 to 400 Ma (Posey et al. 1983; Grant et al. 1984; Stein and Kish 1985). This lack of consistent success in dating the southeast Missouri ores has discouraged attempts to date other MVT districts by radiometric means. Although MVT districts do not normally contain fluorite or quartz, those that do may be directly dated using the Sm-Nd method on fluorite or indirectly using the Rb-Sr on inclusions in paragenetically related quartz (Shepherd et al. 1982; Chesley et al. 1991). Also, Hearn et al. (1987) have used $^{40}Ar/^{39}Ar$ to date authigenic feldspar intergrown with sphalerite in the east Tennessee district to obtain an age of 322 Ma. Nevertheless, over the past three decades several hypotheses have been proposed to model the formation of MVT deposits. Each may be related to a stage in the sedimentologic or tectonic history of a sedimentary basin but, until the deposits can be reliably dated, there is no way to relate the formation of MVT ores to a particular stage in basinal history, thereby precluding selection of the correct deposit model for MVT ores. Our project, therefore, was designed to develop the palaeomagnetic method as a reliable, and universally applicable method of dating this type of deposit.

2 Palaeomagnetic Methods

2.1 Previous Studies

During the 1970s, F.W. Beales and others at the University of Toronto, recognizing the need to successfully date MVT deposits, were the first to try to use palaeomagnetism for this purpose (Beales et al. 1974; Wu and Beales 1981). Unfortunately, for several reasons, they experienced limited success in their attempts. Firstly, the cryogenic magnetometer they used was not sufficiently sensitive or efficient. Secondly, the apparent polar wander path (APWP) was too poorly defined to use for dating their measured pole positions. Finally, they were unlucky in selecting the Pine Point district as one of their two main research targets because, as our own research has revealed, these ores have an unusually weak remanent magnetization compared with those from most other MVT districts.

Wisniowiecki et al. (1983) determined an Early Permian pole for both the Pb-Zn ores and Upper Cambrian host rocks in the Viburnum Trend district of southeast Missouri, confirming with more precise data the results published earlier by Beales et al. (1974) and Wu and Beales (1981). Wisniowiecki et al. (1983) argued, however, that the ores were emplaced some time after the Permian magnetization event because magnetite spheroids appeared to carry the remanence both in breccia blocks predating ore and in the altered host

Fig. 1. Locations of MVT deposits on which palaeomagnetic dating studies have been carried out (*filled triangles*, published; *filled circles*, in progress by present authors). *1* Southeast Missouri (Wisniowiecki et al. 1983); *2* east Tennessee (Bachtadse et al. 1987); *3* northern Arkansas (Pan et al. 1990); *4* Tri-State (Pan et al. 1990); *5* central Missouri (Symons and Sangster 1991); *6* Polaris (Symons and Sangster 1992); *7* Nanisivik; *8* Daniels Harbour (Newfoundland Zinc); *9* Pine Point; *10* Gays River; *11* southern Illinois; *12* central Tennessee; *13* Silesia, Poland

rock, and so must predate sulphide deposition based on geochemical considerations.

Bachtadse et al. (1987) measured an Early Permian pole for ores and Ordovician host rocks in the east Tennessee district. Because they also obtained a marginally negative fold test and because internal sediments in the ore are conformable with folded host strata, these authors suggested the ore predated folding, whereas the magnetization postdated folding.

From the above brief review of previous palaeomagnetic studies of MVT ores, it is apparent that two main factors contributed to render these attempts inconclusive: (1) it was not until recently that Van der Voo (1990) evaluated the North American and European palaeomagnetic databases and published reliable Palaeozoic APWPs; (2) previous authors could not show that the magnetization and mineralization ages were coeval, and so could not demonstrate that the actual mineralization event had been dated. With these factors in mind, the present authors have paid particular attention to developing tests to show that the magnetization and mineralization events are coeval.

2.2 Sampling

Sampling is usually done by collecting four oriented blocks from each site. The decision to collect blocks, rather than drilling cores, was necessary because most mines do not permit gasoline-powered motors underground. It is also easier and safer to hand sample the ore on fractured mine and pit walls rather than trying to drill them. Later, in the laboratory, three to five, 2.5 cm diameter, specimens are drilled from each block. Since most collections for a given MVT district are represented by 20 to 40 sites, the initial collection typically contains from 250 to 500 specimens.

Sites are preferentially located in mines, pits, or deep road cuts where the effects of weathering are absent or minimal. Because the magnetic intensities of ores and carbonate host rocks are so low, even small amounts of modern meteoric weathering will contaminate the material by deposition of iron oxide minerals. Ideally, about an equal number of sites are located in ore, surrounding alteration (usually dolomite), and in correlative host rock several kilometres removed from ore.

Where feasible, attempts are also made to include sites for which it is possible to conduct traditional palaeomagnetic field stability tests (i.e. on the limbs of macro- or microfolds, across dyke or vein contacts, or from rotated conglomerate or breccia clasts) and for more specialized tests (i.e. massive and disseminated ore, aggregates of associated minerals, large single crystals of the ore stage and associated gangue minerals, especially if they can be oriented). Further details on these tests will be presented later.

2.3 Nature Remanence

Specimens with natural remanent magnetization (NRM) intensities $\leq 2 \times 10^{-5}\,\mathrm{A\,m^{-1}}$ are regarded as being too weak and are routinely rejected. In addition, only the three specimens from each block with the largest NRM intensities are subsequently measured.

Six of the seven MVT collections measured thus far have median NRM intensities of between 1×10^{-4} and $2 \times 10^{-4}\,\mathrm{A\,m^{-1}}$ with nearly all specimens in the 5×10^{-5} to $5 \times 10^{-4}\,\mathrm{A\,m^{-1}}$ range. The seventh collection, from Pine Point, has a median NRM intensity of $3 \times 10^{-5}\,\mathrm{A\,m^{-1}}$, and even after rejecting the weakest half of the nearly 500-specimen collection, it increased to only $6 \times 10^{-5}\,\mathrm{A\,m^{-1}}$. Thus, a good cryogenic magnetometer is required to measure the remanent magnetization in MVT ores. The one used in our studies is an automated two-axis CTF magnetometer with the capability of reliably measuring the moment per unit volume down to about $3 \times 10^{-6}\,\mathrm{A\,m^{-1}}$.

2.4 Step Demagnetization

Most commercial alternating field (AF) demagnetizers will work well with MVT specimens. For our studies, a Sapphire Instruments SI-4 unit with a 200 mT capability is used although the remanence of most specimens is reduced to the noise level of the magnetometer by about 80 to 120 mT. AF step demagnetization usually removes the modern viscous remanent magnetization (VRM) acquired in the laboratory or in the Earth's present magnetic field by about 30 mT (Fig. 2), and thereafter isolates a single characteristic remanent magnetization (ChRM) in the ore, alteration zone, and host rock specimens. More complicated signatures have been encountered only in weathered specimens. Note, however, that the VRM component commonly comprises 30 to 60% of the NRM intensity, so that the geologically significant ChRM is being defined in the low $10^{-5}\,\mathrm{A\,m^{-1}}$ to high $10^{-6}\,\mathrm{A\,m^{-1}}$ range. This substantial VRM component means that it is desirable to store the specimens in a shielded low-ambient magnetic field for 1 month or more before measurements begin, and that it is very desirable to work entirely within a shielded room. Measurements for our MVT studies are conducted in a three-layer transformer steel encased room with an ambient field of about 0.2% of that of the Earth.

Thermal step demagnetization is done sparingly on MVT collections because the sulphide minerals begin to oxidize severely at about 400 °C, giving off noxious fumes. Since it typically takes 300 to 350 °C to remove the VRM, thermal demagnetization is useful for barren alteration zone and host rock specimens only (Fig. 3). Except where surface weathering is evident, demagnetization above 600 °C has not yet been necessary. Both Schonstedt TSD-1 and Magnetic Minerals MMTD thermal demagnetizing units have been used to do this testing and, to date, only magnetite Curie temperatures have been found in all unweathered ores and host rocks.

2.5 Statistical Analysis

The Kirschvink (1980) method is used to identify the magnetic components and all mean angular deviation values accepted to date have been $<10^{0}$. Routinely, the VRM component is found to dominate the measured NRM and to align

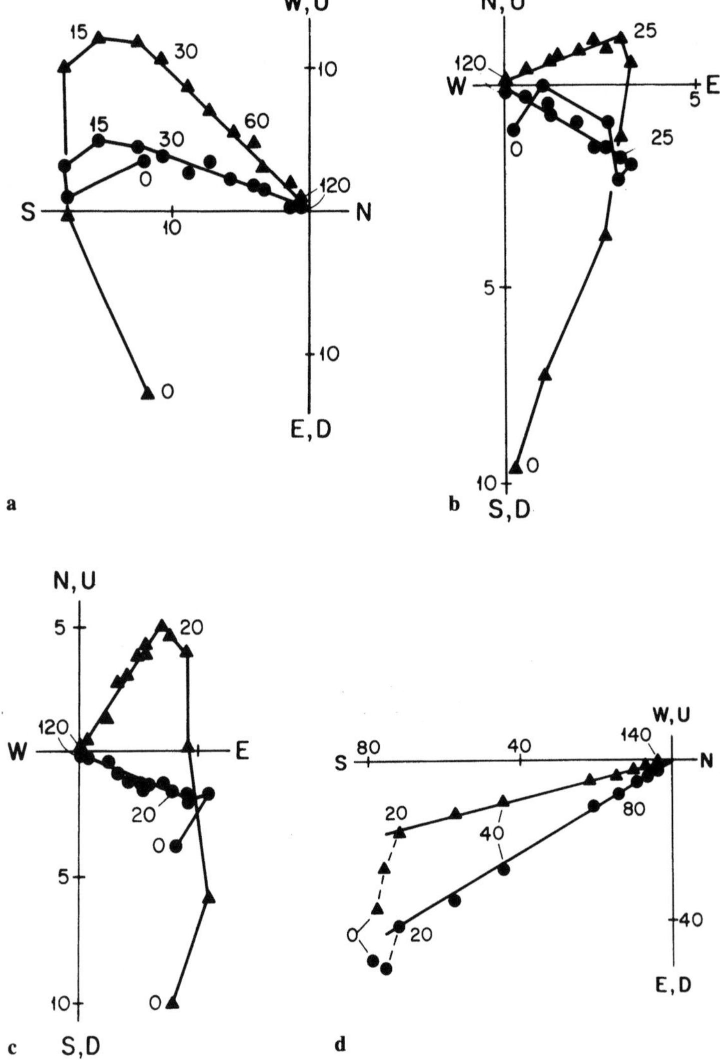

Fig. 2a–d. Orthogonal AF demagnetization diagrams showing examples of specimens from: **a** Polaris host rock limestone; **b,c** Polaris ore with approximately 30% sphalerite and galena; **d** central Missouri massive barite. The axes are north (*N*), east (*E*), south (*S*), west (*W*), up (*U*), and down (*D*). The axial units are in $10^{-5}\,\mathrm{A\,m^{-1}}$. *Circles* and *triangles* denote projections onto the horizontal and vertical planes, respectively. The AF step intensities are in millitesla (mT)

near the present Earth's magnetic field direction (Fig. 4). The ChRM component is obtained by anchoring its direction on the origin because the noise level is normally the limiting factor rather than a residual high coercivity or unblocking temperature component. The ChRM directions are combined using Fisher (1953) statistics to determine site mean directions. For a few specimens,

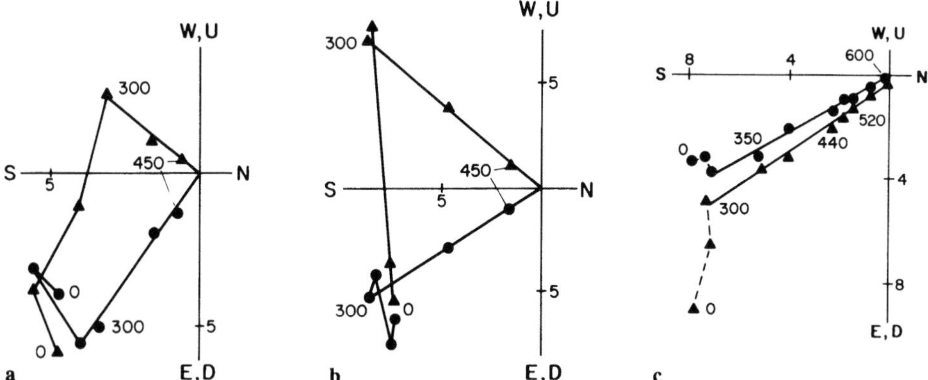

Fig. 3a–c. Orthogonal thermal demagnetization diagrams showing examples of specimens from: **a** limestone with about 10% sphalerite and galena; **b** massive sphalerite and galena from Polaris; **c** massive barite from central Missouri. Conventions as in Fig. 2 except step temperatures are in °C

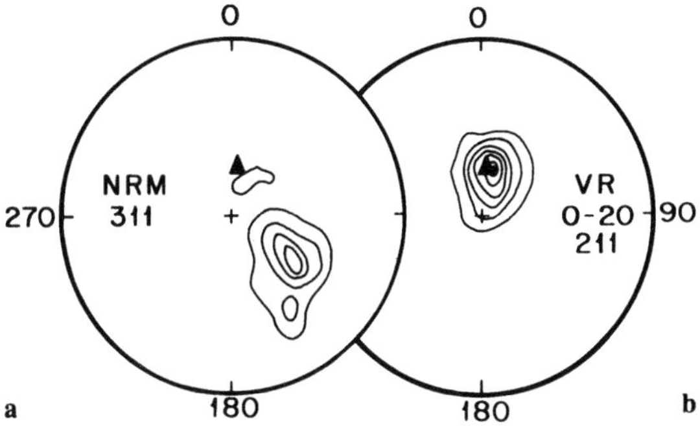

Fig. 4a,b. Contour plot of the percentage density of measured vectors from northern Arkansas samples plotted on the lower hemisphere of an equal-area stereogram showing: **a** the natural remanent magnetization (NRM) for 311 specimens; and **b** the vector removed between NRM and 20 mT on AF demagnetization for 211 specimens, aligned with the Earth's present magnetic field direction (*triangle*)

their data are incorporated using the remagnetization circle method of Bailey and Halls (1984).

In studies thus far, well-defined and consistent ChRM directions have been found at the specimen, site, and collection levels for disseminated and massive sphalerite ore from the east Tennessee and Polaris districts (Fig. 5; Bachtadse et al. 1987; Symons and Sangster 1992); galena ore from the southeastern Missouri and northern Arkansas districts (Wisniowiecki et al. 1983; Pan et al. 1990); barite ore from the central Missouri district (Fig. 6a) (Symons and Sangster 1991); and fluorite ore from the southern Illinois district (unpubl. data). Thus, good palaeomagnetic data have been obtained from all of the major mineralogic types found in MVT deposits. In addition, the hydrothermal

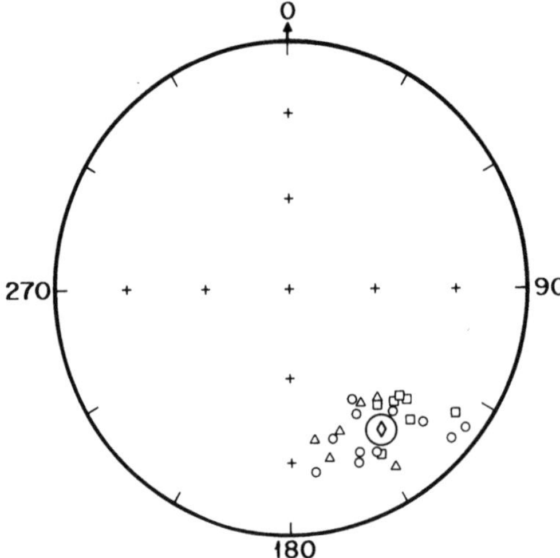

Fig. 5. Equal-area stereogram of the upper hemisphere for the mean remanence directions from Polaris for limestone host rock (*triangles*), dolomite alteration envelope (*circles*) and sphalerite-galena ore (*squares*) sites. The overall mean direction (*diamond*) is surrounded by its cone of 95% confidence

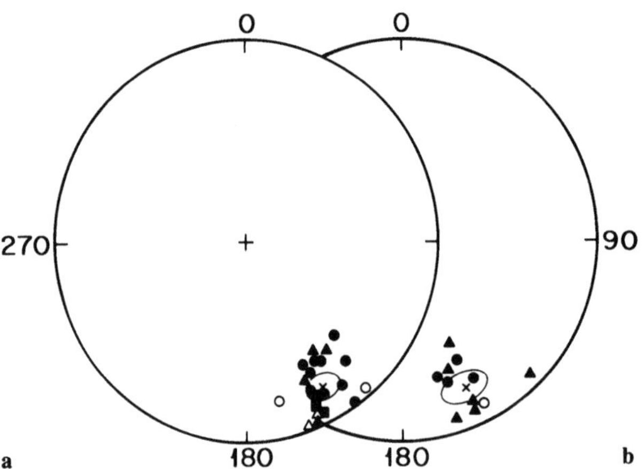

Fig. 6a,b. Equal-area stereograms with down (up) vectors as filled solid (*unfilled open*) symbols showing the ChRM directions from central Missouri for: **a** barite sites from three mining pits (Blackman *circles*, Goller *triangles*, Stenbergen *squares*); **b** remagnetized dolostone host rocks (*circles*) and breccia clasts (*triangles*).

(sparry) alteration dolomite found in many deposits (Figs. 5, 6b) and the siliceous alteration ("jasperoid") associated with the Tri-State and northern Arkansas districts have, in every case, given statistically identical mean ChRM directions to those of their respective ores. Conversely, the remote host rocks may (Wisniowiecki et al. 1983; Bachtadse et al. 1987; Symons and Sangster

1991, 1992) or may not (Pan et al. 1990) be remagnetized to yield the same mean ChRM directions as the ores they host.

2.6 Field Tests

Although field tests for remanence stability are commonly precluded by the geologic characteristics of MVT deposits, there are examples of their successful use. Fold tests have been used in the east Tennessee district where Bachtadse et al. (1987) showed that the ore probably acquired its ChRM after folding of the host strata and in the Polaris district where Symons and Sangster (1992) showed that the ore was prefolding in age. The Nanisivik deposit is cut by a diabase dyke that appears to give a positive contact test (unpubl. data). Similarly, conglomerates associated with the Daniels Harbour (Newfoundland Zinc) deposit yield a negative conglomerate test, indicating that the host rocks were remagnetized since deposition (H. Pan, pers. comm., 1992). Many MVT deposits or their host rocks contain breccias with clasts that have been clearly rotated relative to nearby undisturbed bedding. These situations can be used for a "conglomerate" test, although, in both reported cases to date, the test has been negative, indicating that the breccia has been remagnetized at some time after its formation (Bachtadse et al. 1987; Symons and Sangster 1991; Fig. 6b).

2.7 Saturation Remanence

As a matter of routine, representative specimens of the ores and host rocks, typically about ten per MVT district/deposit, were subjected to saturation isothermal remanent magnetization (SIRM) testing using a Sapphire Instruments SI-4 pulse magnetizer. Specimens were magnetized in direct field steps up to 0.9 or 1.2 T, then AF demagnetized in steps to 50 mT (Fig. 7). Again excluding weathered specimens, results to date have shown that the SIRM acquisition and demagnetization curves are characteristic of single domain (SD) to pseudosingle domain (PSD) magnetite. This indicates that very fine-grained magnetite approaching SD size is the only significant remanence carrier in MVT ore minerals (i.e. sphalerite, galena, barite, fluorite), alteration assemblages, and carbonate host rocks.

2.8 Microscopy

To better characterize the magnetic mineralogy, six to eight extracts of the magnetic material in representative specimens were prepared following methods outlined by Pan et al. (1990). The extracts were examined with a scanning electron microscope with energy dispersive analysis (SEM-EDA). Magnetite spheroids are the common magnetic carrier in Palaeozoic carbonates of mid-continent USA (McCabe and Elmore 1990) and have been shown to be magnetite replacements of pyrite framboids as a result of postdepositional alteration.

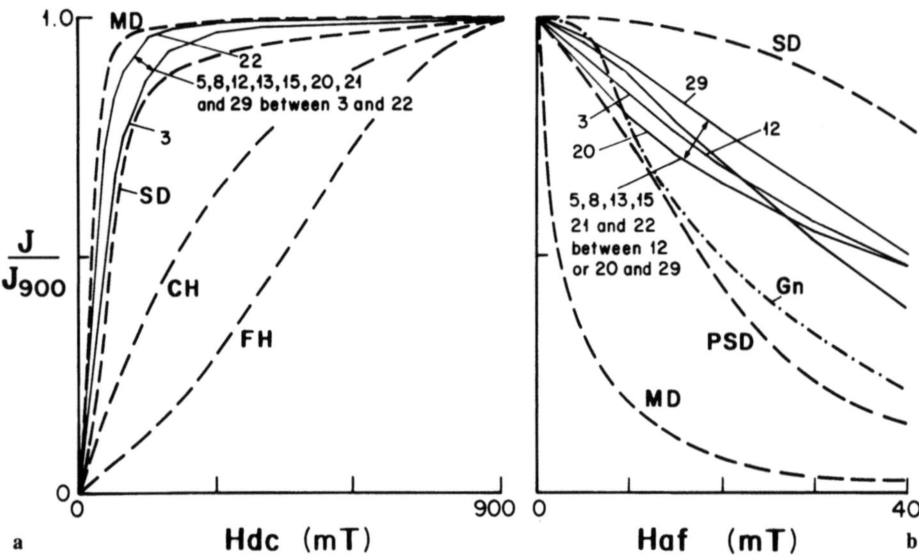

Fig. 7a,b. Saturation isothermal remanent magnetization (SIRM) curves for host rock, alteration zone, and ore specimens from Polaris showing: **a** acquisition of SIRM; **b** demagnetization of SIRM. J/J_{900} is the ratio of the measured remanence intensity to the SIRM intensity at 900 mT and H is the applied direct (*dc*) or alternating (*af*) field in millitesla (mT). *Arrowed bar* locates additional SIRM curves falling between the indicated curves. The type curves are for single domain (*SD*), pseudosingle domain (*PSD*), and multidomain (*MD*) magnetite and for fine-grained (*FH*) and coarse-grained (*CH*) hematite. *Gn* is an artificial specimen of powdered galena in plaster of Paris which tracks specimen curve No. 22 in **a**

Although similar spheroids were found in a few specimens from the northern Arkansas and central Missouri districts, none were found in Polaris specimens. In all three districts, however, small blocky magnetite crystals were found to be the most abundant magnetic mineral in the extracts, indicating that this crystalline form of magnetite, rather than replacements of spheroidal pyrite, is the main magnetic carrier. Occasionally, discrete crystals of sphalerite, galena, and marcasite have been identified in the magnetic extracts, indicating that these minerals also contain a magnetic phase or magnetite inclusions.

Fluid inclusions in MVT ore-stage minerals commonly show homogenization temperatures in the range 70 to 140 °C and this range is in agreement with conodont alteration indices of the host rocks (Sangster 1989). These data demonstrate that the blocky, crystalline magnetite is chemical in origin, that its remanence is a chemical remanent magnetization (CRM) acquired upon formation, and that the remanence has not been overprinted during a later thermal event.

3 Mineralization = Magnetization In Age

3.1 Introduction

In using palaeomagnetic methods to determine the age of MVT mineralization, it is critical to demonstrate that the magnetic component being measured (i.e. magnetite) was deposited at the same time as the non-magnetic ore component (e.g. sphalerite, galena, etc.) whose age is sought. The following discussion describes some of the tests used to confirm that MVT mineralization and magnetism are coeval.

3.2 Massive Ore

Many of the drill specimens were massive "monomineralic" barite, sphalerite, or fluorite from central Missouri, Polaris, and southern Illinois, or of dolomite or chert from alteration zones. Measurements of specific gravity were used to show that these specimens were close to being monomineralic. In each case, the specimens gave remanence directions and intensities statistically identical to those of the ChRM of all other specimens from the same site or the comparable population of monomineralic sites. Further, these specimens have shown the same AF and thermal step demagnetization behaviour, SIRM behaviour, and SEM-EDA characteristics. These observations show that the magnetite ChRM in these "monomineralic" specimens is either coeval with mineralization or possibly postmineralization from secondary intercrystalline magnetite.

3.3 Single Crystals

To check for the possible presence of secondary magnetite, large unoriented single crystals of barite, fluorite and sphalerite from central Missouri and of fluorite from southern Illinois were selected. Cores drilled from these were subjected to AF step demagnetization and SIRM analysis. Thermoremanent analysis was not possible because of distintegration of the crystals during heating. As before, these single crystal specimens yielded magnetization characteristics that are indistinguishable from those of the MVT ore material (Fig. 8). These results are interpreted to indicate that the magnetic mineral properties of the ore mineral aggregates are the same as its constituent minerals and that neither earlier unreplaced nor later secondary interstitial magnetite is responsible for the ChRM.

At the time of preparation of this study, large oriented ore-stage single crystals from the central Tennessee MVT district were collected for determination of their ChRM direction. If these tests show that the crystal ChRM directions are the same as that of the ore, then magnetization and mineralization will be regarded as coeval.

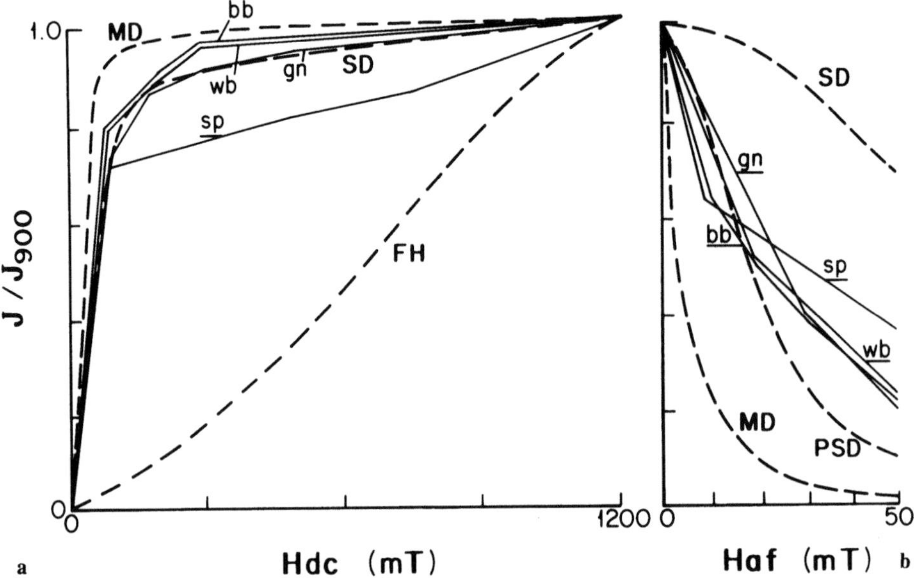

Fig. 8. SIRM curves for single crystals from central Missouri of white barite (*wb*), blue barite (*bb*), and sphalerite (*sp*), and of galena powder from a single crystal dispersed in plaster of Paris (*gn*). Other conventions as in Fig. 7

3.4 Artificial Specimens

Specimens from northern Arkansas and Polaris that carry disseminated galena respond well to AF demagnetization and give good ChRM directions and intensities. Several single crystals of galena were measured and found to have strong NRM intensities of about $10^{-3}\,\mathrm{A\,m^{-1}}$. These very conductive crystals cannot be AF demagnetized or subjected to SIRM testing because of the skin effect, i.e. AF fields of up to 200 mT at 300 Hz do not penetrate and reduce remanence intensity by only about 15%. To test this galena, a portion of the crystal was crushed, mixed with a plaster of Paris or ceramic slurry, and allowed to set in the Earth's ambient magnetic field. These artificial specimens carry about 5 wt% galena and acquire a detrital remanent magnetization parallel to the Earth's field. Blank plaster of Paris and ceramic specimens were also prepared in the same manner for comparison. Artificial and blank specimens were then subjected to AF and thermal demagnetization and to SIRM testing. When the blank specimen values were subtracted from the artificial specimen values, the residual galena coercivity and blocking temperature spectra correspond to SD-PSD magnetite (Fig. 9).

Although the galena crystals have been closely examined with SEM-EDA, magnetite inclusions or phases have not yet been observed. It is evident from the measured remanence intensities of galena and barren host rock, however, that the magnetism carried by the galena is a significant contributor to the ChRM in galena-bearing MVT ores. When disseminated, the galena grains

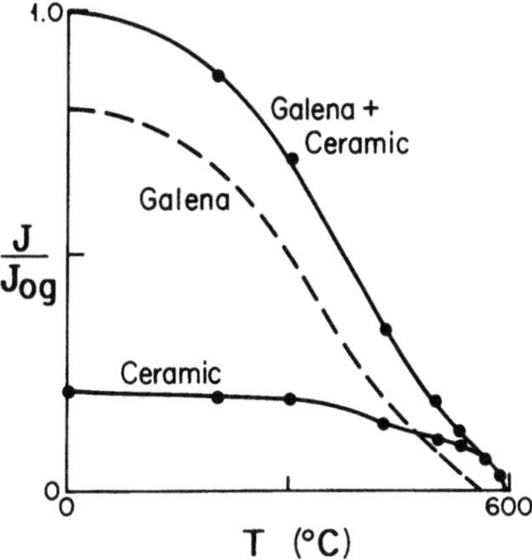

Fig. 9. Decay of remanence intensity on thermal step demagnetization of artificial specimens of ceramic alone and of ceramic plus about 6% galena from a single crystal from Polaris. The galena curve was obtained by subtraction to give a Curie temperature of about 550 °C, indicative of magnetite. The initial remanence intensity of the ceramic plus galena specimen (J_{og}) is $4.7 \times 10^{-2}\,\mathrm{A\,m^{-1}}$

normally possess diameters of less than the skin depth for 300 Hz AF demagnetization. Similarly, they respond well to SIRM testing.

4 Discussion of Results and Geological Models

Three points are important from the preceding discussion. First, MVT deposits commonly occur in a stable tectonic environment so that post-ore rotation of their palaeomagnetic pole is normally not a factor. Second, most MVT ores and altered host rocks have a readily measurable ChRM using current palaeomagnetic methods. Third, all our evidence to date confirms that this ChRM is coeval with the mineralizing event. With this in mind it is instructive to examine palaeomagnetic results from the five MVT districts for which data have been published.

The southeast Missouri, northern Arkansas, Tri-State, and central Missouri MVT districts all give late Pennsylvanian to Early Permian palaeomagnetic ages (Wisniowiecki et al. 1983; Pan et al. 1990; Symons and Sangster 1991) acquired during the Kiaman reversed superchron. Furthermore, we believe that Wisniowiecki et al. (1983) *did* date the age of mineralization in southeastern Missouri and, noting that magnetite inclusions are commonly observed in sulphides and vice versa, we reject the authors' geochemical argument that the magnetite could not have co-precipitated with the sulphides. Thus, all four districts, regardless of host rock age, yield mineralization ages that are within the same range as that of the Ouachita orogeny that uplifted the southern portion of the Arkoma basin in adjacent central Arkansas, thereby supporting the gravity-driven fluid flow model. Leach and Rowan (1986) further suggested that the position of the four districts within the foreland carbonate platform probably coincides with the position of the foreland bulge created

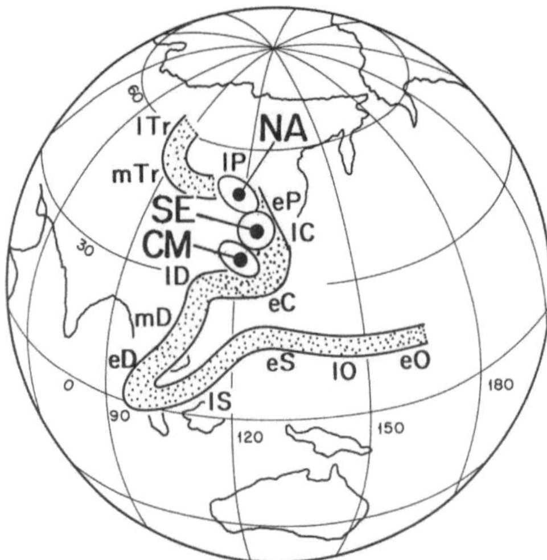

Fig. 10. Apparent polar wander path for the North American craton from Van der Voo (1990): Ordovician (*O*), Silurian (*S*), Devonian (*D*), Carboniferous (*C*) Permian (*P*), Triassic (*Tr*), early (*e*), middle (*m*), late (*l*); with the pole positions for northern Arkansas and Tri-State (*NA*), southeast Missouri (*SE*), and central Missouri (*CM*) MVT districts

as a result of isostatic adjustment associated with the orogeny. Relaxation of stress following an orogeny has been suggested to result in progressive migration of the foreland bulge *toward* the orogenic front (Quinlan and Beaumont 1984). The age difference between northern Arkansas-Tri-State, central Missouri, and southeast Missouri ores, implicit in their palaeopole positions (Fig. 10), is statistically significant at the 95% confidence level. Thus, the district furthest from the Ouachita orogenic front is the oldest and the northern Arkansas-Tri-State districts the youngest. A southward migrating foreland bulge, as proposed by Quinlan and Beaumont (1984), may be responsible for the southward younging of the four MVT districts in the central interior region of the United States.

The palaeomagnetic results from the remaining two MVT districts lead to similar genetic conclusions. The Polaris deposit occurs in Ordovician host rocks, yet gives a Late Devonian age corresponding to the adjacent Ellesmerian orogeny (Symons and Sangster 1992). Similarly, the east Tennessee district is hosted in Ordovician strata, yet gives a Pennsylvanian age corresponding to the early stages of the adjacent Alleghenian orogeny (Bachtadse et al. 1987). This mineralization age does not conflict with the geological observation that mineral deposition must pre-date folding of the host strata (Kendall 1960), nor with the 322 Ma (early Pennsylvanian) age for K-feldspar intergrown with sphalerite (Hearn et al. 1987). Thus, for both Polaris and east Tennessee, the age of mineralization correlates with an adjacent orogenic event, thereby supporting the gravity-driven fluid flow model.

As a general conclusion, the palaeomagnetic method appears to be a valid and generally applicable method for dating MVT mineralization; results to date support an orogeny-induced gravity-driven fluid flow model for ore genesis.

Acknowledgements. The authors take this opportunity to acknowledge, with thanks, three colleagues who have made important contributions to one or more of our published studies: H. Pan, D.L. Leach, and R. Randell. The senior author was supported by funding provided by three E.M.R. Research Agreements, a University of Windsor Research Grant, and, in part, NSERC Operating Grants.

References

Anderson GM, Macqueen RW (1982) Ore deposit models – 6. Mississippi Valley-type lead-zinc deposits. Geosci Can 9:107–117

Bachtadse V, Van der Voo R, Haynes FM, Kesler SE (1987) Late Paleozoic magnetization of mineralized and unmineralized Ordovician carbonates from east Tennessee: evidence for a post-ore chemical event. J Geophys Res 92:165–176

Bailey RC, Halls HC (1984) Estimate of confidence in paleomagnetic directions derived from mixed remagnetization circle and direct observational data. J Geophys 54:174–182

Beales FW, Jackson SA (1966) Precipitation of lead-zinc ores in carbonate reservoirs as illustrated by the Pine Point ore field, Canada. Trans Inst Min Metall Sect B 75:278–285

Beales FW, Carracedo JC, Strangway DW (1974) Paleomagnetism and the origin of Mississippi Valley-type ore deposits. Can J Earth Sci 11:211–223

Bethke CM (1985) A numerical model of compaction-driven groundwater flow and heat transfer and its application to the paleohydrology of intracratonic sedimentary basins. J Geophys Res 80:6817–6828

Cathles LM, Smith AT (1983) Thermal constraints on the formation of Mississippi Valley-type lead-zinc deposits and their implications for episodic basin dewatering and deposit genesis. Econ Geol 78:983–1002

Chesley JT, Halliday AN, Scrivener RC (1991) Samarium-neodymium direct dating of fluorite mineralization. Science 252:949–951

Fisher RA (1953) Dispersion on a sphere. Proc R Soc Lond Ser A 217:295–305

Garven G, Freeze RA (1984a) Theoretical analysis of the role of groundwater flow in the genesis of stratabound ore deposits. 1. Mathematical and numerical model. Am J Sci 284:1084–1124

Garven G, Freeze RA (1984b) Theoretical analysis of the role of groundwater flow in the genesis of stratabound ore deposits. 2. Quantitative results. Am J Sci 284:1125–1274

Grant NK, Laskowski TE, Foland KA (1984) Rb-Sr and K-Ar ages of Paleozoic glauconites from Ohio and Missouri, U.S.A. Isotope Geosci 2:217–239

Hearn PP Jr, Sutter JF, Belkin HE (1987) Evidence for Late-Paleozoic brine migration in Cambrian rocks of the central and southern Appalachians: implications for Mississippi Valley-type sulfide mineralization. Geochim Cosmochim Acta 51:1323–1334

Kendall DL (1960) Ore deposits and sedimentary features, Jefferson City mine, Tennessee. Econ Geol 55:985–1003

Kirschvink JL (1980) The least-squares line and plane and the analysis of paleomagnetic data. R Astron Soc Geophys J 62:669–718

Lange S, Chaudhuri S, Clauer N (1983) Strontium isotopic evidence for the origin of barites and sulfides from the Mississippi Valley-type ore deposits in southeast Missouri. Econ Geol 78:1255–1261

Leach DL, Rowan EL (1986) Genetic link between Ouachita foldbelt tectonism and the Mississippi Valley-type lead zinc deposits of the Ozarks. Geology 14:931–935

Leach DL, Sangster DF (in press) Mississippi Valley-type deposits. In: Kirkham RV, Duke JM, Sinclair WD, Thorpe RI (eds) Mineral deposit modelling. Geol Assoc Can, Spec Pap

McCabe C, Elmore RD (1990) The occurrence and origin of late Paleozoic remagnetization in the sedimentary rocks of North America. Rev Geophys 27:471–494

McGinnis LD (1968) Glaciation as a possible cause of mineral deposition. Econ Geol 58:1145–1156

Noble EA (1963) Formation of ore deposits by water of compaction. Econ Geol 58:1145–1256

Oliver J (1986) Fluids expelled tectonically from orogenic basins: Their role in hydrocarbon migration and other geologic phenomena. Geology 14:99–102

Pan H, Symons DTA, Sangster DF (1990) Paleomagnetism of the Mississippi Valley-type ores and host rocks in the northern Arkansas and Tri-State districts. Can J Earth Sci 27:923–931

Posey HH, Stein HJ, Fullagar PD, Kish SA (1983) Rb-Sr isotopioc analysis of Upper Cambrian glauconites, southern Missouri: implications for movement of Mississippi Valley-type ore fluids in the Ozark region. In: Kisvarsanyi G, Grant SK, Pratt WP, Koenig JW (eds) Int Conf on Mississippi Valley-type lead-zinc deposits Proc Vol. University of Missouri, Rolla, MO, pp 166–173

Quinlan GM, Beaumont C (1984) Appalachian thrusting, lithospheric flexure, and the Paleozoic stratigraphy of the eastern interior of North America. Can J Earth Sci 21:973–996

Rowan EL, Leach DL (1989) Constraints from fluid inclusions on sulfide precipitation mechanisms and ore fluid migration in the Viburnum Trend lead district, Missouri. Econ Geol 84:1948–1965

Sangster DF (1989) Thermal comparison of MVT deposits and their host rocks. Geol Soc Am Annu Mect Abstr Prog 21(6):A7

Shepherd TJ, Darbyshire DPF, More GR, Greenwood DA (1982) Rare earth element and isotope geochemistry of the North Pennine ore deposits. Bur Rech Géol Min Bull (2) II 4:371–377

Stein HJ, Hish SA (1985) The timing of ore formation in southeast Missouri: Rb-Sr glauconite dating at the Magmont mine, Viburnum Trend, southeast Missouri. Econ Geol 80:739–753

Symons DTA, Sangster DF (1991) Paleomagnetic age of the central Missouri barite deposits and its genetic implications. Econ Geol 86:1–12

Symons DTA, Sangster DF (1992) Late Devonian paleomagnetic age for the Polaris Mississippi Valley-type Zn-Pb deposit, Canadian Arctic Archipelago. Can J Earth Sci 29:15–25

Van der Voo R (1990) Phanerozoic paleomagnetic poles from Europe and North America and comparisons with continental reconstructions. Rev Geophys 28:167–208

Wisniowiecki MJ, Van der Voo R, McCabe C, Kelly WC (1983) A Pennsylvanian paleomagnetic pole from the mineralized Late Cambrian Bonneterre Formation, southeast Missouri. J Geophys Res 88:6540–6548

Wu Y, Beales FW (1981) A reconnaissance study by paleomagnetic methods of the age of mineralization along the Viburnum Trend, southeast Missouri. Econ Geol 76:1879–1894

York D, Maslevic A, Hall CM, Kuybida P, Kenyon WJ, Spooner ETC, Scott SD (1981) The direct dating of ore minerals. Ont Geol Surv Misc Pap 98:334–340

Galena and Sphalerite Associated with Coal Seams

D.J. Swaine

Abstract

Occurrences of galena and sphalerite associated with coal seams are described. The sources of lead, zinc and sulphur and mechanisms for the formation of galena and sphalerite associated with coals, either dispersed or in cleats and the like, are discussed. It is postulated that anomalous high concentrations of lead and zinc, and ipso facto of galena and sphalerite, are the result of precipitation from zinc- and lead-bearing brines in a reducing sulphur-rich environment during coalification. Hence, high concentrations of galena and more especially of sphalerite in coal may indicate anomalous metal-rich brines and may be an exploration tool for sediment-hosted Zn-Pb deposits.

1 Introduction

Lead and zinc are present in most coals, usually in low concentrations varying from a few parts per million (ppm) to a few hundred ppm (Swaine 1990). Comparative values for coals, the upper continental crust, terrigenous shale and soil are given in Table 1 for elements selected for their chalcophilic tendencies. The ranges of values for coals and soils are those for "most coals" (Swaine 1990) and for most soils (Swaine 1955). In general, coals have lower contents than shales for most elements and their ranges of values are similar to or less than those for soils. The main exception is selenium which is often higher in coals than in the crust, shale or soils. As well as the chalcophilic elements, reduced sulphur compounds are found in sediments under anaerobic conditions, and hence conditions for the formation of metal sulphides were present during some stages of coalification. There are some situations where local, abnormally high concentrations are found (Table 2) and these are interesting because of their possible relationship to nearby areas where mineralization may have occurred. As examples, the association of galena and sphalerite with coal seams will be dealt with in some detail.

Some information on galena and sphalerite in coals is given by Swaine (1984, 1990). The occurrences are in high rank coals, mostly bituminous, but Finkelman (1982) found them also in four lignite samples from the USA.

CSIRO Division of Coal and Energy Technology, PO Box 136, North Ryde, NSW, 2113, Australia

Spec. Publ. No. 10 Soc. Geol. Applied to Mineral Deposits
Fontboté/Boni (Eds.), Sediment-Hosted Zn-Pb Ores

Table 1. Contents of selected elements in coals, the earth's crust, shale and soil (as ppm)

Element	Coals					Upper continental crust[d]	Terrigenous shale[d,e]	Soil[f]
	World[a]	Australia[b]		USA[c]				
	Range	Range	Mean	Range	Mean	Mean	Mean	Range
Antimony	0.05–10	<0.01–2	0.5	<0.1–7	1	0.2	1.5[e]	0.2–10[g]
Arsenic	0.5–80	0.2–9	1.5	<1–80	15	1.5	13[e]	1–50
Cadmium	0.1–3	0.01–0.20	0.08	<0.01–8	0.4	0.098	0.22[e]	0.02–10[a]
Cobalt	0.5–30	0.6–20	4	0.8–80	8	10	23	1–40
Copper	0.5–50	6–30	15	3–120	18	25	50	2–100
Lead	2–80	2–40	10	<1–60	20	20	20	2–200
Mercury	0.02–1	0.01–0.25	0.10	0.01–1.8	0.2	–	0.18[e]	0.01–0.5[g]
Moblydenum	0.1–10	0.3–4	1.5	0.1–15	3	1.5	1.0	0.2–5
Nickel	0.5–50	3–50	15	1.4–100	16	20	55	5–500
Selenium	0.2–10	0.2–1.6	0.8	< 0.6–20	3.5	0.05	0.5[e]	0.1–2
Silver	0.02–2	<0.02–1	<0.5	0.01–0.48	0.05	0.05	0.07[e]	0.01–8[g]
Thallium	<0.1–1	<0.1–1	–	0.1–2.5	0.4	0.075	1.2[e]	0.1–0.8[g]
Zinc	5–300	10–70	25	2–100	30	71	85	10–300

[a] Swaine (1990).
[b] Based on Clark and Swaine (1962) and Swaine (1977).
[c] Based on Zubovic et al. (1980).
[d] Taylor and McLennan (1985).
[e] Wedepohl (1969–1974).
[f] Swaine (1955).
[g] Bowen (1979).

Table 2. Abnormally high contents of some trace elements in coal (in ppm on a coal basis)

Element	Rank of coal	Content	Location	Reference
Antimony	–	124	China	Cited by Swaine (1990)
	Bituminous	188	South Wales	Cited by Swaine (1990)
Arsenic	Lignite	344	Turkey	Cited by Swaine (1990)
	Bituminous	2 170	Alabama, USA	Finkelman and Brown (1989)
Cadmium	Bituminous	170	Interior Province, USA	Zubovic et al. (1979)
Cobalt	Bituminous	930	Appalachian region, USA	Zubovic et al. (1979)
Copper	Lignite	433	Texas, USA	Cited by Swaine (1990)
		1 668	USA	Finkelman et al. (1990)
Lead	–	900	South Scotland	Taylor (1973)
		1 856	USA	Finkelman et al. (1990)
Molybdenum	Lignite	985	Russia	Cited by Swaine (1990)
	Bituminous	409	South Wales	Cited by Swaine (1990)
Nickel	Bituminous	580	Interior Province, USA	Zubovic et al. (1979)
	–	990	USA	Finkelman et al. (1990)
Selenium	Bituminous	150	Appalachian region, USA	Finkelman et al. (1990)
Silver	Bituminous	66	Texas, USA	Finkelman and Brown (1989)
Zinc	–	32 300	USA	Finkelman et al. (1990)

2 Early Reports

There are some early reports of galena and sphalerite associated with coal seams. More recent reports will be dealt with in later sections.

2.1 Galena

Briggs and Kemp (1928–1929) noted that Percy (1875) stated that galena had been seen in coals from Warwickshire and Yorkshire, England, and Phillips (1896) mentioned the presence of galena in coal. Galena, associated with a fault, was found in the roof of the Eureka seam at the Netherseal colliery, Leicestershire (Binns and Harrow 1897). Another galena-rich vein found in the Maudlin seam, Wearmouth colliery, Yorkshire, also terminated in the roof (Louis 1901–1902). Galena was reported as occurring in coal from central Missouri, USA (Hinds 1912). Ankerite, probably from a Leicestershire coal, was found to contain some galena associated with pyrite and sphalerite (Crook 1913).

2.2 Sphalerite

Wheeler (1895) found sphalerite embedded in lignite from north St. Louis, Missouri, and Jenney (1903) mentioned other occurrences of sphalerite associated with Missouri coals. Binns and Harrow (1897) found sphalerite occurring "as small crystals in cracks, passing through bands and nodules of clay-ironstone, which are found at numerous horizons of the coal-measures, not only in the neighbourhood of faults, but among regular beds" in the Netherseal Colliery, Leicestershire. Ankerite layers in coal, probably from Leicestershire, had sphalerite associated with them, in one case as a thin layer about 0.25 mm thick (Crook 1913). It was noted that the sphalerite "was present in comparatively much larger quantity than the galena". Dove (1921) found sphalerite in pyrite-marcasite from the Pan Handle mine, Bicknell, Indiana, the occurrence being mainly in "the shale above the coal, and between the shale and the coal". The sphalerite grains (1–10 mm) were found as "a filling between the larger granules of pyrite-marcasite". Selected specimens contained about 20% sphalerite, but the overall content was about 0.3%.

3 Occurrences of Galena and Sphalerite

In general, galena is a very rare constituent of coals, whereas sphalerite is commonly found, albeit usually in relatively low concentrations. The highest recorded values for lead in coal are 1856 ppm Pb in a coal from the USA (Finkelman et al. 1990), 900 ppm Pb in a sample of coal from the south of Scotland (Taylor 1973) and 590 ppm Pb in a sample of coal from the Interior Province, USA (Zubovic et al. 1979), whereas values of 4050, 5350, 14 900 and 32 300 ppm Zn were reported for coals from Poland (Tomza 1987), from Knox county, Illinois (Hatch et al. 1976a), from Illinois (Cobb et al. 1980) and from the USA (Finkelman et al. 1990), respectively, Values of greater than about 300 ppm are uncommon for lead and greater than about 1000 ppm are uncommon for zinc (Swaine 1990). In general, zinc is higher than lead in most coals, values being mostly up to 300 ppm for zinc and up to 100 ppm for lead.

3.1 Galena

There is not much detailed information on the occurrences of galena associated with coal seams. It is listed as "rare" by Bouska (1981) and Mackowsky (1982), where "rare" is given as 1–5%, but most reported findings of galena would be equivalent to less than 1%. It is well to recall that quantitative values for minerals at low concentrations are difficult, if not impossible, to achieve. In most cases, estimates are given on the basis of observations by optical microscopy. Lead may occur in coal as galena, clausthalite, associated with pyrite and associated with some barium minerals (Finkelman 1981, 1988). Clausthalite (PbSe) was found first by Finkelman (1978, 1981) in Appalachian coals and in two "foreign" coals, and later in a coal from the Vale of Belvoir, Leicestershire (Cressey and Cressey 1988). However, in most coals galena is probably the main site of lead. Occurrences of galena associated with coals from North America, Europe and Australia will be discussed below.

3.1.1 North America

Finkelman (1981) carried out an in-depth study of mineral matter in coal using scanning electron microscopy (SEM) equipped with an energy dispersive X-ray detector and other techniques (Finkelman 1986). He studied 23 samples of bituminous coals from the Appalachian region, 35 coals (18 bituminous, 5 sub-bituminous, 4 lignite, 8 anthracite) from 5 of the coal provinces in the USA and 21 foreign coals from 14 countries (mostly bituminous), and found that galena was present in 9, 20 and 16% of the samples, respectively. The Appalachian coals were exceptional as 57% of them had clausthalite as the predominant lead mineral. Based on a study of 79 coals, Finkelman (1981) found that the frequency of occurrence of galena was about 25%. An SEM photomicrograph shows micrometre-sized particles of galena on the edges of pyrite grains (Finkelman et al. 1979). In one sample the galena was associated with kaolinite in the pore structure of inertinite and in another sample it was associated with sphalerite in siderite. It may also be associated with pyrite. Galena has been reported as a minor constituent of the Herrin coal, Illinois Basin (Chou and Harvey 1983) and Illinois Basin coals (Ruch et al. 1974).

3.1.2 Europe

Based probably on studies of German coals, Mackowsky (1982) stated that "most coal seams contain small quantities of sphalerite, galena and chalcopyrite". She reported that sometimes "concentric pyritic concretions occur which are composed of alternating layers of pyrite, sphalerite and galena". Bouska (1981) refers to various occurrences of galena associated with Czechoslovakian coal seams, for example, from the Kladno seam, the Ronna, Max and Vanek mines, the Ostrave-Karvina coalfield, and the districts of Radnice, Nyrany, Mirosov, Svatonovice, Zacler, and several mines including Julius, Ferdinand, Antonin and Nosek. The galena often occurs as very fine grains and is sometimes deposited on carbonate minerals. Figure 1 shows galena with pyrite and

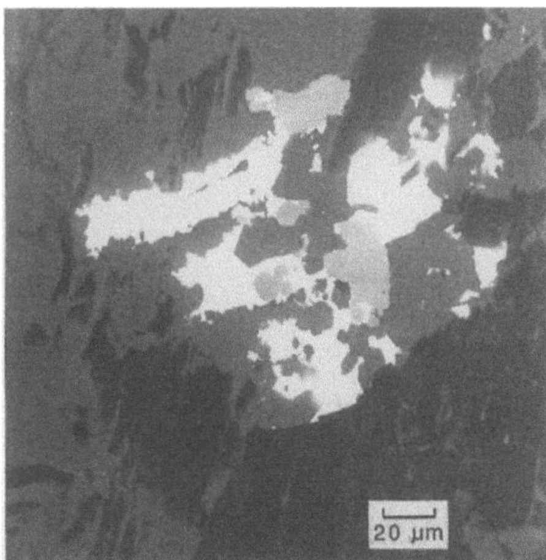

Fig. 1. Galena (*white*), pyrite (*light grey*) and quartz (*dark grey*) filling a cavity in a Czechoslovakian bituminous coal (R.B. Finkelman 1991, pers. comm.)

quartz filling a cavity in a Czechoslovakian bituminous coal (R.B. Finkelman 1991, pers. comm.).

In a study of cleat minerals in four seams (Eight Feet, Park, Yard, Shallow) from the Cannock coalfield, West Midlands, England, Spears and Caswell (1986) found galena in two seams "in monomineralic vitrain cleat". On the basis of the diagenetic sequence proposed for the cleat minerals, "it may be suggested that the cleat minerals developed at a maximum temperature of about 110 °C" (Spears and Caswell 1986). Samples of coal from the Asfordby Hydro Borehole, Vale of Belvoir Prospect, Leicestershire, were examined by Cressey and Cressey (1988) using X-ray powder diffraction and SEM techniques. Galena was identified in coal, shale bands and fracture filling in the coal, and micrographs show the galena in veins with kaolinite and sphalerite, the crystal sizes being generally 10–100 μm.

3.1.3 Australia

Taylor and Warne (1960) found galena in Australian bituminous coals, "though rarely and in insignificant amount". The occurrence is "most often as narrow veins, but also as infillings in fusite, and occasionally as euhedral crystals in shaly coal". It has only been found in coals from the southern area of the Sydney Basin (Kemezys and Taylor 1964) and in one seam from the Collinsville area, Queensland (Taylor 1967).

3.2 Sphalerite

Although sphalerite is rarely found in concentrations exceeding about 1%, it is frequently reported. It is classed as "rare", i.e. 1–5%, by Mackowsky (1982) and "less frequent" by Bouska (1981). Most observations are based on optical microscopy, but in some cases X-ray diffraction and SEM have been used, especially after prior separation of zinc-rich grains. In most high rank coals zinc is predominantly present as sphalerite. Occurrences of sphalerite associated with coals from North America, Europe and Australia will be discussed below.

3.2.1 North America

A consideration of the results for zinc contents in coals from several states in the USA (Table 3) shows that exceptionally high values are found in some coals from the Illinois Basin, Missouri, Iowa and Indiana. Hence, it is not surprising that relatively high concentrations of sphalerite have been found in some coal seams from these areas. These occurrences have been studied in detail and more is known about them than about any other occurrences (Hatch et al. 1976a,b; Cobb et al. 1980; Cobb 1981; Hatch et al. 1984; Whelan et al. 1988). The usual fine grains of sphalerite found in most coals are surpassed in some seams of the Illinois and Forest City basins by sphalerite up to 5 cm wide in cleat fillings (J.R. Hatch pers. comm. to Whelan et al. 1988) and up to 5 cm

Table 3. Zinc contents in coals from several states in the USA (in ppm)

State	Range of values	Mean value	Reference
West Virginia	1–240	14	Zubovic et al. (1980)
Virginia	2–100	15	Zubovic et al. (1980)
Pennsylvania	3–150	24	Zubovic et al. (1980)
Ohio	7–210	32	Zubovic et al. (1980)
Maryland	9–94	36	Zubovic et al. (1980)
Indiana	7–1600	85	Zubovic et al. (1980)
Illinois Basin	10–5300	250	Gluskoter et al. (1977)
Missouri	8–5270	899	Wedge et al. (1976)
Iowa	7–18000	1100	Hatch et al. (1984)

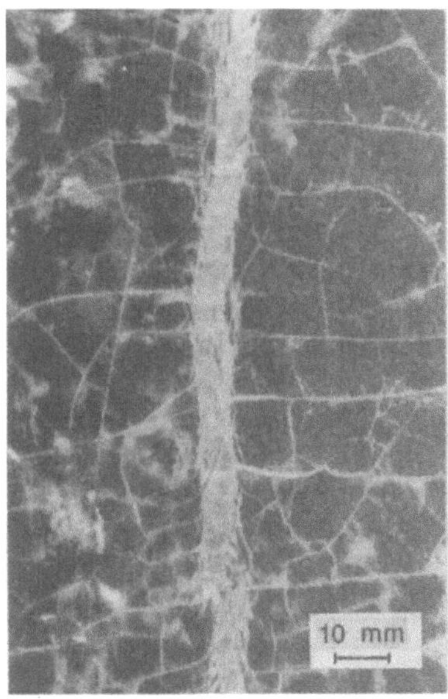

Fig. 2. A sphalerite vein in the Springfield (No. 5) coal, Illinois, USA. (Cobb et al. 1980)

in the longest dimension in some clastic dykes cutting the coal (Cobb 1981). Early studies of coal from the Illinois Basin delineated areas, especially in the northwest, where high zinc coals were common. The zinc was present as sphalerite, usually associated with pyrite, quartz, kaolinite and calcite as fillings in cleats. The term cleat is used to designate vertical fractures in coal. Further investigation by Cobb (1981) and Cobb et al. (1980) showed that the sphalerite-rich coals were from four counties, namely, Fulton, Knox, Peoria and Stark, and that the coals were Herrin No. 6, Danville No. 7, Springfield No. 5 and Colchester No. 2. The main occurrence was "an open-space filling in fractures in the coals" (Cobb et al. 1980) as shown in Fig. 2. The sites included cleats, shears, tension gashes and cell lumens of fusinite. Associations with pyrite nodules and crystal aggregates in clastic dykes were also found. Details of these occurrences, illustrated well by photomicrographs, are given by Cobb et al. (1980).

Fig. 3. Sphalerite (*white*) and calcite (*grey*) vein fillings in a bituminous coal from the Fort Worth basin Texas, USA (R.B. Finkelman 1991, pers. comm.)

The first identification of sphalerite in an Illinois Basin coal was by Zubovic (1960) in a sample of Herrin No. 6 coal and it was also reported by Rao and Gluskoter (1973), Ruch et al. (1974) and Chou and Harvey (1983) from areas where zinc concentrations were not unusually high.

Finkelman (1981) made a detailed study of about 80 coals (mostly bituminous) from various parts of the USA, but including 21 foreign coals. Most of the sphalerite particles were very fine grained, say up to a few micrometre and hence refined techniques, for example, SEM (Finkelman 1981, 1988), were needed. Sphalerite was found in about 35% of the samples. The main occurrence was in the pores of inertinite, as shown by various photomicrographs, for example, a crystal in a crushed inertinite particle (Finkelman 1988), a crystal in semifusinite (Finkelman et al. 1979), pore infilling in a fragment of semifusinite (Finkelman 1978) and globules in sclerotinite (Finkelman and Stanton 1978). It is found occasionally in vitrinite (Finkelman 1981) and there is an interesting photomicrograph of a sphalerite grain with a closely attached smaller grain of PbSe (Finkelman 1986). Sphalerite was not found in association with pyrite. A recent example from a bituminous coal in the Fort Worth basin, Texas (Fig. 3) shows sphalerite grains in a fracture (R.B. Finkelman 1991, pers. comm.).

The multi-technique approach, especially the use of SEM used by Finkelman, provided valuable information about sphalerite and other minerals in the micrometre-size range, not only grain size and morphology, but also the associations with other mineral grains and macerals. Further studies on other coals should be carried out by these techniques.

Wedge et al. (1976) found some high values for zinc (up to 5265 ppm Zn) in coals from several counties in Missouri and inferred from statistical analysis of the results that sphalerite was present. Sphalerite was found as a rare mineral in Seelyville Coal III, Indiana, by Boctor et al. (1976), where "it occurs exclusively in fusinite and is occasionally associated with pyrite as a filling mineral in cleats". A photomicrograph shows crystal aggregates of sphalerite enclosed in a veinlet of pyrite in fusinite. The filling of cell cavities in fusinite is also reported. A study of Indiana Coal Bed V (Springfield) showed that sphalerite only occurs in cellular structures of semifusinite where it is associated with pyrite (Parratt and Kullerud 1979). No sphalerite was found in cleats. In coals from the Cherokee Group, south-central and southeastern Iowa, sphalerite is found along cleats and fractures in the coal in association with pyrite, calcite, kaolinite and barite (Hatch et al. 1976b).

3.2.2 *Europe*

Sphalerite has been identified in German coals. Occasionally, "concentric pyritic concretions occur which are composed of alternating layers of pyrite, sphalerite and galena or also of chalcopyrite" (Mackowsky 1982). Sphalerite has been reported in coals from several areas in Czechoslovakia, namely, the Kladno seam, the Ronna, Max and Vanek mines, the Ostrava-Karvina coalfield and the districts of Radnice, Nyrany, Mirosov, Svatonovice, and several mines including Julius, Ferdinand, Antonin and Nosek (Bouska 1981). The sphalerite is mostly very fine grained, but also occurs in fractures and occasionally deposited on carbonate minerals. Spears and Caswell (1986) found sphalerite in cleats in two of four coal seams studied (Eight Feet, Park) from the Cannock coalfield, West Midlands, England. The ions needed to form the cleat minerals were provided by the coal and associated sediments at temperatures not exceeding about 110 °C. Sphalerite was identified in samples of coal from the Vale of Belvoir Prospect, Leicestershire, by Cressey and Cressey (1988) using modern techniques. It was seen in fracture-fill material, as grains in dull coal and as a coating on a lamination in dull coal. One photomicrograph shows sphalerite and galena in adjacent channels in fibrous coal.

3.2.3 *Australia*

Taylor and Warne (1960) found that sphalerite in Australian coals "is fairly common but occurrences are small". It occurs "in isolated anhedral grains up to 100 microns across in both coal and clay-rich layers". It is also found in veins commonly 10–20 μm wide, often associated with quartz (chalcedony), and as an infilling in cell lumens of semifusinite. An unusual occurrence is as nodules in vitrinite shown on a photomicrograph (Kemezys and Taylor 1964). Sphalerite was observed as grains, veins and in semifusinite lumens in coals from the Tomago Coal Measures, New South Wales (Smyth 1968) and in samples of coal from 12 seams in the Liddell district, New South Wales (CSIRO 1967). Other occurrences in New South Wales were in coals from the Fassifern, Wallarah and Wongawilli seams (Corcoran 1979). "Sphalerite was unusually common (although the concentration was still well under 1 per cent) in some parts" of the Big seam, Blair Athol, Queensland (CSIRO 1960). It was found as small crystals, "Often filling cell lumens in semifusinite" and in those parts of the seam where carbonate minerals were present. Several samples of coal from the Collinsville, Blackwater, Baralaba and Moura/Kianga areas of the Bowen Basin, Queensland had traces of sphalerite (Taylor 1967). Smyth (1966) found that sphalerite in Queensland coals was found only associated with vitrinite and semifusinite, more commonly the former. Sphalerite was also reported from seams in the Callide and Ipswich districts (Corcoran 1979). In the Leigh Creek coalfield, South Australia, sphalerite occurred in coal and clay layers "as rare isolated grains and crystals, generally 50 μm or less in size" (CSIRO 1964).

4 Associations with Macerals, Minerals and Impurities

Various associations with macerals and minerals were referred to previously (Sect. 3). For example, Finkelman (1988) found that sphalerite tended to be preferentially associated with inertinite, although it was also found with vitrinite and exinite (now termed liptinite). It would seem that sphalerite is most likely to be found associated with inertinite or similar macerals (semi-fusinite, sclerotinite), but other associations are also found. Galena and sphalerite occur, at least partly, as very fine-grained minerals embedded in the organic matrix and hence they cannot be separated by ordinary float-sink methods. Cressey and Cressey (1988) found galena in veins with kaolinite and sphalerite. In the high sphalerite coals in some Illinois seams, sphalerite was associated with pyrite, quartz, kaolinite and calcite in fillings in cleats. It would seem unwise to generalize about expected associations of galena and sphalerite with other minerals, because these surely depend on subtle changes in conditions, for example, in pH and Eh, during coalification.

As expected, impurity ions are found in or closely associated with galena and sphalerite in coal. Degens (1965) stated that "it appears that sedimentary sulfides do not allow as much lattice substitution as the comparable high temperature counterpart, for the simple reason that low temperature environments do not favour initial lattice expansion". Hence, there is probably only limited substitution in coal sulphides. There are several reports of pyrite with relatively high contents of lead. Pyrite from the Young Wallsend seam, Sydney Basin, Australia, had about 0.3% Pb, which was probably present as admixed galena (Brown and Swaine 1964). Lead and zinc are associated with pyrite in coal from the Donets Basin, Russia (Dvornikov 1967; Dvornikov and Tikhonenkova 1968), from the Ruhr district, Germany (Pickhardt 1989) and from Bulgaria (Eskenazy 1989). In three Bulgarian coals Eskenazy (1989) found that 8–24% of the total lead and 24–39% of the total zinc in the coal were associated with pyrite. It is feasible that the lead and zinc are present as admixed galena and sphalerite. The thallium found in sulphide inclusions, mainly pyrite, in several coals from Russia (Voskresenskaya 1968) may be associated with admixed galena where isomorphous substitution of thallium for some lead is possible. About 1% zinc was found in a sulphide mineral, mainly marcasite, from the Four Feet seam, Ipswich coalfield, Queensland, and presumed to be admixed sphalerite (Brown and Swaine 1964). The association of cadmium with sphalerite in several coals from Illinois was explained by Gluskoter and Lindahl (1973) as the cadmium being "in solid solution, replacing zinc in the mineral sphalerite (ZnS)". It is likely that sphalerite in coal is the host for cadmium in most, if not all, coals. It is probable that the statement that "it is often difficult to distinguish between matter in true solid solution in the pyrite structure and impurities contained in discrete minerals" (Morse et al. 1987) applies generally to sulphide minerals in coals.

5 Formation of Galena and Sphalerite

Two trends in ore genesis are designated by Amstutz (1964), namely, "causes from within" and "causes from without" and there is an analogous situation for sulphide minerals in coal "where the sources of elements for mineral formation were both within the coal swamp and also introduced from outside" (Swaine 1984). Galena and sphalerite are found as fine, micrometre-sized particles usually intimately dispersed in the coaly matter, and as coarse particles, usually found as epigenetic deposits in cracks, fractures and cleats. In general, these may be up to a few millimetres wide, but sphalerites up to 5 cm wide occur in some cleats (J.R. Hatch, pers. comm. to Whelan et al. 1988). Galena and sphalerite are authigenic and have not been found as detrital minerals (Finkelman 1981).

The formation of fine-grained particles of galena and sphalerite is regarded as taking place during the early stages of coalification. A mechanism for this will be proposed based on information and suggestions in Mackowsky (1982) and Swaine (1975, 1984). Although lead and zinc derived from decaying plant matter could have been the sources for galena and sphalerite, this is most unlikely and the preferred source is aqueous solutions reaching the peat or coal swamp from nearby areas. The lead and zinc may form complexes with organic matter, especially during the early stages of coalification when there are carboxylic acid (—COOH) and phenolic hydroxyl (—OH) groups for reaction to form chelates. There may be a preferential chelation effect which may vary with the stage of diagenesis, the pH and the degree of reduction, as postulated for some Recent marine sediments (Nissenbaum and Swaine 1976). Under conditions of impeded drainage with consequent reducing conditions as found for lead and zinc in soils (Swaine and Mitchell 1960), lead and zinc would be mobilized. Sulphate ions in the presence of suitable organic matter and under relevant pH and anoxic conditions could have been reduced to give H_2S or HS^- for reaction to form galena and sphalerite, assuming that the metal chelates are less stable than the metal sulphides.

For the formation of coarse-grained galena and sphalerite in cracks, fractures and cleats at a later stage of coalification, the source of lead and zinc is probably as chlorocomplexes in solutions entering the voids in the coal. Perhaps the sulphur could come from sulphate accompanying the metals in solution, but it is most likely to have been derived from the decomposition of organic sulphur in the coal, possibly mercapto (—SH). This latter mechanism is supported by sulphur isotope values for sphalerite and pyrite in cleats in coals from the Illinois and Forest City Basins (Whelan et al. 1988). According to Skinner (1967), the slow degradation of organic matter gives a slow release of hydrogen sulphide. Support is also given by Spears and Caswell (1986) who stated that "it is difficult to envisage preferential sulfate reduction in the coal compared with other organic rich sediments in the sequence", presumably including coal. Examination of fluid inclusions in sphalerites from cleats in Illinois coals indicated homogenization and freezing temperatures from 82 to 102 °C in keeping with 75–113 °C found by Cobb (1981) for similar coals, and that the sphalerite was deposited from a sodium chloride-rich brine (Roedder 1979).

The state of lead and zinc in solutions draining from nearby areas into coal swamps is not known, but some speculation is worthwhile. Amongst the possibilities are simple compounds, for example sulphates or carbonates, adsorbed cations on clays and chlorocomplexes, in keeping with the brine suggested by Roedder (1979).

6 Galena and Sphalerite in Coal in Relation to Nearby Mineralization

It seems reasonable to regard lead and zinc in galena and sphalerite in coal as being derived from rocks in nearby areas which drain into the coal swamp. Perhaps the usually relatively low levels of lead and zinc in coals, and ipso facto of galena and sphalerite, mean that the incoming waters are also low in lead and zinc, because the rocks are also low. However, relatively high levels of galena and more usually of sphalerite could be derived from higher levels of lead and zinc coming from areas where there is some lead and zinc mineralization. The relevance of ore deposits to lead and zinc in coal is cited by Bouska (1981) for an area near Bytom, Upper Silesia. According to Voskresenskaya (1968), the concentrations of thallium in various Russian coals depend entirely on the composition of the source rocks.

As a consequence of the above postulates, it seems possible that the levels of galena and sphalerite in coals which are relatively high in these minerals may be useful as guides to mineralization in nearby areas. Some support for this suggestion is given by Hatch (1983) who found that the sphalerite in cleats and fractures in some midcontinent USA coals was "associated with calcite, pyrite, kaolinite and barite, a mineral assemblage common to Mississippi Valley lead-zinc deposits". Increased contents of antimony, arsenic, lead and zinc in pyrite from coals from the central Donets Basin, Russia, may be useful in delineating sectors where new hydrothermal lead-zinc deposits may be found (Dvornikov 1967). Following their investigation of sphalerite associated with coals of the Illinois Basin, Hatch et al. (1976a) suggested "that the primary control of the amount of sphalerite in coal is the proximity to areas of mineralisation". Coal as "a geochemical indicator of mineralization" has been proposed also by Finkelman and Brown (1989) and Finkelman et al. (1990).

The high sphalerite-bearing coals of parts of Illinois are unique and have been considered as a potential source of zinc and cadmium. Cobb et al. (1980) estimated that coals from four counties, where high sphalerite occurs, contained 3–7 million t of zinc and 30000–70000 t of cadmium.

7 Concluding Remarks

Most of the lead and zinc in coals occurs in galena and sphalerite. Although galena is regarded as rare, sphalerite has been reported fairly frequently from various coals, except brown coals. Early formed galena and sphalerite are usually fine-grained and disseminated in the coaly organic matter, whereas the later formed minerals tend to be coarse-grained and to occur in cleats and related voids. In the first case the sulphur is probably derived from the bacterial reduction of sulphate, whereas in the second case, the preferred source is the breakdown of organic sulphur in the coal. The source of the cations is solutions draining into the coal swamp from the nearby area.

When the levels of galena and more notably of sphalerite in coals are relatively high, it is suggested that they may be indicators of mineralization in a nearby area. It must be stressed that such high concentrations are rare, and that most coals would gain lead and zinc from areas where the relevant concentrations are not high.

Acknowledgements. I am most grateful to Dr. R.B. Finkelman, US Geological Survey, for giving me Figs. 1 and 3 and for alerting me to two of his recent papers which also espouse the relevance of certain coals to mineralization.

References

Amstutz GC (1964) Introduction. In: Amstutz GC (ed) Sedimentology and ore genesis. Developments in sedimentology, vol 2. Elsevier, Amsterdam, pp 1–7

Binns GJ, Harrow G (1897) On the occurrence of certain minerals at Netherseal colliery, Leicestershire. Trans Inst Min Eng 13:252–255

Boctor NZ, Kullerud G, Sweany JL (1976) Sulfide minerals in Seelyville Coal III, Chinook mine, Indiana. Mineral Depos 11:249–266

Bouska V (1981) Geochemistry of coal. Elsevier, Amsterdam, 284 pp

Bowen HJM (1979) Environmental chemistry of the elements. Academic Press, London, 333 pp

Briggs H, Kemp CN (1928–1929) A note on the mineralogy of coal, as suggested by X-ray examination. Trans Inst Min Eng 78:5

Brown HR, Swaine DJ (1964) Inorganic constituents of Australian coals. J Inst Fuel 37:422–440

Chou C-L, Harvey RD (1983) Composition and distribution of mineral matter in the Herrin coal of the Illinois Basin. Proc Int Conf Coal Sci Pittsburgh, International Energy Agency, pp 373–376

Clark MC, Swaine DJ (1962) Trace elements in coal. CSIRO Div Coal Res Tech Commun 45, 109 pp

Clarke FW (1916) The data of geochemistry. US Geol Surv Bull 616, 821 pp

Cobb JC (1981) Geology and geochemistry of sphalerite in coal. Thesis, University of Illinois, Urbana, 204 pp

Cobb JC, Steele JD, Treworgy CG, Ashby JF (1980) The abundance of zinc and cadmium in sphalerite-bearing coals in Illinois. Ill State Geol Surv Ill Miner Note 74, 28 pp

Corcoran JF (1979) Mineral matter in Australian steaming coals – a survey. Colloq combustion of pulverised coal: the effect of mineral matter. University of Newcastle, Newcastle, L4-1–L4-16

Cressey BA, Cressey G (1988) Preliminary mineralogical investigation of Leicestershire low-rank coal. Int J Coal Geol 10:177–191

Crook T (1913) On the frequent occurrence of ankerite in coal. Mineral Mag 16:219–223

CSIRO (1960) Studies of the characteristics of the Big seam, Blair Athol, Queensland. CSIRO Div Min Chem Tech Commun 39, 94 pp

CSIRO (1964) Characteristics of coals from Lobe C, Leigh Creek coalfield, South Australia. CSIRO Div Coal Res Loc Rep 329, 47 pp

CSIRO (1967) Characteristics of coals from the Liddell district Northern coalfield, New South Wales, and comparison with Great Northern seam coal from Newvale No. 1 colliery. CSIRO Div Coal Res Tech Commun 49, 73 pp

Degens ET (1965) Geochemistry of sediments – a brief survey. Prentice-Hall, New Jersey, 342 pp

Dove LP (1921) Sphalerite in coal pyrite. Am Mineral 6:61
Dvornikov AG (1967) Mercury distribution in disulfides of iron from coal seams of the Central Donets Basin. Dokl Akad Nauk SSSR 172:211–213
Dvornikov AG, Tikhonenkova EG (1968) Distribution and composition of iron disulfides from coals of the Donets Basin. Zap Vses Mineral Ova 97:309–320
Eskenazy GM (1989) Modes of occurrence of trace elements in Bulgarian coals. J Coal Qual 8:102–109
Finkelman RB (1978) Determination of trace element sites in the Waynesburg coal by SEM analysis of accessory minerals. Scanning Electron Microsc 1:143–148
Finkelman RB (1981) Modes of occurrence of trace elements in coal. US Geol Surv Open-file Rep OFR-81–99, 301 pp
Finkelman RB (1982) The origin, occurrence, and the distribution of the inorganic constituents in low-rank coals. In: Schobert HH (ed) Low-rank coal basic coal science workshop. Proc CONF-811268 US Dept Energy. DOE, Washington DC, pp 70–89
Finkelman RB (1986) Characterization of the inorganic constituents in coal. Proc Mater Res Soc Symp 65:71–76
Finkelman RB (1988) The inorganic geochemistry of coal: a scanning electron microscopy view. Scanning Microsc 2(1):97–105
Finkelman RB, Brown RD (1989) Mineral resource and geochemical exploration potential of coal that has anomalous metal concentrations. US Geol Surv Circ 1039:18–19
Finkelman RB, Stanton RW (1978) Identification and significance of accessory minerals from a bituminous coal. Fuel 57:763–768
Finkelman RB, Stanton RW, Cecil CB, Minkin JA (1979) Modes of occurrence of selected trace elements in several Appalachian coals. Am Chem Soc Div Fuel Chem Prep 24(1):236–241
Finkelman RB, Bragg LJ, Tewalt SJ (1990) Byproduct recovery from high-sulfur coals. In: Markuszewski R, Wheelock TD (eds) Processing and utilization of high-sulfur coals III. Elsevier, Amsterdam, pp 89–96
Gluskoter HJ, Lindahl PC (1973) Cadmium-mode of occurrence in Illinois coals. Science 188:264–266
Gluskoter HJ, Ruch RR, Miller WG, Cahill RA, Dreher GB, Kuhn JK (1977) Trace elements in coal: occurrence and distribution. Ill State Geol Surv Circ 499, 154 pp
Hatch JR (1983) Geochemical processes that control minor and trace element composition of United States coals. In: Shanks WC (ed) Cameron volume on unconventional mineral deposits. Soc Econ Geol, New York, pp 89–98
Hatch JR, Gluskoter HJ, Lindahl PC (1976a) Sphalerite in coals from the Illinois Basin. Econ Geol 71:613–624
Hatch JR, Avcin MJ, Wedge WK, Brady LL (1976b) Sphalerite in coals from southeastern Iowa, Missouri and southeastern Kansas. US Geol Surv Open-file Rep 76–796, 26 pp
Hatch JR, Avcin MJ, van Dorpe PE (1984) Element geochemistry of Cherokee Group coals (Middle Pennsylvanian) from south-central and southeastern Iowa. Iowa Geol Surv Tech Pap 5, 108 pp
Hinds H (1912) The coal deposits of Missouri. Mo Bur Geol Mines Ser 2 (11) 503 pp; cited by Wedge et al. (1976)
Jenney WP (1903) The mineral crest or the hydrostatic level attained by the ore depositing solutions in certain mining districts of the Great Salt Lake Basin. Trans Am Inst Min Eng 33:46
Kemezys M, Taylor GH (1964) Occurrence and distribution of minerals in some Australian coals. J Inst Fuel 37:389–397
Louis H (1901–1902) Note on a mineral vein in Wearmouth colliery. Trans Inst Min Eng 22:127–129
Mackowsky M-Th (1982) Minerals and trace elements occurring in coal. In: Stach E. Mackowsky M-Th, Teichmüller M, Taylor GH, Chandra D, Teichmüller R (eds) Stach's textbook of coal petrology, 3rd edn. Borntraeger, Berlin, pp 153–171
Morse JW, Millero EJ, Cornwell JC, Rickard D (1987) The chemistry of the hydrogen sulfide and iron sulfide systems in natural waters. Earth-Sci Rev 24:1–42
Nissenbaum A, Swaine DJ (1976) Organic matter-metal interactions in Recent sediments: the role of humic substances. Geochim Cosmochim Acta 40:809–816
Parratt RL, Kullerud G (1979) Sulfide minerals in coal bed V, Minnehaha mine, Sullivan County, Indiana. Mineral Depos 14:195–206

Percy J (1875) Metallurgy: introduction, refractory materials and fuel. Murray, London, 596 pp
Phillips JA (1896) A treatise on ore deposits, 2nd edn (rewritten and enlarged by H. Louis). Macmillan, London, 950 pp
Pickhardt W (1989) Trace elements in minerals of German bituminous coals. Int J Coal Geol 14:137–153
Rao CP, Gluskoter HJ (1973) Occurrence and distribution of minerals in Illinois coals. Ill State Geol Surv Circ 476, 56 pp
Roedder E (1979) Fluid inclusion evidence on the environments of sedimentary diagenesis, a review. SEPM Spec Publ 26:89–107
Ruch RR, Gluskoter HJ, Shimp NF (1974) Occurrence and distribution of potentially volatile trace elements in coal. Environ Geol Notes Ill State Geol Surv 72, 96 pp
Skinner BJ (1967) Precipitations of Mississippi Valley type ores: a possible mechanism. In: Brown JS (ed) Genesis of stratiform lead-zinc-barite-fluorite deposits. Econ Geol Monogr 3:363–370
Smyth M (1966) The association of minerals with macerals and microlithotypes in some Australian coals. CSIRO Div Coal Res Tech Commun 48, 35 pp
Smyth M (1968) The petrography of some New South Wales Permian coals from the Tomago Coal Measures. Australas Inst Min Metall Proc 225:1–9
Spears DA, Caswell SA (1986) Mineral matter in coals: cleat minerals and their origin in some coals from the English Midlands. Int J Coal Geol 6:107–125
Swaine DJ (1955) The trace-element contents of soils. Commonw Bur Soil Sci Tech Commun 48, 157 pp
Swaine DJ (1975) Trace elements in coals. In: Tugarinov AI (ed) Recent contributions to geochemistry and analytical chemistry. Wiley, New York, pp 539–550
Swaine DJ (1977) Trace elements in coal. In: Hemphill DD (ed) Trace substances in environmental health XI. University of Columbia, Missouri, pp 107–116
Swaine DJ (1984) Sulfide minerals in coal with emphasis on Australian occurrences. In: Wauschkuhn A, Kluth C, Zimmermann RA (eds) Syngenesis and epigenesis in the formation of mineral deposits. Springer, Berlin Heidelberg New York, pp 120–129
Swaine DJ (1990) Trace elements in coal. Butterworths, London, 290 pp
Swaine DJ, Mitchell RL (1960) Trace-element distribution in soil profiles. J Soil Sci 11:347–368
Taylor CJ (1973) Mercury and other potentially toxic trace elements in British coals. NCB Yorkshire Regional Lab Rep NCB/YRL/Misc 1, 21 pp
Taylor GH (1967) Coals of the Bowen Basin, Eastern Queensland. CSIRO Div Min Chem Tech Commun 50, 51 pp
Taylor GH, Warne SStJ (1960) Some Australian coal petrological studies and their geological implications. Proc Int Comm Coal Petrol 3:75–87
Taylor SR, McLennan SM (1985) The continental crust: its composition and evolution. Blackwell, Oxford, 312 pp
Tomza U (1987) Trace element content of some Polish hard and brown coals. J Radioanal Nucl Chem Lett 119:387–396
Voskresenskaya NT (1968) Thallium in coal. Geochem Int 5:158–168
Wedepohl KH (ed) (1969–1974) Handbook of geochemistry 6 vols. Springer, Berlin Heidelberg New York
Wedge WK, Bhatia DMS, Rueff AW (1976) Chemical analyses of selected Missouri coals and some statistical implications. Mo Dep Nat Resour Div Geol Land Surv Rep Invest 60, 33 pp
Wheeler HA (1895) Trans Acad Sci St Louis 7: 123; cited by Clarke (1916)
Whelan JF, Cobb JC, Rye RO (1988) Stable isotope geochemistry of sphalerite and other mineral matter in coal beds of the Illinois and Forest City Basins. Econ Geol 83:990–1007
Zubovic P (1960) Minor element content of coal from Illinois beds 5 and 6 and their correlatives in Indiana and western Kentucky. US Geol Surv Open-file Rep, 79 pp
Zubovic P, Hatch JR, Medlin JH (1979) Assessment of the chemical composition of coal resources. UN Symp World Coal Prospects Rep TCD/NRET/AC 12/EP/15, 24 pp
Zubovic P, Oman CL, Bragg LJ, Coleman SL, Rega NH, Lemaster ME, Rose HJ, Golightly DE, Puskas J (1980) Chemical analysis of 659 coal samples from the eastern United States. US Geol Surv Open-file Rep 80-2003, 513 pp

North American Deposits

Dolostone-Hosted Sulfide Occurrences in Silurian Strata, Appalachian Basin of New York

G.M. Friedman

Abstract

During the middle Silurian, at about 420 Ma, the Lockport strata (mostly dolostone) formed under peritidal conditions in an arid evaporitic setting. The environment of deposition was that of a shoaling-upward sequence from subtidal to intertidal and, subsequently, a supratidal setting followed by subaerial emergence. Under subaerial conditions now mineral-filled former vugs and cavities, and solution breccias, resulted from dissolution of evaporites by surface and near-surface percolating freshwaters. Vugs, cavities and solution-collapse breccias formed under subaerial conditions, representing an unconformity setting.

Subsequent subsidence depressed the strata to a burial depth of 5 km at which the strata were fractured and filled with sulfide, carbonate, and sulfate minerals, including saddle dolomite, sphalerite, galena, marcasite, pyrite, fluorite, anhydrite, calcite, quartz, barite and celestite. The fluids from which the minerals precipitated were hot and saline; the temperatures were at about 150 °C. Their salinity exceeded that of seawater three- to sevenfold. The period of time between original deposition, subsidence to 5 km, and uplift and erosion took 170 to 220 million years. Uplift coincided with the Allegheny orogeny.

1 Introduction

Minor occurrences of sphalerite, galena, and pyrite are widespread in the northern Appalachian Basin. Among the earliest producing mines in the United States are Mississippi Valley-type (MVT) deposits. Among these, the deposits in the Shawangunk Formation (Silurian) compare with northern European Hercynian-age lead-zinc deposits and are distributed in clastic rocks (Friedman and Mutschler 1987).

The Lockport Formation (middle Silurian) of western New York is a dense, massive, fine-grained dolostone that hosts a suite of minerals which has been well known to mineral collectors for over 200 years. These mineral occurrences were first described by Eaton (1842) and Robinson (1825). The

Department of Geology, Brookyn College and the Graduate School of the City University of New York, Brooklyn, NY 11210 and Northeastern Science Foundation, affiliated with Brooklyn College of the City University of New York, Rensselaer Center of Applied Geology, P.O. Box 746, Troy, NY 12181-0746, USA

Spec. Publ. No. 10 Soc. Geol. Applied to Mineral Deposits
Fontboté/Boni (Eds.), Sediment-Hosted Zn-Pb Ores

minerals include sphalerite, galena, pyrite, marcasite, anhydrite, gypsum, aragonite, barite, celestite, calcite, fluorite, quartz, and sulfur. They fill vugs, geodes, cavities, and fractures. When the Erie Canal was built to connect Lake Erie with the Hudson River, piles of rocks from the Lockport Formation along the canal's banks provided early mineralogists, such as Beck (1842), with an exceptional suite of minerals (Jensen 1978).

2 Lithology and Depositional Setting of the Host Bed Rock

In New York State the Lockport Formation of Niagaran (middle Silurian) age (about 420 Ma) is about 60 m thick and exposed in an east-west outcrop belt that dips gently to the south (Fig. 1). The lithology is predominantly dolostone. In outcrop the dolostone is gray, brownish-gray, and buff and massive to evenly bedded. It is a fine- to medium-crystalline dolostone and contains intraclasts, ooids, peloids, and skeletal fragments, particularly those of trilobites, brachiopods, corals, stromatoporoids, mollusks, and echinoderms. In addition to mineralized vugs and fracture fills (Jensen 1942; Bassett and Kinsland 1973), stylolites and carbonaceous partings are common. The dolostone exudes an odor of crude oil, when broken. Pyrobitumen and natural asphaltum are locally present. Zenger (1965) and Shukla and Friedman (1983) described the Lockport Formation and interpreted its depositional and diagenetic pattern. Figure 2 shows the inferred depositional environment of the Lockport Formation. A subtidal to intertidal depositional environment for the Lockport Formation is indicated by ooids, reef-building organisms, and grain-support fabric. The topmost part of the formation shows evidence of hypersaline deposition in a supratidal zone. The Lockport Formation is a shallowing-upward sequence of carbonate rocks (Fig. 2; Shukla and Friedman 1983) whose

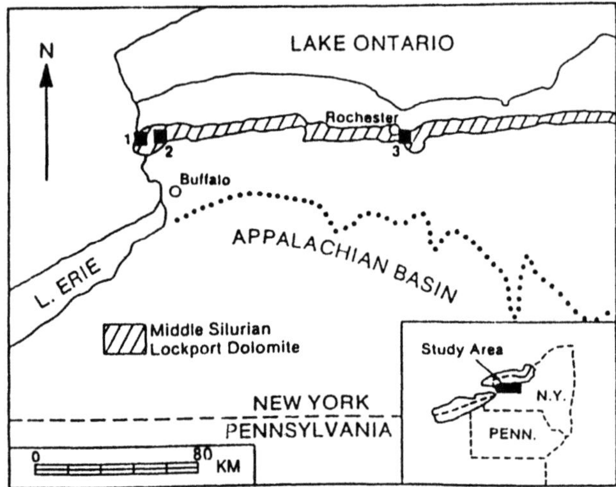

Fig. 1. Location map showing distribution of sampled mineral occurrences in the Lockport Formation with *inset map* of New York, Pennsylvania and eastern Great Lakes. Sampled locations include *1* Pekin; *2* Frontier; *3* Penfield

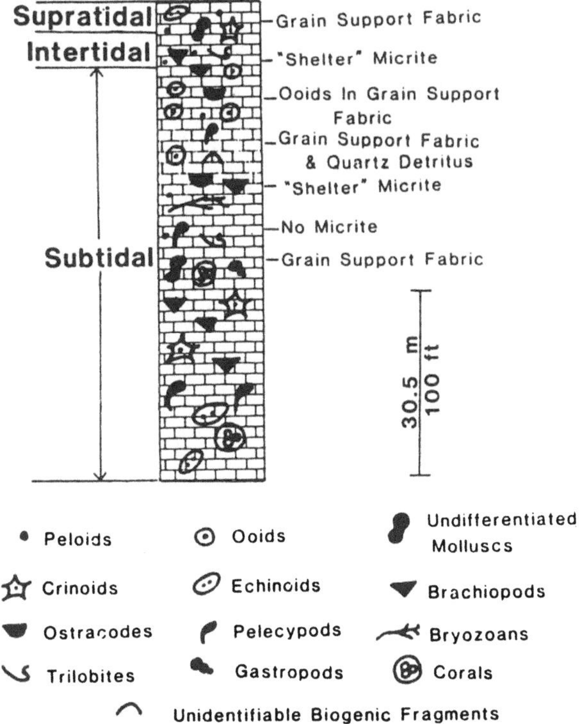

Fig. 2. Vertical column showing interpreted depositional environments of the Lockport Formation at a site where the bed rock occurs as a limestone

peritidal setting was followed by subaerial emergence. Textures and fabrics of quartz crystals commonly indicate the former presence of evaporite minerals (Friedman and Shukla 1980). Under subaerial conditions now mineral-filled former vugs and cavities, including solution breccias, resulted from dissolution of evaporites by surface and near-surface percolating freshwaters. Former vugs, now geodes, and solution-collapse breccias are recognized as karst features. Interestingly, an environmental or climatic change is implied in the formation of original gypsum and anhydrite, followed by a period of freshwater dissolution under conditions of emergence leading to karstification. Variations in dolostone occur in its trace-element chemistry and its crystal morphology. The dolostones were stained for Fe^{2+}. Where present in sufficent concentration, the ferrous ion helps delineate crystal-growth stages. The earlier stage is marked by high concentrations of Fe^{2+} in almost opaque dolomite crystal cores (Fig. 3; Shukla and Friedman 1983).

3 Mineral Assemblages

3.1 Lead-Zinc Minerals

Galena occurring as cubes modified by octahedron and dodecahedron faces (Giles 1920) fills fractures, solution cavities, channels, and even caves. Galena

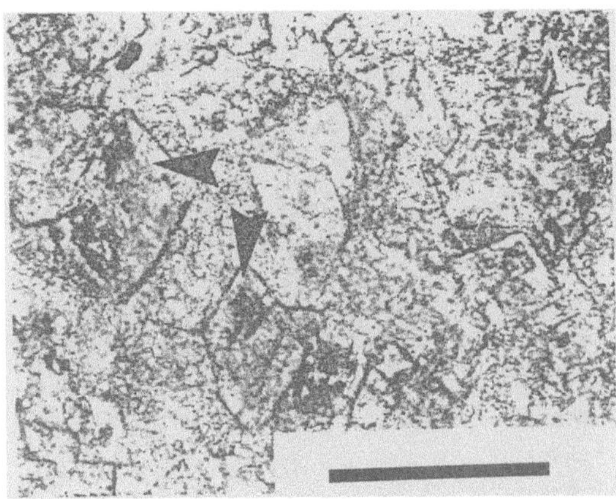

Fig. 3. Photomicrograph of dolomite host rock. Cores of some dolomite crystals are rich in Fe^{2+} and nearly opaque (*arrowheads*), whereas peripheries are poor in Fe^{2+}. Crossed nicols; *scale bar*- 40 μm

crystals weighing more than 100 kg were discovered in a cave-like cavity during construction of the Erie Canal (Hall 1843).

Sphalerite also fills fractures and vugs (Fig. 4), and even the molds of former fossil fragments. The color of sphalerite varies from reddish-brown to yellow and translucent to transparent.

Although uncommon, marcasite occurs as tiny, bladed crystals in cavities lined by saddle dolomite. Pyrite is common as cubic crystals modified by the octahedron, pyritohedron, and dodecahedron. Like the other sulfides it fills vugs.

3.2 Carbonate Minerals

Coarse crystals of white to pink dolomite having curved faces, known as saddle dolomite (Radke and Mathis 1980), fill vugs, channels, and fractures (Fig. 5). Microprobe analysis of saddle dolomite indicates CaO 30.2%, MgO 21.7%, FeO 0.3%, MnO 0.1%, and CO_2 47.7%. Strontium was not detected in any of the samples. The composition of saddle dolomite averages ($Ca_{0.52}$, $Mg_{0.47}$, $Fe_{0.01}$, $Mn_{<0.001}$) $(CO_3)_2$.

White or yellow scalenohedra of calcite are common as dog-tooth spar and occur in vugs in which they are cemented to crystals of saddle dolomite, and hence are younger than the saddle dolomite.

3.3 Sulfate Minerals

White massive to crystalline gypsum and clear selenite crystals, fill vugs and cavities (Fig. 6). Some of the selenite is of optical grade (Jensen 1942). In places selenite crystals occupy parts of saddle dolomite-rimmed geodes. Various

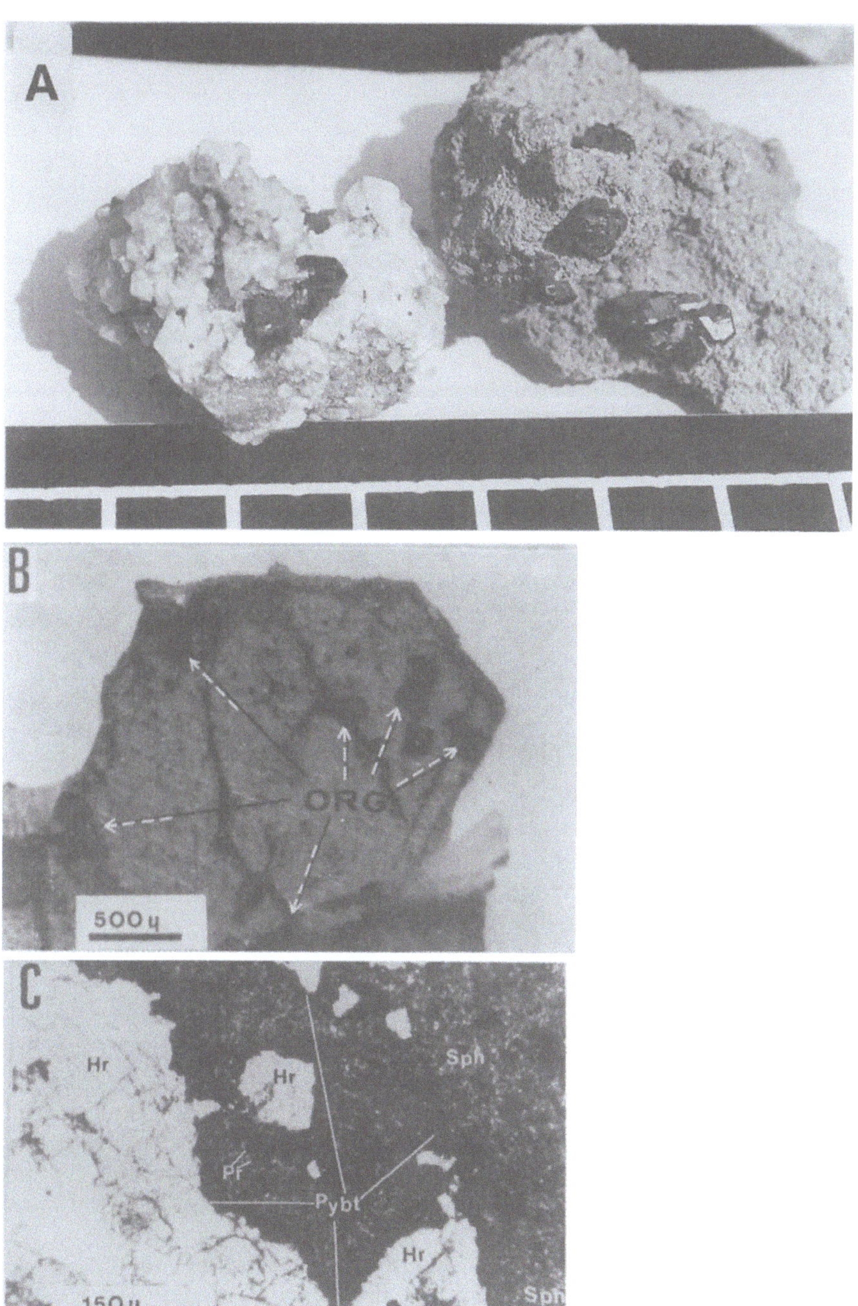

Fig. 4. A Sphalerite and saddle dolomite; crystals fill vug. *Scale bar* in centimeters. **B** Photomicrograph showing bitumen (*arrows*) as inclusions in coarse-crystalline sphalerite. **C** Photomicrograph of sphalerite (*Sph*) vein in dolostone host rock (*Hr*). Sporadic pyrite (*Pr*) and pyrobitumen (*Pybt*) are scattered in vein

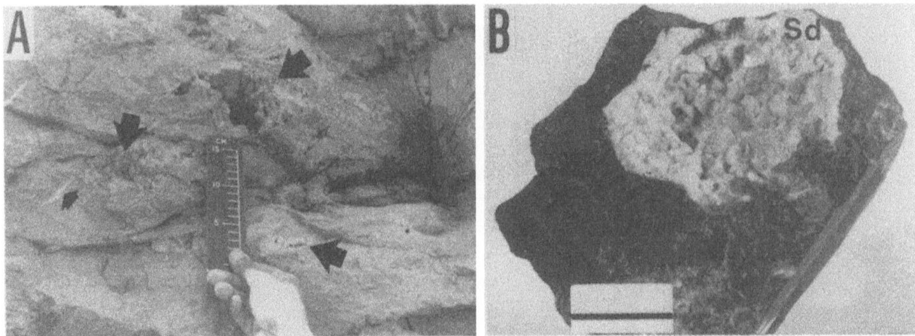

Fig. 5. **A** Outcrop and **B** hand-specimen photographs. Host rock dolostone in which very coarse crystalline saddle dolomite (*Sd*) fills vugs. *Arrows* in **A** point to saddle dolomite

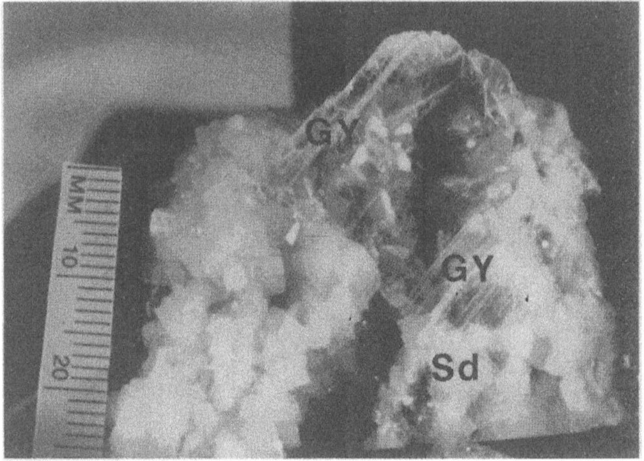

Fig. 6. Gypsum (*GY*) and saddle dolomite (*Sd*) filling vug

crystal morphologies and crystal aggregates of authigenic quartz euhedra fill voids created by dissolution of sulfates, especially gypsum. Isolated anhedra are equant, prismatic or lath-shaped; aggregates of quartz crystals after gypsum are either fan-shaped or form clusters composed of euhedra radiating out in all directions.

In vugs and other cavities, bluish-white to white crystals of anhydrite grade into massive gypsum. In places, blue anhydrite crystals occupy the core of vugs. Some anhydrite is nodular and has an incomplete rind of megaquartz. White to bluish-white barite fills vugs and fractures and is locally embedded in selenite crystals or attached to celestite crystals. Celestite occurs in vugs as fibrous to lamellar masses or as crystals up to 15 cm in length, in places embedded in selenite (Jensen 1978). Gypsum must have formed at the expense of anhydrite during or after uplift which coincided with the Allegheny orogeny.

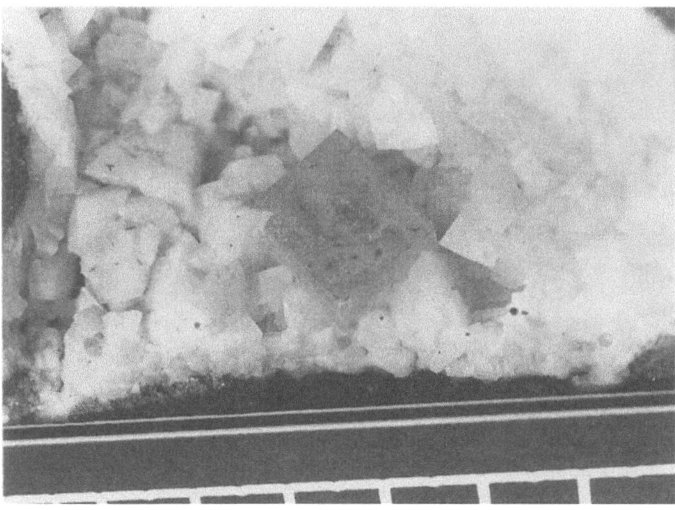

Fig. 7. Crystal of fluorite in vug attached to saddle dolomite. *Bars* on scale are 1 cm apart

3.4 Fluorite

Crystals of fluorite are present as cubes (Fig. 7), in places modified by the octahedron, and occur in the core of vugs, commonly embedded in selenite crystals or attached to saddle dolomite. The color of fluorite ranges from colorless to blue or purple.

4 Thermal and Salinity History of Mineral Assemblage

4.1 Fluid-Inclusion Analyses

Two-phase (aqueous liquid plus vapor), primary and secondary fluid inclusions were analyzed in saddle dolomite, calcite, sphalerite, and fluorite (Fig. 8). The fluid inclusions discussed in this study are primary in origin; this determination was made following the distinction between primary and secondary inclusions of Roedder (1984).

An approximation of the salt concentration in solution may be obtained by freezing fluid inclusions and observing the depression of the freezing temperature. The eutectic melting temperatures (initial melting temperatures) of fluid inclusions in the saddle dolomite, sphalerite, and fluorite crystals considered range from −45 to −55 °C, indicating the presence of other divalent salts, such as Ca^{2+} and Mg^{2+}, in addition to NaCl (Crawford 1981).

Freezing temperatures (final melting temperatures) of fluid inclusions in saddle dolomite, sphalerite, and fluorite were generally in the range of −4 to −21 °C corresponding to salinities ranging from 6 to 23 wt% equivalent NaCl, which is three to seven times the salinity of seawater. Homogenization tem-

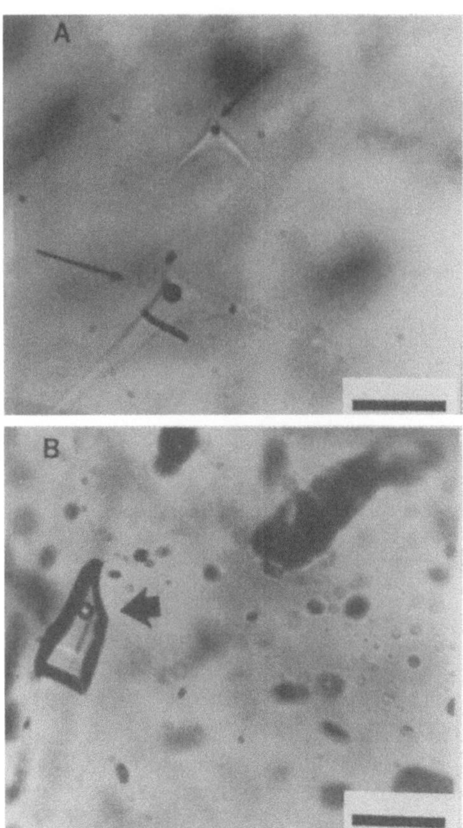

Fig. 8A,B. Photomicrograph showing two-phase (aqueous liquid plus vapor) primary fluid inclusions (*arrows*). **A** Fluid inclusions in fluorite. **B** Fluid inclusions in sphalerite. *Bar* = 104 μm

peratures of two-phase, aqueous liquid-vapor inclusions in late-stage minerals, uncorrected for excess ambient pressures during inclusion entrapment, range from 110 to 200 °C (Fig. 9). These homogenization temperatures provide an estimate of minimum trapping temperatures, as reported by Roedder (1984). Other data on mineralization temperatures from the studied area, also deduced from fluid inclusions, are reported by Kinsland (1977) and Tillman and Barnes (1983), and fall into this same temperature range.

Estimating the pressure at the time of inclusion formation and adding an appropriate correction to the homogenization temperatures is another way of obtaining approximate formation temperatures. If the Lockport dolomites were buried to depths between 1000 and 4000 m at the time inclusions formed (Friedman 1987b), and using the pressure correction data of Potter et al. (1978), the estimated pressure correction to be added to the determined homogenization temperatures is between 10 and 40 °C. Therefore, in this study, homogenization temperatures, uncorrected for pressure, may yield close estimates of the actual mineralization temperatures. Fluid inclusions trapped during the growth of saddle dolomite, sphalerite, fluorite, and calcite (Fig. 8)

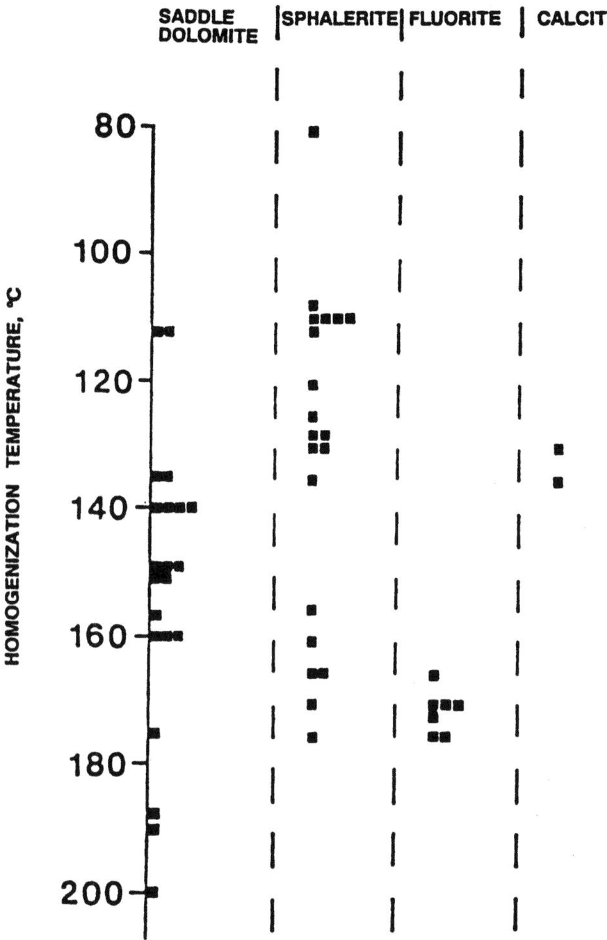

Fig. 9. Frequency-distribution diagram of homogenization temperature determinations of fluid inclusions in saddle dolomite, sphalerite, fluorite, and calcite crystals (not pressure corrected)

indicate that the precipitating fluids were hot (Fig. 9): mean homogenization temperature of saddle dolomite were ≈150 °C, sphalerite showed a range of temperatures around 80–179 °C clustering at 130–140 °C, fluorite indicated a temperature range of 167–177 °C, and calcite a range of 137–139 °C.

4.2 Data from Stable Isotopes

The $\delta^{18}O$ values for saddle dolomite range from −9.75 to −11.11‰ PDB (mean −10.14‰), which may be interpreted (Table 1) to indicate that the temperatures of the precipitating fluids were in the range of 101–193 °C.

The value of $\delta^{13}C$ PDB of CO_2 trapped in a fluid inclusion in saddle dolomite was −23.2‰. This value indicates enrichment in the light carbon

Table 1. Oxygen isotope and fluid inclusion homogenization temperatures, saddle dolomite

Sample	$\delta^{18}O$ SMOW (‰)	$\delta^{18}O$ PDB (‰)	Isotopic temp. (°C)	Mean homogenization temp. (°C)
1	+19.4	−11.11	120–193	200
2	+20.8	0–9.75	107–172	110
3	+20.6	0–9.75	109–175	160
4	+20.3	−10.24	112–180	135
5	+20.6	0–9.95	109–175	175
6	+19.9	−10.63	115–185	190
7	+20.8	0–9.75	107–172	140
8	+20.0	−10.53	114–184	140
9	+19.6	−10.92	118–190	150
10	+20.1	−10.42	113–182	150

Note: SMOW = standard mean ocean water; PDB = belemnite standard from Peedee Formation (Cretaceous) in North Carolina. Temperatures calculated by using the relation for dolomite (Northrop and Clayton 1966):

$$10^3 \ln = 3.2 \times 10^6 T^{-2}(K) - 3.3; 10 \ln = \delta^{18}O_{dolomite} - \delta^{18}O_{water}$$

The $\delta^{18}O$ values are those of water assumed to be in the range of 2 to 8‰ SMOW, which corresponds to saline to highly saline waters. High salinity values for saddle dolomite-precipitating fluids in this study were also obtained from fluid-inclusion freezing temperatures.

isotope which could have resulted from the inorganic oxidation of methane. In this reaction the CO_2 retains the same signature as the original methane. The water of the inclusion in saddle dolomite has a δD_{SMOW} of −59‰.

5 Deep-Burial Mineralization

Studies of fluid-homogenization temperatures, oxygen isotopes, fission-track analysis, and vitrinite reflectance of samples from surface-exposed strata in undeformed areas of the Appalachian Basin (Friedman 1987a,b; Lakatos and Miller 1983) imply that these strata have been heated to temperatures that suggest a great burial depth and unexpectedly large amounts of uplift and erosion, ranging from 4 to 7 km (Friedman and Sanders 1983; Friedman 1987a, b). There is no evidence that the paleotemperatures deduced from these various techniques are related to high thermal anomalies, or to a high thermal gradient, such as a shallow heat source in the upper mantle or possible intrusions, or advective heat transport (Friedman and Sanders 1983; Friedman 1987 a,b).

The studied mineral assemblage which fills fractures, vugs, and various cavities formed under conditions of deep burial at the temperature ranges indicated in Table 1 and Fig. 9. These temperatures were in the range of 150 ±

50 °C. The same range is supported by the data of Harris et al. (1978), Harris (1979), and Kinsland (1977). The average geothermal gradient measured today in eastern and midcontinent regions of North America ranges from 20 to 30 °C km^{-1} (mean 25 °C km^{-1}) (Harrison et al. 1983; Friedman 1987a,b). A paleogeothermal gradient of 26 °C km^{-1} was inferred for the northern Appalachian Basin for the Middle to Late Paleozoic (Friedman and Sanders 1982). If this gradient existed at the time of formation of the mineral assemblage, then the Lockport strata would have been buried to a depth of 5 km. Because the strata are undeformed, vertical uplift and erosion must have brought the inferred, formerly deeply buried strata to their present surface exposure. The period of time between original deposition, subsidence to 5 km, and uplift and erosion took 170 to 220 million years (Friedman 1987a,b).

6 Conclusions

During the middle Silurian, at about 420 Ma, the Lockport strata (mostly dolostone) formed under peritidal conditions in an arid evaporitic setting. The environment of deposition was that of a shoaling-upward sequence from subtidal to intertidal and, subsequently, a supratidal setting followed by subaerial emergence. Under subaerial conditions now mineral-filled former vugs and cavities, including solution breccias, resulted from the dissolution of evaporites by surface and near-surface percolating freshwaters.

A mineral assemblage, consisting for the most part of lead-zinc sulfides, saddle dolomite, and sulfate minerals such as anhydrite, barite, and celestite, is interpreted to have precipitated under deep-burial conditions of approximately 5 km. The fluids from which the minerals precipitated were hot and saline. At the time of the Allegheny orogeny (296 Ma), this mineral assemblage was uplifted to shallower depth and ultimately to the surface.

Acknowledgements. Thanks are extended to D.L. Leach, L. Fontboté, and M. Boni for reviewing this paper.

References

Bassett WA, Kinsland GL (1973) Mineral collecting at Penfield Quarry. In: Hewitt PC (ed) New York State Geol Assoc, Guidebook, pp H1–H9

Beck LC (1842) Mineralogy of New York. White and Vissher, Albany

Crawford ML (1981) Phase equilibria in aqueous inclusions. In: Hollister LS, Crawford ML (eds) Short course in fluid inclusion: applications to petrology. Mineral Assoc Can, pp 75–99

Eaton A (1824) A geological and agricultural survey of the district adjoining the Erie Canal, in the State of New York. Packard and Van Benthuysen, Albany

Friedman GM (1987a) Vertical movements of the crust: case histories from the northern Appalachian Basin. Geology 15:1130–1133

Friedman GM (1987b) Deep-burial diagenesis: its implications for vertical movements of the crust, uplift of the lithosphere and isostatic unroofing – a review. Sediment Geol 50:67–94

Friedman GM, Sanders JE (1982) Time-temperature-burial significance of Devonian anthracite implies former great (≈6.5 km) depth of burial of Catskill Mountains, New York. Geology 10:93–96

Friedman GM, Sanders JE (1983) Reply to on "Time-temperature-burial significance of Devonian anthracite implies former great (≈6.5 km) depth of burial of Catskill Mountains, New York". Geology 11:123–124

Friedman GM, Shukla V (1980) Significance of authigenic quartz euhedra after sulfates: example from the Lockport Formation (middle Silurian) of New York. J Sediment Petrol 50:1299–1304

Friedman JD, Mutschler FE (1987) New geophysical, geochemical, and geological investigations of northern Appalachian zinc-lead sulfide deposits of New York State. US Geol Survey Research on Mineral Resources. US Geol Surv Circ 995:21–22

Giles AW (1920) Minerals in the Niagara limestone of western New York. Rochester Acad Sci Proc 6:57–72

Hall J (1843) The natural history of New York. Carroll and Cook, Printers to the Assembly, Albany, New York, 525 pp

Harris AG (1979) Conodont color alteration: an organo-mineral metamorphism index and its application to Appalachian Basin geology. Soc Econ Paleontol Mineral Spec Publ 26:3–16

Harris AG, Harris LD, Epstein JB (1978) Oil and gas data from Paleozoic rocks in Appalachian basin: maps for assessing hydrocarbon potential and thermal maturity (conodont color alteration isograds and overburden isopachs). US Geol Surv Misc Invest Map 1-917E scale 1:2500000 (4 sheets)

Harrison WE, Luza KV, Prater ML, Cheung PK (1983) Geothermal resource assessment in Oklahoma. Okla Geol Surv Spec Publ 83

Jensen DE (1942) Minerals of the Lockport dolomite in the vicinity of Rochester. Rock Minerals Mag 17:199–203

Jensen DE (1978) Minerals of New York State. Ward's Natural Science Establishment, Rochester

Kinsland GL (1977) Formation temperature of fluorite in the Lockport dolomite in upper New York State as indicated by fluid inclusion studies with a discussion of heat sources. Econ Geol 72:849–854

Lakatos S, Miller DD (1983) Fission-track analysis of apatite and zircon defines a burial depth of 4 to 7 km for lowermost Upper Devonian Catskills, New York. Geology 11:103–104

Northrop DA, Clayton RN (1966) Oxygen isotope fractionation in systems containing dolomite. J Geol 74:174–196

Potter RW, Clynne MA, Brown DL (1978) Freezing point depression of aqueous sodium chloride solutions. Econ Geol 73:284–285

Radke BM, Mathis RL (1980) On the formation and occurrence of saddle dolomite. J Sediment Petrol 50:1149–1168

Robinson S (1825) A catalogue of American minerals with their localities. Boston (priv publ)

Roedder E (1984) Fluid inclusions. In: Ribbe PH (ed) Reviews in mineralogy. Mineral Soc America 12, 644 pp

Shukla V, Friedman GM (1983) Dolomitization and diagenesis in a shallowing-upward sequence: the Lockport Formation (middle Silurian), New York State. J Sediment Petrol 53:703–717

Tillman JE, Barnes HL (1983) Deciphering fracturing and fluid migration histories in the northern Appalachian Basin. Bull Am Assoc Petrol Geol 67:692–705

Zenger DH (1965) Stratigraphy of the Lockport Formation (Silurian) in New York State. N Y State Mus Sci Serv Bull 104, 210 pp

Mississippi Valley-Type Deposits in Continental Margin Basins: Lessons from the Appalachian-Caledonian Orogen

S.E. Kesler

Abstract

One of the major questions concerning Mississippi Valley-type (MVT) deposits at continental margins is the mechanism by which mineralizing brines were expelled from their source basins. Possible mechanisms, listed in order of the intensity of regional deformation with which they are associated, include basinal overpressuring, regional tilting, hinterland recharge of foreland basins, and destruction of source basins by regional thrusting. MVT mineralization in the Caledonian-Appalachian orogen provides an excellent opportunity to compare the effectiveness of these four brine expulsion processes. In general, the largest MVT deposits are found in areas of most intense regional thrusting, suggesting that hinterland recharge or basin deformation is the important brine expulsion mechanism. The fact that widespread MVT mineralization in the East Tennessee area of the southern Appalachians is adjacent to overlapping Taconic and Acadian foreland basins and metamorphic maxima suggests that basin deformation is the dominant process for brine expulsion.

1 Introduction

Although it is clear that Mississippi Valley-type (MVT) deposits are formed by basinal brines (Anderson and Mcqueen 1982; Sverjensky 1986; Hanor 1987), less is known about processes by which these brines travel to sites of ore deposition. Possible processes that can cause brines to flow within and beyond their parent basins include sedimentary overpressuring, regional tilting, hinterland recharge, and regional thrusting (Sharp 1978; Cathles and Smith 1983; Garven 1985; Bethke 1986; Oliver 1986; Sharp and Kyle 1988; Bethke and Marshak 1990). In the order listed, these processes occur at successively later stages in basin evolution and involve increasing degrees of tectonic deformation (Fig. 1). For instance, sedimentary overpressuring can form without regional deformation, and regional tilting might not involve significant folding of basinal sediments. In contrast, hinterland recharge and compression beneath thrust sheets require that parts or all of the basin actually undergo deformation. Oliver (1986) has suggested that the last of these processes, regional thrusting, is necessary to force large volumes of fluids from continental margin basins

Department of Geological Sciences University of Michigan Ann Arbor, MI 48109, USA

Spec. Publ. No. 10 Soc. Geol. Applied to Mineral Deposits
Fontboté/Boni (Eds.), Sediment-Hosted Zn-Pb Ores

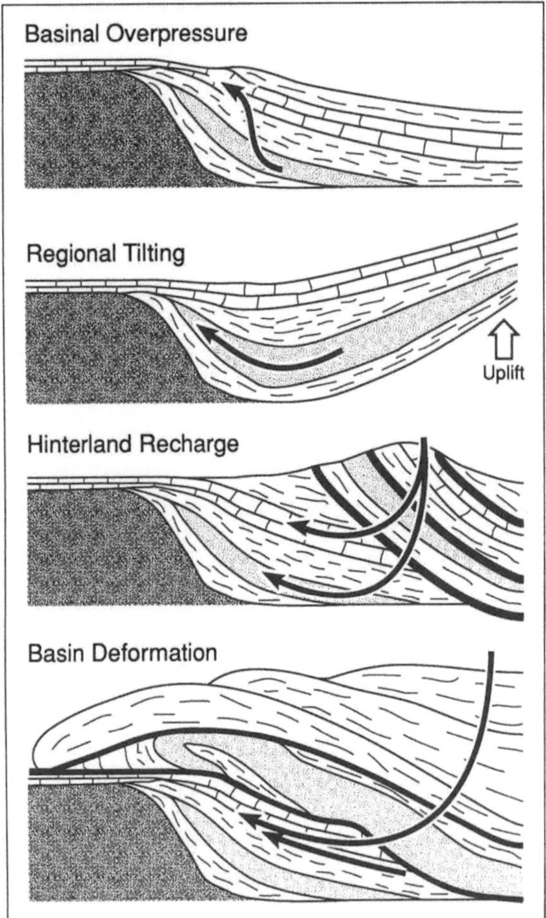

Fig. 1. Schematic illustration of four possible processes for expelling brines from the deeper parts of their host sedimentary basins

and that such continent-scale fluid expulsion events are necessary to form hydrocarbon and basin-related mineral deposits.

MVT deposits in continental margin basins provide an opportunity to assess the relative importance of these processes and, in particular, to evaluate the importance of regional thrusting. The Appalachian-Caledonian orogen (Fig. 2), which is one of the most extensive and longest-lived orogens in the Phanerozoic record (Williams 1984), is particularly well suited to such a test. It has undergone several important regional thrusting events and is one of the largest MVT provinces in the world (Fig. 2). As shown below, regional thrusting appears to be the dominant control on the development of MVT mineralization; even more important, apparently, than the presence of favorable host rocks for ore deposition.

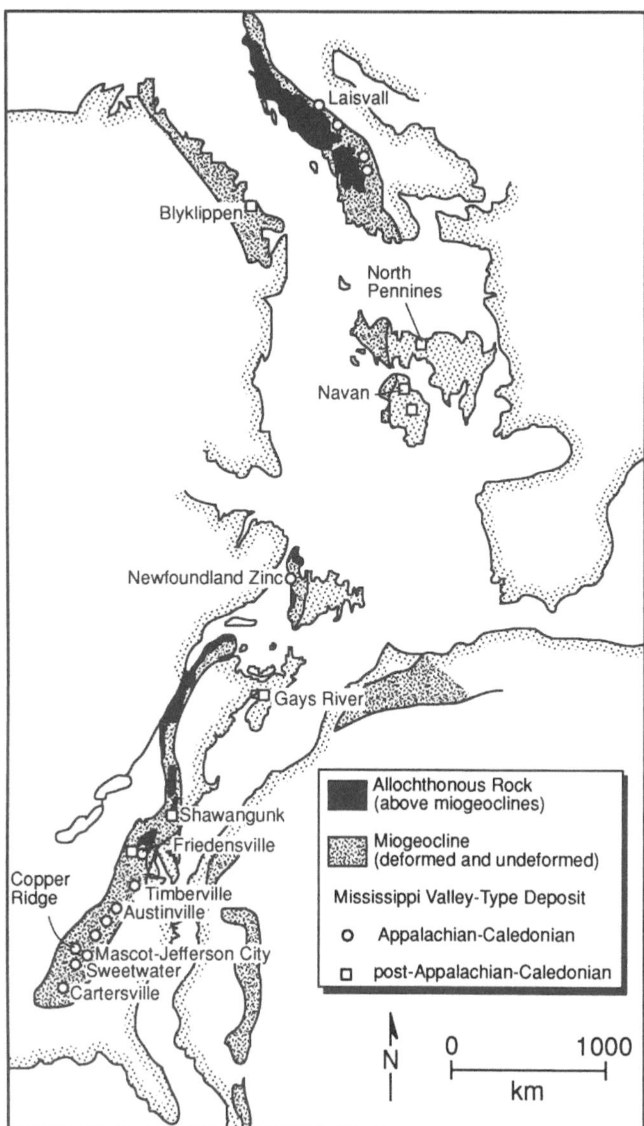

Fig. 2. Distribution of important Mississippi Valley-type (MVT) deposits in the Appalachian-Caledonian orogen. Deposits are divided into those that appear to be part of the Appalachian-Caledonian collisional margin orogen (*open circles*) and those that are related to later tectonic cycles (*open squares*). With only a few exceptions, such as the Shawangunk deposits in New York (Wilbur et al. 1990), deposits related to the Appalachian-Caledonian orogen are hosted by sediments of the Cambro-Ordovician miogeocline. Appalachian-Caledonian orogen is from Williams (1984) and locations of MVT deposits are from Bjørlykke and Sangster (1981), Clark (1989), Howe (1981), Sawkins (1966), Smith (1977), Sorenson et al. (1978), Ravenhurst et al. (1989), Wilbur et al. (1990), Williams and McArdle (1978)

2 Regional Geology of the Appalachian-Caledonian Orogen

The Appalachian-Caledonian orogen (Fig. 2) is exposed along the margin of the present North Atlantic Ocean in Greenland, Great Britain, Scandinavia, eastern North America, and western Africa (Williams 1984; Gee and Sturt 1985; Rast 1989). One of the most important features of the orogen is its extensive prism of Late Precambrian to Ordovician miogeoclinal sediments. This sequence, which is present throughout most of the orogen, consists of early clastic sediments overlain by carbonate bank deposits and coeval deeper water deposits consisting of shales, sandstones and carbonate breccias (Williams 1984; Bovan and Read 1987). Highly porous sandstones in the clastic rocks, and karst and other breccias in the carbonate rocks in this sequence provided numerous favorable zones for MVT mineralization (Hoagland 1976; Bjørlykke and Sangster 1981).

Late Precambrian and Early Paleozoic sediments in the Appalachian-Caledonian orogen underwent several periods of deformation associated with opening and closing of the proto-Atlantic (Iapetus) Ocean. Although there is considerable overlap in timing of individual events from place to place, deformation in the orogen is commonly divided into three major episodes (Williams 1984; Roberts and Gee 1985; Rast 1989). The earliest phase, which is known as Finnmarkian, Grampian, and Taconic from north to south in the orogen, took place in Ordovician-Silurian time. It was followed by a Silurian-Devonian orogenic phase (Caledonian, Scandian, Cymrian, Acadian), and a Pennsylvanian-Permian phase (Hercynian, Variscan, Alleghanian). These orogenic events did not affect all of the orogen equally, but the sum of these processes produced the orogen that is exposed today along the margins of the North Atlantic Ocean.

About half of the Early Paleozoic miogeoclinal sediments in the Caledonian-Appalachian orogen have been strongly deformed by thrusting, and large nappes have been carried onto the continents on both sides of the Atlantic Ocean (Fig. 2). It is apparent from the distribution of these nappes that thrusting associated with the closing of the Iapetus was not equally distributed on both sides of the Atlantic, however. In the northernmost part of the orogen, for instance, large segments of the Iapetus ocean floor were thrust eastward across the Baltic craton, whereas west-directed thrusting in Greenland just across the Iapetus involved largely miogeoclinal sediments (Henriksen 1985; Hurst et al. 1985). To the south, the polarity of intense thrusting was reversed, with translation of large oceanic nappes westward across North America and limited ophiolite emplacement in adjacent Africa (Williams 1984; Rast 1989).

3 Geologic Setting and Age of MVT Mineralization in the Appalachian-Caledonian Orogen

3.1 Introduction

MVT mineralization is widespread in the Appalachian-Caledonian orogen and appears to have formed at several different times during the tectonic evolution of the region. It is important in this study to distinguish between deposits that are actually related to the Appalachian-Caledonian collisional margin and other deposits that formed after the orogen had closed. As noted below, although deposits related to compressional margin tectonics are widespread throughout the orogen, a province of deposits related to later rifting is found in the northern Appalachians and British Isles (Fig. 2). In general, the division between these two groups of deposits is reflected in the age of their host rocks. Deposits hosted by the Late Precambrian to Ordovician sedimentary prism formed during compressional margin tectonics, whereas those in younger host rocks formed during the rifting that caused deposition of their host sediments.

3.2 Scandinavia and Greenland

The Scandinavian Caledonides lack conventional carbonate-hosted MVT mineralization, but have important sandstone-hosted lead deposits and related vein deposits (Bjørlykke and Sangster 1981). They are found largely as disseminations in upper Precambrian to Lower Cambrian quartzitic sandstones that form the basal part of the Scandinavian Paleozoic marine transgression (Bjørlykke 1978; Willdén 1980). At the largest of these deposits, Laisvall, Sweden, fluid inclusion studies show that the mineralizing fluid was a basinal brine similar to that in carbonate-hosted MVT deposits (Lindblom 1986), and it is likely that other Scandinavian sandstone lead deposits formed from similar fluids. These deposits, which include Dorotea, Vassbo and Osen, consist of galena with lesser amounts of sphalerite and essentially no pyrite (Rickard et al. 1979; Bjørlykke and Sangster 1981).

Host rocks for the deposits are in and below the Caledonide thrust sheets, with larger deposits such as Laisvall in autochthonous sediments beneath the thrust sheet and smaller deposits in fractures in either the basement or overlying thrust sheets (Rickard et al. 1979; Bjørlykke and Sangster 1981). The ore minerals have clearly been affected by thrusting, indicating that MVT mineralization preceded regional deformation in their immediate vicinity. Deformation took place in this area during mid-Ordovician time and again at the end of the Silurian (Sturt 1978; Williams 1984; Roberts and Gee 1985; Torsvik et al. 1991). The earlier of these events appears to have completely dismembered the Cambro-Ordovician sedimentary sequence (Chapman et al. 1985; Ramsay et al. 1985). Thus, unless ore-forming brines came from Devonian or younger successor basins on the Caledonian thrust sheets (Steel et al. 1985), which seems unlikely, Scandinavian MVT mineralization was emplaced by the end of Ordovician time. No isotopic measurements relating to the age of these deposits have been made.

In contrast to the abundance of basin-related mineralization in Scandinavia, MVT deposits are scarce in Greenland. The only significant deposit of this type is Blyklippen, which is hosted in Upper Carboniferous to Lower Triassic clastic sediments (Fig. 2). These host rocks are younger than the Caledonian orogen in this area, suggesting that mineralization here is related to later tectonic or intrusive activity (Sorenson et al. 1978). In this same area, Stendal and Ghisler (1984) describe stratiform copper deposits in Late Proterozoic clastic rocks, which might be cupriferous equivalents of the Laisvall mineralization. MVT mineralization is also present in adjacent parts of the Canadian Arctic, including

Nanisivik on Baffin Island (McNaughton and Smith 1986) and Polaris on Little Cornwallis Island (Jowett 1977; Symons and Sangster 1991), although they do not appear to be directly related to the Appalachian-Caledonian orogen.

3.3 British Isles

MVT and other basin-related deposits in the British Isles are hosted by younger Paleozoic rocks that are largely unrelated to Caledonian collisional margin tectonics. In England, the best-known MVT mineralization, which is located in the Pennines (Fig. 1), is hosted by Lower Carboniferous sedimentary rocks around a basement dome (Sawkins 1966; Jones et al., this Vol.). Although individual deposits in this district are relatively small, total production of lead has amounted to over 4 million t (Sawkins 1966), indicating that the mineralizing system was large. Possibly coeval vein deposits are found throughout northern England and Scotland (Dunham et al. 1978). The Pennine deposits have yielded Late Triassic-Early Jurassic isotopic ages of about 200 Ma (Shepherd et al. 1982; Halliday et al. 1989; Table 1).

Conventional MVT mineralization is also lacking in Ireland; in its place are the well-known Irish Pb-Zn deposits, including Navan, Tynagh and Silvermines (Evans 1976; Andrew and Ashton 1985). These deposits are also hosted by rocks of Lower Carboniferous age (Fig. 2) and exhibit similarities to sedimentary exhalative deposits, which constrain their age to that of the host sediments. Lower Carboniferous isotopic ages have also been obtained from the Gortdrum deposit (Table 1).

Table 1. Isotopic age measurements on Mississippi Valley-type deposits in the Appalachian-Caledonian orogen. As noted in the text, deposits in England, Ireland and Nova Scotia appear to have formed after closing of the Appalachian-Caledonian collisional margin in their respective locations

Deposit/location	Age (Ma)	Type of measurement	Mineral[a]	Reference
North Pennines/ England	~200	Sm-Nd	FL	Halliday et al. (1990)
	206 ± 9	Rb-Sr	FI	Shepherd et al. (1982)
Gortdrum/Ireland	359 ± 26	Pb-Pb	PB	Duane and deWit (1988)
	340 +25/−20	U-Pb	PB	
Newfoundland Zinc/ Newfoundland, Canada	350–370	Ar/Ar	FS	Hall et al. (1989)
	200	Ar/Ar	FS	Hall et al. (1989)
Gays River, Brookfield, Smithfield, Upper Brookside/Nova Scotia, Canada	322 ± 17	FT	ZR	Ravenhurst et al. (1989)
	300 ± 6	Rb/Sr	IL-AB	Ravenhurst et al. (1989)
	315–365	K/Ar	CL	Ravenhurst et al. (1989)
Coy/Mascot-Jeff. City East Tennessee, USA	377 ± 29	Rb-Sr	SP	Nakai et al. (1990)

[a] Mineral/feature dated: FL = fluorite; FI = fluid inclusions; PB = galena; FS = feldspar; ZR = zircon; IL = illite; AB = albite; CL = chlorite; SP = sphalerite.

3.4 Northern Appalachians

MVT mineralization is widespread in the northern Appalachians and appears to represent at least two MVT systems of different ages. Deposits of the older system which appear to be part of the Appalachian-Caledonian orogen, are found largely in the Lower Ordovician St. George Formation in Newfoundland (Collins and Smith 1975), and include the Newfoundland Zinc mine (Lane 1989). These deposits consist almost entirely of sphalerite and sparry dolomite and they exhibit strong megascopic similarities to MVT ores in East Tennessee (Kesler and van der Pluijm 1990). Smaller prospects, some containing lead, fluorite and barite, are found in younger sedimentary units in Newfoundland (D.F. Sangster, pers. comm., 1989). The age of these deposits is constrained by geologic relations reported in the Newfoundland Zinc mine area (Lane 1989), as well as by laser $^{40}Ar-^{39}Ar$ ages of 350 to 370 Ma on authigenic feldspar related to mineralization in the area (Hall et al. 1989; Table 1). The basin to which this mineralization was related was deformed and obscured by thrusting during Acadian time, making reconstruction of basinal morphology or fluid flow patterns difficult (Kesler and van der Pluijm 1990; Quinn 1991).

The other large group of deposits in the northern Appalachians is hosted by Lower Carboniferous (Mississippian) Windsor Group carbonate rocks (Akande and Zentilli 1984). These deposits are associated with the Maritimes Basin, which consists of a series of interconnected grabens containing thick terrestrial red beds and minor marine sediments that formed during Middle Devonian to Early Permian time (Williams 1984; Sangster and Vaillancourt 1990). This basin overlies and post-dates deformation of the Appalachian-Caledonian miogeocline in this area. In contrast to the Cambro-Ordovician-hosted deposits, the Windsor-hosted deposits contain a wide range of minerals. The large Gays River deposit, which is the best-known example of this group, consists of massive sphalerite and galena with sparry dolomite (Akande and Zentilli 1984). Other deposits, including Walton and Yava, range from limestone replacement to sandstone-hosted lead, and contain a complex mineral suite including sphalerite, galena, barite, fluorite, and celestite (Boyle et al. 1976; Sangster and Vaillancourt 1990). It is not clear how many of these deposits, other than Gays River, are of MVT type (see discussion in Sangster and Vaillancourt 1990), although all of them appear to be related to the hydrothermal evolution of their host basins. The formation of these deposits appears to be constrained to uppermost Lower Carboniferous time or slightly younger by isotopic ages (Ravenhurst et al. 1989; Table 1).

3.5 Southern and Central Appalachians

MVT deposits and prospects in the southern and central Appalachians (Fig. 2) are found in rocks ranging in age from Cambrian to Pennsylvanian (Hoagland 1976; Clark 1989; Kesler and van der Pluijm 1990), although most economically important deposits are in the Cambrian to Ordovician prism. Cambrian-hosted MVT deposits, which are in the Shady Dolomite and related units (Fig. 2), are largely restricted to the southern Appalachians and include the Austinville-Ivanhoe (Virginia) zinc-lead district (Brown and Weinberg 1968; Foley et al. 1981) and the Cartersville (Georgia) barite district (Kesler 1950). Ordovician-hosted MVT mineralization is much more widespread and economically important, however. The largest of these deposits are in East Tennessee, where ore is hosted by solution collapse breccias in the upper part of the Upper Cambrian-Lower Ordovician Knox Group (Harris 1971; Haynes and Kesler 1989). Districts in East Tennessee include Mascot-Jefferson City and Copper Ridge, which are remarkable for their high Zn:Pb ratios (Sangster 1983) and Sweetwater, which contains barite, fluorite and limited sphalerite (Hoagland 1976; Kesler et al. 1989). Other important deposits include Timberville and Friedensville which are found in the Lower Ordovician Beekmantown Formation in Virginia and Pennsylvania, respectively (Callahan 1968). Although MVT mineralization is found in younger rocks ranging from Silurian to Pennsylvanian in age, no other large deposits are known in the area (Smith 1977; Howe 1981; Wilbur et al. 1990).

Although considerable controversy exists about the age of brine expulsion in the southern and central Appalachians, growing evidence suggests that important MVT mineralization took place during Acadian deformation (Kesler and van der Pluijm 1990). The controversy has been fueled by paleomagnetic poles obtained on MVT ores and isotopic ages from authigenic minerals, which record a major fluid expulsion event during Late Paleozoic time (Bachtadse et al. 1987; Elliot and Aronson 1987; Hearn and others 1987; Jackson et al. 1988; Miller and Kent 1988). This event, which apparently reflects expulsion of fluids during Alleghanian deformation associated with the final consolidation of the orogen (Oliver 1986), has led to suggestions that MVT mineralization in this part of the orogen formed at the end of Paleozoic time. However, hydrocarbon deposits and MVT deposits are actually associated with pre-Alleghanian features (Haynes and Kesler 1989; Wilbur et al. 1990). The largest MVT deposits, those in East Tennessee, have yielded an Acadian (Devonian) Rb-Sr age of 377 ± 29 Ma on sphalerite from the Coy mine in the Mascot-Jefferson City zinc district (Nakai et al. 1990; Table 1). Kesler et al. (1988, 1989) have obtained essentially identical Sr isotope model ages for this mineralization. No other isotopic ages have been measured for deposits in this segment of the orogen, and their age is limited only by the fact that most deposits have been affected by Late Paleozoic Alleghanian deformation (Hoagland 1976).

3.6 Western Africa

Curiously, MVT mineralization is almost as scarce in western Africa as it is common in adjacent parts of North America. Although small lead-zinc prospects are present in this area, most are hosted by older rocks and none have been significant producers (Stam 1960). It is not clear whether MVT-forming processes operated in this region prior to formation of the numerous, important deposits that are hosted by Mesozoic of northern Africa.

4 Relation Between MVT Deposits and Appalachian-Caledonian Evolution

It is clear from the relations shown in Fig. 2 that significant MVT mineralization in the Appalachian-Caledonian orogen is restricted to those parts of the orogen that have undergone the largest amount of regional thrusting. This suggests that processes stronger than basinal overpressuring and regional tilting are required for large-scale brine expulsion and MVT genesis. Whether this brine expulsion was actually caused by hinterland recharge during thrusting or by more extensive deformation of the basin (Fig. 1) must be determined from the relation between mineralization and deformation-related features such as foreland basins and hinterland metamorphism.

Foreland basins record the progress of thrust sheets associated with regional deformation (Bradley and Kusky 1986; Jamieson and Beaumont 1988) and, as such, can be useful indicators of brine expulsion centers. It is not likely that these basins act as MVT brine sources because of their small volume and lack of evaporites (Kesler et al. 1989). However, they show where the greatest amount of sediment was loaded onto underlying strata, as well as the probable location of the thickest parts of the thrust sheet, where maximum recharge might have occurred. Extensive cover of thrust sheets in the Scandinavian segment of the orogen has prevented delineation of passive margin or foreland basins that might have been associated with mineralizing brines at Laisvall and related sandstone lead deposits (Bassett 1985; Bergstrom and Gee 1985; Bruton et al. 1985; Foyn 1985). Similar relations prevail in the northern Appalachians, where only small remnants of Early Paleozoic foreland basins remain (Quinn 1991). The only area in which the relation between mineralization and foreland basins has been documented is the southern Appalachians, where thrusting was widespread but has not completely covered the foreland, leaving large thicknesses of orogen-related sediment preserved (Colton 1971). Kesler et al. (1989) have shown, for instance, that there is a close spatial relation between MVT districts of East Tennessee and the Taconic-age Sevier foreland basin (Fig. 3). Average fluid inclusion temperatures in the deposits even appear to decrease westward away from the basin. It is unlikely that this indicates a genetic relation between MVT mineralization and the Sevier basin because the Sevier basin is not large enough to have supplied the necessary volume of brines and because the deposits appear to be of Devonian age, considerably younger than the basin. In fact, recent faunal and geologic relations suggest that a sizeable basin of Devonian (Acadian) age was present just west of the present location of the Sevier basin (Tull and Groszos 1990; Unrug and Unrug 1990), and it is possible that the MVT isotherms shown in Fig. 3 reflect brine expulsion related to this basin.

These observations appear to be corroborated by the relation between degree of hinterland metamorphism and MVT mineralization in the adjacent foreland basins in the Appalachians. This relation is obscured in most parts of the Caledonides by extensive thrusting and multiple phases of deformation. In Scandinavia, for instance, the present distribution of metamorphic grades reflects the overlapping effects of three phases of metamorphism as well as

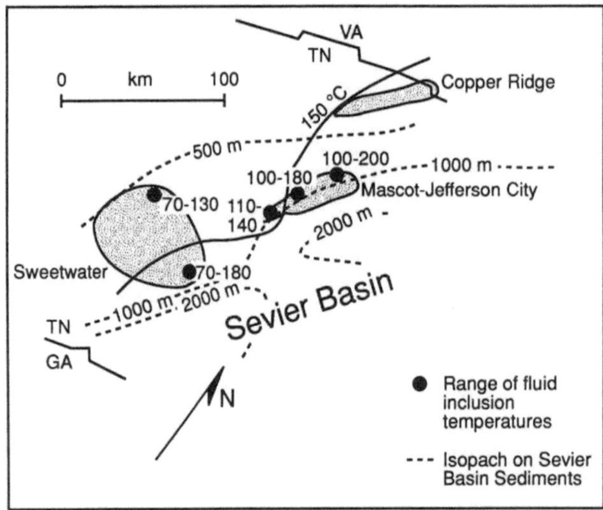

Fig. 3. Relation between Lower Ordovician-hosted MVT mineralization in East Tennessee and isopachs of the Sevier basin (in palinspastic reconstruction) (after Kesler et al. 1989), with approximate 150 °C contour on fluid inclusion temperatures (data from Zimmerman and Kesler 1981; Taylor et al. 1983). As noted in the text, this relation probably reflects the presence of a highly deformed Acadian-age(?) basin that overlapped the Sevier basin on the east. (Tull and Groszos 1990; Unrug and Unrug 1990)

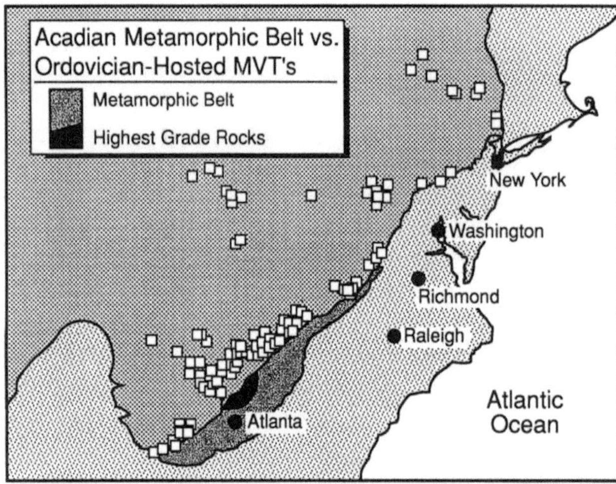

Fig. 4. Relation between MVT mineralization in Orovician rocks and zones of Acadian metamorphism in the southern and central Appalachians. Note that the eastern Tennessee deposits are immediately adjacent to the zone of highest Acadian metamorphic grade. Isotopic ages of 460 to 320 Ma have been obtained for this metamorphism compared to the age of 377 ± 29 Ma obtained by Nakai et al. (1990) for sphalerite in the East Tennessee deposits

cratonward translation of previously metamorphosed thrust sheets (Bryhni and Andreasson 1985). In the southern Appalachians, however, the area of highest grade Acadian metamorphic rocks is immediately east of the MVT deposits (Fig. 4), where it overlaps a similar high-grade area of Taconic metamorphism. If, as seems likely, metamorphism is most intense in the hinterland of foreland basins, this suggests that an Acadian foreland basin was present just east of the present location of the East Tennessee MVT deposits.

These relations suggest that large-scale thrusting is more important than hinterland recharge in the formation of MVT deposits. For instance, during formation of the Sevier basin in Taconic time, there was probably ample opportunity for hinterland recharge to have expelled brines to form MVT deposits in the southern Appalachians. Instead, MVT deposits for which dates are available appear to have formed only during late Acadian regional thrusting and basin development. Furthermore, the fact that the Scandinavian MVT deposits are found in sandstones rather than carbonate rocks, which are more common MVT hosts (Sangster 1983), suggests that once a basinal brine was expelled, it deposited MVT ore in the most favorable location that was available in its flow path. This suggests further that the most important single factor in the formation of MVT deposits is brine expulsion; although MVT mineralization is clearly most common in carbonate host rocks, it will probably occur in favorable settings, regardless of lithology, as long as brines can be expelled to form the deposits.

5 Conclusions

This review of the regional setting of MVT mineralization in the Caledonian-Appalachian orogen supports the following conclusions.

1. Significant mineralization is closely associated with large-scale regional thrusting. Where such thrusting is present, economically important MVT deposits are present; where it is absent, MVT deposits are also absent or small.
2. Inasmuch as large MVT deposits in the Caledonian-Appalachian orogen are in both carbonate and sandstone host rocks, brine expulsion was probably more important to the formation of MVT deposits than host rock lithology.
3. These observations lend strong support to the hypothesis of Oliver (1986) that large-scale thrusting is the most important mechanism for expulsion of basinal brines and hydrocarbons, and suggest that the timing of basin deformation can be determined by further dating of MVT deposits.

Acknowledgements. Support for field and laboratory studies related to this proposal has been provided by National Science Foundation Grant (EAR 8305909) and by the Turner Fund of the University of Michigan. Discussions and field sessions with A. Bjørlykke, J. Gesink, V. Greene, A. Halliday, D. Harper, F. Haynes, H. Jones, R. Kettler, S. Lindblom, E. McCormick, D. Sangster, B. van der Pluijm, F. Vokes, and M. Willdén have been of great help in forming the ideas presented here.

References

Akande SO, Zentilli M (1984) Geologic, fluid inclusion, and stable isotope studies of the Gays River lead-zinc deposit, Nova Scotia, Canada. Econ Geol 79:1187–1211

Anderson GM, Macqueen RW (1982) Ore deposit models – 6. Mississippi Valley-type lead-zinc deposits. Geosci Can 9:108–117

Andrew CJ, Ashton JH (1985) The regional setting, geology and metal distribution of the Navan orebody, Ireland. Trans Inst Min Metall Sect B 94:66–93

Bachtadse V, Van der Voo R, Haynes FM, Kesler SE (1987) Late Paleozoic magnetization of mineralized and unmineralized Ordovician carbonates from East Tennessee: evidence for a post-ore chemical event. J Geophys Res 92:14165–14176

Bassett MG (1985) Silurian stratigraphy and facies development in Scandinavia. In: Gee DG, Sturt BA (eds) The Caledonide orogen – Scandinavia and related areas. Wiley, New York, pp 283–292

Bergstrom J, Gee DG (1985) The Cambrian in Scandinavia. In: Gee DG, Sturt BA (eds) The Caledonide orogen – Scandinavia and related areas. Wiley, New York, pp 247–272

Bethke CM (1986) Hydrologic constraints on the genesis of the Upper Mississippi Valley district from Illinois Basin brines. Econ Geol 81:233–249

Bethke C, Marshak S (1990) Brine migration across North America – the plate tectonics of groundwater. Annu Rev Earth Planet Sci 18:287–315

Bjørlykke A (1978) The eastern marginal zone of the Caledonide orogen in Norway. In: Caledonian-Appalachian orogen of the North Atlantic region. Geol Surv Can Pap 78–13:49–56

Bjørlykke A, Sangster DF (1981) An overview of sandstone lead deposits and their relation to red-bed copper and carbonate-hosted lead-zinc deposits: Econ Geol 75th Anniversary Vol: 179–213

Bovan JA, Read JF (1987) Incipiently drowned facies within a cyclic peritidal ramp sequence, Early Ordovician Chepultepec interval, Virginia Appalachians. Geol Soc Am Bull 98:714–727

Boyle RW, Wanless RK, Stevens RD (1976) Sulfur isotope investigation of the barite, manganese and lead-zinc-copper-silver deposits of the Walton-Cheverie area, Nova Scotia, Canada. Econ Geol 71:749–762

Bradley DW, Kusky TM (1986) Geologic evidence for rate of plate convergence during the Taconic arc-continent collision. J Geol 94:667–681

Brown WH, Weinberg EL (1968) Geology of the Austinville-Ivanhoe district, Virginia. In: Ridge JD (ed) Ore deposits of the United States, 1933–1967, vol 1. AIME, New York, pp 169–186

Bruton OL, Lindstrom M, Owen AW (1985) The Ordovician of Scandinavia. In: Gee DG, Sturt BA (eds) The Caledonide orogen – Scandinavia and related areas. Wiley, New York, pp 273–282

Bryhni I, Andresson P-G (1985) Metamorphism in the Scandinavian Caledonides. In: Gee DG, Sturt BA (eds) The Caledonide orogen – Scandinavia and related areas. Wiley, New York, pp 763–781

Callahan WH (1968) Geology of the Friedensville zinc mine, Lehigh County, Pennsylvania. In: Ridge JD (ed) Ore deposits of the United States 1933–1967 (Graton-Sales Volume). AIME, New York, pp 95–107

Cathles LM, Smith AT Jr (1983) Thermal constraints on the formation of Mississippi Valley-type lead-zinc deposits and their implications for episodic basin dewatering and deposit genesis. Econ Geol 78:983–1002

Chapman TJ, Gayer RA, Williams GD (1985) Structural cross-sections through the Finnmark Caledonides and timing of the Finnmarkian event. In: Gee DG, Sturt BA (eds) The Caledonide orogen – Scandinavia and related areas. Wiley, New York, pp 593–610

Clark SHB (1989) Metallogenic map of zinc, lead, and barium deposits and occurrences in Paleozoic sedimentary rocks, east-central United States. US Geol Surv Map I-1773

Collins JA, Smith L (1975) Zinc deposits related to diagenesis and intrakarstic sedimentation in the Lower Ordovician St. George Formation, western Newfoundland. Bull Can Petrol Geol 23:393–427

Colton GW (1971) The Appalachian basin – its depositional sequences and their geologic relationships. In: Fisher GW, Pettijohn FJ, Reed JC Jr (eds) Studies of Appalachian geology: central and southern. Wiley, New York, pp 4–47

Duane MJ, de Wit MJ (1988) Pb-Zn ore deposits of the northern Caledonides: products of continental-scale fluid mixing and tectonic expulsion during continental collision. Geol 16:999–002 .
Dunham K, Beer KE, Ellis RA, Ballagher MJ, Nutt MJC, Webb BC (1978) United Kingdom. In: Bowie SHU, Kvalheim A, Haslam HW (eds) Mineral deposits of Europe, vol 1. Northwest Europe. Institution of Mining and Metallurgy, London, pp 263–346
Elliot I, Aronson JL (1987) Alleghanian episode of K-bentonite illitization in southern Appalachian basin. Geol 15:735–739
Evans AM (1976) Genesis of Irish base-metal deposits. In: Wolf KH (ed) Handbook of stratbound and stratiform ore deposits, vol 5. Elsevier, Amsterdam, pp 231–255
Foley NK, Sinha AK, Craig JR (1981) Isotopic composition of lead in the Austinville-Ivanhoe Pb-Zn district, Virginia. Econ Geol 76:2012–2017
Foyn S (1985) The Late Precambrian in northern Scandinavia. In: Gee DG, Sturt BA (eds) The Caledonides Orogen – Scandinavia and related areas. Wiley, New York, pp 233–246
Garven G (1985) The role of regional fluid flow in the genesis of the Pine Point deposit, western Canada sedimentary basin. Econ Geol 80:307–324
Gee DG, Sturt BA (eds) (1985) The Caledonide orogen – Scandinavia and related areas. Wiley, New York, 1514 pp
Hall CM, York D, Saunders CM, Strong DF (1989) Laser 40Ar/39Ar dating of Mississippi Valley-type mineralization western Newfoundland. Int Geological Congr Abstr Vol, pp 2–10
Halliday AN, Shepherd TJ, Nakai S, Chesley J, Dickin AP (1989) Sm-Nd and Rb-Sr dating of MVT deposits. GSA Abstr Prog 7:175
Hanor JS (1987) Origin and migration of subsurface sedimentary brines. Soc Econ Paleontol Mineral Lect Notes Short Course 21, 247 pp
Harris LD (1971) A Lower Paleozoic paleoaquifer – the Kingsport Formation and Mascot Dolomite of Tennessee and southwest Virginia. Econ Geol 66:735–743
Hatcher RD (1987) Tectonics of the southern and central Appalachian internides. Annu Rev Earth Planet Sci 15:337–362
Haynes FM, Kesler SE (1989) Pre-Alleghanian (Pennsylvanian-Permian) hydrocarbon emplacement along Ordovician Knox unconformity, eastern Tennessee. Am Assoc Petrol Geol Bull 73:289–297
Hearn PP Jr, Sutter JF, Belkin HE (1987) Evidence for Late-Paleozoic brine migration in Cambrian carbonate rocks of the central and southern Appalachian: implications for Mississippi Valley-type sulfide mineralization. Geochim Cosmochim Acta 51:1323–1334
Henriksen N (1985) The Caledonides of central East Greenland 70°–76°N. In: Gee DG, Sturt BA (eds) The Caledonide orogen – Scandinavia and related areas. Wiley, New York, pp 1095–1114
Hoagland AD (1976) Appalachian zinc-lead deposits. In: Wolfe KH (ed) Handbook of stratabound and stratiform ore deposits, vol 6. Elsevier, Amsterdam, pp 495–534
Howe SS (1981) Mineralogy, fluid inclusions, and stable isotopes of lead-zinc occurrences in central Pennsylvania. MSc Thesis, State College, Pennsylvania State University, 155 pp
Hurst JM, Jepsen HF, Kalsbeek F, McKerrow WS, Peel JS (1985) The geology of the northern extremity of the East Greenland Caledonides. In: Gee DG, Sturt BA (eds) The Caledonide orogen – Scandinavia and related areas. Wiley, New York, pp 1047–1064
Jackson M, McCabe C, Ballard MM, Van der Voo R (1988) Magnetite authigenesis and diagenetic paleotemperatures across the northern Appalachia basin. Geology 16:592–595
Jamieson RA, Beaumont C (1988) Orogeny and metamorphism: a model for deformation and pressure-temperature-time paths with application to the central and southern Appalachians. Tectonics 7:417–445
Jowett C (1977) Nature of the ore forming fluids of the Polaris lead-zinc deposit, Little Cornwallis Island, NWT, from fluid inclusion studies. Can Min Metall Bull (March):23–31
Kesler SE, van der Pluijm BA (1990) Relation of Lower Ordovician-hosted MVT mineralization to Appalachian orogenic events. Geology 18:1115–1118
Kesler SE, Jones LM, Ruiz J (1988) Strontium isotopic geochemistry of Mississippi Valley-type deposits, East Tenessee: implications for age and source of mineralizing brines. Geol Soc Am Bull 100:1300–1307 (reply to discussion in 102:1600–1602)
Kesler SE, Gesink JA, Haynes FM (1989) Evolution of mineralizing brines in the east Tennessee Mississippi Valley-type ore field. Geology 17:466–469

Kesler TL (1950) Geology and mineral deposits of the Cartersville district, Georgia. US Geol Surv Prof Pap 224, 97 pp

Lane T (1989) Sphalerite/dolomite stratigraphy and the tectonic origin of an MVT deposit, Daniel's Harbour, Newfoundland, Canada. Geol Soc Am Abstr Prog 21:A8

Lindblom S (1986) Textural and fluid inclusion evidence for ore deposition in the Pb-Zn deposit at Laisvall, Sweden. Econ Geol 81:46–64

McNaughton K, Smith TE (1986) A fluid inclusion study of sphalerite and dolomite from the Nanisivik lead-zinc deposit, Baffin Island, Northwest Territories, Canada. Econ Geol 81:713–720

Miller JD, Kent DV (1988) Regional trends in the timing of Alleghanian remagnetization in the Appalachians. Geology 16:588–591

Nakai S, Halliday AN, Kesler SE, Jones HD (1990) Rb-Sr dating of sphalerite and genesis of MVT deposits. Nature 346:354–357

Oliver J (1986) Fluids expelled tectonically from orogenic belts: their role in hydrocarbon migration and other geologic phenomena. Geology 14:99–102

Quinn L (1991) Ordovician foredeep sandstones of the Goose Tickle Group, western Newfoundland. Geol Assoc Can Prog Abstr 16:A103

Ramsay DM, Sturt BA, Jansen O, Andersen TB, Sinhan-Roy S (1985) The tectonostratigraphy of western Porsangerhalvoya, Finnmark, north Norway. In: Gee DG, Sturt BA (eds) The Caledonide orogen – Scandinavia and related areas. Wiley, New York, pp 611–622

Rast N (1989) The evolution of the Appalachian chain. In: Bally AW, Palmer AR (eds) The geology of North America, vol A. The geology of North America – an overview. Geological Society of America, Boulder, CO, pp 323–348

Ravenhurst CE, Reynolds PH, Zentilli M, Krueger HW, Blenkinsop J (1989) Formation of Carboniferous Pb-Zn and barite mineralization from basin-derived fluids, Nova Scotia, Canada. Econ Geol 84:1471–1488

Rickard DT, Willden MY, Narinder N-E, Donnely TH (1979) Studies on the genesis of the Laisvall sandstone lead-zinc deposit, Sweden. Econ Geol 74:1255–1285

Roberts D, Gee DG (1985) An introduction to the structure of the Scandinavian Caledonides. In: Gee DG, Strut BA (eds) The Caledonide orogen – Scandinavia and related areas. Wiley, New York, pp 56–68

Sangster DF (1983) Mississippi Valley-type deposits: a geological melange. In: Kisvarsanyi G, Grant SK, Pratt WP, Koenig JW (eds) International Conference on Mississippi Valley-type Lead-Zinc Deposits, Proc Vol. University of Missouri, Rolla, MO, pp 7–19

Sangster DF, Vaillancourt PD (1990) Geology of the Yava sandstone-lead deposit, Cape Breton Island, Nova Scotia. In: Sangster AL (ed) Mineral deposit studies in Nova Scotia, vol 1. Geol Surv Can Pap 90-8:203–244

Sawkins FJ (1966) Ore genesis in the North Pennine orefield, in the light of fluid inclusion studies. Econ Geol 61:385–401

Sharp JJ Jr (1978) Energy and momentum transport model of the Ouachita basin and its possible impact on the formation of economic mineral deposits. Econ Geol 73:1057–1068

Sharp JM, Kyle JR (1988) The role of groundwater processes in the formation of ore deposits. In: Back W, Rosensheim JS, Seaber PR (eds) The geology of North America, hydrogeology, vol O-2. Geological Society of America, Boulder, CO, pp 461–483

Shepherd TJ, Darbyshire DPF, Moore GR, Greenwood DA (1982) Age of Pennine mineralization. Bull Bur Mech Gites Min 11:371–377

Smith RC II (1977) Zinc and lead occurrences in Pennsylvania. Pa Geol Surv Mineral Resour Rep 72, 318 pp

Sorenson H, Nielsen BL, Jacobsen FL (1978) Denmark and Greenland. In: Bowie SHU, Kvalheim A, Haslam HW (eds) Mineral deposits of Europe, vol 1. Northwest Europe. Institution of Mining and Metallurgy, London, pp 251–261

Stam JC (1960) Some ore occurrences of the Mississippi Valley-type in equatorial Africa. Econ Geol 55:1708–1715

Steel R, Siedlecka A, Roberts D (1985) The Old Red Sandstone basins of Norway and their deformation. In: Gee DG, Sturt BA (eds) The Caledonide orogen – Scandinavia and related areas. Wiley, New York, pp 293–316

Stendal H, Ghisler M (1984) Strata-bound copper sulfide and non strata-bound arsenopyrite and base metal mineralization in the Caledonides of east Greenland – a review. Econ Geol 79:1574–1585

Sturt BA (1978) The Norwegian Caledonides. Geol Surv Can Pap 78-13:13–16
Sverjensky DA (1986) Genesis of Mississippi Valley-type lead-zinc deposits. Annu Rev Earth Planet Sci 14:177–199
Symons DTA, Sangster DF (1991) Paleomagnetism of the Mississippi Valley-type Polaris Zn-Pb deposit: genetic consequences of a Late Devonian age. Geol Assoc Can Prog Abstr 16:A121
Taylor M, Kesler SE, Cloke PL, Kelly WC (1983) Fluid inclusion evidence for fluid mixing, Mascot-Jefferson City zinc district, Tennessee. Econ Geol 78:1425–1439
Torsvik TH, Ryan PD, Trench A, Harper DAT (1991) Cambrian-Ordovician paleogeography of Baltica. Geology 19:7–10
Tull JF, Groszos M (1990) Nested Paleozoic "successor" basins in the southern Appalachian Blue Ridge. Geology 18:1046–1049
Unrug R, Unrug S (1990) Paleontological evidence of Paleozoic age for the Walden Creek Group, Ocoee Supergroup, Tennessee. Geology 18:1041–1045
Wilbur JS, Mutschler FE, Friedman JD, Zartman RE (1990) New chemical, isotopic and fluid inclusion data from zinc-lead-copper veins, Shawangunk Mountains, New York: Econ Geol 85:182–196
Williams CE, McArdle P (1978) Ireland. In: Bowie SHU, Kvalheim A, Haslam HW (eds) Mineral deposits of Europe, vol 1. Northwest Europe. Institution of Mining and Metallurgy, London, pp 319–346
Williams H (1984) Miogeoclines and suspect terranes of the Caledonian-Appalachian orogen: tectonic patterns in the North Atlantic region. Can J Earth Sci 21:887–901
Willdén MU (1980) Paleoenvironment of the autochthonous sedimentary rock sequence at Laisvall, Swedish Caledonides. Acta Univ Stockholm, Stockholm Contrib Geol 33, 100 pp
Zimmerman RK, Kesler SE (1981) Fluid inclusion evidence for solution mixing, Sweetwater (Mississippi Valley-type) district, Tennessee. Econ Geol 76:134–142

Genesis of the Ozark Mississippi Valley-Type Metallogenic Province, Missouri, Arkansas, Kansas, and Oklahoma, USA

D.L. Leach[1]

Abstract

The Ozark region of the United States midcontinent is host to the largest Mississippi Valley-type (MVT) lead-zinc province in the world. The region includes the world-class districts of the Old Lead Belt, Viburnum Trend, and Tri-State districts and the smaller northern Arkansas, central Missouri, and southeast Missouri barite districts. This metallogenic province was the product of an enormous hydrothermal system that affected more than 350 000 km^3 of rock. The hydrothermal fluids migrated into the Ozark region in late Paleozoic in response to convergent plate tectonics in the Ouachita foldbelt.

Ore districts were localized by a variety of paleohydrological controls on the migration of the hot basinal brines. Rock-water interactions in lithologically different aquifers produced ore districts with distinct metal ratios and isotopic compositions.

Reaction path modeling of quantitative fluid inclusion compositions shows that each district has its own depositional mechanism. The dominant ore depositional mechanisms are fluid mixing and wall rock reaction. The near-isothermal effervescence of CO_2 from the ore fluid during migration to lower confining pressures produced widespread sparry dolomite and trace sulfides.

1 Introduction

It is increasingly clear that most North American Mississippi Valley-type (MVT) deposits formed from enormous hydrothermal systems in which the fluid drive was provided by deformation of foredeeps and uplift of foreland thrust belts during collisional tectonics (e.g. Leach and Rowan 1986; Oliver 1986; Bethke and Marshak 1990; Leach and Sangster 1994). The model that best explains the regional hydrothermal systems is topographically driven fluid flow in which groundwater, recharged in the uplifted orogenic flank of a foredeep, migrates through deep portions of the basin, acquiring heat and dissolved components, and which is discharged along the basin's cratonic flank (Bethke and Marshak 1990). These regional hydrothermal systems produced MVT districts with broad fringing areas of trace mineralization, extensive but subtle hydrothermal alteration, broad thermal anomalies, and regional deposition of sparry dolomite.

[1] United States Geological Survey MS 973 Box 80225 Federal Center Denver, CO 80225, USA

Spec. Publ. No. 10 Soc. Geol. Applied to Mineral Deposits
Fontboté/Boni (Eds.), Sediment-Hosted Zn-Pb Ores

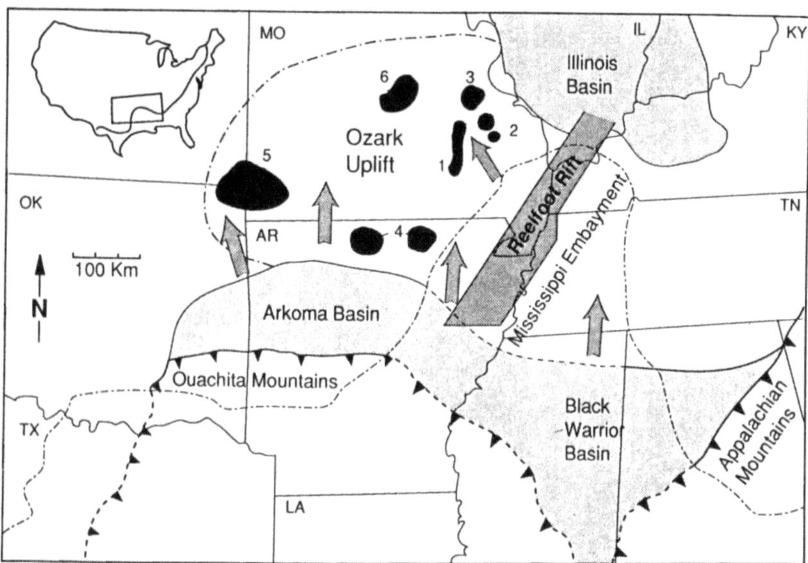

Fig. 1. Map showing locations of mining districts, major tectonic features, sedimentary basins, and likely ore fluid flow paths in the Ozark Region. MVT districts are: *1* Viburnum Trend subdistrict of southeast Missouri; *2* Old Lead Belt subdistrict of southeast Missouri; *3* southeast Missouri barite; *4* northern Arkansas; *5* Tri-State; *6* central Missouri

The regional nature of ore-forming processes for MVT deposits is best documented in the Ozark region of the US midcontinent. This region is located north of the Ouachita foldbelt and covers more than 240000 km^2 (Fig. 1). Ore districts include the world-class Viburnum Trend, Old Lead Belt, and Tri-State districts, as well as the smaller southeast Missouri barite, northern Arkansas, and central Missouri districts. Trace and minor occurrences of sphalerite and sparry dolomite are common throughout most of the stratigraphic section in the Ozark region (Leach 1979; Erickson et al. 1981; Coveney and Goebel 1983). Leach and Rowan (1986) presented evidence that brine migration from the Arkoma foredeep (Fig. 1) in response to Late Pennsylvanian-Early Permian orogenesis in the Ouachita foldbelt was responsible for the formation of MVT deposits in the Ozark region. This chapter expands on the Leach and Rowan (1986) concept, summarizes new information, and presents a genetic model for this important metallogenic province.

1.1 Regional Tectonic and Geologic Setting

MVT deposits in the Ozark region occur in relatively flat-lying and slightly deformed platform carbonate rocks of the Ozark uplift. Uplift of the Ozark dome (Fig. 1) occurred in Early Devonian, late Mississippian, and post-Pennsylvanian. The most significant uplift of the Ozark dome together with widespread gentle folding and normal faulting can be attributed to late Paleozoic orogenies along the continental margins.

Faults in the Paleozoic rocks of the Ozark region are generally normal faults that trend predominantly northwest-southeast together with a less well-developed secondary northeast to southwest trend. This rectilinear pattern of faulting in the Ozarks probably reflects the reactivation of preexisting faults in the Precambrian basement (McCracken 1971). Most of these faults have normal displacements less than several tens of meters but some have displacements of up to 125 m. In northern Arkansas and Tri-State districts, normal faults are commonly spatially associated with ore deposits (McKnight 1935; McKnight and Fischer 1970). In the Old Lead Belt in southeast Missouri, extensive normal faults together and fractures were important ore controls. Some major faults in the Old Lead Belt have considerable left-lateral displacement. The pattern of faulting in southeast Missouri as well as throughout the Ozark region is attributed to reactivation of basement structures during plate collision in the Ouachita foldbelt and uplift of the Ozark dome in late Paleozoic (Viele 1983; Kaiser and Ohmoto 1988).

In southeast Missouri, the Ozark platform carbonate sequence thickens dramatically into the graben formed as result of subsidence within the Reelfoot rift zone. As much as 7 km of Paleozoic rocks, including more than 1 km of arkosic Cambrian sandstones, occurs within the deeper part of the rift complex (Schwalb 1982). The Reelfoot rift is a failed continental rift that formed during the late Precambrian to Early Cambrian continental breakup (Hinze et al. 1980). The Paleozoic was marked by continued subsidence along the rift complex and infilling with marine sediments. In the late Paleozoic, the southern Reelfoot rift zone was overridden by thrusting during the Ouachita orogeny. After a period of uplift, erosion, and intrusions along the margins of the rift in the Mesozoic, subsidence of the Reelfoot rift complex associated with opening of the Gulf of Mexico resulted in Cretaceous and Tertiary sedimentation that comprises the Mississippi Embayment.

East of the Ozark uplift is the intracratonic Illinois basin (Fig. 1), containing more than 2.5 km of Paleozoic sediments. Early Paleozoic sedimentation was related to subsidence of the Reelfoot rift complex that forms the southerm margin of the basin. Subsidence of the Illinois basin continued through most of the Paleozoic for reasons which are not well understood.

In north-central Arkansas, Paleozoic rocks stratigraphically equivalent to the host rocks of the MVT deposits in the Ozark region are overlain by as much as 12 km of Carboniferous flysch of the Arkoma foredeep (Lillie et al. 1983). To the southeast into Tennessee and northern Alabama, they are overlain by an unknown but probably comparable thickness of flysch comprising the Black Warrior foredeep. The Arkoma and Black Warrior foredeeps are part of a series of foredeeps bordering the northern flank of the Ouachita foldbelt, formed in response to crustal downwarping and rapid sedimentation accompanying the Ouachita orogeny (Graham et al. 1975; Houseknecht 1986). Development of the Ouachita orogeny is generally accepted to be the result of closure of a Paleozoic ocean basin. South-dipping subduction of the North American plate beneath a magmatic arc or continental plate (Llanoria) formed an accretionary wedge and culminated in plate collision and suturing (Viele 1979; Nelson et al. 1982). As discussed below, the migration of fluids responsible for the Ozark MVT deposits is believed to coincide with final plate suturing and uplift of the tectonic flanks of the foredeeps in Late Pennsylvanian and Early Permian.

The tectonic effects of the Ouachita orogeny on the North American craton include reactivation of older basement faults, uplift of the Pascola arch in southern Illinois, and the development of numerous normal and strike-slip faults in southeast Missouri (Viele 1983; Kaiser and Ohmoto 1988). Small-scale faults, fractures, and gentle folds that are important ore controls within the ore districts in the Ozark region can be related to Ouachita orogeny (Kaiser and Ohmoto 1985).

1.2 Geology and Mineralogy of the Ore Districts

Mississippi Valley-type deposits occur in Upper Cambrian rocks in the Viburnum Trend, Old Lead Belt and southeast Missouri barite districts, and in Ordovician through Pennsylvanian rocks in the northern Arkansas, Tri-State, and central Missouri districts (Fig. 2). These districts have many features in common; however, each can be distinguished on the basis of stratigraphic ore horizons, metal ratios, geologic controls, and isotopic and trace element compositions. These differences are believed to be related to paleohydrological controls on ore-fluid migration, fluid-rock interactions in lithologically different aquifers, and precipitation mechanisms that varied from district to district (Viets and Leach 1990; Leach and Sangster 1994; Plumlee et al. 1994). Table 1 gives a highly generalized description of ore, gangue and alteration mineral assemblages together with the principal host rocks for each district.

1.2.1 Tri-State District

The Tri-State district, long recognized as one of the great mining districts of the world, covers more than 1800s km^2 in southwestern Missouri, northeastern Oklahoma, and southeastern Oklahoma (Fig. 1). The ore consists predominantly of sphalerite and galena (Zn/Pb atomic ratio is 16 for the district) with small amounts of chalcopyrite and pyrite. Important gangue minerals are dolomite and quartz in the form of jasperoid replacement of the host carbonate rocks. Jasperoid is a term for epigenetic silica replacement of previously lithified host rock and excludes syngenetic or diagenetic forms of silica such as chert (Lovering 1972). Mineralization occurs in Ordovician through Pennsylvanian rocks, however, ore deposits are generally restricted to the Mississippian Keokuk and Warsaw Limestones. Most of the deposits are in breccias and broad mineralized zones bordering the breccia. The most common deposits are elongate "runs" and stratiform "blanket" breccias associated with steeply dipping fractures and faults (McKnight and Fischer 1970). Although the origin of these breccias is uncertain, the map view of these breccias is similar to a karst system (McKnight and Fischer 1970). The typical alteration pattern for an ore run in the Tri-State district (McKnight and Fischer 1970) is a central dolomite core surrounded by a jasperoid zone, which grades into unaltered limestone. Sulfides are disseminated throughout the dolomite core and

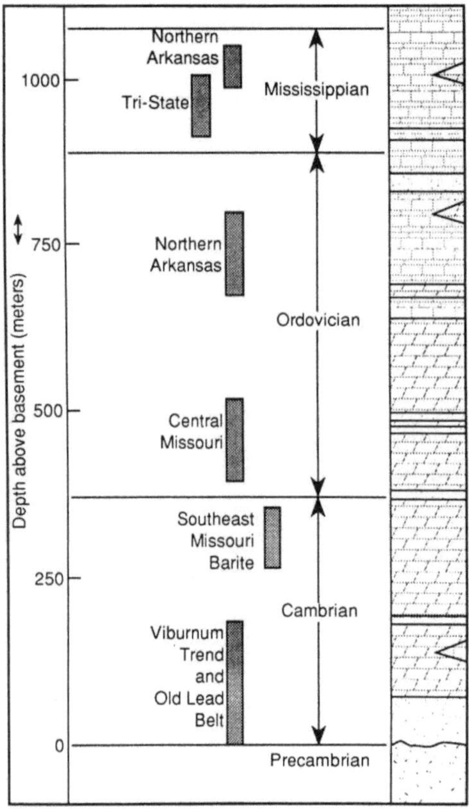

Fig. 2. Highly simplified illustration of the stratigraphic position and depth to basement rocks for Ozark MVT districts. Shown are the most important stratigraphic ore horizons for each district. However, minor mineralization can be found throughout the stratigraphic section

jasperoid, but are typically concentrated in a narrow zone between the dolomite core and jasperoid. Although most sulfides occur as replacement ore, sulfides and sparry dolomite commonly fill available porosity, fractures, and solution collapse features (McKnight and Fischer 1970). More detailed descriptions are in Hagni and Grawe (1964), Brockie et al. (1968), McKnight and Fischer (1970), and Hagni (1976).

1.2.2 Northern Arkansas District

Many small deposits occur in northern Arkansas on the southern flank of the Ozark uplift. The ore is in the Cotter and Powell Dolomites and Smithville and Everton Formations of Ordovician age and in the Boone Formation and Batesville Sandstone of Mississippian age. Ore deposits are developed along faults and solution collapse breccias. The most common forms of breccias are linear runs and stratiform or blanket veins (McKnight 1935), developed along

Table 1. Mississippi Valley-type districts in the Ozark region. A highly generalized description of the mineralogy and alteration assemblage together with principal host rocks for each district. Listed for each district are the most important minerals present and those in bold are the most abundant. Although barite is abundant in the central Missouri district, it is considered to have been deposited much later than sphalerite and galena (Leach 1980). The sulfide minerals listed occur both as replacement of host rocks and as open-space fillings. In the Viburnum Trend, Old Lead Belt, northern Arkansas, and Tri-State districts, replacement ores are the most common sulfide occurrence. Host rocks given are the principal ore horizons but sulfide minerals do occur in other formations

District	Mineralogy	Alteration	Host rocks
Southeast Missouri lead district: Viburnum Trend and Old Lead Belt subdistricts	**galena**, **sphalerite**, **pyrite**, **marcasite**, **dolomite**; minor copper-nickel-cobalt sulfides, quartz	Recrystallization of early dolostone, minor silicification, alteration of feldspar, and deposition of dickite and illite	Cambrian Bonneterre dolostone, minor occurrence in Cambrian Lamotte Sandstone
Tri-State	**Sphalerite**, **galena**, **dolomite**; minor pyrite, marcasite, chalocopyrite	Local dolomitization of limestone and silicification of carbonate rocks (jasperoid)	Mississippian Keokuk and Warsaw Limestones
Northern Arkansas	**Sphalerite**, **dolomite**; minor quartz, galena, pyrite, marcasite	Local dolomitization of limestone and extensive silicification of carbonate rocks (jasperoid)	Ordovician Everton Formation, Mississippian St. Joe and Boone Formations (all interbedded dolostones and limestones)
Central Missouri	**Sphalerite**, **galena**, **barite**	None	Ordovician Jefferson City and Gasconade Dolomites
Southeast Missouri barite	**Sphalerite**, **galena**, **barite**	None	Cambrian Potosi and Eminence dolomites

minor folds and steeply dipping fractures and faults similar to those in the Tri-State district. The sulfide ore is dominantly sphalerite (Zn/Pb atomic ratio of about 50); important gangue minerals are sparry dolomite and quartz, usually in the form of jasperoid. Host rocks in the northern Arkansas district have been altered to dolomite and jasperoid, similar to the Tri-State district. More detailed descriptions of the ore deposits are in McKnight (1935).

1.2.3 Central Missouri District

More than 250 small, high-grade deposits of barite, galena, and sphalerite comprise the central Missouri district. Historical production data from the district are incomplete but it is known that more than 350 000 t of barite, 17 000 t of galena, and 6000 t of sphalerite were mined. Deposits are located on a prominent northwest structural knob of the Ozark uplift in a gently dipping, relatively undisturbed series of rocks that range in age from Late Cambrian through Pennsylvanian. Although all the formations are mineralized, deposits are mainly in the Jefferson City and Gasconade Dolomites of Early Ordovician age. The deposits are in fractures, faults, and solution collapse breccia zones. The most productive deposits occur in domal solution collapse breccias. The central Missouri district differs from the Tri-State and northern Arkansas districts in that jasperoid and quartz are absent in the deposits, sparry dolomite is uncommon, ore minerals typically fill open space, replacement is uncommon, and barite is an abundant late mineral. Barite is believed to represent a distinctly later mineralizing event than the sulfides (Leach 1980).

1.2.4 Southeast Missouri Lead District

The southeast Missouri lead district includes the major subdistricts of the Viburnum Trend and Old Lead Belt (Fig. 1) and is the world's largest lead mining district. The Viburnum Trend and Old Lead belt are distinct from the other Ozark MVT districts in that the ores are lead-dominant, with Zn/Pb atomic ratios less than 0.4, and they contain minor but economically recoverable Cu-Co-Ni-Ag sulfides in a variety of mineral phases. The bulk of the galena has an octahedral crystal habit and occurs predominantly as replacement of host dolostone; hydrothermal dissolution and collapse brecciation of host dolostone accompanied and postdated deposition of octahedral-galena ore. Late-stage cubic galena crystals, occurring largely in open spaces, accounts for a minor part of the lead ore. Also, in contrast to Tri-State and northern Arkansas, extensive silicification is absent in southeast Missouri. The most obvious host rock alteration in the southeast Missouri lead districts is recrystallization of dolostone. Locally, dickite and illite are dispersed through the ore, and when mineralization occurs in the Lamotte Sandstone (Upper Cambrian), feldspars are altered to dickite.

The ore deposits are restricted to Upper Cambrian rocks, which unconformably overlie the Precambrian crystalline rocks that crop out in the St. Francois Mountains that define the crest of the Ozark uplift. About 95% of the ore has been produced from dolostones in the Bonneterre Formation (Upper Cambrian) and the remainder from the basal Lamotte Sandstone. Ore controls are both varied and complex and include pinch-outs of the Lamotte Sandstone and the Bonneterre Formation against buried Precambrian knobs, abrupt facies changes, sedimentary breccias, algal reef complexes, faults, and fractured rocks. In the Viburnum Trend, the most important hosts for ore are extensive linear solution collapse breccias. Detailed descriptions of the mines of the Viburnum Trend are given in a series of papers in Economic Geology (vol. 72, No. 3, 1977). Other important papers include Snyder and Gerdemann (1968), Gerdemann and Meyers (1972), and Sverjensky (1981).

1.2.5 Southeast Missouri Barite District

The southeast Missouri barite district is northwest of the Viburnum Trend and Old Lead Belt. Barite, minor galena and sphalerite occur in fractures, faults, solution channels, and along bedding in the Potosi and Eminence Dolomites of Late Cambrian age, and in the Gasconade Dolomite, Roubidoux

Formation, and the Jefferson City and Cotter Dolomites of Early Ordovician age. The district produced more than 1 million t of barite, largely from residual deposits derived from the weathering of the Potosi and Eminence Dolomites. Minor but unknown amounts of lead and zinc were produced from this district. For more detailed descriptions, see Leach (1979) and Kaiser et al. (1987).

2 Evidence for a Regional, Interconnected Hydrothermal System

Many small MVT deposits of sulfides occur outside the limits of the main ore districts and trace occurrences of MVT minerals commonly occur throughout the stratigraphic section in coal mines, rock quarries, and outcrops in the Ozark region. Geochemical studies of a drill core in the Ozark region by Erickson et al. (1981, 1983, 1988) demonstrated the widespread occurrence of anomalous concentrations of MVT metals in insoluble residues from carbonate rocks. Erickson et al. (1981) interpreted these anomalies to indicate the passage of metal-bearing fluids through extensive parts of the stratigraphic section. Leach (1979) suggested that the apparent areal continuity of minor MVT mineralization with the main ore districts suggests a common genetic relationship.

Sulfide ore minerals in the Ozark MVT districts are associated with coarse, saddle-shaped, sparry dolomite that paragenetically spans the deposition of ore minerals. Although sparry dolomite is closely associated with sulfide minerals, the coprecipitation of dolomite and sulfides in the ore deposits was rare; rather, dolomite was generally precipitated before or after deposition of paragenetic stages of sulfide minerals. In the Ozark region, this sparry dolomite occurs as fracture and vug lining within secondary or secondarily enhanced porosity and as replacement of limestones. Distant from the ore districts, sparry dolomite may occur without associated sulfides, but it is commonly interlayered with trace sulfides in ore-barren rocks studied by Erickson et al. (1981, 1988).

A cathodoluminescent (CL) microstratigraphy was defined in the sparry dolomite that spanned main stages of Viburnum Trend mineralization and correlated throughout the district by Voss and Hagni (1985). Rowan (1986) extended this CL microstratigraphy southwest from the Viburnum Trend to sparry dolomite in a drill core near the northern Arkansas and Tri-State districts. The regional correlation of the CL microstratigraphy suggests extensive circulation of the mineralizing fluids through hydrologically continuous formations in the Ozark region.

The author observed that sparry dolomite cement in the Ozark region decreases in abundance from northern Arkansas to central Missouri and is absent in northern Missouri. This suggests that the fluids which deposited sparry dolomite may have been derived from the Arkoma basin. In addition, recent studies by Gregg and Shelton (1989a,b) presented data for trace and minor elements in sparry dolomite (both replacement and cement) which indicate a regional fluid movement north from the Arkoma basin, in agreement with Leach and Rowan (1986). The minor element pattern in the sparry dolomite is more complex in the Viburnum Trend area where the presence of more than one fluid was postulated (in agreement with Goldhaber and Mosier 1989; Rowan and Leach 1989; Brannon et al. 1991; Viets and Leach 1990).

The apparent areal continuity of the MVT districts, the CL microstratigraphic zonation in sparry dolomites, and the metal-enriched insoluble residues from carbonate rocks throughout the stratigraphic section in the Ozarks argue strongly for an interconnected hydrothermal system on a regional scale. This hydrothermal system affected more than 350 000 km^3 of rock.

3 Age of Mineralization

In the Tri-State and northern Arkansas districts, the occurrence of ore along Pennsylvanian faults constrains mineralization to Pennsylvanian or younger. Disseminated sphalerite in rocks as young as mid-Pennsylvanian places similar constraints on the timing of mineralization in the central Missouri and Tri-State districts. The ore districts in southeast Missouri lack any such geological constraints on the timing of mineralization. However, the observed correlation of mineral paragenesis of sulfides and sparry dolomite in the Viburnum Trend with ores in the Tri-State and northern Arkansas districts argues strongly for contemporaneous mineralization by the same regional, interconnected, hydrothermal system. Therefore, the timing of mineralization in southeast Missouri is younger than mid-Pennsylvanian.

Recent attempts to date MVT mineralization have been mainly by paleomagnetic methods using apparent polar wandering paths. By this technique, the Viburnum Trend, northern Arkansas, Tri-State, and central Missouri districts have been examined (Wisniowiecki et al. 1983; Pan et al. 1990; Symons and Sangster 1991; Symons and Sangster, this Vol.). The paleomagnetic results yield a regionally consistent age of Late Pennsylvanian to Permian.

Consistent with the paleomagnetic ages, several radiometric studies in the US midcontinent region indicate an episode of regional fluid migration in Permian time. Hay et al. (1988) determined a Permian age (265 Ma) for illite from bentonites on the Mississippi River arch in Missouri. A 250 Ma (mid-Permian) age for postmineralization illite in the Viburnum Trend gives an upper bound on the age of mineralization (Aronson, reported in Rothbard 1983). As discussed below, fluids responsible for the Ozark MVT deposits are believed to have migrated from foredeep basins bordering the Ouachita foldbelt during closing stages of Ouachita orogeny in Late Pennsylvanian to Permian. Consistent with this hypothesis is evidence for widespread fluid migration in the Ouachita foldbelt and adjacent Arkoma foredeep. Desborough et al. (1985) found Permian fission-track ages in detrital zircon from the Pennsylvanian of the Arkansas River valley in north-central Arkansas. Also, Bass and Ferrara (1969) determined an Rb/Sr age of 283 ± 4 Ma for adularia from the Ouachita mountains. Shelton et al. (1986) documented a K-Ar age of 262 ± 10 Ma for Ba-rich adularia veins from the Ouachita Mountains and suggested a link to hydrothermal mineralization in southeast Missouri. A Late

Pennsylvanian to mid-Permian age of mineralization is consistent with fluid migration during the closing stages of Ouachita orogeny.

In contrast to the Late Pennsylvanian to Permian ages, some radiometric studies give ages that cluster in Middle Paleozoic. Stein and Kish (1985, 1991) reported Late Devonian to Early Mississippian ages for glauconites in the Upper Cambrian Bonneterre Formation in southeast Missouri. In view of the other age dating studies and the field evidence that indicate a late Pennsylvanian to Permian age for mineralization, this older date (Devonian-early Mississippian) may represent the age of regional dolomitization in the Ozark region prior to the MVT mineralization event, as suggested by paleomagnetic studies (Pan et al. 1990).

4 Thermal Regime in the Ozark Region During Mineralization

Fluid inclusion studies of MVT deposits in the Ozarks show that the ore fluid was a brine at temperatures predominantly between 80 and 140 °C (see references in Leach and Rowan 1986; Rowan and Leach 1989; Shelton et al. 1992). Assuming hydrostatic pressures and a simple H_2O-NaCl fluid, the resulting pressure corrections to the homogenization temperatures would not exceed approximately 15 °C (Leach and Rowan 1986; Rowan and Leach 1989). A few fluid inclusion homogenizaton temperatures, largely from sparry dolomite, are between 140 and 180 °C (Rowan and Leach 1989; Shelton et al. 1992). As pointed out by Rowan and Leach (1989, p. 1955), "The nature of inclusions in dolomite is such that there is substantial 'noise' in the homogenization temperature distribution, far exceeding the precision of individual measurements, due most likely to necking-down". Therefore, the extreme low and high homogenization temperatures in sparry dolomite may not be representative of the actual thermal regime during dolomite precipitation.

Fluid inclusion studies show that the ore districts were not thermally anomalous during ore deposition (Leach and Rowan 1986; Rowan and Leach 1989). Furthermore, widespread minor and trace occurrences of sphalerite in the region yield homogenization temperatures and salinities identical to those in adjacent ore districts (Leach 1979; Coveney and Goebel 1983) which suggests regional heating of the stratigraphic section to at least 100 °C. Independent evidence for a broad heating of the Ozark region is shown by the anomalous thermal maturity of Pennsylvanian rocks (Houseknecht et al. 1991) and conodont alteration studies in rocks with reset paleomagnetic poles (Sangster 1989). These results indicate that the mineralizing fluids were in thermal equilibrium with an enormous volume of rock.

On the basis of reconstruction of the stratigraphic section, the estimated maximum possible cover is about 1 km for Ordovician- and Mississippian-hosted deposits and up to 2 km for deposits in Cambrian rocks. If the ore fluids were heated by conduction from the basement, unreasonable thermal gradients of at least 60 to 100 °C/km would be required in the relatively shallow shelf carbonate sequences that host the MVT deposits. A more reasonable explanation for the broad distribution of saline fluid inclusions with higher temperatures relative to model temperatures based on basement heat-flow-controlled

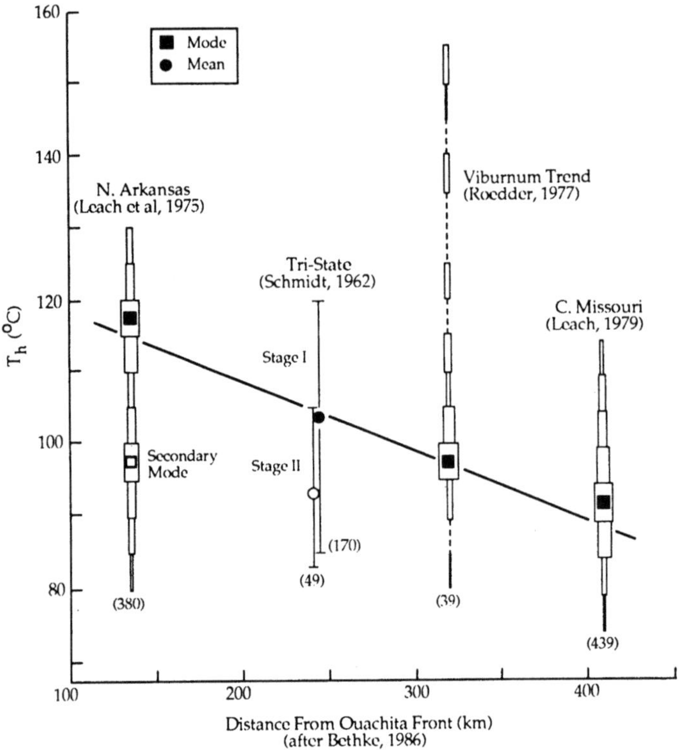

Fig. 3. Distribution of fluid inclusion homogenization temperatures in sphalerite from four principal MVT districts in the Ozark region vs. distance from the Ouachita foldbelt (after Bethke 1986). Lateral thermal gradient is approximately 0.1 °C/km. Number of homogenization temperatures for each district are given in parentheses

geothermal gradients is the advective heat transport via rapid, large-scale flow of brines originating at greater depths in nearby basins.

Despite the absence of thermal gradients away from the ore deposits, there is an apparent south to north trend of decreasing fluid inclusion homogenization temperatures; values for sphalerite from the region demonstrate this trend (Fig. 3). Although there were multiple stages of ore deposition spanning a range of temperatures in each district, the distribution means and modes systematically decrease from the edge of the Arkoma basin northward to central Missouri. This distribution of fluid inclusion homogenization temperatures reflects a lateral thermal gradient of about 0.1 °C/km of a slowly cooling brine moving northward from the Arkoma basin. This regional trend in fluid inclusion temperatures is consistent with the northward decrease in rank of Pennsylvanian coals in this region (Damberger 1974).

This apparent decrease in temperature away from the Arkoma basin has recently been questioned by Shelton et al. (1992) largely on the basis of their observed homogenization temperatures in sparry dolomite from the Ozark region. As mentioned previously, the inherent scatter or "noise" in homogenization temperatures for sparry dolomite in the Ozark region limits the precision in defining thermal gradients. It is the author's experience that homogenization temperatures obtained from sphalerite in the Ozark region have less scatter or noise relative to dolomite and therefore provide a more precise definition of thermal gradients.

5 Composition of the Ore Fluids

5.1 Salinity

Final melting temperatures of fluid inclusions in samples throughout the Ozarks (Leach et al. 1975; Leach 1979, 1980; Leach and Rowan 1986; Rowan and Leach 1989) range mainly from −10 to −27 °C with a well-defined mode at −21 °C, corresponding to 23 equivalent NaCl wt%. None of these studies show any systematic relationship between homogenization and final melting temperatures. These studies show a range of final melting temperatures with a small, secondary mode at around −10 to −15 °C, corresponding to salinities of 14 to 18 equivalent NaCl wt%. Rowan and Leach (1989) proposed that this secondary mode of final melting temperatures is evidence for another more dilute brine present in the region during mineralization, consistent with Bauer and Shelton (1989) and Shelton et al. (1992). Recent studies by Bauer and Shelton (1989) and Shelton et al. (1992) on sparry dolomite from the Ozark region indicate a larger range in homogenization and final melting temperatures than have been previously reported. The large amount of "scatter" in the data presented by Bauer and Shelton (1989, Fig. 1, p. A3) may reflect necking-down of the inclusions in sparry dolomite after fluid entrapment as commonly observed by Rowan and Leach (1989).

5.2 Ion Ratios

The ion ratios of Na, Ca, Mg, K, Br, and Cl in fluid inclusions in ore and gangue minerals from ore districts and minor occurrences were determined by Viets and Leach (1990). They identified two distinct brine compositions; one brine was present in all districts, referred to as the regional brine and a second brine, referred to as the VT brine, was present only in the main ore-stage (octahedral-galena stage) minerals in the Viburnum Trend. The VT brine is enriched in K, Mg, and Br relative to the regional brine, consistent with a study by Crocetti and Holland (1989). Viets and Leach (1990) proposed that the difference in composition between the two brines was controlled by the lithology of the principal aquifers for brine migration. The enrichment of K in the VT brine probably reflects fluid reaction with arkosic sandstones in the basal Lamotte Sandstone which is generally accepted as the principal aquifer for the deposits in southeast Missouri. However, Viets and Leach (1990) did not preclude the possibility that the distinct brine composition of the VT brine may be related to a bittern-derived fluid, perhaps with an origin in either the Arkoma or Black Warrior foredeeps. The regional brine was the ore fluid for

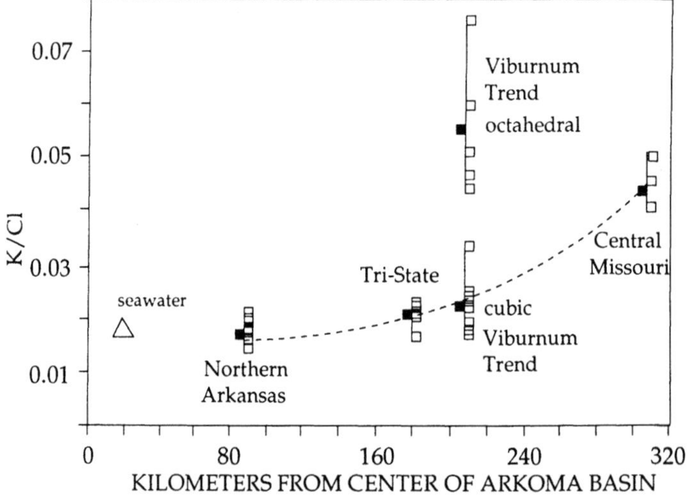

Fig. 4. Potassium to chloride atomic ratios of inclusion fluids extracted from hydrothermal minerals of the Ozark MVT districts vs. the distance of their geographic center to the center of the Arkoma basin. (Viets and Leach 1990)

Tri-State, northern Arkansas, and central Missouri, and the cubic-stage ores in the Viburnum Trend and their composition were ascribed by Viets and Leach (1990) to fluid migration through carbonate aquifers.

Shown in Fig. 4 is the K/Cl composition of the fluid inclusions plotted against distance from the Arkoma basin. The regional brine shows a systematic increase in K content away from the proposed source of brines in the Arkoma basin. This systematic increase in K content was interpreted by Viets and Leach (1990) to be due to progressive dissolution of K-rich minerals by the ore fluids, such as K-feldspar or glauconites, during fluid migration. Also evident in Fig. 4 is the distinctly different K/Cl ratios of the regional and VT brines (octahedral-stage minerals) in the Viburnum Trend. Other cations show small variations throughout the region which reflect local variations in the lithology of the host rocks.

Changes in composition of the ore fluid through mineral paragenesis were observed only in samples from the Viburnum Trend. Figure 5 presents the temporal changes in K/Cl and Ca/Mg through mineral paragenesis in the Viburnum Trend. Viets and Leach (1990) interpreted this variation to reflect the presence of two brine compositions during ore formation in the Viburnum Trend.

5.3 Gas Composition

Composition of gases in fluid inclusions in samples of ore and gangue minerals from te Ozark region was determined by quadrupole mass spectrometry (Hofstra et al. 1989; Landis and Hofstra 1991). Water-dominant fluid inclusions typically contain less than 1–2 mol% gases; CO_2 is the major gas, and there are lesser

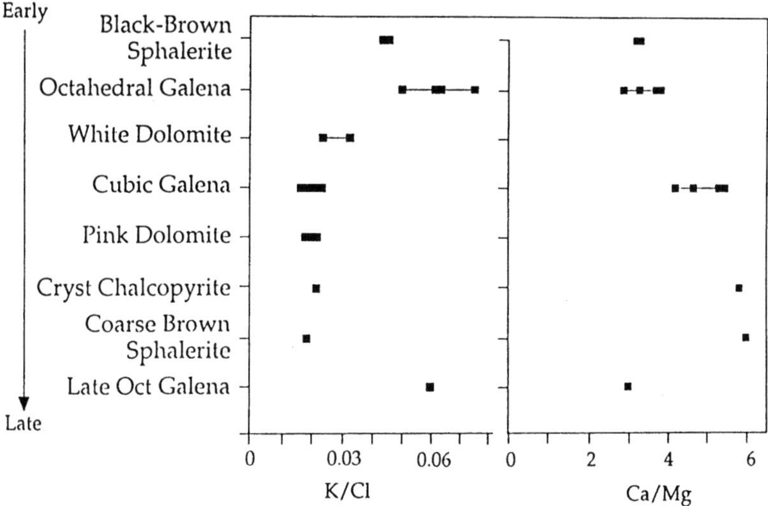

Fig. 5. Temporal changes in K/Cl and Ca/Mg atomic ratios of inclusion fluids for a generalized paragenesis of ore-stage minerals in the Viburnum Trend. (Viets and Leach 1990)

amounts of CH_4, H_2S, N_2, and short-chain hydrocarbons. Gas analyses of fluid inclusions also show that both ore and gangue minerals contain gas-dominant (CO_2-rich) inclusions and H_2O-dominant inclusions (Hofstra et al. 1989; Landis and Hofstra 1991). The CO_2-rich inclusions indicate that a separate CO_2-rich gas phase was present during deposition of sparry dolomite and sulfides and that the fluids were saturated with respect to CO_2. The data indicate that CO_2 effervescence was a widespread phenomena in the Ozark region, occurring in ore zones during sulfide mineralization, and along fluid migration pathways during deposition of sarry dolomite. The CO_2 effervenscence resulted from two processes: (1) pressure decrease associated with migration of basinal brines to shallower crustal levels and lower confining pressures; (2) P_{CO2} increase associated with the dissolution of carbonates by acid generated during sulfide deposition (Leach et al. 1991).

The gas compositions of inclusion fluids are remarkably similar across the Ozark region with only very small variations apparent from district to district. In the Viburnum Trend, the gas compositions of fluid inclusions in octahedral-stage and later cubic-stage galena, sphalerite, and dolomite and essentially the same. Fluid inclusions in sulfides contain more H_2S (~0.2 mol%) and CH_4 ± hydrocarbons (~0.2 mol%) than fluid inclusions in gangue minerals, ($H_2S \leqslant 0.002$ and $CH_4 \leqslant 0.1$ mol%). The higher conceentrations of CH_4 and H_2S in fluid inclusions in sulfides indicate that a reduced fluid enriched in these components interacted with the ore fluid during periods of sulfide deposition.

5.4 Modal Composition of the Regional and VT Ore Fluids

The total salinity and absolute concentrations of individual salt components can be computed using a thermodynamic model relating the ice melting temperature

Table 2. Reference compositions for ore-forming fluids used in reaction path calculations. Regional fluid found in regional hydrothermal dolomite, the cubic galena stage of the Viburnum Trend, and the main stages of all other districts are presented. The main-stage Viburnum Trend fluid (VT) is that found in the main octahedral galena stage of the Viburnum Trend. Concentrations are given as molality (m) or parts per million (ppm). (Plumlee et al, 1994)

Parameter	Regional fluid	VT fluid	Range	Determined by
Temperature (°C)	120	120	80–180	FI[a]
pH	4.55	4.53	4.3–5.0	Dolomite saturation
Ionic strength (m)	4.0	4.0	3.3–4.5	Calculated
Saturation pressure (bar)	148.3	148.0	118–250	Calculated
Cl(m)	4.9	4.9	3.1–5.3	Charge balance
S_{ox} (m)	2.5×10^{-2}	2.5×10^{-2}	2.4×10^{-2}–2.5×10^{-2}	FI
S_{red} (m)	1.4×10^{-3}	1.4×10^{-3}	1.5×10^{-2}–1.4×10^{-4}	FI
CO_2 (m)	3.5×10^{-1}	3.5×10^{-1}	2.0×10^{-2}–8.0×10^{-1}	FI
CH_4 aq (m)	5.6×10^{-3}	5.6×10^{-3}	5.6×10^{-3}	FI
Ca (m)	5.7×10^{-1}	5.5×10^{-1}	2.0×10^{-2}–4.5×10^{-1}	FI
Mg (m)	1.02×10^{-1}	1.73×10^{-1}	2.0×10^{-2}–1.9×10^{-1}	FI
Ca/Mg	5.6	3.2	3.0–6.0	FI
K (m)	1.02×10^{-1}	2.54×10^{-1}	8.2×10^{-2}–1.2×10^{-1}	FI
Na (m)	3.53	3.53	2.8–4.0	FI
SiO_2 (m)	5.6×10^{-4}	5.6×10^{-4}	5.6×10^{-4}–9.3×10^{-4}	Quartz saturation
Al (m)	1.2×10^{-6}	1.1×10^{-6}	1.0×10^{-6}–1.9×10^{-6}	Muscovite saturation
Fe (ppm)	1.7×10^{-4}	0.6	2.5×10^{-6}–1.0	Pyrite saturation
Zn (ppm)	3.8×10^{-1}	0.1	6.1×10^{-2}–54	Sphalerite saturation
Pb (ppm)	9.0×10^{-2}	11.4	7.3×10^{-2}–12	Galena saturation

FI parameter measured through fluid inclusion studies. All other values calculated by SOLVEQ assuming charge balance or saturation with various minerals, with the exception of Zn for the VT fluid (which was fixed at a level approximately 3 orders of magnitude undersaturated with sphalerite).

(Tm) of fluic inclusions to ratios of ions in the system Na-K-Ca-Mg-Cl-SO_4-H_2O (Spencer et al. 1990). For both the regional and VT brines, the concentrations of these components calculated for the Tm mode at −21 °C are given in Table 2. Also shown are the concentrations of CO_2, CH_4, and H_2S determined by quadrupole mass spectrometry. Plumlee et al. (1994) use the chemical speciation program, SOLVEQ (Reed 1982) to calculate the fluid chemistry for the regional and VT brines using the measured fluid inclusion temperature, solute, and gas data, and assuming saturation with various hydrothermal and inferred aquifer minerals (Table 2).

5.5 Fluid pH Values

Due to their high CO_2 contents, the calculated pH values of dolomite-saturated Ozark fluids are moderately acidic (from 4.25 to 5) and relatively insensitive to variations in temperature, Ca/Mg ratios, or ionic strength (Plumlee et al. 1994). Calculated pH values for calcite-saturated fluids (flowing through limestone rather than dolostone host rocks) are in the same general range as those of dolomite-saturated fluids. These low pH values are consistent with the abundant hydrothermal marcasite in the Ozark system, because marcasite requires pH values less than 4.5–5 to form (Murowchick and Barnes 1986; Goldhaber and Stanton 1987; Stanton and Goldhaber 1991); they are also consistent with observed aluminosilicate alteration mineralogies (Sverjensky 1984).

5.6 Saturation Pressures

The calculated saturation pressures of the Ozark basinal fluids vary greatly depending upon the dissolved gas contents, salinities and temperatures (Table 2). Calculated saturation pressures using measured modal fluid inclusion gas contents and salinities generally fall between 120 and 200 bar, whereas extreme gas contents and salinities yield pressures significantly outside this range. The maximum stratigraphic cover in the region at the time of ore deposition was probably no more than 1500 to 2000 m, which corresponds to a maximum hydrostatic pressure of 150 to 200 bar. These calculations indicate that the Ozark basinal brines were either saturated or close to saturation with a vapor phase, consistent with fluid inclusion gas data indicative of two-phase trapping.

6 Source of the Ore Components

6.1 Source of Metals

Virtually every rock in the stratigraphic section has been proposed as the original source of the metals. Metals in the ore fluids are attributed to the source sedimentary basin or to leaching of metals from aquifers through which the brines migrated. Metals can be derived from the source basin for the brines; however, evidence suggests the metal character of the ore fluid, and therefore of the MVT deposits, is directly related to the lithology of the aquifer through which the fluids migrated. As noted by Gustafson and Williams (1981), Bjørlykke and Sangster (1981), and Sangster (1983, 1990), the Zn/Pb ratios of MVT deposits throughout the world are related to host rock lithology; sandstone-hosted ore deposits are lead-rich, whereas MVT deposits, hosted by carbonate rocks, are typically zinc-rich. This observation, in itself, may suggest that the lead and zinc in the Ozark MVT deposits were derived from the local stratigraphic sequence. The southeast Missouri MVT deposits appear to be the exception, because these largely carbonate-hosted deposits are lead-rich. However, an important aquifer for the southeast Missouri deposits is believed to be the basal sandstone (e.g. Doe and Delevaux 1972). Recent studies support the concept that the lead, copper, nickel, and cobalt contents of the southeast Missouri deposits are due to fluid interaction with the thick red clastic sediments in the Reelfoot rift complex (Diehl et al. 1989; Viets and Leach 1990). Another plausible source for these metals in southeast Missouri could be the mafic and ultramafic rocks within the Reelfoot rift (Horrall et al., in press).

6.2 Sulfur and Lead Isotopes

Lead and sulfur isotopic studies of the MVT districts agree that these components were derived from crustal sources: beyond this point there is little

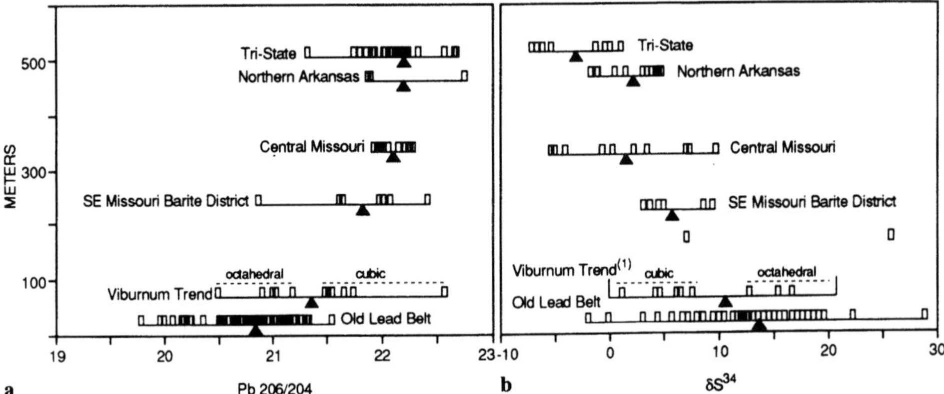

Fig. 6. **a** Ranges of $^{206}Pb/^{204}Pb$ ratios of galena for the MVT districts in the Ozark region vs. their approximate distance from the basal sandstone. *Triangles* indicate district means (Viets and Leach 1990). Original sources of data are listed in Viets and Leach (1990). **b** Sulfur isotope compositions of ore minerals for the MVT districts in the Ozark region vs. their approximate distance above the basal sandstone. *Triangles* indicate district means. *1* Shown for the Viburnum Trend is the range of sulfur isotopes from octahedral galena indicated by the *vertical line* at each end of the range; data are from Hennigh et al. (1990), Burstein et al. (1991), and Shelton et al. (1991). Other sources of data are given in Viets and Leach (1990)

agreement on specific sources for each district. Each district has a range of sulfur isotope values which has been interpreted to result from multiple sources of sulfur and various mechanisms of producing reduced sulfur from sedimentary or seawater sulfate (Heyl et al. 1974). Lead isotopes are radiogenic with $^{206}Pb/^{204}Pb$ greater than 20.

Figure 6 A,B shows the range of sulfur and lead isotope values for each district in the Ozarks versus distance above the basal sandstone. The distribution of sulfur and lead isotopes of the Ozark region shows a rough correlation with stratigraphic position; sulfur isotopes become lighter and lead becomes more radiogenic higher in the stratigraphic section (Goldhaber 1989; Goldhaber and Mosier 1989; Leach and Viets 1990). This suggests a more *local* host rock contribution of *lead* and *sulfur* rather than a distal basin source. As pointed out by Viets and Leach (1990), cubic galena in the Viburnum Trend is isotopically similar to Tri-State, northern Arkansas, southeast Missouri barite, and central Missouri districts and therefore probably shared a common brine source. This conclusion is consistent with ion microprobe mass spectrometry of a single crystal of galena from the Tri-State district showing that the isotopic composition of sulfur and lead varied together and therefore both elements were probably transported together in the same ore fluid (Deloule et al. 1986).

The lead deposits of southeast Missouri have a wide range in both sulfur and lead isotopes. A trend from isotopically heavier sulfur and less radiogenic lead in early and main-stage octahedral galena to isotopically lighter sulfur and more radiogenic lead in the cubic-stage galenas throughout the Viburnum Trend is well documented (Sverjensky et al. 1979; Sverjensky 1981; Crocetti et al. 1988). Ion microprobe analysis of individual galena crystals (Deloule et al. 1986; Crocetti and Holland 1989) shows significant and commonly independent variations in sulfur and lead isotopes at the micron scale. The variation in lead and sulfur through mineral paragenesis, together with the micron-scale and independent isotopic variations and fluid inclusion chemistry, was interpreted by Crocetti and Holland (1989), Goldhaber (1989), Goldhaber and Mosier (1989), and Viets and Leach (1990) to indicate mineralization in the Viburnum Trend was by at least two chemically and isotopically distinct ore fluids. Strontium isotopic studies of ore minerals in the Viburnum Trend are also consistent with mineralization from at least two isotopically distinct fluids (Brannon et al. 1991).

Recent sulfur isotope studies of Goldhaber and Mosier (1989), Hennigh et al. (1990), Burstein et al. (1991) and Shelton et al. (1991) indicate a significantly larger range of $\delta^{34}S$ values for main-stage octahedral galena (-1 to $+27$ per mill) in the Viburnum Trend than has been previously reported. Burstein et al. (1991) suggested that the paragenetically early copper and zinc ore is enriched in ^{32}S and may reflect a fluid moving through the Precambrian crystalline basement or sediments derived from these rocks. The observed shift to more ^{34}S-enriched sulfides in the main lead-zinc ore indicates the introduction of another basinal fluid. The Goldhaber and Mosier (1989) study of sulfur isotopes in iron sulfides within the Bonneterre Formation indicates that portions of the Bonneterre Formation were the likely proximal source for the heavy sulfur in the main-stage ores. The return to ^{32}S-enriched sulfide deposition in the main-stage marcasite and late-stage galena mineralization may represent renewed dominance of the early fluid source (Burstein et al. 1991). These new studies of sulfur isotopes clearly indicate the presence of multiple, metal-specific fluids with distinct sulfur isotope signatures.

Figure 7 presents the combined sulfur and lead isotopic data for the Ozark MVT districts which show that the lead and sulfur isotopic systematics observed in the Viburnum Trend are similar for the entire Ozark region. Viets and Leach (1990) interpreted this linear array as indicating that the cubic galena stage in the Viburnum Trend, Tri-State, northern Arkansas, and central Missouri formed from a brine with more radiogenic lead traveling through a carbonate aquifer. The main-stage octahedral galena of the Viburnum Trend and Old Lead Belt was interpreted to have formed from a brine with less radiogenic lead that traveled through the basal sandstone aquifer.

7 Migration of the Brines

Any fluid-driven mechanism for the Ozark hydrothermal system must be consistent with the multiple lines of evidence for regional mineralization at temperatures exceeding normal geothermal gradients and the absence of thermal anomalies associated with the ore districts. These thermal observations constrain possible hydrological models.

Bethke et al. (1988) evaluated the possible role of compaction-driven fluid flow from the Arkoma basin. They determined that fluid discharge, even when optimum conditions are assumed to have occurred during development of overpressured zones in a rapid depocenter such as the Arkoma basin, *could not* transport adequate heat to account for the wide distribution of fluid inclusions

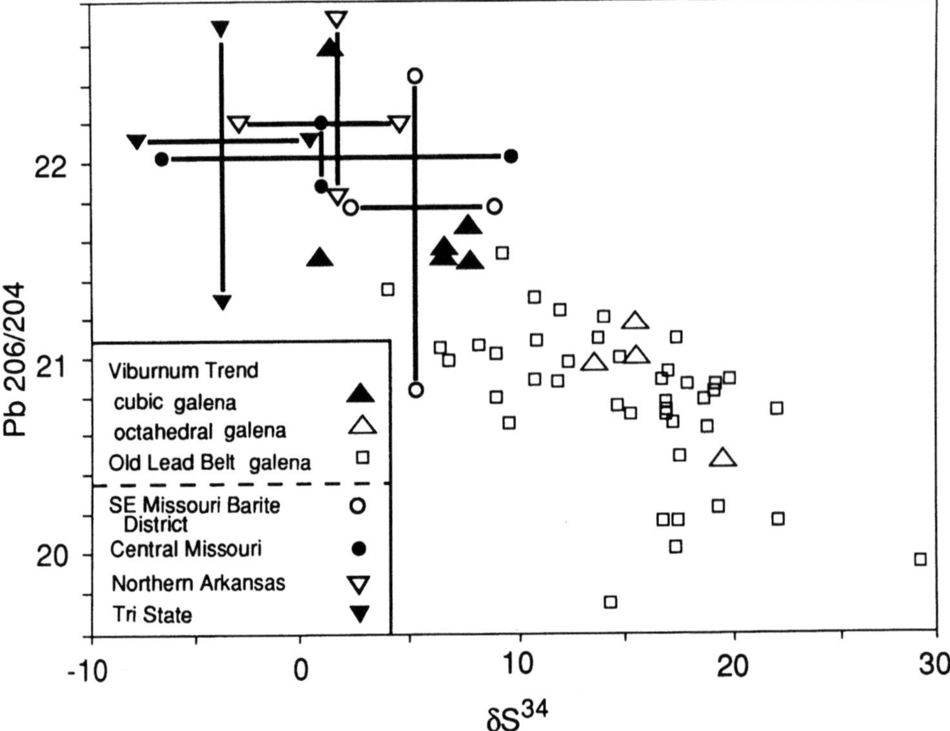

Fig. 7. Combined $^{206}Pb/^{204}Pb$ ratios and $\delta^{34}S$ values for galena from MVT districts in the Ozark region (Viets and Leach, 1990). The Old Lead Belt and Viburnum Trend have both lead and sulfur data for individual galena samples, whereas other districts are shown with *bars* indicating the total range in sulfur and lead isotopes reported for galena. Sulfur isotope compositions of sphalerite from the districts lie within the range of values reported for galena. Original sources of data are given in Viets and Leach (1990)

with high temperatures in the Ozark region. The "tectonic squeezing" of fluids away from advancing thrust sheets (Oliver 1986) is predicted to yield flow rates *too slow* to account for the regional distribution of hot fluid inclusion temperatures (Ge and Garven 1989; Bethke and Marshak 1990).

The model that best explains regional MVT mineralization in the Ozarks is *topographically driven fluid flow* (Garven and Freeze 1984a,b; Garven 1985; Bethke 1986; Bethke et al. 1988). Bethke et al. (1988) modeled fluid flow from the Arkoma basin in response to uplift of the basin's tectonic flank during the Ouachita orogeny. Their calculations show that high flow rates (m/year) could be achieved by uplift of the Ouachita foldbelt. These high flow rates best explain the high temperatures recorded in shallow strata and high fluid/rock ratios inferred from diagenetic changes in the rocks. In this model, groundwater, recharged in the uplifted orogenic flank of the basin, migrates through the deep portions of the basin, thereby acquiring heat and dissolved components, and is discharged through an enormous hydrothermal system affecting broad areas along the basin's cratonic flank (Fig. 8).

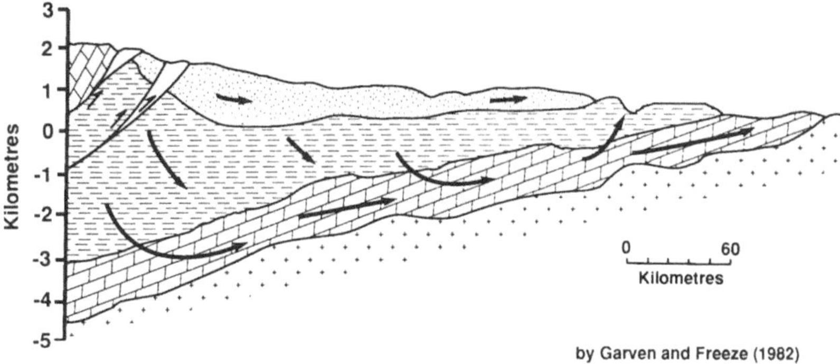

Fig. 8. Conceptual model of topographically driven fluid flow in foreland basins (Garven and Freeze 1984a). Groundwater, recharged in the uplifted orogenic flank of the basin, migrates through deep portions of the basin, acquiring heat and dissolved components, and is discharged along the basin's cratonic flank

Within this regional hydrothermal system, it is reasonable that certain stratigraphic units were principal aquifers for the ore fluids. The aquifer that may have been an important pathway for the regional brine of Viets and Leach (1990) is the post-Bonneterre, Cambrian-Ordovician regional aquifer which today is an important groundwater resource. Many trace element anomalies occur in this aquifer (Erickson et al. 1981, 1983) and abundant sparry dolomite is present (Rowan 1986). Viets and Leach (1990) suggest that this aquifer may have been the principal flow path for the ore fluids for the northern Arkansas, Tri-State, and central Missouri districts.

It is generally accepted that the basal Lamotte Sandstone was an important aquifer for the southeast Missouri districts. Recent studies show than an important component of fluid flow for the lead-rich deposits in the Viburnum Trend and Old Lead Belt was out of the Reelfoot rift zone into southeast Missouri (Fig. 1). Farr (1989) found sparry dolomite in the Bonneterre Formation became less radiogenic in strontium northwesterly from the Reelfoot rift and toward the ore districts. Erickson et al. (1988) found a lead-rich geochemical zone in the Bonneterre Formation with a map pattern consistent with fluid flow from the Reelfoot rift zone. Coincident with this geochemically enriched zone, Diehl et al. (1989) observed that feldspar in the basal Lamotte Sandstone was significantly leached. Leaching of feldspars in the Lamotte Sandstone may have mobilized lead to produce the observed lead enrichment in the overlying Bonneterre Formation.

8 District Controls

MVT deposits of the Ozark region are the products of an enormous hydrothermal system that left trace mineralization over a wide area, and the controls that localized economic concentrations of ore minerals can be considered within a hydrogeologic framework. Table 3 presents the important

Table 3. Geologic controls for Ozark MVT deposits

District	Shale edges	Limestone/ dolomite	Reef and barrier complex	Solution-collapse breccias	Basement topography	Fault/ fractures	Unconformity features
Viburnum Trend	●	●	●	●	●	●	
Old Lead Belt	●	●	●		●	●	
Southeast Missouri barite						●	●
Central Missouri	○			●		○	
Tri-State	●	•		●		●	
Northern Arkansas	●	•		●		●	

● major; ○ less important; • minor

geologic features that localized ore in each district. Each geologic feature can be viewed as a fundamental control on fluid transmissivity, either at the district or mine scale, which allows the ascent and focusing of fluid flow. Abrupt changes in either lateral or vertical transmissivity of fluids permit relatively rapid changes in temperature and pressure of the ore fluid as well as providing opportunities for fluid mixing or fluid-rock reactions to precipitate ore. All ore-controlling features typically have their unmineralized counterpart in each district, suggesting that *several* controls must have been necessary in localizing ore. The various controls listed in Table 3 are generally interrelated. For example, shale depositional edges, limestone to dolostone transitions, and reef complexes are parts of sedimentary facies tracts, all of which may be related to basement topography. In addition, the development of karst features is commonly controlled by fractures and shale edges.

8.1 Shale Edges

Shales, shaley carbonates, and dense limestones, acting as aquitards within the stratigraphic sequence, provided an important constraint on the migration of the ore fluids (Palmer and Hayes 1989). At places where the aquitards are breached, such as through-going faults, or where facies changes result in the absence of the aquitard (e.g. shale edges), fluids could ascend and deposit ore through subsequent cooling, fluid mixing, or wall rock alteration. The relationship between a shale edge and localization of MVT deposits in the Tri-State district is shown by the occurrence of ore only beyond the subcrop edge of the Chattanooga (Upper Devonian and Lower Mississippian) and Northview (Lower Mississippian) Shales, both of which stratigraphically underlie the main ore-bearing carbonate units in the district (Siebenthal 1916; Brockie et al. 1968). Palmer and Hayes (1989) extended this concept to other Ozark MVT districts and found similar relationships. In the Viburnum Trend, ore deposits occur east of a transition of shaley limestone to dolostone in the Bonneterre Formation. However, ore deposits are uncommon in this district where shale facies of the overlying Davis Formation is absent. In this latter situation, the shale may have acted as an aquitard to confine fluid flow within the ore zone.

8.2 Limestone to Dolostone Transitions

In the Viburnum Trend and Old Lead Belt, ore deposits are located near a transition of limestone to dolostone where the dolostone is clearly pre-ore. In the Ozark region, limestones commonly behave like an aquitard (Palmer and Hayes 1989). Therefore, at transitions of limestone to dolostone, ore fluids migrating below a limestone aquitard have an opportunity to ascend and deposit ore (Palmer and Hayes 1989). In addition, dolostones generally have significantly greater intergranular and fracture-controlled permeability relative to limestones. This inherently greater permeability of dolostones over limestones may account for the preference of MVT deposits to occur in dolostones.

8.3 Carbonate Reef Complex

The Viburnum Trend and Old Lead Belt are classic examples of ore deposition related to a carbonate reef complex. In the Viburnum Trend, little ore is actually found within reef rock; rather, ore is located in permeable carbonate facies and breccia zones fringing the reef. In the Old Lead Belt, much of the ore is localized in carbonate-sand bars, sedimentary breccias, and other highly permeable facies related to a barrier reef complex. Both reef and carbonate barrier complexes within a stratigraphic sequence are found where abrupt changes in sedimentary facies occur that result in dramatic permeability contrasts. Such abrupt lateral and vertical changes in fluid transmissivity create opportunities for fluid mixing and steep physicochemical gradients in the migrating ore fluid to produce sulfide deposition.

8.4 Basement Topography

The best example of basement-controlled migration of MVT ore fluids is in the southeast Missouri lead districts. In the Viburnum Trend and Old Lead Belt, an important ore control is the pinch-out of the Lamotte Sandstone against Precambrian knobs (Snyder and Gerdemann 1968). At sandstone pinch-outs, ore fluids migrating in the Lamotte Sandstone were forced to rise into permeable reef facies that host most of the ore.

8.5 Faults

In all the districts, faulting provided important ground preparation and enhanced permeabilities for fluid migration through development of fault breccias and dilational fractures. One of the most important ore controls for the Ozark MVT deposits is the coincidence of extensively faulted ground with one or more of the other fluid-focusing features given in Table 3. For example, faulted ground may have allowed ore fluids to ascend into the karst features in northern Arkansas, Tri-State, and southeast Missouri barite districts. In the Old Lead Belt, the coincidence of highly faulted ground with permeable carbonate reef facies was an important ore control.

8.6 Karst Breccias

Preexisting solution-collapse breccias and related carbonate dissolution features are important hosts for all MVT districts in the Ozarks, with the exception of the Old Lead Belt. These early solution-collapse features, always located beneath an unconformity or disconformity, have generally been enlarged by dissolution and collapse during deposition of the sulfides. Pre-ore solution-collapse breccias are important hosts for ore because they provide extremely focused fluid flow in carbonate terranes, they serve as common pathways for cross-formational fluid flow, and they provide opportunities for a variety of ore depositional processes to operate.

9 Precipitation Mechanisms

Plumlee et al. (1994) evaluated possible depositional mechanisms for MVT deposits in the Ozark region using the reaction path program, CHILLER (Reed 1982). Their study used quantitative fluid inclusion composition and chemical speciation of the regional and VT brines (Table 2) as starting compositions for the reaction path calculations. A number of different modifications of the regional and VT brine compositions (Table 2) were evaluated in the calculations to cover the range of observed fluid inclusion data. Modifications to these brine compositions included variations in temperature, salinity, Ca/Mg ratios, CO_2, H_2S, and metal contents. The fluids were initially constrained to be saturated with dolomite, quartz, pyrite, sphalerite, galena, and muscovite (approximating illite). Together, more than 150 different reaction paths were evaluated. The suitability of each of the proposed depositional mechanisms was evaluated by comparing the hydrothermal mineral assemblages predicted by the reaction path calculations with those observed in the Ozark MVT districts. The main depositional mechanisms considered were (1) isothermal boiling of a single basinal fluid in response to decreasing pressure; (2) boiling of a single fluid with both decreasing pressure and temperature; (3) isobaric cooling of a single fluid; (4) thermochemical sulfate reduction; (5) reaction of a metal-bearing fluid with H_2S and CH_4 (approximating sour gas); (6) reaction of a single fluid with limestone host rocks; (7) mixing of two fluids with different temperatures, salinities, CO_2, H_2S, and metal contents; and (8) isothermal boiling of fluids followed by mixing with fluids of different gas and metal contents.

The results from Plumlee et al. (1994) show that no single depositional process can reproduce all the hydrothermal mineral assemblages observed throughout the Ozark region; rather, specific districts require their own dis-

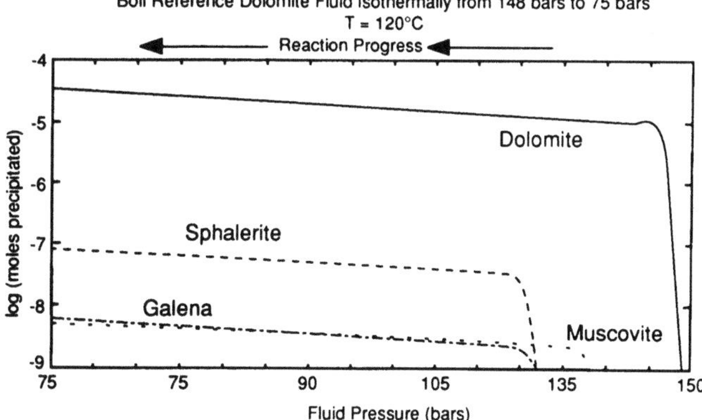

Fig. 9. Reaction path modeling isothermal boiling of the regional fluid (Table 2) from 120 to 75 bar. In this and all subsequent reaction path diagrams, the reaction proceeds from *right* to *left* in the figure. The fluid is assumed to be initially saturated with dolomite, sphalerite, galena, quartz, pyrite, and muscovite. Shown on the vertical axis is log moles of minerals precipitated. This process yields an open-space filling mineral assemblage of hydrothermal dolomite and two orders of magnitude less of sulfides. A component of fluid cooling produces a quartz-dominated assemblage, not observed in the Ozarks. (After Plumlee et al. 1994)

tinctive depositional mechanism(s) that reflect the local host rock composition, structural setting, and hydrology. In Figs. 9–13 (after Plumlee et al. 1994) the reaction path calculations for the dominant precipitation mechanisms for the most important MVT mineral assemblages in the Ozarks are presented. All reaction paths are flow-through models, where solids (and/or gases) formed at each reaction step are removed from the system after the step. For the reaction path diagrams presented here, reaction progress is from *right to left*. The X-axis shows *what parameter is changing* (e.g. P, T, amount of mineral, gas, and/or fluid added). The Y-axis indicates the *amount* of mineral precipitated at each step in log moles. Where only fluids are involved, the best way to view these graphs is as *mineral paragenesis diagrams*; where rock is reacted with fluid, the graphs are best viewed as *mineral zonation diagrams*.

9.1 Regional Trace MVT Metals and Sparry Dolomite

Chemical reaction paths that predict mineral assemblages most similar to the regional sparry dolomite occurrences (saddle dolomite with traces of base sulfides) are those in which a single fluid boils isothermally (CO_2 effervescence) or nearly isothermally in response to pressure decreases from 120 down to 75 bar (Fig. 9). These pressure decreases correspond to reasonable decreases in the hydrostatic confining pressure for a fluid migrating from northern Arkansas to central Missouri. Due to the relatively low hydrothermal temperatures and the high dissolved gas contents, the gas phase separated during boiling is CO_2-dominant rather than water-dominant. Nearly all fluid compositions evaluated produce the same general mineral assemblages dominated by dolomite with two to three orders of magnitude less sphalerite, galena, and muscovite. Isothermal boiling of a dolomite-saturated fluid will result in precipitation of dolomite with trace sulfides in open spaces but does *not* produce sulfides in amounts that may yield an ore deposit. Therefore, other mechanisms are necessary to account for the sulfide-dominated assemblages in the ore deposits.

9.2 Central Missouri

Central Missouri ore deposits are largely open-space fillings of sphalerite plus minor galena in dolostone karst features localized along a broad anticline. Hydrothermal solution collapse during ore deposition was a minor process, indicating that dolomite was slightly undersaturated during ore deposition. No silicification and only minor sparry dolomite is present in the ore deposits. Late-stage barite cements the sphalerite and galena.

The reaction path that best explains the features of the central Missouri deposits is the near-isothermal mixing of two dolomite-saturated fluids with different H_2S and metal contents (Fig. 10). Paleokarst features may have allowed the regional brine to rise stratigraphically and mix with fluids with locally derived H_2S. The formation of the late barite was not modeled; however, Leach (1980) proposed that fluid mixing was the deposition mechanism on the basis of fluid inclusion data.

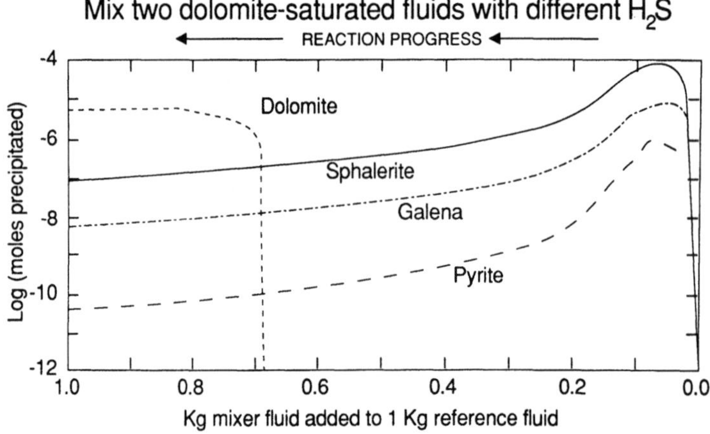

Fig. 10. Reaction path diagram showing minerals precipitated during the isothermal mixing of the regional dolomite-saturated fluid with a dolomite-saturated brine having the same salinity and Ca/Mg ratio, but different H_2S and metal contents. The regional fluid has H_2S concentrations of 1.4×10^{-3} molal, whereas the mixer fluid has H_2S concentrations of 1.5×10^{-2} molal. Metal concentrations for each brine are controlled by sphalerite, galena, and pyrite saturation. This process yields predominantly an open-space filling of sphalerite, galena, and pyrite. Dolomite is slightly undersaturated in the first part of the reaction path which leads to minor replacement of the dolomite host by sulfides. This process reproduces the sphalerite-galena ores in central Missouri and southeast Missouri barite districts. (After Plumlee et al. 1994)

9.3 Tri-State and Northern Arkansas

Both northern Arkansas and Tri-State districts are localized by normal faults that probably allowed fluids, travelling predominantly in deeper Cambrian-Ordovician dolostones, to rise into carbonate sequences dominated by limestones. In northern Arkansas, jasperoid perferentially replaced limestones in the mixed dolostone-limestone sedimentary sequence.

Reaction paths that best describe the ore and alteration assemblage for the Tri-State and northern Arkansas districts are ones where an initially dolomite-saturated brine flows into a cooler limestone (Figs. 11 and 12). Adjacent to fluid conduits, where water/rock ratios were the highest, limestone was replaced by dolomite (Fig. 11). As the fluids moved outward into cooler limestone, jasperoid and sulfide replaced limestone (Fig. 12). Isothermal boiling of the ore fluids may have produced open-space filling of sparry dolomite with minor sulfides in breccia and fault zones (Fig. 10). Local mixing of the regional brine with locally derived sulfur also undoubtedly played a role in the development of open-space filling in sulfide-rich ore runs.

9.4 Viburnum Trend and Old Lead Belt

Several distinctive features of the ore deposits in the Viburnum Trend and Old Lead Belt provide important constraints on reaction path calculation. First, dolomite was undersaturated during sulfide deposition. Evidence for this includes the following: (1) sulfide ore occurs mainly as replacement of host dolostone and open-space filling by sulfides is minor; (2) paragenetic studies

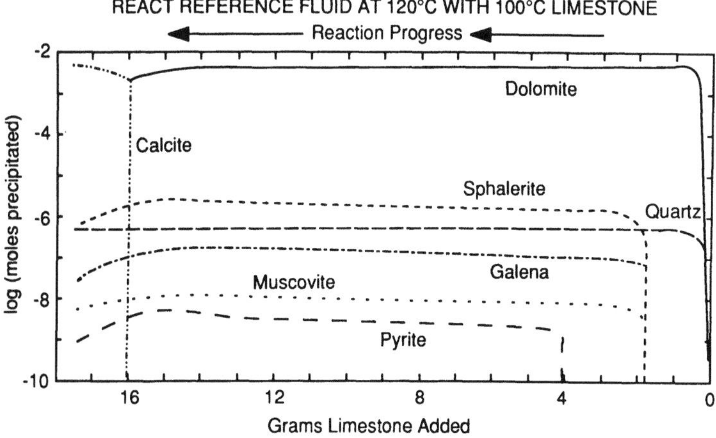

Fig. 11. Reaction path diagram showing minerals precipitated during the progressive addition of 100 °C limestone (as grams of calcite) to 1 kg of 120 °C regional fluid. Limestone is dolomitized adjacent to fluid conduits (e.g. Tri-State breccia ore runs). Here, the system is fluid-dominated so that cooling is due only to the mass of limestone reacting with the ore fluid. This process duplicates the dolomite cores of ore deposits in Tri-State and northern Arkansas. (After Plumlee et al. 1994)

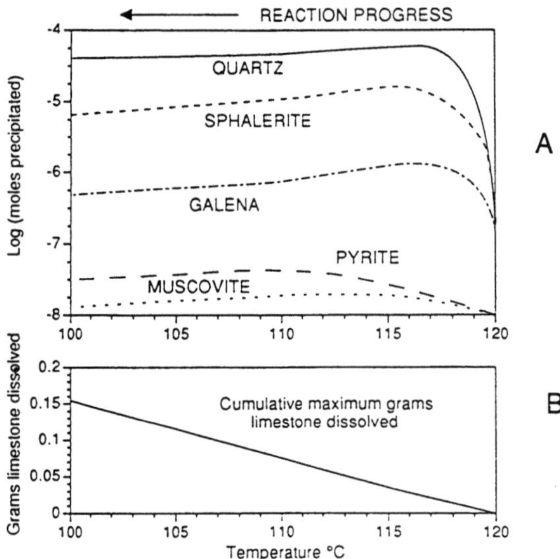

Fig. 12. Reaction path diagrams showing the progressive addition of limestone (as grams of calcite) at 100 °C to an initial 1 kg of the regional fluid at 120 °C, assuming additional heat loss due to conductive cooling. The reaction path was carried out by repeated cycles in which the fluid was cooled over a small temperature interval (5 °C) and then allowed to react with limestone until nearly saturated with dolomite. If the fluid was allowed to completely react to limestone saturation after each cooling step, then a small amount of limestone would be dolomitized until calcite saturation was reached (after Plumlee et al. 1994). **A** Minerals precipitated. This mineral assemblage describes the jasperoid-sulfide replacement ore deposits in Tri-State and northern Arkansas districts. **B** Cumulative maximum amounts of limestone leached along the reaction path prior to reaching dolomite saturation

(Heyl 1983; Hagni 1986) show that sulfide and dolomite generally did not coprecipitate, rather sparry dolomite is interspersed with episodes of sulfide deposition; (3) CL zones in sparry dolomite are commonly corroded in ore zones; (4) ore-related hydrothermal brecciation of the host dolostone is common. Secondly, the precipitation model must account for a lead-dominant ore assemblage. Finally, silicification of the host dolostone is minor and only small amounts of late-stage quartz are present locally in the ore deposits.

The reaction path which best reproduces the broad features of the Viburnum Trend and Old Lead Belt ores is one in which a dolomite-saturated, lead-rich, zinc- and H_2S-poor VT brine (Table 2) mixes with a cooler, less saline, H_2S-rich brine (Fig. 13). The initial mixing steps of this reaction path best reproduce the galena-rich replacement ores in the Viburnum Trend. The VT brine was initially chosen to be several orders of magnitude undersaturated with sphalerite; greater Zn concentrations led to precipitation, upon mixing, of sphalerite-dominant mineral assemblages which are less common in the Viburnum Trend. Cooling associated with fluid mixing may have contributed to sulfide deposition.

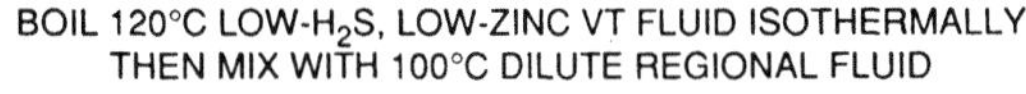

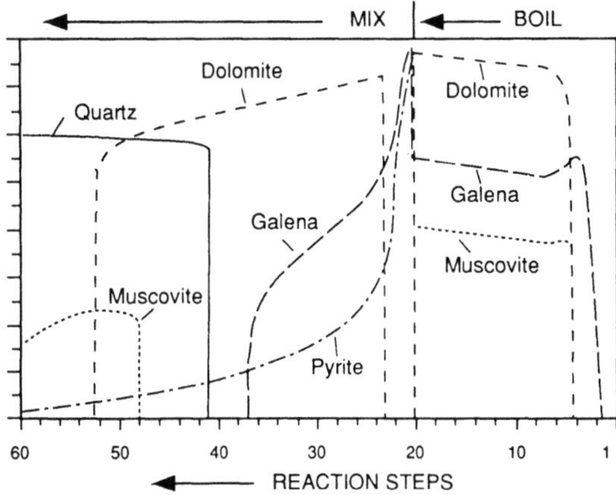

Fig. 13. The first part of this reaction path describes the migration of the VT ore fluid to the Viburnum Trend and Old Lead Belt districts followed by mixing with another fluid in the ore zone. The first part shows the minerals precipitated during isothermal boiling of a lead-rich, zinc-poor, H_2S-poor Viburnum Trend fluid (120 °C, 10^{-5} molal H_2S, tens of ppm Pb, several hundred ppb Zn) from 138 to 75 bar. The second part describes the consequences of mixing the VT brine with a cooler, dilute H_2S-rich regional fluid (100 °C, 1.5×10^{-2} molal H_2S; total salinity one-half that of VT brine). Reaction steps along the boiling portion of the path are equivalent to pressure drops of 1 bar each; reaction steps along the mixing path are each equivalent to the addition of 50 g of mixer fluid to an initial 1 kg of 120 °C regional fluid. The initial mixing steps of this reaction path best reproduce the galena-rich replacement ores observed in the Viburnum Trend. The VT fluid was chosen to be several orders of magnitude undersaturated with sphalerite initially, because greater Zn concentrations led to the precipitation, upon mixing, of sphalerite-dominant mineral assemblages which is only locally important in some mines in the district. Note that the deposition of dolomite occurs again after the third mixing step (150 g total mixer solution added); if correct, this path suggests that ore deposition could have occurred in only limited zones of fluid mixing. (After Plumlee et al. 1994)

9.5 General Comments on the Reaction Path Calculations

The reaction path models reproduce only the most general features of the mineral assemblages for each district; the actual ore-forming event was certainly more complex as demanded by the complex mineralogy, repeated episodes of overlapping mineral deposition, and dissolution of earlier deposited sulfides and/or dolomite. Temporal or transient changes in the temperature or composition of the fluids flowing into ore deposition sites were not addressed.

In the reaction path calculations, iron sulfides are always subordinate in mass to sphalerite and galena. These proportions are correct for much of the main Pb-Zn generations, but do not reproduce the monomineralic pyrite/marcasite stages present in the Viburnum Trend and other districts. It is likely that the reaction path calculations do not predict FeS_2-rich assemblages because they assume redox equilibrium between aqueous sulfate and sulfide and do not account for metastable intermediate-valency species such as polysulfides which are known to enhance FeS_2 precipitation (Murowchick and Barnes 1986). The botryoidal marcasite generation, which occurs between the octahedral and cubic-galena mineralization stages in the Viburnum Trend (Heyl 1983), has very light sulfur isotopes (as light as −25‰ $\delta^{34}S$), which further indicates that disequilibrium played an important role in marcasite deposition (Plumlee and Rye 1994).

In addition to predicting the dominant dopositional process(es) for each district, geochemical modeling also shows that some processes could not have played a dominant role in ore deposition. For example, sulfate reduction at the site of ore deposition, commonly postulated as the sulfide precipitation mechanism (e.g. Anderson and Garven 1987), produces either a dolomite- or a sulfide-dominant mineral assemblage with limited dissolution of the host carbonate rocks. However, most MVT ore deposits in the Ozark region occur as sulfide replacement of the host carbonates and ore-stage dissolution and brecciation of the host rocks attest to significant carbonate dissolution during ore formation. Therefore, sulfate reduction at the site of ore deposition could not have played a dominant role in sulfide precipitation. Simple cooling of the ore fluid predicts a quartz-dominated mineral assemblage that fills open space which is also inconsistent with Ozark MVT deposits.

10 Summary of the Genetic Model

Plate convergence along the southern margin of the North American craton set the stage for the formation of the largest Mississippi Valley-type lead-zinc metallogenic province in the world. Convergence of Llanoria with North American in late Paleozoic created a series of foredeeps along the southern margin of North American. The largest and deepest of these foredeeps was the Arkoma basin, which has a poorly understood connection with the Black Warrior foredeep to the east. By Late Pennsylvanian to Permian time, final plate suturing and uplift of the southern tectonic flank of the Arkoma and Black Warrior basins established an enormous hydrothermal system that affected at least 350 000 km^3 of rocks in the Ozark region. Effects of this large hydrothermal system include the formation of MVT lead-zinc districts with widespread trace sulfides and sparry dolomite, extensive but subtle hydrothermal alteration, reset paleomagnetic and radiometric dates, and regional thermal anomalies.

The model that best explains the regional hydrothermal system is topographically driven fluid flow. Meteoric water, recharged in the uplifted orogenic flank of the foredeep, migrates through deep portions of the basin, thereby acquiring heat and dissolved components, and is discharged through broad areas on the craton.

Although fluid flow was extensive throughout the stratigraphic section, certain regional aquifers were the principal pathway for the main ore districts. Rock-water interactions in lithologically different aquifers led to deposits with different metal and isotopic compositions. Migration of the ore fluids through carbonate rocks, primarily the post-Bonneterre Cambrian-Ordovician regional aquifer, produced the zinc-dominant deposits in the northern Arkansas, Tri-State, central Missouri, and southeast Missouri barite districts. In the southeast

Missouri lead districts, two chemically and isotopically distinct fluids were present during ore deposition. The main-stage (octahedral galena-stage) ores were produced by a potassium-rich brine containing less radiogenic lead that migrated through the basal sandstone aquifer; this brine had a significant migration pathway through thick arkosic Cambrian sandstones in the Reelfoot rift complex. A second fluid containing more radiogenic lead, and chemically and isotopically similar to the regional brine that formed other Ozark MVT districts, produced the late cubic galena-stage ores.

The locations of the Ozark MVT districts were determined by paleohydrological features that allowed stratigraphic ascent and focused fluid flow. Important paleohydrological controls included shale depositional edges, limestone to dolostone transitions, sedimentary facies associated with carbonate reefs, faults, and basement topography. These features created an opportunity for fluid mixing, cooling, and fluid-rock reactions to precipitate ore.

Reaction path modeling of quantitative fluid inclusion compositions shows that no one depositional process can reproduce all the hydrothermal mineral assemblages observed throughout the Ozark region. Rather, specific districts require their own distinctive depositional mechanism(s) that reflect host rock composition, structural setting, and hydrology. The regional occurrences of sparry dolomite with trace sulfides were formed by isothermal or near-isothermal effervescence of CO_2 in response to a decrease in the confining pressure during fluid migration to shallower depths. The central Missouri and southeast Missouri barite districts are best explained by the mixing of two dolomite-saturated brines with different H_2S and metal contents. Northern Arkansas and Tri-State districts can largely be explained by the reaction of a dolomite-saturated brine with cool limestone. The lead-dominant replacement ores in the southeast Missouri lead districts require the mixing of a dolomite-saturated, lead-rich, and H_2S-poor brine with a cooler, more dilute, and H_2S-rich brine. Reaction path calculations also show that cooling, CO_2 effervescence, and thermochemical sulfate reduction could not have played a dominant role in ore deposition.

Acknowledgements. This paper incorporates the work of so many scientists that it is not possible to acknowledge them all. However, I am especially indebted to the following colleagues who were particularly influential in the development of the model presented here: John V. Viets, Elizabeth Rowan, D.F. Sangster, Tim Hayes, Albert Hofstra, Joel Leventhal, Martin Goldhaber, Craig Bethke, Herb Rumi, and Dwight Bradley. The important geochemical modeling was performed by Geoff Plumlee. Jimmy-Carter Borden and Jon Wesley Powell helped in the preparation of the manuscript. I appreciate the helpful reviews of Warren Day, Joel Leventhal, Kevin Shelton, Lluis Fontboté, and an anonymous reviewer that significantly improved the manuscript. I am also indebted to the many mine geologists who graciously allowed access to the ore deposits.

References

Anderson GM, Garven G (1987) Sulfate-sulfide-carbonate associations in Mississippi Valley-type lead-zinc deposits. Econ Geol 82:482–488

Bass MN, Ferrara G (1969) Age of adularia and metamorphism, Ouachita Mountains, Arkansas. Am J Sci 257:491–498

Bauer RM, Shelton KL (1989) Fluid inclusion studies of regionally extensive epigenetic dolomites, Bonneterre Dolomite, S.E. Missouri: evidence of multiple fluids during Pb-Zn ore mineralization. Geol Soc Am Abstr Prog 21:A3

Bethke CM (1986) Hydrologic constraints on the genesis of the Upper Mississippi Valley mineral district from Illinois basin brines. Econ Geol 81:233–249

Bethke CM, Marshak S (1990) Brine migrations across North America – the plate tectonics of groundwater. Annu Rev Earth Planet Sci 18:228–315

Bethke CM, Harrison WJ, Upson C, Altaner SP (1988) Supercomputer analysis of sedimentary basins. Science 239:261–267

Bjørlykke A, Sangster DF (1981) An overview of sandstone-lead deposits and their relationship to red-bed copper and carbonate-hosted lead-zinc deposits. Econ Geol 75th Anniv Edn: 179–213

Brannon JC, Podosek FA, Viets JG, Leach DL, Goldhaber M, Rowan EL (1991) Strontium isotopic constraints on the origin of ore-forming fluids of the Viburnum Trend, southeast Missouri. Geochim Cosmochim Acta 55:1407–1419

Brockie DC, Hare EH Jr, Dingess PR (1968) The geology and ore deposits of the Tri-State district of Missouri, Kansas, and Oklahoma. In: Ridge JD (ed) Ore deposits of the United States, 1933–1967. Am Inst of Min, Metal, and Petrol Eng, New York, pp 400–430

Burstein IB, Shelton KL, Hagni RD, Mavrogenes JA (1991) District-wide sulfur isotope systematics of MVT ores in the world-class southeast Missouri Pb-Zn-Cu deposits: sulfur isotope anomalies indicative of the simultaneous presence of multiple fluids. Geol Soc Am Abstr Prog 24:414

Coveney RM, Goebel ED (1983) New fluid inclusion homogenization temperatures for sphalerite from minor occurrences in the midcontinent area. In: Kisvarsanyi G, Grant SK, Pratt WP, Koenig JW (eds) Proc Int Conf Mississippi Valley type lead-zinc deposits. University of Missouri, Rolla Press, Rolla, pp 234–242

Crocetti CA, Holland HD (1989) Sulfur-lead isotope systematics and the composition of fluid inclusions in galena from the Viburnum Trend, Missouri. Econ Geol 84:2196–2216

Crocetti CA, Holland HD, McKenna LW (1988) Isotopic compositions of lead in galenas from the Viburnum Trend, Missouri. Econ Geol 83:355–376

Damberger HH (1974) Coalification patterns of Pennsylvanian coal basins of the eastern United States. In: Dutcher RR, Hacquebard PA, Schopt JM, Simon JA (eds) Carbonaceous materials as indicators of metamorphism. Geol Soc Am Spec Pap 153:53–73

Deloule E, Allègre C, Doe B (1986) Lead and sulfur isotope microstratigraphy in galena crystals from Mississippi Valley-type deposits. Econ Geol 81:1307–1321

Desborough GA, Zimmerman RA, Elrick M, Stone C (1985) Permian thermal alteration of Carboniferous strata in the Ouachita region and Arkansas River Valley, Arkansas. Geol Soc Am Abstr Prog 17:155

Diehl SF, Goldhaber MB, Mosier EL (1989) Regions of feldspar precipitation and dissolution in the Lamotte Sandstone, Missouri – implications for MVT ore genesis. US Geol Surv Open File Rep 89-169:5–7 (Abstr)

Doe BR, Delevaux MH (1972) Sources of lead in southeast Missouri galena ores. Econ Geol 67:409–425

Erickson RL, Mosier EL, Odland SK, Erickson MS (1981) A favorable belt for possible mineral discovery in subsurface Cambrian rocks in southern Missouri. Econ Geol 76:921–933

Erickson RL, Mosier EL, Viets JG, Odland SK, Erickson MS (1983) Subsurface geochemical exploration in carbonate terrane-midcontinent U.S.A. In: Kisvarsanyi G, Grant SK, Pratt WP, Koenig JW (eds) Proc Int Conf Mississippi Valley type lead-zinc deposits. University of Missouri, Rolla Press, Rolla, pp 575–583

Erickson RL, Chazin B, Erickson MS, Mosier EL, Whitney H (1988) Tectonic and stratigraphic control of regional subsurface geochemical patterns, midcontinent, U.S.A. In: Kisvarsanyi G, Grant SK (eds) Proc Tectonic control of ore deposits and the vertical and horizontal extent of ore systems. University of Missouri, Rolla Press, Rolla, Missouri, pp 435–446

Farr MR (1989) Compositional zoning characteristics of late dolomite cement in the Cambrian Bonneterre Formation, Missouri: implications for parent fluid migration pathways. Carbonates Evaporites 4:177–194

Garven G (1985) The role of regional fluid flow in the genesis of the Pine Point deposit, western Canada sedimentary basin. Econ Geol 80:307–324

Garven G, Freeze RA (1984a) Theoretical analysis of the role of groundwater flow in the genesis of stratabound ore deposits. 1. Mathematical and numerical model. Am J Sci 284:1084–1124

Garven G, Freeze RA (1984b) Theoretical analysis of the role of groundwater flow in the genesis of stratabound ore deposits. 2. Quantitative results. Am J Sci 284:1125–1174

Ge S, Garven G (1989) Tectonically induced transient groundwater flow in foreland basins. In: Price RA (ed) The origin and evolution of sedimentary basins and their energy and mineral resources. Geophys Monogr 48:145–157

Gerdemann PE, Meyers HE (1972) Relationships of carbonate facies patterns to ore distribution and to ore genesis in the southeast Missouri lead district. Econ Geol 67:426–433

Goldhaber M (1989) Contrasting sulfur sources for the lower versus upper midcontinent Mississippi Valley-type ores – implications for ore genesis. In: Schindler K (ed) USGS research on mineral resources – 1989. Program with Abstracts. US Geol Surv Circ 1035:23

Goldhaber M, Mosier E (1989) Sulfur sources for southeast Missouri MVT ores – implications for ore genesis. US Geol Sur-Mo Geol Surv Symp, Mineral-resource potential of the midcontinent program and Abstracts. US Geol Surv Open File Rep 89–169:10–11

Goldhaber M, Stanton MR (1987) Experimental formation of marcasite at 150–200 °C; implications for carbonate-hosted Pb-Zn deposits. Geol Soc Am Abstr Prog 19(7):678 (Abstr)

Graham SA, Dickinson WR, Ingersoll RV (1975) Himalayan-Bangal model for flysch dispersal in the Appalachian-Ouachita system. Geol Soc Am Bull 86:273–286

Gregg JM, Shelton KL (1989a) Minor- and trace-element distribution in the Bonneterre Dolomite (Cambrian), southeast Missouri: evidence for possible multiple-basin fluid sources and pathways during lead-zinc mineralization. Geol Soc Am Bull 101:221–230

Gregg JM, Shelton KL (1989b) Geochemical and petrographic evidence for fluid sources and pathways during dolomitization and lead-zinc mineralization in southeast Missouri: a review. Carbonates Evaporites 4:153–175

Gustafson LB, Williams N (1981) Sediment-hosted stratiform deposits of copper, lead and zinc. In: Skinner B (ed) Econ Geol 75th Anniversary Vol: 139–178

Hagni RD (1976) Tri-State ore deposits: the character of their host rocks and their genesis. In: Wolf KH (ed) Handbook of strata-bound and stratiform ore deposits, vol 6. Elsevier, Amsterdam, pp 457–494

Hagni RD (1986) Paragenetic sequence of the lead-zinc-copper-cobalt-nickel ores of the southeast Missouri lead district, U.S.A. In: Craig JR, Hagni RD, Kiesel W, Lange IM, Petrovskaya NV, Shadlum TN, Udubasa G, Augustithis SS (eds) Mineral paragenesis. Theophrastus Publications, Athens, Greece, pp 90–132

Hagni RD, Grawe OR (1964) Mineral paragenesis in the Tri-State district, Missouri, Kansas, Oklahoma. Econ Geol 59:449–457

Hay RL, Lee M, Kolata DR, Matthews JC, Morton JP (1988) Episodic potassic diagenesis of Ordovician tuffs in the Mississippi Valley area. Geology 16:743–747

Hennigh QT, Shelton KL, Hagni RD (1990) Sulfur isotope studies of early copper (bornite pod) and zinc mineralization in the Viburnum Trend, MO: do ^{32}S-enriched sulfides indicate basement sulfur and metal sources? Geol Soc Am Abstr Prog 22:A134

Heyl AV (1983) Geologic characteristics of three Mississippi Valley-type districts. In: Kisvarsanyi G, Grant SK, Pratt, WP, Koenig JW (eds) Proc Int Conf Mississippi Valley type lead-zinc deposits. University of Missouri, Rolla Press, Rolla, Missouri, pp 27–30

Heyl AV, Landis GP, Zartman RE (1974) Isotopic evidence for the origin of Mississippi Valley-type mineral deposits: a review. Econ Geol 69:992–1006

Hinze WJ, Braile LW, Keller GR, Lidiak EG (1980) Models for midcontinent tectonism. In: Burchfiel BS, Oliver JE, Silver LT (eds) Continental tectonics. Natl Acad Sci, Washington DC, pp 73–83

Hofstra AH, Leach DL, Landis GP, Viets JG, Rowan EL, Plumlee GS (1989) Fluid inclusion gas geochemistry as a monitor of ore depositional processes in Mississippi Valley-type deposits in the Ozark region. In: Pratt WP, Goldhaber MB (eds) US Geological Survey-Missouri Geological Survey Symp – mineral resource potential of the midcontinent. US Geol Surv Circ 1043:11–12

Horrall KB, Hagni RD, Kisvarsanyi G (1992) Mafic and ultramafic plutons associated with the New Madrid Rift – a possible major source for the copper-cobalt-nickel mineralization of southeast Missouri. Econ Geol (in press)

Houseknecht DW (1986) Evolution from passive margin to foreland basin: the Atoka Formation of the Arkoma basin, south-central U.S.A. In: Allen PA, Homewood P (eds) Foreland basins. International Association of Sedimentologists Spec Publ 8. Blackwell, London, pp 327–345

Houseknecht DW, Hathon LA, McGilvery DA (1991) Thermal maturity of Paleozoic strata in the Arkoma basin. Okla Geol Surv Circ 91:72 pp

Kaiser CJ, Ohmoto H (1985) A kinematic model for tectonic structures hosting North American Mississippi Valley-type mineralization: implications for timing and hydrology. Geol Soc Am Abstr Prog 17:622

Kaiser CJ, Ohmoto H (1988) Ore-controlling structures of Mississippi Valley-type mineralization on the North American midcontinent as products of Late Paleozoic convergent plate tectonism. In: Kisvarsanyi G, Grant SK (eds) Proc Tectonic control of ore deposits and the vertical and horizontal extent of ore systems. University of Missouri, Rolla Press, Rolla, Missouri, pp 424–430

Kaiser CJ, Kelly WC, Wagner RJ, Shanks WC (1987) Geologic and geochemical controls of mineralization in the southeast Missouri barite district. Econ Geol 82:719–734

Landis GP, Hofstra AH (1991) Fluid inclusion gas chemistry as a potential minerals exploration tool: case studies from Creede, CO, Jerritt Canyon, NV, Coeur d'Alene district, ID and MT, Alaska Juneau Gold Belt, AK, and mid-Continent MVT's. J Geochem Exp 42:25–59

Leach DL (1979) Temperature and salinity of the fluids responsible for minor occurrences of sphalerite in the Ozark region of Missouri. Econ Geol 74:931–937

Leach DL (1980) Nature of mineralizing fluids in the barite deposits of central and southeast Missouri. Econ Geol 75:1168–1180

Leach DL, Rowan EL (1986) Genetic link between Ouachita foldbelt tectonism and the Mississippi Valley-type lead zinc deposits of the Ozarks. Geology 14:932–935

Leach DL, Sangster DF (1994) Mississippi Valley-type (MVT) lead-zinc deposits. Ore deposit models. Spec Publ Geol Assoc Can (in press)

Leach DL, Nelson RC, Williams D (1975) Fluid inclusion studies in the northern Arkansas zinc district. Econ Geol 70:1084–1091

Leach DL, Plumlee GS, Hofstra AH, Landis GP, Rowan EL, Viets JB (1991) Origin of late dolomite cement by CO_2-saturated deep basin brines: evidence from the Ozark region, USA. Geology 19:348–351

Lillie RJ, Nelson KD, de Voogd B, Brewer JA, Oliver JE, Brown LD, Kaufman S, Viele GW (1983) Crustal structure of the Ouachita Mountains, Arkansas: a model based on integration of COCORP reflection profiles and regional geophysical data. AAPG Bull 67:907–931

Lovering TG (1972) Jasperoid in the United States – its characteristics, origin, and economic significance. US Geol Surv Prof Pap 710, 164 pp

McCracken MH (1971) Structural features of Missouri. Rep Inv 49. MO Geol Surv Rolla, MO, 99 pp

McKnight ET (1935) Zinc and lead deposits of northern Arkansas. US Geol Surv Bull 853

McKnight ET, Fischer RP (1970) Geology and ore deposits of the Pitcher Field, Oklahoma and Kansas. US Geol Surv Prof Pap 558, 165 pp

Murowchick JB, Barnes HL (1986) Marcasite precipitation from hydrothermal solutions. Geochim Cosmochim Acta 50:2615–2629

Nelson KD, Lillie RJ, deVoogd B, Brewer JA, Oliver JE, Kaufman S, Brown L, Viele GW (1982) COCORP seismic reflections profiling in the Ouachita Mountains of western Arkansas: geometry and geologic interpretations. Tectonics 1:413–430

Oliver J (1986) Fluids expelled tectonically from orogenic belts: their role in hydrocarbon migration and other geologic phenomena. Geology 14:99–102

Palmer JR, Hayes TS (1989) Late Cambrian lithofacies and their control on the Mississippi Valley-type mineralizing system in the Ozark region. In: Schindler KS (ed) United States geological survey research on mineral resources – 1989. Program and abstracts. US Geol Surv Circ 1035:51–53

Pan H, Symons DTA, Sangster DF (1990) Paleomagnetism of the Mississippi Valley-type ores and host rocks in the northern Arkansas and Tri-State districts. Can J Earth Sci 27:923–931

Plumlee GS, Rye RO (1994) Botryoidal pyrite, marcasite, and other distinctive characteristics of the fringes of hydrothermal systems: the perithermal environment. Econ Geol (in press)

Plumlee GS, Leach DL, Hofstra AH, Landis GP, Rowan EL, Viets JB (1994) Chemical reaction path modelling of ore deposition in Mississippi Valley-type Pb-Zn deposits of the Ozark region, US midcontinent. Econ Geol in press

Reed MH (1982) Calculation of multicomponent equilibria and reaction processes in systems involving minerals, gases, and an aqueous phase. Geochim Cosmochim Acta 46:513–528

Roedder E (1977) Fluid inclusion studies of ore deposits in the Viburnum Trend, southeast Missouri. Econ Geol 72:474–479

Rothbard DR (1983) Diagenetic history of the Lamotte sandstone, southeast Missouri. In: Kisvarsanyi G, Grant SK, Pratt WP, Koenig JW (eds) Proc Int Conf Mississippi Valley-type lead-zinc deposits. University of Missouri, Rolla Press, Rolla, pp 385–392

Rowan EL (1986) Cathodoluminescent zonation in hydrothermal dolomite cements: relationship to Mississippi Valley-type Pb-Zn mineralization in southern Missouri and northern Arkansas. In: Hagni RD (ed) Process mineralogy VI. Metallurgical Society 631, Warrendale, Pennsylvania, pp 69–87

Rowan EL, Leach DL (1989) Constraints from fluid inclusions on sulfide precipitation mechanisms and ore fluid migration in the Viburnum Trend lead district, Missouri. Econ Geol 84:1948–1965

Sangster DF (1983) Mississippi Valley-type deposits: a geological melange. In: Kisvarsanyi G, Grant SK, Pratt WP, Koenig JW (eds) Proc Int Conf Mississippi Valley type lead-zinc deposits. University of Missouri, Rolla Press, Rolla, pp 7–19

Sangster DF (1989) Thermal comparison of MVT deposits and their host rocks. Geol Soc Am Abstr Prog 21(6):A7

Sangster DF (1990) Mississippi Valley-type and sedex deposits: a comparative examination. Trans Inst Mining Metall Sect B 99:B21–B42

Schmidt RA (1962) Temperature of mineral formation in the Miami-Pitcher district as indicated by liquid inclusions. Econ Geol 57:1–20

Schwalb HR (1982) Paleozoic geology of the New Madrid area. Ill State Geol Surv NUREG/CR-2909

Shelton KL, Reader JM, Ross LM, Viele GW, Siedeman DE (1986) Ba-rich adularia from the Ouachita Mountains, Arkansas: implications for a postcollisional hydrothermal system. Am Min 71:916–923

Shelton KL, Burstein IB, Hagni RD (1991) Sulfur isotope geochemistry of detailed mineral parageneses, Viburnum Trend MVT deposits, S.E. MO: evidence of multiple basement and basinal sulfur sources. Geol Soc Can Prog Abstr 16:A133

Shelton KL, Bauer RM, Gregg JM (1992) Fluid inclusion studies of regionally extensive epigenetic dolomites, Bonneterre Formation (Cambrian), southeast Missouri: evidence of multiple fluids during dolomitization and lead-zinc mineralization. Geol Soc Am Bull 104:675–683

Siebenthal CE (1916) Origin of the zinc and lead deposits of the Joplin region, Missouri, Kansas, and Oklahoma. US Geol Surv Bull 606, 283 pp

Snyder FG, Gerdemann PE (1968) Geology of the southeast Missouri lead district. In: Ridge JD (ed) Ore deposits of the United States, 1933–1967. American Institute of Mining, Metallurgy, and Petroleum Engineers, New York, pp 326–358

Spencer RJ, Moller-Weare N, Weare JH (1990) The prediction of mineral solubilities in natural waters: a chemical equilibrium model for the Na-K-Ca-Mg-Cl-SO_2-H_2O system at temperatures below 25 °C. Geochim Cosmochim Acta 54:326–358

Stanton, Goldhaber MB (1991) Experimental studies of the synthesis of pyrite and marcasite (FeS_2) from 0 to 200 °C and summary of results. US Geol Surv Open-File Rep 91–310

Stein HJ, Kish SA (1985) The timing of ore formation in southeast Missouri: Rb-Sr glauconite dating at the Magmont Mine, Viburnum Trend. Econ Geol 80:739–753

Stein HJ, Kish SA (1991) The significance of Rb-Sr glauconite ages, Bonneterre Formation, Missouri; Late Devonian-early Mississippian brine migration in the midcontinent. J Geol 99:468–481

Sverjensky DA (1981) The origin of a Mississippi Valley-type deposit in the Viburnum Trend, southeast Missouri. Econ Geol 76:1848–1872

Sverjensky DA (1984) Oil field brines as ore-forming solutions. Econ Geol 79:23–37

Sverjensky DA, Rye DM, Doe BR (1979) The lead and sulfur isotopic compositions of galena from a Mississippi Valley-type deposit in the New Lead Belt, southeast Missouri. Econ Geol 74:149–153

Symons DTA, Sangster DF (1991) Paleomagnetic age of the central Missouri barite deposits and its genetic significance. Econ Geol 86:1–12

Viele GW (1979) Geological map and cross section, eastern Ouachita Mountains, Arkansas. Geol Soc Am Map and Chart Series MC-28F, scale 1:250000, 1 sheet, 8 pp

Viele GW (1983) Collision effects on the craton caused by the Ouachita orogeny. Geol Soc Am Abstr Prog 16:712

Viets JV, Leach DL (1990) Genetic implications of regional and temporal trends in ore fluid geochemistry of Mississippi Valley-type deposits in the Ozark region. Econ Geol 85:842–861

Voss RL, Hagni RD (1985) The application of cathodoluminescent microstratigraphy for sparry dolomite from the Viburnum Trend, southeastern Missouri. In: Hausen DM, Kopp OC (eds) Mineralogy – applications to the mineral industry. Proc, Paul F Kerr Memorial Symp, American Institute of Mining Engineers, New York, pp 51–58

Wisniowiecki MJ, Van der Voo R, McCabe C, Kelly WC (1983) A Pennsylvanian paleomagnetic pole from the mineralized Late Cambrian Bonneterre Formation, southeast Missouri. J Geophys Res 88:6540–6548

Relations Between Diapiric Salt Structures and Metal Concentrations, Gulf Coast Sedimentary Basin, Southern North America

H.H. Posey[1], J.R. Kyle[2] and W.N. Agee[3]

Abstract

Salt dome-hosted mineral systems are a distinct class of sediment-hosted metalliferous deposits that have received detailed study only within the last decade. Known deposits of this class are of low tonnage (10 to 15 million t, typically) and modest grade (5 to 15% combined Zn + Pb). However, exploration for these deposits has been limited, particularly in the USA, and the average deposit size yet may be found to be greater. Mineralogically, salt dome-hosted mineral deposits are similar to many carbonate- and shale-hosted mineral deposits as they contain pyrite, marcasite, sphalerite, galena, and barite; many deposits contain high concentrations of pyrrhotite or celestite. The host rocks are remarkably different from major sediment-hosted mineral deposits as they consist mainly of anhydrite and bacteriogenic limestone.

Salt dome mineral deposits apparently begin forming during the earliest phases of salt diapirism and continue, perhaps episodically, as long as oil- or methane-bearing brines circulate in their vicinity. Some of the earliest minerals to form are sulfides that precipitate in equilibrium with anhydrite. As crude oil and thermogenic methane come in contact with the anhydrite in low temperature environments (below about 70 °C), calcite may form; in this environment, sulfides, barite, and celestite may also form at the same time as or after the calcite.

Anhydrite cap rocks form from top to base, by underplating, as salt dissolves at the tops of diapirs. Calcite cap rocks form also from top to base, but by progressive replacement of the underlying anhydrite. Metal sulfides that form syngenetically with either the anhydrite or calcite cap rocks show isotopic and major element compositions which vary with depth. The range of isotopic values may record chemical changes in the upward migrating metal-bearing brines or variations in local chemical conditions during sulfide precipitation. The range in relative metal concentrations most likely reflects changes in brine composition due either to brine mixing or to evolution of brine compositions.

The rate of diapirism appears to be controlled mainly by sediment loading. Periods of rapid diapirism generally accompany major periods of fluvioclastic

---

[1] Colorado Department of Natural Resources, Division of Minerals and Geology, 1313 Sherman Street, Denver, CO 80203, USA

[2] Department of Geological Sciences, The University of Texas at Austin, Austin, TX 78712, USA

[3] Department of Geological Sciences, The University of Texas at Austin, Austin, TX 78712, USA; present address: Fluor Daniel, 12790 Merit Drive, Suite 200, Dallas, TX 75251, USA

Spec. Publ. No. 10 Soc. Geol. Applied to Mineral Deposits
Fontboté/Boni (Eds.), Sediment-Hosted Zn-Pb Ores

sedimentation; halokinesis is also active during periods of marine carbonate loading. However, diapirism may be induced or at least aided by fluids that invade the salt column and interact with materials contained therein, particularly anhydrite. The source of these invasive fluids during the early stages of diapirism is enigmatic. The fluids come either from beneath the salt stocks after passing through incredible thicknesses of evaporites, or from laterally adjacent beds. Another problem regarding fluid sources is that calcite cap rock formed from a combination of oilfield brine and meteoric water but the source of the meteoric water, its flow path, and the flow mechanisms are yet unknown.

1 Introduction

Salt diapirs and salt dome cap rocks are widely known as sources of halite and native sulfur and for their association with oil and gas traps. What is less known, however, is that iron, zinc, lead, and silver sulfides, barite, and celestite are common in salt dome cap rocks. In southern North America salt domes occur both onshore and offshore in the Gulf of Mexico. These are obvious targets for sulfur and hydrocarbon exploration, and although they are not exploited in this region, these domes are also targets for metal exploration.

To understand the origin of salt dome metal deposits researchers have combined several features of sedimentary basin evolution and found what appear to be strongly dependent relationships. Oil and gas maturation and migration, fluid-rock interaction chemistry, isotope geochemistry, basinal fluid migration, salt diapirism, and the physics, hydrology and chemistry of cap rock accumulation and formation all seem to affect salt dome mineralization. Fluids which contribute to salt dome mineralization are similar in composition to those that form both Mississippi Valley-type deposits and shale-hosted sedimentary exhalative deposits (see Sangster 1990). Thus, considering these similarities, it follows that by resolving questions about the origins of salt dome mineral occurrences, we will strengthen our understanding of the other, more economically important deposits.

In this chapter we discuss the origins of salt dome mineral deposits, particularly those we have studied in the US Gulf Coast. Insofar as they relate to our understanding of the origin of the metals in cap rocks, we highlight studies of cap rock formation and native sulfur.

2 Origin of Salt Diapirs and Cap Rocks

2.1 Gulf Coast Geology

During the Triassic, continental rifting and crustal subsidence along the former junction of South America and North America created the proto-Gulf of Mexico. North of that, early incipient rifting created smaller basins in the continental interior (Fig. 1). By mid-Jurassic the Gulf area had become a broad shallow shelf or, perhaps, a restricted sea accumulating evaporites. By late Jurassic, following deposition of the extensive Louann Formation evaporites,

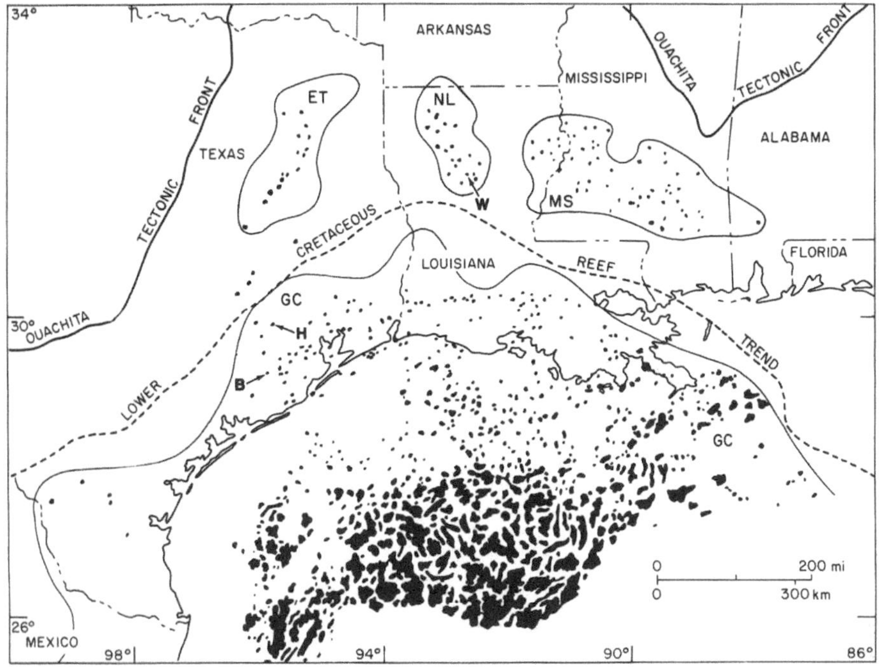

Fig. 1. General geologic setting of the Gulf Coast showing location of salt basins and selected salt domes. *ET* East Texas Basin; *NL* North Louisiana Basin; *MS* Mississippi Salt Basin; *GC* Gulf Coast Basin; *B* Boling dome; *H* Hockley dome; *W* Winnfield dome. (After Martin 1978)

the area had become a shallow marine environment accumulating carbonate sediments. With sediment loading, the upper surface of the Louann evaporites began to deform, and salt structures started developing. Diapirism continued throughout the Tertiary, a period dominanted by clastic sedimentation that continues today, especially in the offshore.

With the exception of the Late Jurassic Smackover and Lower Cretaceous Edwards marine carbonate formations, the Gulf Coast section is largely one of prograding fluvio-deltaic clastic complexes (Fig. 2). These units dip gently gulfward and are interrupted by down-to-basin growth faults (Fig. 3). As will be demonstrated, the salt diapirs and growth faults are principal routes of fluid migration.

2.2 Diapirism

Halite diapirs form in response to loading of the mother evaporite by younger sediments (Trusheim 1960; Seni and Jackson 1983). From deformation that creates broad anticlines, halokinesis progresses into discrete high-amplitude diapirs. Although halite diapirism is commonly considered a piercement process, in fact, diapirs in the Gulf Coast generally do not pierce significant thicknesses of overlying strata. Seismic cross sections, which show thicker

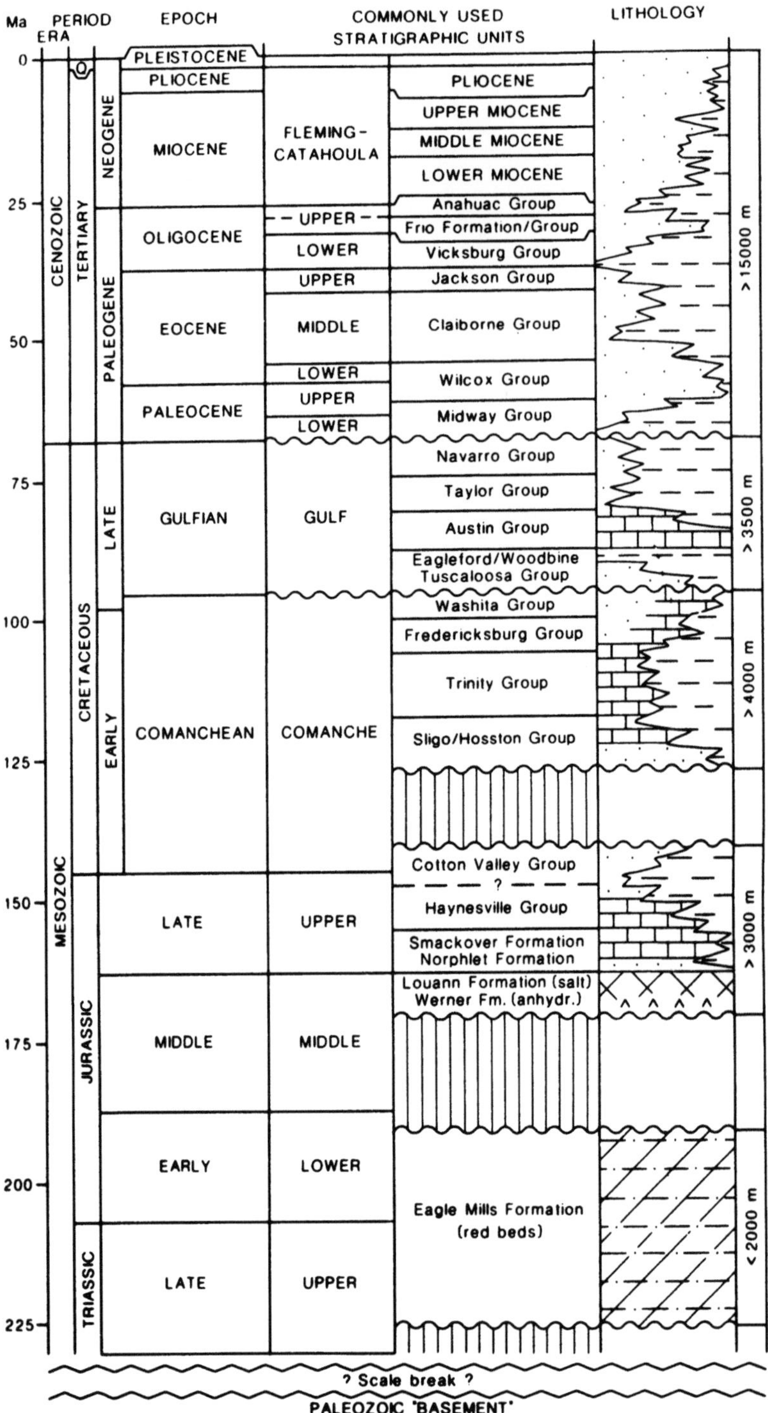

Fig. 2. Stratigraphic column for the Mesozoic and Cenozoic strata of the Gulf Coast Basin. (Curtis 1987)

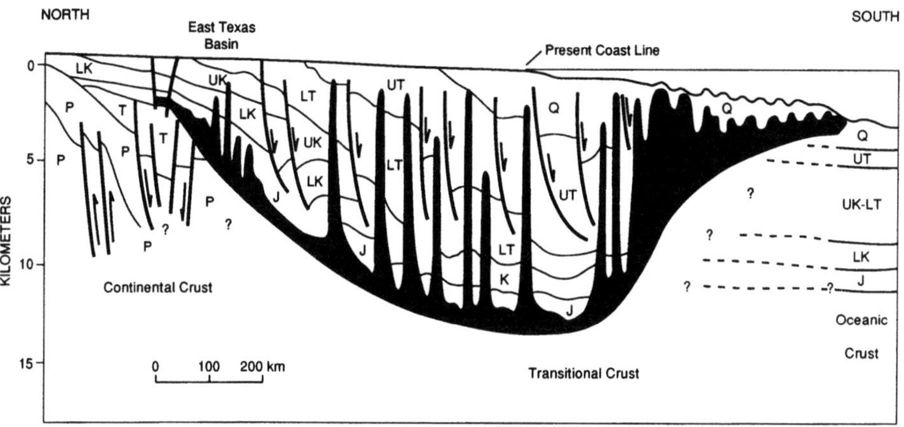

Fig. 3. Generalized north-south geologic cross section of the northern Gulf Coast Basin. *P* Undifferentiated Paleozoic; *T* Triassic red beds; *J* Middle (?) – Upper Jurassic; *LK* Lower Cretaceous; *UK* Upper Cretaceous; *LT* Lower Tertiary; *UT* Upper Tertiary; *Q* Quaternary. *Solid pattern* represents in-place and deformed Middle Jurassic evaporites of the Louann Formation. (After Salvador and Buffler 1982; Curtis 1987)

sedimentary units adjacent to diapir stems than away from them, indicate that diapirs are continuously fed by halite from below, and grow upward, more or less apace of sedimentation (Posey and Kyle 1988). The subtle amplitudes of domed surface layers – which reach only about 50 m onshore and 150 m offshore – indicate that the amount of piercement is minor compared with the amplitudes of the diapirs, most of which exceed several kilometers in height. Sediment layers alongside the diapirs are deformed by dragging or by vertical thrusting of the rising diapir and by subsidence of the sediment layers into rim synclines that develop as diapiric salt evulses from the mother salt. Disruption of the bedded units along diapir margins creates avenues for vertical fluid movement.

A note to consider with regard to mineralization is that the cap rock-forming environment, i.e. the top of the salt mass, occurs near the seafloor. Active marine calcite cap rock-forming sites have been identified over shallow salt diapirs on the present Gulf Coast seafloor (Kennicutt et al. 1985; Roberts et al. 1990a,b). These sites are coincident with petroleum seeps, and calcite that is present – including some mollusk shells – reflects a crude oil carbon source in its isotopic signature (Kennicutt et al. 1985). Thus, it follows that cap rock mineralization may take place in sediments just below the seafloor and, under specific circumstances, mineralizing brines may be "exhaled" onto the seafloor to form brine pools (Kyle and Price 1986). Considering that methane and brines are seeping today into offshore sediments, and that sulfide smokers accompany some of these seeps (see, for instance, Martens et al. 1991), we speculate that some salt dome mineralization may be taking place at present.

In a diapir field or salt dome basin, diapirs that form earliest in a succession occupy the basin edge where fluvial systems first begin releasing their sediment load to the marine environment. With prograding sedimentation, diapirism moves toward the basin center. This occurs both at the scale of individual salt basins in the Gulf Coast as well as the entire Gulf (Fig. 1). Because sediments were dumped first near the margins of the craton, domes in the interior basins – away from the Gulf of Mexico basin – formed before those in or near the present-day Gulf. A significant part of the Louann Formation in the offshore is allochthonous; it has migrated up-section and toward the center of the Gulf in response to progradational sediment loading (Jackson and Cornelius 1987).

During halokinesis, halite within the stems of diapirs interacts with fluids from adjacent sedimentary units. Citing the onset of diapirism soon after Louann deposition and the presence of high temperature minerals in calcite cap rocks, Light et al. (1987) concluded that high temperature, hydrocarbon-bearing fluids must have arrived at the site of cap rock formation soon after Louann deposition. They concluded that the diapir-invading fluids must have come from beneath the Louann, perhaps from the Triassic Eagle Mills clastics or from older crustal rocks. Anhydrite, which makes up a few percent of halite diapirs, apparently recrystallizes and inherits isotopic ratios that are a mixture of the original anhydrite and the invading fluids (Posey 1986; Posey et al. 1987a; Land et al. 1988). Halite also recrystallizes and, in the process, loses trace elements (notably bromine) that were incorporated while the halite precipitated originally from seawater (Land et al. 1988).

Although some metal sulfides occur in the salt diapirs themselves, these do not appear to be related to sulfides in the cap rocks. Cap rock mineralization is epigenetic relative to the primary marine evaporites, but is in part syngenetic with its cap rock hosts.

2.3 Cap Rock Formation

Salt dome cap rocks consist of variable amounts of anhydrite, calcite, and gypsum. Virtually all cap rocks contain anhydrite, but gypsum and calcite may or may not be present. The characteristic, complete vertical sequence consists of an upper calcite zone, a central gypsum-rich zone, and a lower anhydrite zone (Fig. 4). Lenses of calcite also occur inside the anhydrite zone. The contact between calcite and anhydrite is generally diffused with gypsum. The gypsum zone is typically a thin, irregular "transition" zone between the calcite and anhydrite cap rocks. Calcite-cemented clastic rocks occur commonly above or alongside diapirs.

Cap rocks commonly contain several accessory minerals, the most abundant being doubly terminated quartz crystals, single euhedral carbonate crystals, and clay minerals. These form prior to the cap rocks and are encased commonly by the younger cap rock minerals. Other common accessories include base metal sulfides, barite, and celestite, which will be discussed in the following section. These form prior to, during, and after their cap rock hosts.

Ghost fabrics of anhydrite in calcite indicate that anhydrite forms prior to calcite in the cap rock sequence (Posey 1986; Prikryl et al. 1988; Prikryl 1989). Similarities in strontium isotopic values of these two minerals indicate that the strontium in calcite and, thus, the calcium as well, are derived from the

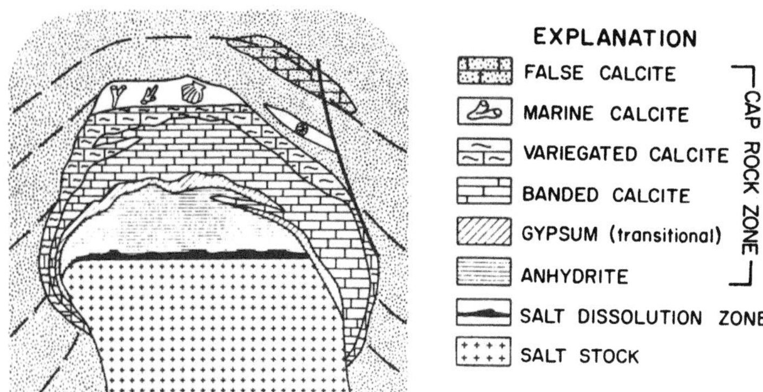

Fig. 4. Schematic cross section of a salt dome cap rock showing an idealized sequence of cap rock lithotypes. (After Posey 1986; Posey and Kyle 1988)

anhydrite (Posey 1986). However, the span of time between each cap rock event, whether they form through two separate sequential events of underplating or at nearly the same time, is not well known. Gypsum forms after both calcite and anhydrite (Werner et al. 1988). Base metal minerals form both during and after the formation of anhydrite and before, during, and after calcite cap rocks.

Anhydrite Cap Rock. Diapiric halite contains a few percent of anhydrite crystals. As halite at the top of a diapir dissolves, there accumulates a residuum of anhydrite crystals. As pressure is exerted by the rising diapir, the anhydrite crystals fuse to form an anhydrite cap. As more halite dissolves, new anhydrite layers fuse to the overlying layers, and an inverted stratigraphy evolves.

Locally, anhydrite underplating causes a type of small-scale internal diapirism in that loosely consolidated layers of anhydrite are intruded from beneath by younger unconsolidated layers (Kyle et al. 1987). This small-scale diapirism probably mimics the larger-scale diapirism.

Calcite Cap Rocks. Calcite cap rock formation is an intricate process controlled by several coincident chemical and physical parameters. Calcium is derived from the dissolution of anhydrite (Feely and Kulp 1957; Kirkland and Evans 1976; Posey 1986). Bicarbonate is derived in some cases through the breakdown of crude oil (Feely and Kulp 1957; Sassen 1980, 1987) and in others by the oxidation of methane (Kirkland and Evans 1976; Posey 1986). The oxidation of crude oil or methane in salt dome environments is a biochemical (bacteriogenic) process that accompanies sulfate reduction, and the entire process takes place at relatively low temperatures.

Calcite carbon is derived from hydrocarbons. The precise hydrocarbon compounds, though probably many, are not well known. Machel (1987, 1989) generalized the bacteriogenic calcite-forming process as follows:

Hydrocarbons + SO_4^{2-} → altered hydrocarbons + bitumen + HCO_3^- + H_2S+CO_2(?) + heat.

Bicarbonate combines with calcium to precipitate calcite which has a light carbon isotope signature inherited from the hydrocarbon presursor. The presence of metals may lead to precipitation of metal sulfides. If the H_2S is trapped and subsequently oxidized, native sulfur may form.

It appears that along the outermost rim of anhydrite cap rock, where the anhydrite is most exposed to pore fluids from surrounding rock and sediment, the earliest calcite forms. Other calcite layers form later, beneath this early calcite, and limestone cap rock – like anhydrite cap rock – grows from top to base (Posey 1986; Posey et al. 1987b; Prikryl et al. 1988).

In some cases, calcite formation involves crude oil and cool, bacteria-laden waters (Feely and Kulp 1957; Sassen 1980, 1987; Sassen et al. 1988). In others, it appears that methane provides carbon for calcite formation (Posey 1986; Posey et al. 1987b; Prikryl et al. 1988). Calcite that forms from the bacterial degradation of crude oil has carbon isotopic compositions ranging from about −10 to −30‰ (PDB), whereas isotopically lighter calcite, −30 to at least −55‰, forms from thermogenic methane. Overall, the carbon isotopes of cap rock limestone (−10 to −55‰) are distinct from marine limestones (+5 to −5‰) and generally distinct from freshwater limestones as well (−5 to −15‰).

Calcite cap rocks form also on the seafloor, and these commonly involve and incorporate biota living there (Baker 1979; Kennicutt et al. 1985; Roberts et al. 1990a,b; Fig. 4). Though minor in abundance, local native sulfur and base metal minerals have also been found in this seafloor unit. Marine salt dome limestones, even the skeletal material, bear the characteristic light carbon isotope values of salt dome calcite cap rocks (Sassen 1980, 1987; Posey 1986). Calcites with this distinct isotopic composition also occur as cements in porous clastic rocks many hundreds of meters away from diapirs (McManus and Hanor 1988).

Gypsum Cap Rock. Petrographic studies show that gypsum forms most commonly following both anhydrite and calcite formation. Spatial considerations suggest that water invades cap rock along the calcite/anhydrite contact, hydrates or replaces part of the anhydrite, and gypsum precipitates in both of the preexisting cap rocks. Studies of gypsum hydration waters indicate that meteoric fluids are responsible for hydrating the anhydrite to gypsum (Werner et al. 1988). Gypsum appears to form both before and after sulfur in various cap rocks.

3 Metallic Mineralization

3.1 General

Our discussion of salt dome mineralization will focus on five topics: (1) host rock and mineral relations, (2) the timing of mineralization, (3) fluid migration mechanisms, (4) the sources of metals, and (5) the sources of sulfur.

Sulfide and barite occurrences have been documented in at least 16 domes in the Gulf Coast (Kyle and Price 1986). In these, the most common metallic

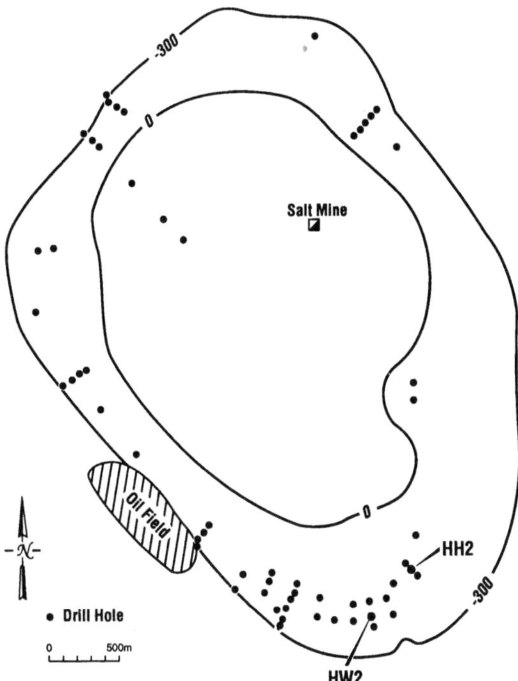

Fig. 5. Structure contour map showing depth to cap rock in meters at Hockley dome, Texas. Also note locations of cap rock metal exploration drill holes, oil field, and salt mine shaft. (After Agee 1990)

minerals, in decreasing abundance order, are: pyrite and marcasite, barite, sphalerite, pyrrhotite, galena, and celestite. Acanthite and strontianite occur as minor minerals along with many other trace minerals (Kyle and Price 1986). Though very common in diapirs of North Africa (Kyle and Posey 1991), celestite is abundant in only a few domes in the Gulf Coast (Saunders et al. 1988).

The base-metal minerals occur as anhydrite crystal cements, open-space fills in anhydrite, and as carbonate and sulfate replacements and open-space fills. The majority of anhydrite-hosted mineralization formed during anhydrite cap rock accumulation, whereas most of that hosted by calcite probably formed prior to or during calcite precipitation (Kyle and Price 1986).

Hockley dome in the Texas Gulf Coast and the Winnfield dome in northwest Louisiana have relatively major occurrences of sulfides and have been the objects of considerable study (Fig. 1). The Hockley dome cap rock rests about 12 m below surface. It is a site of active underground salt mining, and oil is produced from its flanks (Fig. 5). The Hockley dome was also the site of a recent mineral exploration program in which 65 core tests were drilled to evaluate the zinc-lead-silver potential of the Hockley cap rock. The Winnfield dome cap rock occurs at the surface. Formerly, it was mined for salt under-

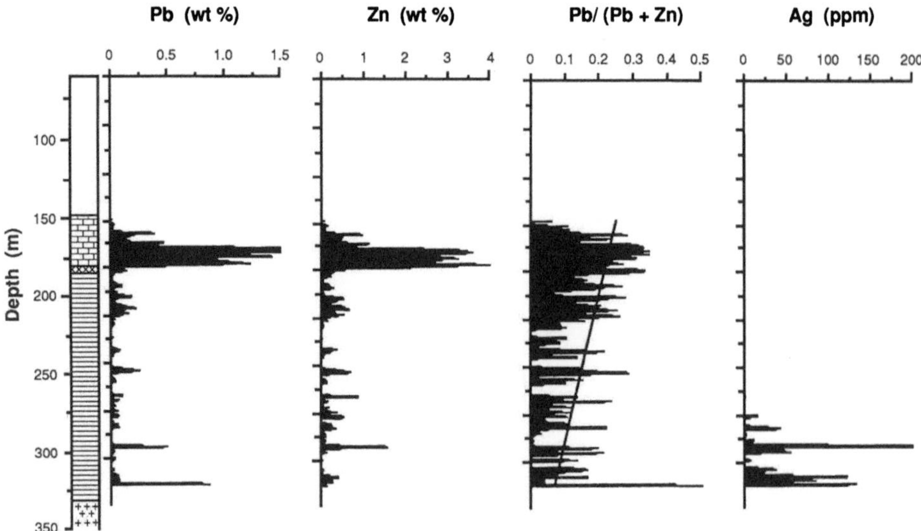

Fig. 6. Cap rock zones and metal intercepts for the HW2 drill hole at Hockley dome; see Fig. 5 for drill hole location. (After Agee 1990)

ground and the cap rock was quarried for limestone and gypsum. Currently, it is being quarried for anhydrite.[1]

The Hockley cap rock is composed of a typical sequence of calcite, gypsum, and anhydrite totaling as much as 275 m in thickness; anhydrite comprises the majority of the cap rock. Although discontinuous, the calcite portion is commonly several meters thick and may exceed 30 m. This zone is highly fractured and brecciated and the vugs in it contain calcite, barite, and sulfur crystals. Petroleum is present locally. Sulfide and barite in the Hockley dome occur mostly around the margins of both the calcite and anhydrite zones (Fig. 5; Price et al. 1983; Kyle and Price 1986; Kyle and Agee 1988; Agee 1990).

In the southern part of the dome is a calcite zone enclosed within the anhydrite cap rock. Mineralization in this lower calcite zone consists primarily of pyrite, marcasite, sphalerite, and galena in replacement and open-space filling modes. Barite occurs exclusively in the calcite cap rock. Mineralization within anhydrite consists mostly of sulfide cements. The richest zones of metallic mineralization in the anhydrite cap rock occur in and adjacent to the lens of calcite along the southern margin of the dome.

The annular zone of metallic sulfides that occurs in the calcite and anhydrite zones of Hockley ranges from depths of 120 to 365 m (Price et al. 1983; Kyle and Price 1986; Fig. 6). In sedimentary units above and adjacent to the cap rock, sulfide and barite are concentrated locally. Marcasite, pyrite,

[1] Anhydrite from Winnfield is used mainly as a road aggregate on farm and oilfield drill roads. Because the anhydrite swells considerably as it hydrates to gypsum it cannot be used beneath concrete or asphalt. However, in the moist climate of northern Louisiana this aggreagate is ideal as a road aggregate where the land is wet or swampy.

sphalerite, and galena range from trace to local massive abundances, and barite concentrations reach as high as 60%. Significant Zn + Pb intercepts include 5.5 m of 7.1% and 12 m of 4.2%. The overall Zn to Pb ratio for the deposit is about 3 to 1. The highest Ag assay, 550 ppm, comes from a 2-m intercept averaging over 200 ppm. Acanthite appears to carry the higher grade silver assays. The most significant silver zones occur deep in the anhydrite cap rock and are not related, spatially, to the major sulfide zones.

Winnfield mineralization consists mostly of iron-sulfides (pyrrhotite, pyrite, and marcasite), sphalerite, galena, and barite (Ulrich et al. 1984; Kyle et al. 1987). High metal concentrations are found in massive sulfide lenses generally in the lowermost portion of the calcite cap rock, above and within the gypsum transition zone. Trace mineralization, most of which is pyrrhotite, occurs throughout the anhydrite cap rock. The metallic mineralization in the Winnfield cap rock appears to be considerably less well developed than at Hockley. The presence of pyrrhotite laminae that accumulated in a progressive sequence was critical for studying timing of cap rock and mineral formation (Gose et al. 1985, 1989) and will be considered later.

3.2 Host Rock and Mineral Relations

Anhydrite-Hosted Mineralization. Most of the secondary mineral concentrations in the anhydrite cap rock section are sulfides of iron, zinc, and lead. Generally, barite and celestite are confined to calcite cap rocks. Much of the anhydrite-hosted sulfide mineralization consists of stratiform layers that encase euhedral anhydrite crystals (Fig. 7A), and thus must have formed at the anhydrite/salt interface soon after salt dissolution (Ulrich et al. 1984; Kyle and Posey 1991).

What happened to this mineralization when the anhydrite was replaced by calcite is not well known. Some calcite replacement textures clearly indicate the former presence of anhydrite-hosted sulfides (Fig. 7B; Kyle and Agee 1988), but these are not common. It is possible that some of the breccia-form massive sulfide zones in calcite cap rock (Fig. 7C; Kyle and Price 1986) are the result of later mineralization overprinting anhydrite-hosted sulfide concentrations.

Calcite-Hosted Mineralization. The dominant types of calcite cap rock (not including false cap and marine cap) are variegated calcite cap rock and true, or banded, calcite cap rock (Fig. 4). Both of these cap rock types are mineralized.

Variegated calcite cap rock is a highly brecciated, polymictic, bacteriogenic limestone (Posey 1986; Posey et al. 1987b; Prikryl et al. 1988; Prikryl 1989). Each clast consists of one or more types of limestone that differ in color due mostly to crystal size and clay abundance. Some of the clay in these limestone clasts was inherited from the precursor anhydrite, but most of it apparently came from enveloping clastic strata which supplied clay to the calcite-forming environment. The upper section of variegated calcite generally contains more clay than the lower, and both have more clay than the banded calcite cap rock. Most of these clasts are well rounded.

Fig. 7A-F. Features of mineralized cap rock from Gulf Coast salt domes. **A** Stratiform sulfide laminae in anhydrite cap rock, Winnfield dome. **B** Disrupted sulfide laminae in calcite cap rock relict from stratiform sulfide laminae in anhydrite; from the interior calcite zone on the southern flank of the Hockley dome. **C** Breccia-form massive sulfides from the interior calcite zone on the southern flank of the Hockley dome. **D** Scanning electron photomicrograph of marcasite (*mc*) spheres intergrown with vug-filling calcite (*ca*) scalenohedra in calcite cap rock, Boling dome. **E** Scanning electron photomicrograph of sulfides along a fracture in calcite cap rock, Boling dome. *ga* Galena; *gy* gypsum; *mc* marcasite; *sl* sphalerite. **F** Reflected light photomicrograph of colloform sulfide aggregate, Hockley dome. *ga* Galena; *mc* marcasite; *sl* sphalerite. Field of view = 3.2 mm

100 µ
mc
ca
d
sl
ga
gy
mc
500 µ
e
ga
mc
sl
mc
f

Overall, we believe that variegated calcite cap rocks form in a CO_2-rich environment. Periodically, when CO_2 builds up to excess, it dissolves part of the previously formed calcite cap rock. The CO_2 may derive from any of several hydrocarbon sources including thermogenic gas, biogenic CO_2 from the oxidation of crude oil, or from the oxidation of thermogenic CH_4. (Biogenic CH_4 does not appear to contribute to salt dome calcite.) As brecciation proceeds, excess CO_2 causes the margins of limestone clasts to dissolve, giving them a rounded, abraded appearance. As CO_2 levels drop, calcite again precipitates, generating new limestone that encrusts the existing limestone clasts. As this process of precipitation and partial dissolution continues, a complex unit of rounded, subrounded, and brecciated limestone with downward-diminishing clay content develops. This process continues until the CO_2 concentration decreases sufficiently to allow the true banded calcite cap rocks to form. Perhaps because of the higher clay content, the variegated calcite becomes an effective seal, whereas the true cap rock below remains more porous and permeable.

True calcite cap rock forms in a repetitious two-stage process of anhydrite replacement and open-space filling (Posey 1986; Posey et al. 1987b; Prikryl et al. 1988). Partial dissolution and replacement of anhydrite takes place at the calcite/anhydrite interface. A layer of replacement limestone, containing acid-insoluble residues from the anhydrite, forms first and leaves a space between the new limestone layer and the overlying one formed previously. Precipitation of calcite in this open space follows, and the process begins again, causing bands of calcite to underplate.

Mineralization in the calcite cap rock formed generally during and after the host calcite (Fig. 7D,E). This mineralization consists of open-space filling minerals, particularly barite, and replacement minerals, mostly sulfides. Sulfide-hosting porosity appears to have been created progressively, a feature that suggests the ore fluids created their own porosity (Kyle and Price 1986). In many instances, sulfides cross-cut earlier sulfides and cap rock units (Fig. 7E). Particularly in the variegated cap rock, sulfide clasts are contained within other sulfide clasts or a sulfide matrix. CO_2 concentration has an effect on sulfide solubilities, but this effect has not been recognized in the mineral textures.

Paragenesis, Sulfide Textures, and Zoning. Banded or colloform textures consisting of extremely fine-crystalline aggregates are common textures exhibited by the iron sulfides and sphalerite (Fig. 7F). Although there are many superposition and cross-cutting mineral textures in the sulfides and barite, only a general paragenesis can be deciphered. That is, for any particular site, iron sulfides generally formed early, galena and sphalerite formed later, and barite, celestite, and sulfur were generally the last to precipitate.

The majority of high-grade mineralization occurs in massive sulfide lenses, both at Hockley and Winnfield. These massive sulfide lenses are vuggy, fractured, and brecciated. At Winnfield, the layers consist mostly of iron sulfides, and the vugs and fractures are commonly lined with sphalerite, galena, barite, and calcite.

A zoning sequence has not been recognized at Winnfield, but Hockley does appear to exhibit mineral zoning (Kyle and Price 1986; Agee 1990). Agee

(1990) was able to demonstrate general trends and correlations in metal concentrations for some drill cores through the Hockley cap rock. Correlations based on metal abundances and ratios between core holes are possible in some cases. These correlations are related to the sequential formation of anhydrite cap rock and stratiform sulfides by underplating. Metal ratios indicate that there is an overall increase, downward, of Zn relative to Pb in the anhydrite cap rock (Fig. 6). For example, the metal concentrations for drill hole HW2 show Pb/(Pb + Zn) values that commonly range from 0.3 to 0.4 in the upper section. Ratios below 0.2 and commonly less than 0.1 are typical of the lower section. Barite occurs exclusively in the Hockley calcite cap rock, and the sulfide minerals are more common in the lower calcite and anhydrite cap rocks (Price et al. 1983; Kyle and Price 1986). Acanthite is most commonly associated with galena and appears to be concentrated low in the anhydrite cap rock stratigraphy, indicating a position late in the mineralization sequence (Fig. 6).

3.3 Timing of Mineralization

Because cap rock mineralization is a complex, evolving process, it is most realistic to view mineral-host rock textural relations as indicators of local or time-specific events. For example, sulfide minerals in the anhydrite cap rock, which occur as stratiform laminae that formed at the cap rock-salt contact, are syngenetic with the enclosing host rock. However, paleomagnetic studies of the rates of diapirism and cap rock formation indicate that similar sulfide laminae formed as the result of fluid events covering several million years (Kyle and Gose 1991). Sulfide minerals are locally intergrown with calcite and thus are contemporaneous with calcite cap rock formation (Fig. 7A). In other cases, calcite cap rock is clearly an overprint on stratiform sulfide laminae that were originally hosted by anhydrite as described above (Fig. 7B). Sulfide minerals also occur within fractures and breccias composed of either cap rock lithology (Fig. 7C,E). Within this complex environment, it is possible, perhaps even probable, that the mineralization system operating at a particular time produced a spectrum of local textural types indicative of "syngenetic" and "epigenetic" mineralization.

Pyrrhotite in the Winnfield dome occurs commonly as a cement that encases euhedral crystals of anhydrite. These anhydrite crystals are identical in form, size, and isotopic composition to those in the salt mass, and leave no doubt that they are a residue from the dissolution of halite (Posey 1986). Encasement of undeformed anhydrite crystals in pyrrhotite laminae, whereas the unmineralized cap rock anhydrite crystals are fused, indicates that the sulfide mineralization formed as the anhydrite crystals were being liberated from the salt, prior to their compaction or fusing by the rising diapir (Ulrich et al. 1984; Kyle et al. 1987).

Following evidence that anhydrite forms top to base by underplating (Ulrich et al. 1984), it was extrapolated that a cap rock sequence containing stratiform pyrrhotite laminae would record changes in magnetic declination, and that the ages of pyrrhotite would vary with stratigraphic position (Gose et al. 1985, 1989; Kyle et al. 1987). Using paleomagnetic evidence, it has been established that pyrrhotite mineralization and anhydrite cap rock formation at the Winnfield dome began at 157 Ma (shortly after deposition of the Louann evaporites) and continued at least until 145 Ma (Gose et al. 1989). Anhydrite cap rock beneath that already sampled probably records even younger ages.

3.4 Fluid Migration Mechanisms

The three most prevalent fluid-driving mechanisms in sedimentary basins are probably convective drive, geopressure drive, and gravity drive. As they relate to mineral deposits, these driving mechanisms are discussed by, among others, Sharp and Kyle (1988). Density drive, a process that takes place around salt domes where fluid density increases as a result of diapiric halite dissolution, is locally an important mechanism for moving fluid (Hanor 1987a,b; Ranganathan and Hanor 1988). Even though salt domes are positive thermal anomalies and, as such, affect fluid convection (Jensen 1983; Wood and Hewett 1984), convective flow is probably not of importance in the Gulf Coast salt dome area. Certainly, the effects of convection-driven flow are not as great as brine-density flow and geopressure-driven flow (Ranganathan and Hanor 1988). The gravity drive and geopressure drive mechanisms are both operative in the Gulf Coast. A discussion of these two systems is in order.

Geopressure-Driven Flow. As sediments are deposited, fluids that occupy the intergranular spaces will equilibrate with respect to pressure, and the value of fluid pressure at any depth will depend on the pressure derived from the weight of the overlying water column, the fluid density, and atmospheric pressure. Hanor (1987a) provided an excellent explanation of fluid pressure derivation. The hydrostatic pressure in a fluid having common density at any depth below the surface is calculated thus:

$$P_z = P_o + \rho g D,$$

where P_z is the hydrostatic pressure (i.e. the fluid pressure at elevation z); P_o is the fluid pressure at some reference point where elevation $z = 0$; ρ is the fluid density; g is the gravitational constant; and D = depth (in positive terms).

Geopressuring occurs if fluids are either buried too quickly for the fluids to equilibrate as described above, or if fluids in one unit are capped by impermeable rocks. Under these conditions, fluid pressure in the deeper section will be subjected to both the weight of the water column (see equation above) and the weight of the overlying rocks. Geopressuring is a common phenomenon and occurs, in general, below depths of about 2 to 3 km in the Gulf Cost.

Typical formation fluid pressure gradients in the Gulf Coastal region show an inflection away from the hydrostatic pressure gradient trend toward the lithostatic pressure gradient line starting at the "top of geopressure" zone. Fluids below this level are said to be "overpressured" and will escape upward if the cap at the top of geopressure is breached.

Saline fluids in the present Gulf Coast are known to be migrating upward, out of the geopressured zone. Most of these fluids move along two principal escape routes: growth faults and salt dome margins. It seems no coincidence that both of these structural conduits are the loci of oil and gas accumulations, as such brines may carry hydrocarbons, metals, H_2S, or other dissolved gases.

Gravity-Driven Flow. Gravity-driven flow, also called "forced convection," can be simply described by plugging off the low-elevation end of a J-shaped pipe, then filling it with water. When the plug is removed, the fluid pressure exerted by the water column on the high-elevation side will cause water to flow out the low-elevation side. Fluids flow from landward areas by gravity flow, and may participate in mineral formation. In this case, meteoric waters move into the subsurface after falling on land and proceed into the deep subsurface by a gravity-driven flow system (see, for example, Garven and Freeze 1984; Bethke 1985, 1986; Garven 1985). In some cases, fluids flow beneath a confining unit into the deeper subsurface. Such a system, if bisected by faults or salt domes, will direct the fluids upward, into a lower pressure regime (Kyle and Posey 1991). During this process, they may participate in mineralization or cap rock formation.

3.5 Source of Metals

To form anhydrite cap rock, little more is required in the way of fluid than halite-undersaturated water. The formation of metallic minerals and calcite is more complicated and is only partly understood.

Arguments about the source of metals are similar in most respects to those about the origin of Mississippi Valley-type mineralizing fluids. The documentation of metal-rich formation waters, or "oilfield brines," in the Gulf Coast was a major advance in the genetic modeling for Mississippi Valley-type deposits (Carpenter et al. 1974). Metal-bearing brines have been identified in many oilfields in the Gulf Coast (Carpenter et al. 1974; Land and Prezbindowski 1981; Kharaka et al. 1987) and are believed to be the source of the metals in salt dome cap rock deposits (Price et al. 1983; Kyle and Price 1986). Dissolved metal contents, particularly iron, zinc, lead, and barium, in the range of tens to hundreds of ppm are not uncommon. The mean Zn to Pb ratio for the Gulf Coast formation waters (n = 110) compiled by Agee (1990) is slightly greater than 4 to 1. This is similar to the 3 to 1 Zn to Pb ratio observed in Hockley dome cap rock sulfide concentrations (Kyle and Price 1986). In general, reservoir fluids in Mesozoic rocks appear to contain higher metal contents than Cenozoic reservoirs, regardless of the reservoir lithologic type, temperature, and salinity. It is clear that not all formation waters, even concentrated brines, are metal-enriched.

Deep basin brines moved updip out of the overpressured deep Gulf of Mexico basin via formation aquifers and major fault systems that served as vertical pathways (Land and Prezbindowski 1981). Sulfide precipitation was probably initiated by the interaction of cool surface waters with warmer metalliferous formation waters that migrated into the cap rock-forming environment

(Kyle and Price 1986; Kyle and Agee 1988). Many of these metal concentrations are stratiform sulfides that were precipitated at the zone of salt dissolution as the anhydrite cap rock accumulated by sequential underplating (Ulrich et al. 1984; Kyle and Price 1986). Metal ratio data suggest that the sulfide-precipitating fluid for the Hockley cap rock-hosted sulfide deposit became more Zn-rich and possibly Ag-rich during anhydrite cap rock accumulation (Agee 1990).

The ultimate source of metals is difficult to determine. Rock-water interaction accompanying burial diagenesis is a general explanation for metal enrichment in formation waters. Using this explanation, some researchers propose that the metals are derived from the alteration of feldspars and clays to provide most of the iron, lead, barium and strontium, and the alteration of carbonates to supply most of the zinc (Land and Prezbindowski 1981; Sverjensky 1981, 1984; Kharaka et al. 1987). Others provide evidence that iron-oxydes-hydroxides in red beds are a likely source of metals (Carpenter et al. 1974).

In the simplest case, metals are derived from sediments in briny solutions that appear to be warm – above about 100 °C. In the case of salt domes, these fluids appear to have traveled both with and apart from hydrocarbons. For instance, there is little evidence that hydrocarbons participated in the production of sulfides in most of the anhydrite cap rocks, yet calcite cap rocks depend on hydrocarbons for their carbon. Thus, because sulfides form during the deposition of both of these cap rock units, it seems that one set of sulfides formed from nil-hydrocarbon fluids, whereas the other formed in the presence of abundant hydrocarbons.

3.6 Source of Sulfur

Cap rock sulfur occurs in the form of anhydrite, gypsum, metal sulfides, metal sulfates, and native sulfur, and the source of each of these sulfur phases is, at least, slightly different. As discussed previously, Jurassic seawater is the source of sulfur in cap rock anhydrite and gypsum. However, the source of sulfur in the other mineral phases is not clearly known.

Kyle and Agee (1988) suggested that metal sulfide sulfur may be derived from two sources, via two processes. A profile of sulfur isotope values for pyrite and marcasite in the mineralized anhydrite section for the HH2 core yields sulfur isotope ratios ranging from about −30 to +4‰ (CDT) (Fig. 8). These extremes were interpreted to represent end members of a two-component sulfur system. The light end member represents biochemical reduction of cap rock-sourced sulfate in low-temperature groundwater. The heavy end member indicates a trend toward some value representing higher temperature thermochemical sulfate reduction of marine sulfate in hydrocarbon reservoirs. Actual values of the two sulfur end members are not known, but may be as light as the lightest native sulfur, which is −35‰, and as heavy as Cretaceous marine sulfate, which is +20‰.

Within this sulfur isotope profile there is a reversal coinciding with a limestone unit within the anhydrite (Fig. 8). This reversal also correlates with

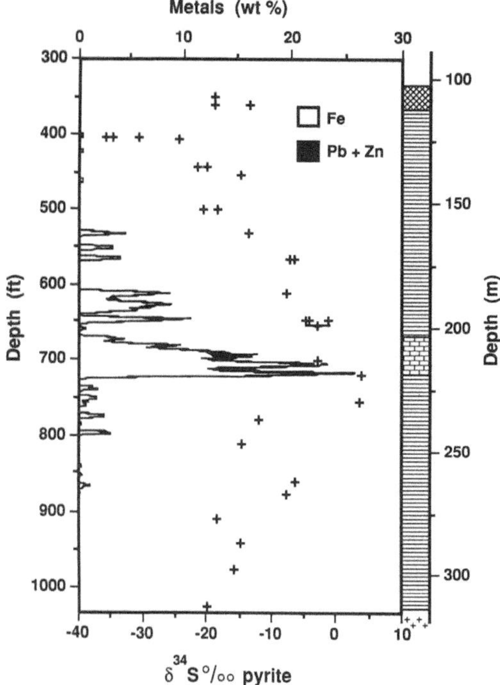

Fig. 8. Vertical profile of metal composition and $\delta^{34}S$ composition of sulfides in drill hole HH2 through the cap rock on the southern margin of Hockley dome; see Fig. 5 for drill hole location. Histograms show assay values indicative of metal sulfide concentrations; iron sulfide minerals are dominant in most intervals. Some of the highest Pb + Zn concentrations are associated with the interior calcite zone between the depths of 205 and 215 m. (After Kyle and Agee 1988; Agee 1990)

the highest metal abundance (iron, zinc, and lead) in the anhydrite section. Therefore, the source of isotopically heavy sulfur was also the source of the metals (Kyle and Agee 1988). Although the two-fluid model seems the simplest explanation, these data might also be explained through variations in oxygen availability, water availability, pH, total sulfur in the system, sulfate/sulfide ratios, or kinetic isotope effects (Kyle and Posey 1991). The coincidence of limestone and heavy sulfur may indicate higher pH, higher sulfur, and lower fO_2, all of which would favor sulfide deposition and greater stability of limestone (Ohmoto and Rye 1979).

Sulfur isotope compositions of barite reach extremely heavy values, ranging from about +20 to +80‰ (Kyle and Price 1986; Posey et al. 1987a). This extreme enrichment in ^{34}S apparently resulted when a finite reservoir of SO_4^{2-}, from the cap rocks, was partially reduced and the reservoir became enriched in the heavier sulfur isotope. Dissolved sulfate mixed periodically with barium from the ore fluids, and the barite incorporated the sulfur isotope composition of the sulfate reservoir at the time of barite precipitation.

4 Discussion

Mineral and fluid diagenesis and low-rank metamorphic reactions in sedimentary basins are controlled, in the simplest sense, by variations in burial temperatures, the primary mineral assemblage, and fluid composition. However, in basins that undergo fluid expulsion as a result of overpressuring, and particularly in basins where oil and gas are generated and moved by overpressuring, these diagenetic and metamorphic reactions respond to more complex controls. Multiple generations of mineral deposition and destruction develop because fluids and hydrocarbon-associated gases are driven from the high-temperature environments of their origin, to shallower, lower temperature strata where, because they are out of equilibrium, they cause minerals to dissolve or to precipitate. In basins where salinities vary widely during burial, such reactions may be even more complex.

Several attempts have been made to account for all of the fluid-rock interactions and fluid pathways that pertain to salt dome mineralization in the Gulf Coast (Price et al. 1983; Light et al. 1987; Kyle and Agee 1988; Agee 1990; Light and Posey 1992). All of these models require, at a minimum, interactions between a deep basin brine, appropriate sediments to supply metals, a source of sulfur, and meteoric water. Modern offshore domes, which have isotopic compositions similar to those onshore and, hence, apparently similar material sources, present a special problem, namely: how to move meteoric water down-dip and down-section in order to participate in salt dome mineralization and cap rock formation.

Salt domes and their associated cap rocks, minerals, and hydrocarbons are normal products of evaporite basin evolution involving either the destruction and transformation of minerals by temperature and chemical variations associated with burial diagenesis, material transfer within and around the diapir, oil and gas maturation and migration, or freshwater intrusion in the shallow environment. Metals appear to be derived by reactions between deep basin brines and clay, feldspar, carbonates or, possibly, red beds, or other iron-manganese oxy-hydroxides. Reduced sulfur travels either with these brines or is reduced at the site of metal sulfide deposition. Sulfur for native sulfur deposits is probably derived through biogenic reduction of anhydrite-sourced sulfate, the unreduced portions of which may form barite and other sulfate minerals.

Diapirs intrude over relatively long periods such that the amount of piercement during any single stage is small compared with the total amount of apparent upward migration of the salt mass. Halite diapirs "intrude" sedimentary cover probably with the aid of warm fluids that invade the evaporite body, probably from below. Such fluids interact with and change the isotopic and trace element compositions of halite and associated anhydrite and probably soften the formation by lowering its shear strength.

Halite diapirism and anhydrite cap rock formation begin soon after deposition of the mother salt and cease either when the supply of halite from the mother salt is exhausted, when the evaporite plug is cut off from the mother salt, or when rapid burial forms too thick a cover for the diapir to penetrate. Calcite cap rock may form any time after anhydrite cap rock, but requires mature hydrocarbons for its formation. Base metal sulfides, barite, and celestite form during the entire cap rock-forming event, so they must involve fluids from both below and above the mother salt. Fluids involved in anhydrite cap rock formation are of unknown salinity, slightly undersaturated with respect to anhydrite, and are slightly reducing (Posey et al. 1987a). Calcite cap rock-forming fluids are mixtures of formation and meteoric waters, and the mixture is probably relatively warm. Gypsum and sulfur, which form late in the cap rock sequence, form at substantially lower temperatures, and in the presence of lower salinity, probably meteoric, fluids.

Mineralization in the anhydrite, as seen in the mineral textures, is primarily syngenetic with respect to the anhydrite cap rock. Whatever the composition of this fluid, it must have been in equilibrium with the anhydrite, otherwise there would have been extensive anhydrite solution. However, the fact that higher concentrations of sulfides occur on the flanks of the anhydrite cap, and that at least part of this mineralization is syngenetic, provides an enigma. Perhaps the margins of the anhydrite cap rock, during its formation, produced a more reducing environment, whereas the core, perhaps richer in dissolved sulfate and therefore less reducing, provided an environment that was less conducive to sulfide formation.

Regarding calcite-hosted mineralization, there is evidence that sulfides formed before, during, and after their calcite host. However, most of the sulfide mineralization within the calcite cap rock apparently formed prior to or synchronously with the host calcite cap rocks, at least for the major concentrations. Like mineralization in the anhydrite section, most of the calcite-hosted sulfide mineralization at Hockley occurs on the flanks of the cap rock. This may argue either for epigenetic mineralization or for physical and chemical differences between cap rock cores and flanks that promoted mineralization in the flank sections while inhibiting mineralization in the cores. It is likely that the high metal sulfide zones are hybrid concentrations involving both syngenetic, diagenetic, and epigenetic mineralizaton, althought the relative contributions of each are imperfectly known.

Fluctuations in metal ratios of iron, zinc, and lead suggest that ore fluids either arrived in pulses (Ulrich et al. 1984; Agee 1990) or that fluid chemistry changed, possibly as a result of fluid mixing, during a single sustained pulse of mineralizing fluid movement. Provided one of the fluids was meteoric groundwater, variations in dissolved oxygen would have caused considerable variation in fluid chemistry and would have affected the relative mineral stabilities.

Exploration geologists interested in salt domes are perhaps most interested in the ability of this system to produce a large tonnage deposit. To date, salt dome-hosted mineral deposits are not known that are comparable in size with typical Mississippi Valley type or sedimentary exhalative deposits – the deposits with which they are linked most closely. Salt dome metal deposits are not known to exceed 15 million t of ore. The Hockley dome apparently has about 13 million t of ore averaging 3.1% combined Zn and Pb (Wessel 1983; Agee 1990). The Bou Grine deposit in Tunisia has 7.3 million t of ore averaging 11.1% combined Zn and Pb (Orgeval et al. 1989). Compared with 100 million t deposits of the Mississippi Valley and sedimentary exhalative types, the known salt dome deposits are not very large.

The absence of a large metal deposit associated with a salt dome is certainly no proof that a large deposit never formed. Perhaps the most encouraging aspect is the presence of very large elemental sulfur deposits in Gulf Coast cap rocks, some of which exceed 100 million t, e.g. Boling dome (Fig. 1). These obviously represent an enormous amount of reduced sulfur, which, if combined with a comparable amount of metal, could produce a huge metal sulfide deposit.

So, could a large deposit develop in salt dome cap rock? To form any fluid-generated ore deposit, there must be the ore components (in this case, metals and sulfur), a fluid-drive mechanism, a fluid-mixing mechanism (if two fluids participate), a precipitation mechanism, and characteristics inherent in these features to provide sustained mineral precipitation.

Metal and sulfur abundances provide the simplest controls. The metals apparently derive from brines, but not all brines in the Gulf Coast contain high metal abundances. Clearly, the brine must be an inherently metal-bearing one. Provided sulfur is derived from anhydrite cap rock, the limit to sulfur abundance would depend not only on the abundance of cap rock but also on the presence of a sustained mechanism for dissolving and reducing the anhydrite mineral sulfate. The fact that all known cap rocks contain, at least, anhydrite indicates that in the Gulf Coast there exists some phenomenon that limits the degree to which anhydrite cap rock can be dissolved, and this phenomenon thus limits the availability of sulfur. Plumbing may be the key control here.

Insofar as calcite and anhydrite cap rocks are concerned, there is ample evidence that the cap rock-forming system is a two-fluid system. Deep basin brines interact with meteoric fluids to form both of these cap rocks. Sulfur isotope evidence suggests that anhydrite-hosted mineralization also results from interaction of these two fluids, but the evidence is more circumspect. Regardless of whether sulfide precipitation requires a single fluid or two, the fact remains that ore grade mineralization does exist, thus a precipitation mechanism must exist. Provided H_2S is derived by bacterial reduction of sulfate, the abundance of H_2S (and, hence, metal sulfides) depends on the abundance of organic matter, whether it be in the form of crude oil or natural gas.

What remains is to determine whether the fluid-drive mechanism is sufficient to maintain sustained fluid flow. Gravity-driven flow can endure as long as a highland meteoric water recharge exists and as long as the basin aquifers remain permeable. If the recharge area is large like, say, a foreland thrust belt, the fluid-drive system can exist for tens of millions of years. However, overpressure-driven flow may be much more limited. Fluids can flow upward out of the geopressured zone only as long as abnormal formation pressures are sustained. Duration of flow in this case is determined by the pressure and by the number of escape avenues. Clearly, a large number of salt domes or growth faults will lead to a shorter duration of flow than a small number of fluid conduits.

A final issue regarding mineralization is the effect of calcite cap rock formation on fluid flow. It appears that the upper portions of calcite cap rock act as a seal. Also, bacteriogenic carbonate cements typically form in the permeable sediments adjacent to the domes. If these rocks and cements ultimately seal the cap rock area, they may actually inhibit fluid flow and the system may clog. In this case, both the cap rocks and the metal sulfides would cease to form.

Considering all of the above, it seems that there are more features restricting mineral deposition in the salt dome system than in MVT or Sedex systems. The salt dome system appears to be self-sealing, the availability of H_2S depends at least in part on hydrocarbon abundance, and overpressure flow is probably of limited duration. This is not to say that a large deposit could never form, but it does appear that such an ore deposit will be the exception in this system. On the other hand, if it can be shown that fluids move by gravity drive and that part of the H_2S is thermogenic and travels either with the metals or is generated locally by the hot metalliferous fluids, then the probability of forming a high tonnage deposit would be far greater.

For the present, salt dome mineral deposits offer an important laboratory for the study of other sediment-hosted mineral deposits, particularly the sedimentary-exhalative and Mississippi Valley-type deposits. Future studies should focus on brine seeps and methane seeps offshore, and on possible relationships between modern offshore mineral deposits (the so-called black smokers) and sulfur derived from ancient evaporites. Given that salt diapirs, worldwide, occur mostly in Cenozoic sedimentary sections, it appears that older diapirs – if they existed at all – were "washed out" of the system because of their highly soluble nature. Considering further that the difference in sulfur isotope compositions of seawater and coexisting sulfate minerals is practically nil, could it be that ancient sedimentary exhalative deposits derived sulfur not from seawater, as is commonly held, but rather from evaporites? If so, could it be that those evaporites were related to salt diapirs? We do not intend to draw conclusions to these questions, but rather to draw attention to potential field of future study.

Acknowledgments. We are pleased to acknowledge many research colleagues who collaborated with us in the study of salt domes and on the geology of Gulf Coast, in particular W.A. Gose, M.P.A. Jackson, T. Jackson, L.S. Land, P.E. Price, J. Prikryl, J.A. Saunders, S.J. Seni, J. Sharp, M.R. Ulrich, and G.R. Wessel. We are grateful for salt dome research support provided by National Science Foundation grants EAR-8407736 and EAR-8709319 and by the Petroleum Research Fund grant 19624-AC2 of the American Chemical Society. S.E. Kesler reviewed an earlier version of the manuscript and made valuable suggestions. Manuscript preparation was supported by the Owen Coates Fund of the Geology Foundation of The University of Texas at Austin.

References

Agee WN (1990) Relation of metal sulfide mineralization to anhydrite cap rock formation at Hockley salt dome, Harris County, Texas. MA Thesis, University of Texas, Austin, 255 pp

Baker HW (1979) General geology of Damon Mound. In: Etter EM (ed) Damon Mound field trip guidebook. Houston Geological Society, pp 10–25

Bethke CM (1985) A numerical model of compaction-driven groundwater flow and heat transfer and its application to the paleohydrology of intracratonic sedimentary basins. J Geophys Res 90:6817–6828

Bethke CM (1986) Inverse hydrologic analysis of the distribution and origin of Gulf Coast-type geopressured zones. J Geophys Res 91:6535–6545

Carpenter AB, Trout ML, Pickett EE (1974) Preliminary report on the origin and chemical evolution of lead and zinc-rich oil field brines in central Mississippi. Econ Geol 69:1191–1206

Curtis DM (1987) The northern Gulf of Mexico Basin. Episodes 10:267–270

Feely HW, Kulp JL (1957) Origin of Gulf Coast salt dome sulfur deposits. Am Assoc Petrol Geol Bull 41:1802–1853

Garven G (1985) The role of regional fluid flow in the genesis of the Pine Point deposit. Econ Geol 80:307–324

Garven G, Freeze RA (1984) Theoretical analysis of the role of groundwater flow in the genesis of stratabound ore deposits. I. Mathematical and numerical model. Am J Sci 284:1085–1124

Gose WA, Kyle JR, Ulrich MR (1985) A paleomagnetic study of the cap rock of the Winnfield salt dome, Louisiana. Trans Gulf Coast Assoc Geol Soc 35:97–106

Gose WA, Kyle JR, Farr MR (1989) Direct dating of salt diapir growth by means of paleomagnetism. Soc Econ Paleon Mineral, Gulf Coast Section, Proc 10th Annu Res Conf, Houston, Texas, December, 1989, pp 48–53

Hanor JS (1987a) Origin and migration of subsurface sedimentary brines. Soc Econ Paleontol Mineral Short Course 21, 247 pp

Hanor JS (1987b) Kilometre-scale thermohaline overturn of pore fluids in the Louisiana Gulf Coast. Nature 327:501–503

Jackson MPA, Cornelius RR (1987) Stepwise centrifuge modeling of the effects of differential sedimentary loading on the the formation of salt structures. In: Lerche I, O'Brien JJ (eds) Dynamical geology of salt and related structures. Academic Press, Orlando, pp 163–259

Jensen PK (1983) Calculations of thermal conditions around a salt diapir. Geophys Prosp 31:481–489

Kennicutt MC, Brooks JM, Bidigare RR, Fay RR, Wade TL, McDonald TJ (1985) Vent-type taxa in a hydrocarbon seep region of the Louisiana slope. Nature 317:351–353

Kharaka YK, Maest AS, Carothers WW, Law LM, Lamothe PJ, Fries TL (1987) Geochemistry of metal-rich brines from central Mississippi Salt Dome Basin. Appl Geochem 2:543–561

Kirkland DW, Evans R (1976) Origin of limestone buttes, Gypsum Plain, Culberson County, Texas. Am Assoc Petrol Geol Bull 60:2005–2018

Kyle JR, Agee WN (1988) Evolution of metal ratios and $\delta^{34}S$ composition of sulfide mineralization during anhydrite cap rock formation, Hockley dome, Texas, USA. Chem Geol 74:37–56

Kyle JR, Gose W (1991) Paleomagnetic dating of sulfide mineralization and cap rock formation in Gulf Coast salt domes. Am Assoc Petrol Geol Bull 75:615

Kyle JR, Posey HH (1991) Halokinesis, cap rock development, and salt dome mineral resources. In: Melvin J (ed) Developments in sedimentology 50. Evaporites, petroleum and mineral resources. Elsevier, New York, pp 413–476

Kyle JR, Price PE (1986) Metallic sulphide mineralization in salt-dome cap rocks, Gulf Coast, USA. Trans Inst Min Metall 95:B6–B16

Kyle JR, Ulrich MR, Gose WA (1987) Textural and paleomagnetic evidence for the mechanism and timing of anhydrite cap rock formation, Winnfield salt dome, Louisiana. In: Lerche I, O'Brien JJ (eds) Dynamical geology of salt and related structures. Academic Press, Orlando, pp 497–542

Land LS, Prezbindowski DR (1981) The origin and evolution of saline formation water, lower Cretaceous carbonates, south-central Texas, USA. J Hydrol 54:51–74

Land LS, Kupecz JA, Mack LE (1988) Louann salt geochemistry (Gulf of Mexico sedimentary basin, USA): a preliminary synthesis. Chem Geol 74:25–35

Light MPR, Posey HH (1992) Diagenesis and its relation to mineralization and hydrocarbon reservoir development: Gulf Coast and North Sea basins. In: Wolf KH, Chilingarian GV (eds) Diagenesis III. Elsevier, New York, pp 511–541

Light MPR, Posey HH, Kyle JR, Price PE (1987) Model for the origins of geopressured brines, hydrocarbons, cap-rocks, and metallic mineral deposits, USA. In: Lerche I, O'Brien JJ (eds) Dynamical geology of salt and related structures. Academic Press, Orlando, pp 787–830

Machel HG (1987) Some aspects of diagenetic sulphate-hydrocarbon redox reaction. In: Marshall JD (ed) Diagenesis of sedimentary sequences. Geol Soc Spec Publ 36. Blackwell, London, pp 15–28

Machel HG (1989) Relationships between sulphate reduction and oxidation of organic compounds to carbonate diagenesis, hydrocarbon accumulations, salt domes, and metal sulphide deposits. Carbonates Evaporites 4:137–151

Martens CS, Chanton JP, Paull CK (1991) Biogenic methane from abyssal brine seeps at the base of the Florida escarpment. Geology 19:851–854

Martin RG (1978) Northern and eastern Gulf of Mexico continental margin: stratigraphic and structural framework. In: Bouma AH, Moore GT, Coleman JM (eds) Framework, facies, and oil-trapping characteristics of the upper continental margin. Am Assoc Petrol Geol, Studies in Geology 7, pp 21–42

McManus KM, Hanor JS (1988) Calcite and iron sulfide cementation of Miocene sediments flanking the West Hackberry salt dome, southwest Louisiana, USA. Chem Geol 74:99–112

Ohmoto H, Rye RO (1979) Isotopes of sulfur and carbon. In: Barnes HL (ed) Geochemistry of hydrothermal ore deposits, 2nd edn. Wiley, New York, pp 509–567

Orgeval JJ, Giot D, Karoui J, Monthel J, Sahli R (1989) The discovery and investigation of the Bou Grine Pb-Zn deposit (Tunisian Atlas). Chron Res Min Spec Iss: 53–68

Posey HH (1986) Regional characteristics of strontium, carbon, and oxygen isotopes in salt dome cap rocks of the western Gulf Coast. PhD Thesis, University of North Carolina, Chapel Hill, 235 pp

Posey HH, Kyle JR (1988) Fluid-rock interactions in the salt dome environment: an introduction and review. Chem Geol 74:1–24

Posey HH, Workman AL, Hanor JS, Hurst SD (1985) Isotopic characteristics of brines from three oil and gas fields, southern Louisiana. Trans Gulf Coast Assoc Geol Socs 35:261–267

Posey HH, Kyle JR, Jackson TJ, Hurst SD, Price PE (1987a) Multiple fluid components of salt diapirs and salt dome cap rocks, Gulf Coast, USA. Appl Geochem 2:523–534

Posey HH, Price PE, Kyle JR (1987b) Mixed carbon sources for calcite cap rocks of Gulf Coast salt domes. In: Lerche I, O'Brien JJ (eds) Dynamical geology of salt and related structures. Academic Press, Orlando, pp 593–630

Price PE, Kyle JR, Wessel GR (1983) Salt dome related zinc-lead deposits. In: Kisvarsanyi G, Grant SK, Pratt WP et al. (eds) Proc Int Conf on Mississippi Valley-type lead-zinc deposits. Univ Missouri-Rolla, pp 558–571

Prikryl JD (1989) Origin of salt dome limestone cap rocks in the U.S. Gulf Coast. MA Thesis, University of Texas, Austin, 206 pp

Prikryl JD, Posey HH, Kyle JR (1988) A petrographic and geochemical model for the origin of calcite cap rock at Damon Mound salt dome, Texas, USA. Chem Geol 74:67–97

Ranganathan V, Hanor JS (1988) Density-driven groundwater flow near salt domes. Chem Geol 74:173–188

Roberts HH, Aharon P, Carney R, Larkin J, Sassen R (1990a) Sea floor responses to hydrocarbon seeps, Louisiana continental slope. Geo-Marine Lett 10:232–243

Roberts HH, Sassen R, Carney R, Aharon P (1990b) The role of hydrocarbons in creating sediment and small-scale topography on the Louisiana continental slope. In: Soc Econ Paleon Mineral, Gulf Coast Section, Proc 9th Annu Res Conf, New Orleans, Louisiana, December, 1990, pp 311–324

Salvador A, Buffler RT (1982) The Gulf of Mexico basin. In: Palmer AR (ed) Perspectives in regional geological synthesis: planning for the geology of North America. Geol Soc Am, Decade of North American Geology, pp 157–162

Sangster DF (1990) Mississippi Valley-type and Sedex lead-zinc deposits: a comparative examination. Trans Inst Min Metall 99:B21–B42

Sassen R (1980) Biodegradation of crude oil and mineral deposition in a shallow Gulf Coast salt dome. Org Geochem 2:153–166

Sassen R (1987) Organic geochemistry of salt dome cap rocks, Gulf Coast Salt Basin. In: Lerche I, O'Brien JJ (eds) Dynamical Geology of salt and related structures. Academic Press, Orlando, pp 631–649

Sassen R, Chinn EW, McCabe C (1988) Recent hydrocarbon alteration, sulfate reduction and formation of elemental sulfur and metal sulfides in salt dome cap rock. Chem Geol 74:57–66

Saunders JA, Prikryl JD, Posey HH (1988) Mineralogic and isotopic constraints on the origin of strontium-rich cap rock, Tatum dome, Mississippi, USA. Chem Geol 74:137–152

Seni SJ, Jackson MPA (1983) Evolution of salt structures, east Texas diapir province, part 1. Sedimentary record of halokinesis. Am Assoc Petrol Geol Bull 67:1219–1244

Sharp JM, Kyle JR (1988) The role of ground-water processes in the formation of ore deposits. In: Back W, Rosenshein JS, Seaber PR (eds) The geology of North America, hydrogeology, vol O-2. Geol Soc Am, Boulder, CO, pp 461–483

Sverjensky DA (1981) The origin of a Mississippi Valley-type deposit in the Viburnum Trend, southeast Missouri. Econ Geol 76:1848–1872

Sverjensky DA (1984) Oil field brines as ore-forming solutions. Econ Geol 79:23–37

Trusheim F (1960) Mechanisms of salt migration in northern Germany. Am Assoc Petrol Geol Bull 44:1519–1540

Ulrich MR, Kyle JR, Price PE (1984) Metallic sulfide deposits in the Winnfield salt dome, Louisiana: evidence for episodic introduction of metalliferous brines during cap rock formation. Trans Gulf Coast Assoc Geol Soc 34:435–422

Werner ML, Feldman MD, Knauth LP (1988) Petrography and geochemistry of water-rock interactions in Richton dome cap rock (southeastern Mississippi, USA). Chem Geol 74:113–136

Wessel GR (1983) Geologic summary of Hockley dome, Texas: unpublished company report, Marathon Mineral Resources, 43 pp

Wood JR, Hewett TA (1984) Reservoir diagenesis and convective fluid flow. In: McDonald DA, Surdam RC (eds) Clastic diagenesis. Am Assoc Petrol Geol Mem 37:47–62

Sulfide Breccia in Fossil Mississippi Valley-Type Mud Volcano Mass at Decaturville, Missouri, USA*

R.A. Zimmermann and M. Schidlowski

Abstract

The fossil sulfide and carbonate breccia of Decaturville, Missouri, is described and the structural position of the breccia mass discussed. The large and the small carbonate rock fragments show remnant textures typical of marine carbonates. The isotopic compositions of the carbonate fragments and the "flow" carbonates of the breccia compare well with those of sedimentary carbonates from the host sequence of Mississippi Valley-type stratiform lead-zinc deposits of southeastern Missouri. Therefore, an involvement of deep-seated (mantle) carbon in the process of breccia formation is not indicated.

1 Introduction

The Decaturville sulfide breccia is unique for its flow features, differing markedly from common karst breccias (Amstutz et al. 1974) or fault breccias. Instead, it compares well to mud volcano breccia as first recognized in 1960 during a field trip to Decaturville by G.C. Amstutz and the senior author with A. Gansser who, at that time, had just completed a study of mud volcanoes and salt domes in the western hemisphere and in the Near East (Gansser 1960). Since then, the peculiar features of the sulfide breccia and the surrounding rocks have attracted the attention of many investigators (Krishnaswamy and Amstutz 1960; Amstutz 1965; Zimmerman and Amstutz 1965; Amstutz and Bubenicek 1967; Zimmerman and Amstutz 1972, 1973, 1979; Zimmerman 1978, 1991; Zimmerman and Spreng 1984).

2 Regional and Local Geological Frame

The Decaturville structure belongs to a family of structures all lying more or less on line east to west in Missouri and neighboring states as shown in Fig. 1 (Singewald and Milton 1930; Tarr and Keller 1933; Rust 1937; Amstutz 1959, 1964, 1965; Snyder and Gerdemann 1965; Zimmerman and Amstutz 1965;

* In honor of Prof. G.C. Amstutz, and in memory of Mr. and Mrs. H.B. Hart of Decaturville.

Mineralogisch-Petrographisches Institut, Universität Heidelberg, Im Neuenheimer Feld 236, 69120 Heidelberg, Germany

Spec. Publ. No. 10 Soc. Geol. Applied to Mineral Deposits
Fontboté/Boni (Eds.), Sediment-Hosted Zn-Pb Ores

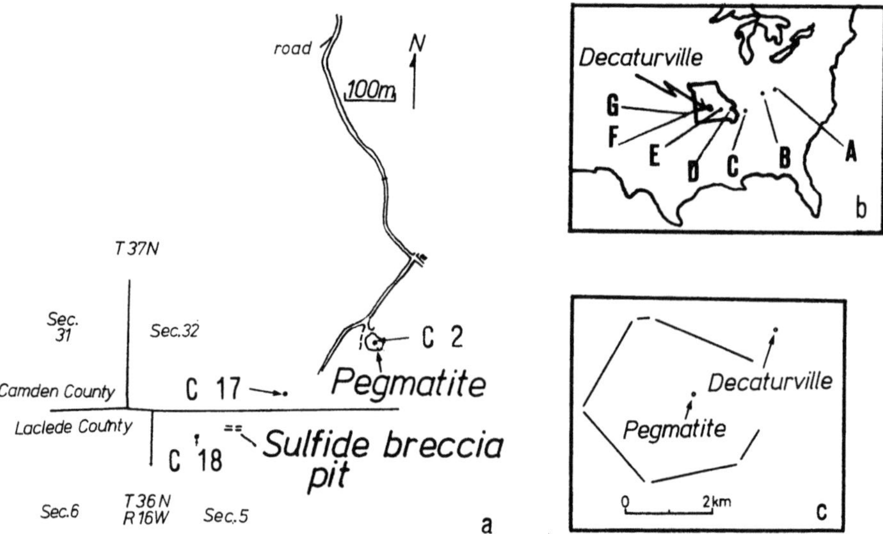

Fig. 1. **a** Location map of outcrops of the tectonically uplifted sulfide breccia (sulfide breccia pit) in the Davis Formation (of U. Cambrian mud volcano origin), and the tectonically uplifted Precambrian pegmatite in the center of the Decaturville polygonal structure. The *capital letters* in the *inset map* **b** (midwest of the United States) denote the locations of the following structural features: *A* Serpent Mound Structure; *B* Jeptha Knob; *C* Hicks Dome; *D* Avon Diatremes; *E* Crooked Creek Structure; *F* Weaubleau Creek Disturbance; *G* Rose Dome; see text (for a map with other ring and polygonal structures in the USA, see Zimmerman and Amstutz 1965). *C 2*, *C 17*, and *C 18* indicate the locations of drill holes in the Decaturville structure (see also Fig. 2 and Table 1). **c** *Inset map*, the major faults forming the polygonal pattern of the Decaturville structure are shown. In **a**, *Sec. 5*, *Sec. 6*, *Sec. 31*, *Sec. 32*, *T36N*, *T37N* and *R16W* are Dept. of the Interior designations for land surveyed in the public domain and appear also on the Stoutland and Macks Creek topographic quadrangle maps. The Camden County and Laclede County line was placed between townships 36 N and 37 N

Heyl 1972). Alnoite, peridotite and kimberlite, rock types of mantle origin, were noted to occur in some of the structures.

The Decaturville breccia is recognized in outcrops (see Fig. 2, inset) and in drill cores (Fig. 1 and Table 1) down to more than 100 m in depth. The breccia contains fragments of sedimentary rocks and sulfides. This mud volcano breccia occurs in sedimentary rocks making up a folded and faulted terrane within the polygonal Decaturville structure, which is about 4 km in diameter (Fig. 1c).

The tectonically uplifted breccia body near the center of the structure just outside of the uplift (Fig. 2, inset) contains up to 15% sulfides. Similar breccia bodies crop out 2.25 km WSW and 2.25 km SE (Zimmerman and Spreng 1984) of the dome center at the pegmatite (Figs. 1, 2), but contain only 1% or less of sulfide.

3 The Mud Volcano Breccia

The mud volcano breccia (Figs. 2, 3 and 4) consists of carbonate rock and sulfide fragments. The size of the carbonate rock pieces ranges from boulder to

Table 1. Parts of the drilled core near the center of the Decaturville structure (see Figs. 1, 2) (Courtesy of Mr. H.B. Hart, proprietor, 1962) (All drill holes are vertical)

Diamond drill core C 17		Diamond drill core C 18		Diamond drill core C 2	
Depth (in ft) and description[a]		Depth (in ft) and description[a]		Depth (in ft) and description[a]	
385	Ls. with dikelets of fine-grained carb.	111	Sulfide breccia; ls., no qtz.	70	Similar to sulfide breccia but with only traces of sulfide fragments; with glauconite and round qtz. grains (Lamotte)
415	Ls. with crenulation cleavage	210	Sulfide breccia; ls.; no qtz.	210	Ls. fragments with extremely fine-grained carb. in matrix; light sulfide breccia; Lamotte Ss. grains
430	Sulfide breccia; no qtz.; stringers of extremely fine-grained ls. breccia			220	Similar to C 2–70 (above) with tiny pockets of round ss. grains in light sulfide breccia
449	Sulfide breccia (heavy) (Bonneterre?)			274	Ss. (Lamotte); round grains; fragments
502	Ls. with diss. sulf. (Bonneterre?)			290	Disrupted ls. with carb. microbreccia in matrix
				298	Fragments of carb. with diss. sulf. in dikelets of fine-grained carb.
				366	Similar to C 2–290 but with stringers of Lamotte Ss. and sulf. in dikelets
				399	Similar to C 2–366 and light sulfide breccia
				428	Ls. fragments; fine-grained matrix; sulf. breccia with fragments of ZnS
				507	Qtz. with undulating extinction; chalcedony; fragmented; sericite; diss. sulf.; Lamotte ss.
				522	Qtz. with undulating extinction; diss. sulf.; some sericite; Lamotte ss.

[a] *Abbreviations*: Carb.: carbonate; Diss.: disseminated; Ls.: limestone; Qtz.: quartz; Ss.: sandstone; Sulf.: sulfides, mostly Fe sulfides.

submillimeter size; the largest sulfide aggregates are of walnut size and the smallest are the size of dust particles. All fragments float in a fine mass which under the microscopy shows the same composition as that just mentioned; yet, the whitish color and the fluidal texture (see Fig. 4, No. 7; Zimmerman 1978)

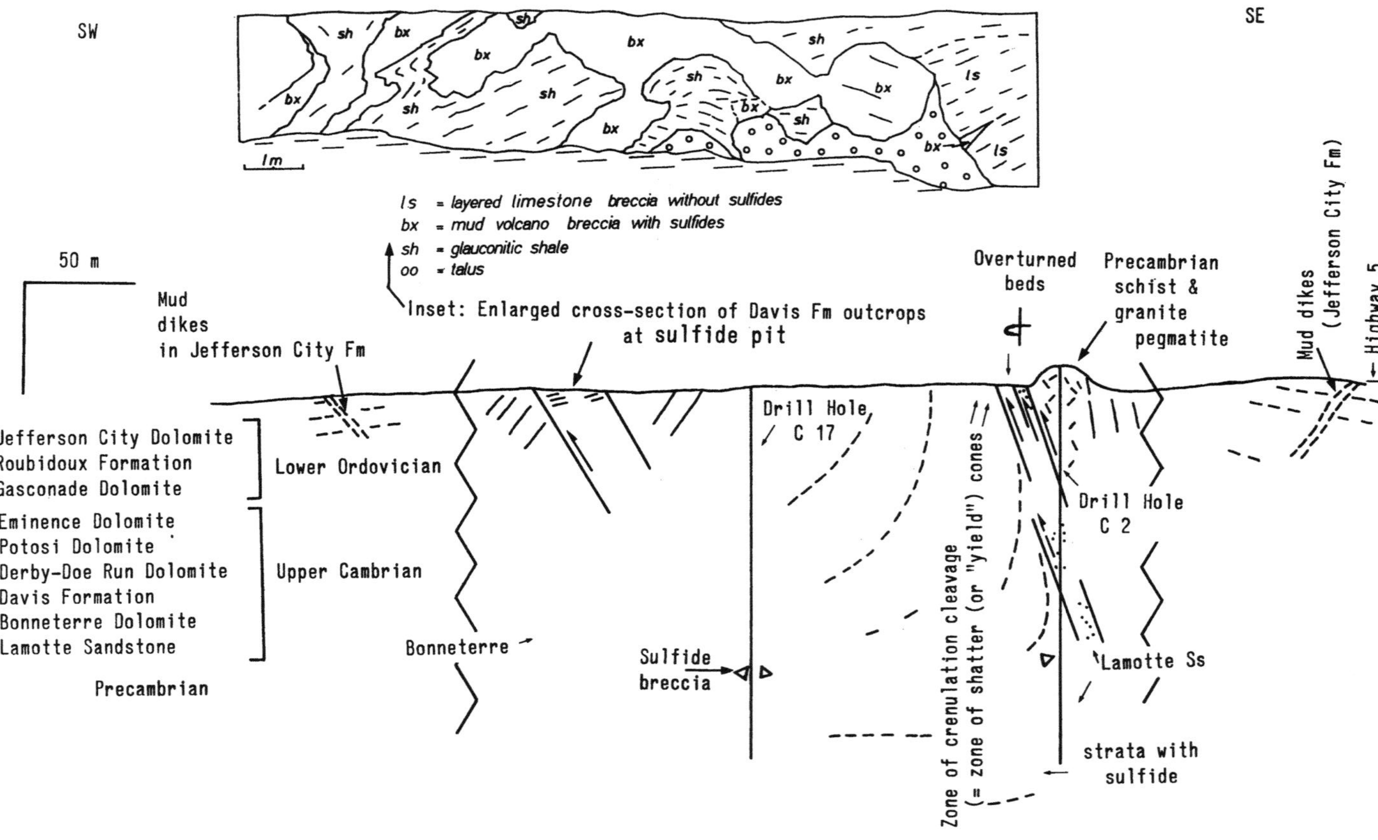
SW
SE
ls = layered limestone breccia without sulfides
bx = mud volcano breccia with sulfides
sh = glauconitic shale
oo = talus
1m
50 m
Mud dikes in Jefferson City Fm
Inset: Enlarged cross-section of Davis Fm outcrops at sulfide pit
Overturned beds
Precambrian schist & granite pegmatite
Mud dikes (Jefferson City Fm)
Highway 5
Jefferson City Dolomite
Roubidoux Formation
Gasconade Dolomite
Lower Ordovician
Eminence Dolomite
Potosi Dolomite
Derby-Doe Run Dolomite
Davis Formation
Bonneterre Dolomite
Lamotte Sandstone
Upper Cambrian
Precambrian
Bonneterre
Drill Hole C 17
Sulfide breccia
Zone of crenulation cleavage
(= zone of shatter (or "yield") cones
Drill Hole C 2
Lamotte Ss
strata with sulfide

suggest that much of this fine mass was not completely crystallized when the mud became "volcanic", i.e. when it intruded the fractured, overlying rock (very likely Davis Shale with limestone layers).

The mud volcano mass also contains locally sandstone layers or "streaks" which must be dislodged Lamotte Sandstone. (No sand facies is known in the Davis Shale or the Bonneterre.) The drawn out texture appears to suggest that these sandstone lenses were not completely indurated when the mud volcano mass was squeezed into the Davis Shale. Moreover, fragments of both carbonate rock and sulfides form indentation features in the sand lenses.

The large and the smaller carbonate rock fragments display remnant bedding at mega- and microscopic scale and yield glauconite. Some of these fragments also contain sulfides (Fig. 4, Nos. 1, 9 and 10). The sulfide fragments consist of unbroken grains and aggregates. The fine matrix of the mud volcano breccia occasionally cemented fissures in the carbonate rock as well as in sulfide fragments. The originally extremely fluid condition of the gray sulfide breccia matrix is also evidenced by thin dikelets of sulfide breccia (Fig. 3) and/or carbonate layers (Zimmerman and Amstutz 1972; Fig. 4). Consequently, it must be concluded that the formation of the sulfides ranged from pre- to postbrecciation.

As described previously (Zimmerman and Amstutz 1973), apart from pyrite and marcasite, sphalerite and galena occur in varying quantities; their intergrowths are illustrated in various parts of Fig. 4. Alternating mineral intergrowths are not rare, and colloform textures known also from recent sulfide-bearing mud volcanoes (Gansser 1960) are common. The same features and an identical sulfide paragenesis are also found in drill cores (see Amstutz 1965).

Fig. 2. Structure section showing position of the Lamotte Sandstone, Davis Formation and sulfide breccia at a depth down to about 150 m (approx. 500 ft), according to drill hole data (*C 2* and *C 17*; see also Table 1), and outcrops in the center of the structure, at the sulfide pit (see enlarged view of cross-sectional surface of Davis Formation outcrops) and in the SW and SE parts of the Decaturville polygonal structure. The relative position of Lamotte Sandstone, as indicated by drill holes, is evidence for the stratigraphic continuity of the bedding: this is an alternative interpretation to the cross-section by Offield and Pohn (1979) where they depict the dome center as an impact structure. Zimmerman and Amstutz (1979) have shown that the fractures in the Lamotte Sandstone flanking the pegmatite could only have formed tectonically by repeated movements. Recurrent mud volcano activity of the sulfide breccia is indicated by the presence of the breccia as a stratigraphic part of the Davis Formation as well as its occurrence higher up in the section in the Jefferson City Formation in the SW and SE parts of the Decaturville polygonal structure

a

b

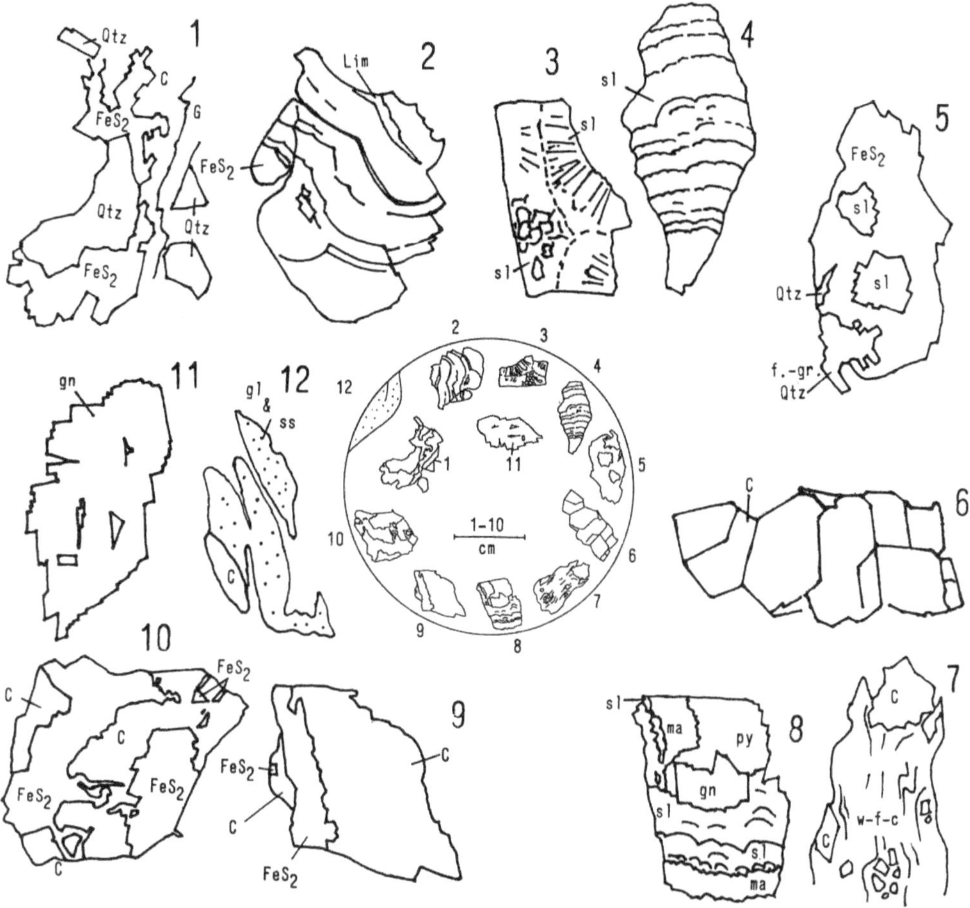

Fig. 4. Spectrum of materials including fragments (**2**, **3**, **4**, **5**, **6**, **8**, **9**, **10**, **11**), carbonate and glauconitic sandstone features (**7**, **12**) and late generation iron sulfides (**1**) contained in the sulfide mud volcano breccia. **1** Detrital quartz with cements of FeS_2, glauconite (*G*) and carbonate (*C*); **2** colloform FeS_2 and limonite (*Lim*); **3**, **4** ZnS (*sl*), honey-yellow, colloform and payement texture; **5** idiomorphic ZnS (*sl*) and fine-grained quartz in FeS_2; **6** carbonate grains (*C*) broken along idiomorphic grain boundaries; **7** fine-grained, white flow-textured carbonate (*w-f-c*) with included carbonate fragments (*C*); **8** marcasite-pyrite (*ma-py*) and ZnS (*sl*) in colloform texture with included PbS (*gn*); **9** "layered" FeS_2 and carbonate (*C*); **10** idiomorphic FeS_2 intergrown with carbonate (*C*); **11** cleavage fragment of PbS (*gn*); **12** glauconitic (*gl*) calcareous sandstone (*ss*) in "flow" lenses; matrix (white area surrounding the 12 features in *circle*) made up of the above materials ranging to microscopic sizes. Magnification: 1 to 5 mm diameter in nature as observed in thin and polished sections; large iron sulfide fragments in the outcrop in walnut sizes; **1**, **6**, **7** and **12** up to boulder sizes

←

Fig. 3. a Outcrop of the body of the sulfide breccia at the sulfide breccia pit (see Fig. 1 for location, and Fig. 2, inset, for outcrop detail). Note the flow structure of the breccia indicated by the elongation of the sandstone layer in the center of the outcrop. Geology students of Prof. Amstutz in 1961, *left to right*: M. Greeley, E. Schot, O. Aguilar, S . Ganthavee, D. Zimmerman, W. Park. **b** "Foot specimen" of the sulfide breccia: *A* sulfide (mostly pyrite and marcasite) fragments; *B* white "flow" carbonate; *C* carbonate breccia fragments; *D* carbonate fragment; *E* fragment of mottled carbonate; *F* Lamotte Sandstone with flow features; *G* fragment of glauconitic caroonate. Length of specimen: 40 cm. Sulfide breccia pit (see Figs. 1 and 2)

Table 2. Carbon and oxygen isotope composition of Upper Cambrian to Ordovician limestones and dolomites from the American midwest, including eight original analyses of carbonates from the Decaturville sulfide breccia pit (Nos. 7–14; see also Figs. 1, 2, 3)

Sample No.	$\delta^{18}O$ (PDB)	$\delta^{13}C$ (PDB)	Rock type	Age	Reference	Remarks
1.	−6.9	+0.3	Dolomite	U. Cambrian	Gregg and Shelton (1990)	Planar dolomite (fine crystalline; replaced cryptalgalaminate); back reef; Bonneterre and Davis Formations; St. Francois Mountains
2.	−7.5	−1.1	Dolomite	U. Cambrian	Gregg and Shelton (1990)	Open-space-filling dolomite cements (Bonneterre and Davis Formations); back reef; St. Francois Mountains
3.	−7.8	−0.1	Limestone	U. Cambrian	Gregg (1985)	Offshore facies, Bonneterre Formation
4.	−8.2[a]	−0.7	Dolomite	U. Cambrian	Hannah and Stein (1984)	Bonneterre Formation; host rock, Viburnum Trend, Magmont mine; sample No. Mag-7
5.	−9.7	−0.1	Limestone	Cambrian	Keith and Weber (1964)	Marine from Cambrian of Texas, Nevada, Utah, Alaska, Montana, Tennessee, Idaho, Oklahoma, and Alabama, USA; BC, Canada
6.	−6.2	0.3	Limestone	Ordovician	Keith and Weber (1964)	Marine from Ordovician of Pennsylvania, New York, Indiana, USA; Quebec and Ontario, Canada; Sweden
7.	−6.0	−0.8	Dolomite	U. Cambrian	This study, Is-1	Fine-grained fragments of dolomite and sulfide (10% sulfide) (sulfide breccia); Decaturville, MO
8.	−6.5	−0.4	Dolomite	U. Cambrian	This study, Is-2	Dolomite fragment (angular) in sulfide breccia; Decaturville, MO; (identical to Nos. 6 and 9 in Fig. 4)
9.	−4.8	−0.0	Dolomite	U. Cambrian	This study, Is-3	White "flow" dolomite similar to Fig. 7 in Zimmerman (1978); consists of broken dolomite fragments floating in a very fine-grained dolomite; entire white carbonate mass in dark gray sulfide breccia (similar to No. 7 in Fig. 4); Decaturville, MO
10.	−5.3	−0.2	Dolomite	U. Cambrian	This study, Is-4	Similar to sample Is-3 (above); (hand specimen No. 101, Mineralogy Institute, Heidelberg); similar to Fig. 8 in Zimmerman and Amstutz (1972); Decaturville, MO

11.	−5.8	−0.7	Dolomite	U. Cambrian	This study, Is-4b	Fine-grained carbonate of the dark gray sulfide breccia around sample Is-4; similar to sample Is-1; Decaturville, MO
12.	−5.1	−0.1	Dolomite	U. Cambrian	This study, Is-5	Similar to samples Is-3 and Is-4, above; hand specimen No. Hart 5; similar to Fig. 8 in Zimmerman and Amstutz (1972)
13.	−5.1	−0.2	Dolomite	U. Cambrian	This study, Is-6	Similar to sample Is-5, above (white "flow" carbonate); hand specimen No. 110
14.	−4.3	−0.1	Dolomite	U. Cambrian	This study, Is-7	Similar to sample Is-3 ("flow" carbonate); hand specimen 1.7.001; Decaturville, MO; samples in this study are from the Bonneterre and Davis Formations

[a] Recalculated from SMOW.

4 Isotopic Composition of the Carbonate Constituents of the Breccia

Carbon and oxygen isotope analyses of Cambrian carbonates from the midwest of the United States (Bonneterre and Davis Formations) have been reported by Hannah and Stein (1984), Gregg (1985) and Gregg and Shelton (1990), with results falling broadly into the range of other marine carbonates of this age group (cf. Keith and Weber 1964) and sedimentary carbonates in general (Schidlowski et al. 1975; Veizer and Hoefs 1976). Table 2 summarizes these data and presents a list of the isotope values of eight carbonate samples from the Decaturville breccia obtained in this investigation, comprising both "fragmental" and "flow type" (see above: whitish, fluidal textured carbonate) dolomite.

With $\delta^{13}C$ values ranging narrowly from ±0 to −0.8‰ and $\delta^{18}O$ values, between −4.3 and −6.5‰ (both vs. PDB), the isotopic compositions of the breccia carbonates adequately match those of the coeval carbonate members of the Bonneterre and Davis Formations, with only $\delta^{18}O$ values moderately shifted in a positive direction by 1 to 4‰ (cf. Table 2). Furthermore, it is interesting to note that the three "fragmental" dolomites analyzed (Nos. 7, 8 and 11) have consistently yielded $\delta^{13}C$ values several tenths of a per mill more negative than the five samples of their "flow-type" counterparts. This minor, but nevertheless, very consistent and distinct difference may be plausibly explained in terms of different stages of diagenetic maturation reached by the two precursor carbonates at the time of formation of the mud volcano. Though diagenetically stabilized carbonate rocks tend to preserve the isotopic composition of their parent carbonate muds within about 1‰ of the original values, it is known that diagenesis shifts $\delta^{13}C$ values slightly in a negative direction (cf. Longstaffe 1989). Therefore, the diagenetically mature fragmental carbonates can indeed be expected to be isotopically lighter than the "flow" variant, which is considered a carbonate mud at the time of its emplacement in the breccia. Perhaps as a result of its sudden "eruption" into a different environment, the $\delta^{13}C$ values of the mobile carbonate slurry may have been diagenetically arrested, whereby the subsequent maturation within the breccia obviously did not close the gap to those carbonates which had already been flushed into the mud volcano as solid fragments.

In any case, it is obvious that both dolomite components of the sulfide breccia bear the isotopic signature of marine carbonates, resembling that of sedimentary carbonate formations from the broader geological frame of the structure. This, together with the fact that the fragments contain sulfides resembling Mississippi Valley-type textures may be taken as evidence that the Pb-Zn ores of the Decaturville deposit were derived from sediment-hosted precursor sulfides of Mississippi Valley type. Moreover, the isotopic evidence clearly excludes any involvement of mantle carbon in the making of the breccia carbonates as $\delta^{13}C$ values of deep-seated carbonate rocks ("carbonatites") have been reported to scatter around −5.9‰ (Kobelski et al. 1973) or, more specifically, they average ca. −5.4 ± 0.2‰ (Deines 1992).

Acknowledgements. The authors are grateful to G. Josten and H. Oberhänsli for help rendered in the laboratory.

References

Amstutz GC (1959) Polygonal and ring tectonic patterns in the Precambrian and Paleozoic of Missouri, USA Eclogae Geol Helv 52:904–913

Amstutz GC (1964) Impact, cryptoexplosion, or diapiric movements? (A discussion of the origin of polygonal fault patterns in the Precambrian and overlaying rocks in Missouri and elsewhere.) Trans Kans Acad Sci 67:343–356

Amstutz GC (1965) Tectonic and petrographic observations on polygonal structures in Missouri. In: Whipple HE (ed) Geological problems in lunar research. Ann N Y Acad Sci 123:876–894

Amstutz GC, Bubenicek L (1967) Diagenesis in sedimentary mineral deposits. In: Larsen G, Chilingar GV (eds) Diagenesis in sediments. Developments in sedimentology, vol 8. Elsevier, Amsterdam, pp 417–475

Amstutz GC, Zimmerman RA, Love LG (1974) Copper deposit at Cornwall, Missouri: observations on the petrology, the sedimentary features and the sulfides (especially the framboidal pyrite). Neues Jahrb Mineral Monatsh 1974:289–307

Deines P (1992) Mantle carbon: concentration, mode of occurrence, and isotopic composition. In: Schidlowski M, Golubic S, Kimberley MM, McKirdy DM, Trudinger PA (eds) Early organic evolution: implications for mineral and energy resources. Springer, Berlin Heidelberg New York, pp 133–146

Gansser A (1960) Über Schlammvulkane und Salzdome. Viertel-jahrsschr Naturforsch Ges Zür 105:1–46

Gregg JM (1985) Regional epigenetic dolomitization in the Bonneterre Dolomite (Cambrian), souutheastern Missouri. Geology 13:503–506

Gregg JM, Shelton KL (1990) Dolomitization and dolomite neomorphism in the Back Reef Facies of the Bonneterre and Davis Formations (Cambrian), southeastern Missouri. J Sediment Petrol 60:549–562

Hannah JL, Stein HJ (1984) Evidence for changing fluid composition: stable isotope analyses of secondary carbonates, Bonneterre Formation, Missouri. Econ Geol 79:1930–1935

Heyl AV (1972) The 38th parallel lineament and its relationship to ore deposits. Econ Geol 67:879–894

Keith ML, Weber JN (1964) Carbon and oxygen isotopic composition of selected limestones and fossils. Geochim Cosmochim Acta 28:1787–1816

Kobelski BJ, Gold DP, Deines P (1979) Variations in stable isotope compositions for carbon and oxygen in some South African and Lesothan kimberlites. In: Boyd FR, Meyer HOA (eds) Kimberlites, diatremes, and diamonds: their geology, petrology, and geochemistry. Proc 2nd Int Kimberlite Conf, vol I. AGU, Washington, DC, pp 252–271

Krishnaswamy DS, Amstutz GC (1960) Geology of the Decaturville disturbance in Missouri. Bull GSA 71:1910 (Abstr)

Longstaffe FJ (1989) Stable isotopes as tracers in clastic diagenesis. In: Hutcheon IE (ed) Short course in burial diagenesis. Mineral Assoc Can, Short Course Handbook 15. Mineal Assoc Can, Montreal, pp 201–277

Offield TW, Pohn HA (1979) Geology of the Decaturville impact structure, Missouri. Prof Pap USGS 1042:1–48

Rust GW (1937) Preliminary notes on explosive volcanism in southeastern Missouri. J Geol 45:48–75

Schidlowski M, Eichmann R, Junge CE (1975) Precambrian sedimentary carbonates: carbon and oxygen isotope geochemistry and implications for the terrestrial oxygen budget. Precambrian Res 2:1–69

Singewald JT Jr, Milton C (1930) An alnoite pipe, its contact phenomena, and ore deposition near Avon, Missouri. J Geol 38:54–66

Snyder FG, Gerdemann PE (1965) Explosive igneous activity along an Illinois-Missouri-Kansas axis. Am J Sci 263:465–493

Tarr WA, Keller WD (1933) A post-Devonian igneous intrusion in southeastern Missouri. J Geol 41:815–823

Veizer J, Hoefs J (1976) The nature of O^{18}/O^{16} and C^{13}/C^{12} secular trends in sedimentary carbonate rocks. Geochim Cosmochim Acta 40:1387–1395

Zimmerman RA (1978) The Cambro-Ordovician fossil mud volcano of Decaturville, Missouri. In: Augustithis SS (ed) Proc vulcanism, vol III. Int Congr on Thermal waters, geothermal energy and vulcanism of the Mediterranean area. Nat Techn Univ, Athens, Greece, Oct 1976, pp 265–279

Zimmerman RA (1991) Crenulation cleavage and the sequence of small-scale internal structures in ("Yield") cones. EOS Trans AGU, Oct 29, 1991 Suppl (Fall Meeting): 458 (Abstr)

Zimmerman RA, Amstutz GC (1965) The polygonal structure at Decaturville, Missouri: new tectonic observations. Neues Jahrb Mineral Monatsh 1965:288–307

Zimmerman RA, Amstutz GC (1972) The Decaturville sulfide breccia–a Cambro-Ordovician mud volcano. Chem Erde 31:253–274

Zimmerman RA, Amstutz GC (1973) Intergrowth and crystallization features in the Cambrian mud volcano of Decaturville, Missouri, USA. In: Amstutz GC, Bernard AJ (eds) Ores in sediments. International Union of Geological Sciences, Series A, Number 3. Springer, Berlin Heidelberg New York, pp 339–350

Zimmerman RA, Amstutz GC (1979) Tectonic features in the polygonal structure of Decaturville, west-central Missouri. Neues Jahrb Mineral Monatsh 1979:453–470

Zimmerman RA, Spreng AC (1984) Sedimentary and diagenetic features in the sulfide-bearing sedimentary dikes and strata of Lower Ordovician dolomites, Decaturville, Missouri, USA. In: Wauschkuhn A, Kluth C, Zimmerman RA (eds) Syngenesis and epigenesis in the formation of mineral deposits, a volume in honor of Professor G.C. Amstutz. Springer, Berlin Heidelberg New York, pp 350–372

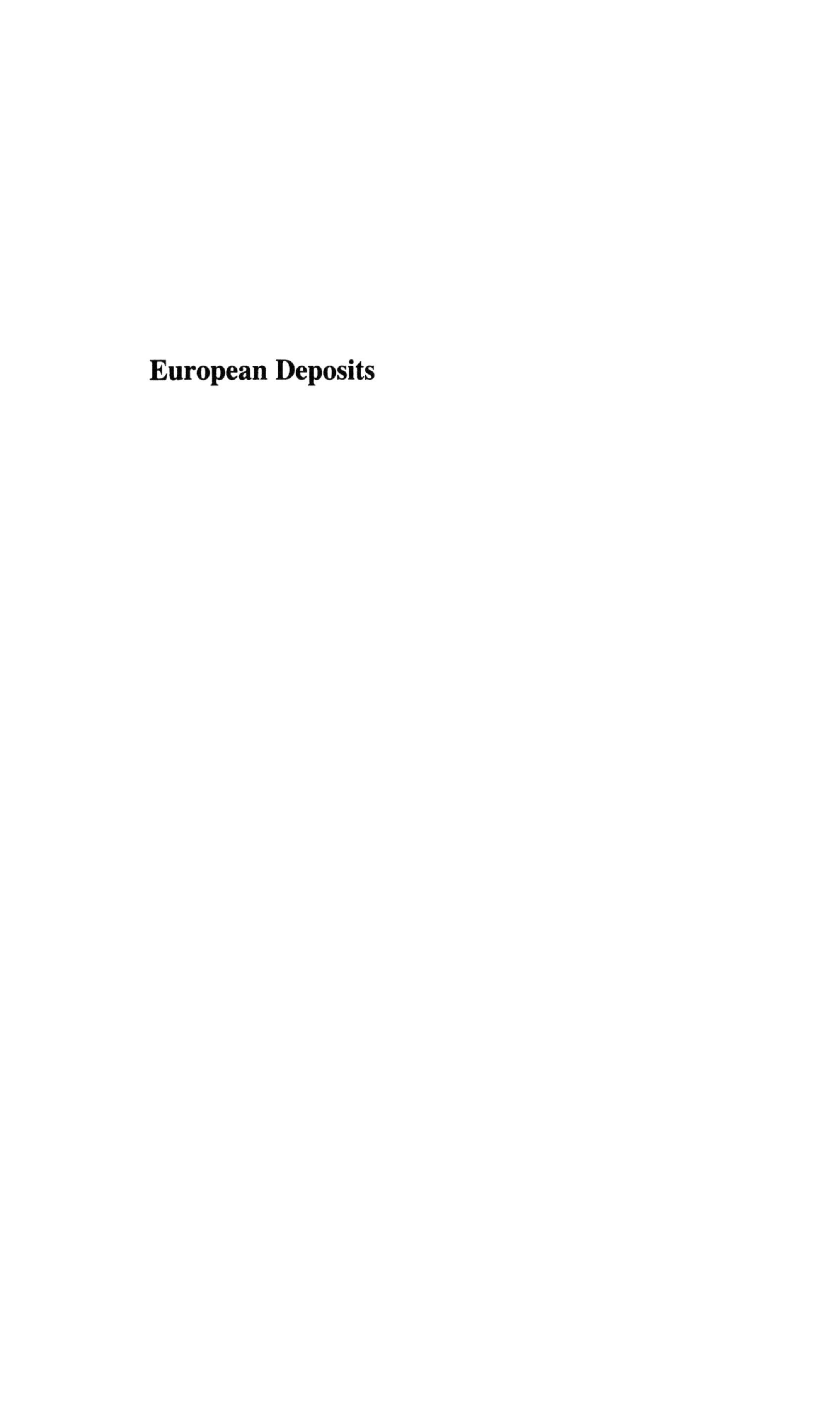

European Deposits

Trace Element Distribution of Middle-Upper Triassic Carbonate-Hosted Lead-Zinc Mineralizations: The Example of the Raibl Deposit (Eastern Alps, Italy)

L. Brigo[1] and P. Cerrato[2]

Abstract

The Raibl deposit (northern Italy) was an economically important carbonate-hosted, fissure-controlled lead-zinc concentration in the Alpine-type Triassic. Trace element analysis of 111 sulphide-rich samples, representing the whole deposit, has revealed a zonal distribution for Ge, Cd, Ga, Tl, As, Sb. The considered trace elements are linked to sphalerite (Ge-Cd-Ga association) and to specific Pb-As-(Sb)-S phases (Tl-As-Sb association). The obtained results suggest that Ge, Tl and As are prominent trace metals in the northern part of the deposit, whereas to the south Cd, and to a lesser extent Sb and Ga, increase conspicuously. The resulting zones partly coincide with distinctive geological and paragenetic features. This overlapping suggests a possible genetic connection between the ore geochemistry and the tectonic and depositional evolution of the deposit.

1 Introduction

The geochemical characterization of the Mississippi Valley-type carbonate-hosted lead-zinc ores (Sangster 1983 and 1990 and references therein) is helpful to constrain the genetic possibilities.

Several stratabound carbonate-hosted mineralizations occur at different stratigraphic positions in the Eastern and Southern Alps: (1) the Middle Triassic Alpine-type high-tonnage lead-zinc concentrations (Brigo et al. 1977), comprising the deposits of Bleiberg and Mežica in the Austroalpine Drava Range and northern Karawanke, and deposits of Salafossa and Raibl in the eastern Southern Alps; (2) the widespread Cu, Sb, Zn, Pb, fluorite and barite occurrences and some minor subeconomic deposits such as M. Avanza and Coccau-Thörl (Brigo and di Colbertaldo 1972) linked to a persistent Devono-Dinantian unconformity (Assereto et al. 1976) of the Palaeocarnic Chain (Southern Alps) and of the southern Karawanke (Fig. 1). Geochemical research on the eastern Alpine ores and host rocks (comprehensive references in Brigo and Omenetto 1985; Köppel and Schroll 1988; Cerny 1989; Schroll 1990; Omenetto and Brigo 1991; Zeeh and Bechstädt, this Vol.) has contributed

[1] Dipartimento di Scienze della Terra, Via Botticelli 23, I-20133 Milano, Italy
[2] I-33012 Cave del Predil, Udine, Italy

Spec. Publ. No. 10 Soc. Geol. Applied to Mineral Deposits
Fontboté/Boni (Eds.), Sediment-Hosted Zn-Pb Ores

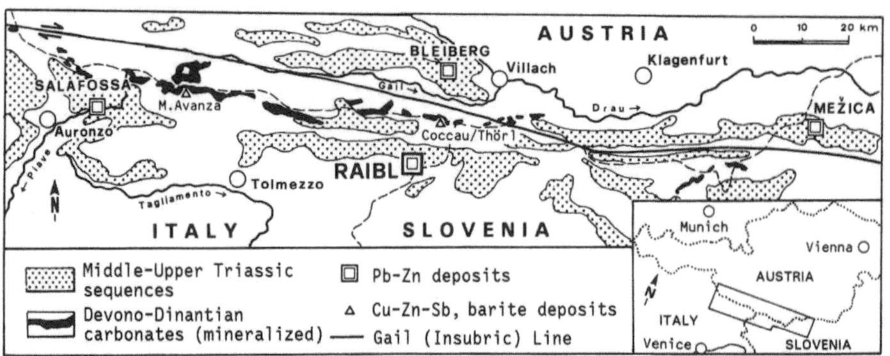

Fig. 1. Location of the major Middle-Upper Triassic Alpine-type carbonate-hosted lead-zinc deposits and the mineralized Devono-Dinantian carbonates of the Eastern and Southern Alps

to smoothen some genetic debates, showing however more differences than similarities among the individual, mineralized districts.

The dominantly fracture-controlled lead-zinc deposit of Raibl represents a particular example of the Alpine-type deposit group. This report focuses on the abundance, distribution and correlation of the trace elements Ge, Cd, Ga, Tl, As, Sb in the mineralization, with particular emphasis on their zoning pattern. Ore zoning is widely recognized (Schroll 1984; Crerar et al. 1985) to depend more on variables such as pH, Eh, degree of saturation, sulphur and oxygen fugacities, buffer capacity, amorphous to crystalline state of sulphide phases, etc. than on the simple effect of temperature.

Trace element zoning overlapping the palaeogeographic-palaeotectonic pattern is an intriguing feature that is difficult to interpret genetically.

2 Geological Setting

The lead-zinc deposit of Raibl is located near the international Italian, Austrian and Slovenian boundary, within the Middle-Upper Triassic lithotypes of the eastern part of the Southern Alps, directly south of the Neogene dextral transpressive Gail and Fella-Sava lines (Doglioni 1988a) of the Insubric fault system (Fig. 1).

On a regional scale the very strong lateral and vertical facies differentiation of the Mid-Triassic period in the Alps was first explained in terms of aborted rifting (Bechstädt et al. 1978) and, more recently, in terms of transcurrent rifting (Brandner 1984; D'Argenio and Horvath 1984; Doglioni 1986; Massari 1986; Lemoine and Trümpy 1987). In the Raibl area the mineralized Middle-Upper Triassic sequence corresponds to a system of carbonate platform (Dolomia Metallifera) and associated basinal sediments (coeval Ladinian carbonatic tuffaceous Buchenstein Formation and superimposed early Carnian highly bituminous marly limestones and black shales of the Calcare del Predil Formation). The stratigraphic and palaeogeographic setting is schematically

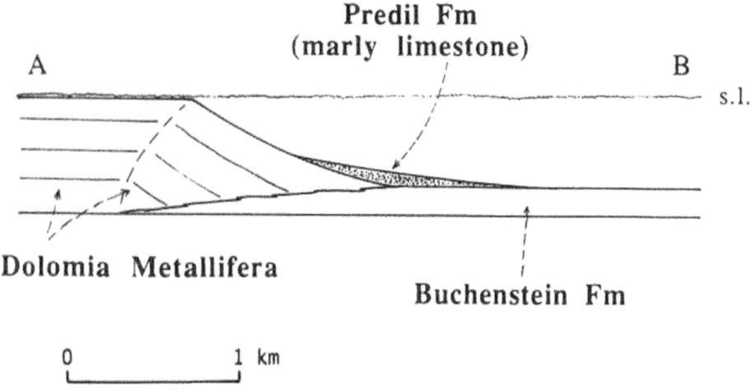

Fig. 2. Stratigraphic and palaeogeographic relationship of the main Middle-Upper Triassic platform-basin formations in the Raibl area (for section location, see Fig. 3a)

illustrated in Fig. 2. The platform comprises an interior domain (cyclic peritidal lagoonal facies) and a southward thinning marginal wedge (slope facies) developing with an overall oblique sigmoidal progradation on the Buchenstein Formation. The Calcare del Predil Formation, during deposition, partly onlapped the slope of carbonatic platform and rested on the basinal sediments of Buchenstein Formation. Lens-shaped talus breccias, solution-collapse breccias and debris flow in the Calcare del Predil (Brigo and Omenetto 1978 and unpubl. data) record emersion and erosion (with karst formation) of the Dolomia Metallifera. Possible reasons for a similar evolution of the platform-basin associations in the Triassic of the Alps are (Bechstädt et al. 1978) eustatic fluctuation of the sea level, selective bioconstruction and erosional activities by currents and synsedimentary tectonics. In the Raibl area Doglioni (1988b) differentiates, with reference to the chronostratigraphic scheme of Haq et al. (1987), into the association of Dolomia Metallifera and the Buchenstein Formation as the product of a high stand phase and into the Calcare del Predil Formation as the basal unit of the low stand wedge. Also, the role of synsedimentary tectonics has been suggested (Assereto et al. 1968) and recently confirmed (Broglio Loriga et al. 1988). Figure 3a shows the schematic palaeogeography of late Ladinian-early Carnian times in the surroundings of the Raibl area, corresponding to an embayment and a restricted basin environment as part of a WNW-ESE trending platform-basin association (Doglioni 1988a). It is conceivable that this palaeogeographic setting was disrupted by synsedimentary faults (Assereto et al. 1968) and by Late Triassic and/or Early Jurassic normal faulting (Fig. 5 in Doglioni 1988a). However, the main deformation took place much later during polyphase Tertiary tectonics: the old zone of weakness was reactivated during Palaeogene (Dinaric phase) and (Fig. 3b) Neogene (South Alpine phase) times in the form of N-S, NE-SW and NW-SE trending strike-slip faults.

The mutual dependence of the platform-basin configuration and its structural evolution provides a useful reference for the present extension of the Raibl deposit and of the associated anoxic sediments of the basal Calcare del Predil Formation, as well as for the distribution of the orebodies (mainly veins, lens-shaped talus breccias and solution-collapse breccias).

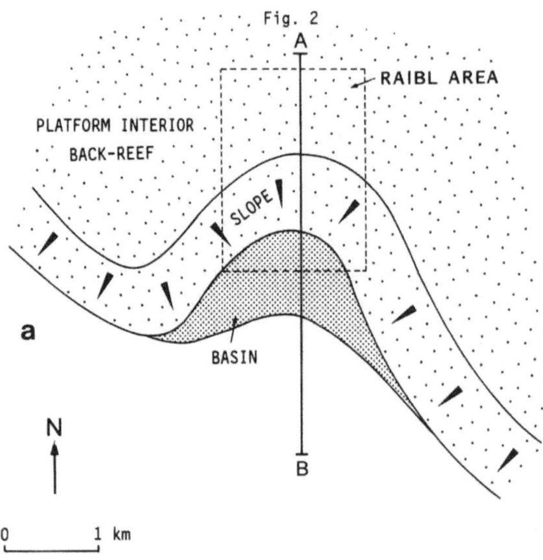

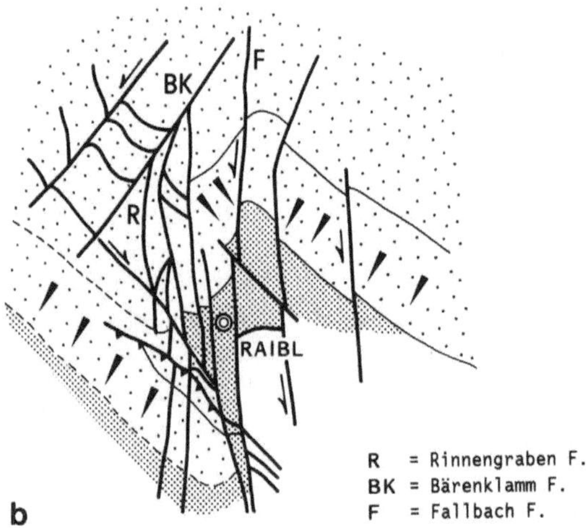

Fig. 3. Schematic palaeogeography and structural record of the Raibl area in **a** late Ladinian-early Carnian and **b** Neogene times. The faults *R*, *BK* and *F* define the ore deposit

Noteworthy is the intensive, mainly post-ore dolomitization (di Colbertaldo 1948) strongly involving the structurally confined deposit area, outside of which limestones prevail (Omenetto and Brigo 1991). The model of fracture- and porosity-controlled thermally driven convection flow of magnesium-rich fluids triggered by Triassic magmatism (Wilson et al. 1990) does not strictly apply to the situation described above, owing to the lack of post-Buchenstein volcanism.

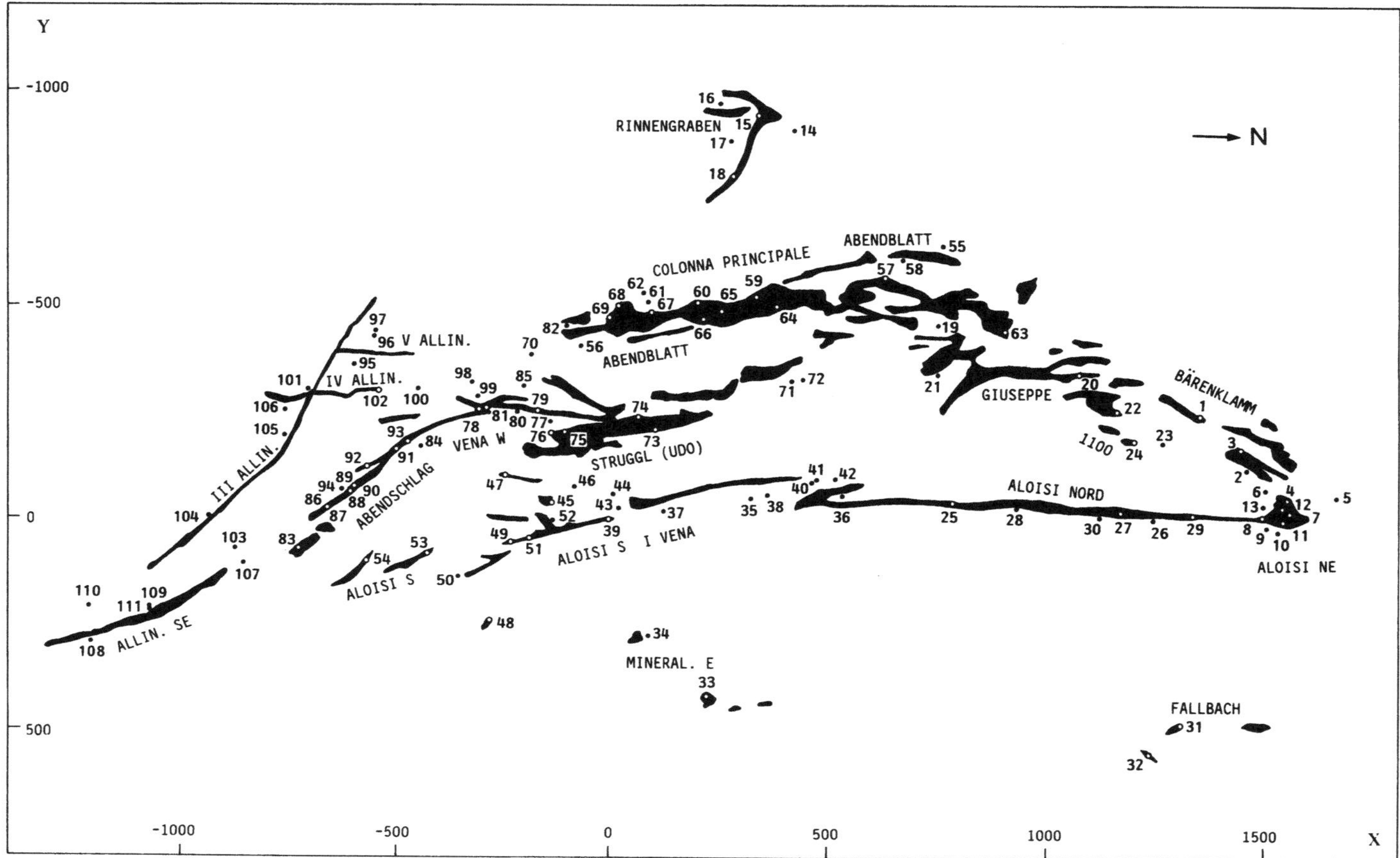

Fig. 4. Schematic mine plan of the Raibl deposit with ore-sample distribution (*1–111*)

3 Analytical Procedures

All the orebodies accessible in the underground workings were systematically sampled. From more than 150 analyzed hand-picked samples and drill cores, 111 were selected to be representative of the whole deposit. The distribution of samples (numbered approximately from north to south) is shown in a schematic mine plant (Fig. 4).

The sulphide-enriched samples selected by hand-picking consist of sphalerite, galena, pyrite-marcasite and dolomite. The analyses of the elements (Zn, Pb, Fe; Ge, Cd, Ga, Tl, As, Sb) of interest were carried out in the laboratories of the Bundesversuchs- und Forschungsanstalt – Arsenal – in Vienna (Austria). Zn, Pb and Ge were analyzed by common methods, Fe, Cd, Ga and Tl by inductively coupled plasma (ICP) and As and Sb by atomic absorption spectrometry (AAS).

In Table 1 the 111 ore samples are listed according to a general north-south trending of orebodies (labelled from A to R; see Fig. 5). For each sample coordinates, original analyses (%) of the major elements (Zn, Pb, Fe) and calculated (ppm) contents of Ge, Cd, Ga with respect to sphalerite and of Tl, As, Sb with respect to bulk sulphides (sphalerite + galena + pyrite-marcasite) are given. Original trace element data can be found in unpublished tables that may be obtained from the authors. Averages, ranges and standard deviations

Table 1. Zn, Pb, Fe (%) and trace element (ppm) contents of 111 selected massive sulphide samples of the Raibl deposit

I. n	Orebodies[a] localization	x	y	z	Zn	Pb	Fe	Ge	Cd	Ga	Tl	As	Sb
A.	Bärenklamm												
1	Sebastiani	1350	−250	142.5	26.21	3.65	4.51	1308	51	25	376	200	90
2	Giuseppe	1480	−115	63.8	24.88	1.21	6.51	2912	178	26	1124	2405	8
3	Zero	1450	−170	0	42.61	4.82	21.51	817	50	15	2895	1132	11
4	XII Clara	1550	−50	−201.5	21.77	0.78	29.57	897	62	30	1397	1319	7
5	XIII Clara	1655	50	−240.7	19.94	0.71	15.31	1373	67	33	828	1041	11
6	XIII Clara	1500	−70	−280	37.55	1.48	5.77	1261	182	18	909	2309	1
7	XV Clara	1550	−15	−321.4	18.31	0.74	23.48	2580	84	37	851	1238	4
B.	Aloisi NE												
8	Zero	1500	−10	0	47.27	1.59	5.76	874	314	14	1378	415	12
9	X Clara	1505	15	−127	42.47	7.94	7.61	1054	355	16	1708	2480	4
10	XII Clara	1530	20	−201.5	50.68	5.15	6.73	862	162	13	1449	1003	11
11	XIII Clara	1535	0	−275	25.41	1.06	5.25	1291	79	69	766	3505	3
12	XIV Clara	1535	−25	−270	32.71	0.81	12.04	1034	139	20	970	791	7
13	XV Clara	1500	−35	−321.4	50.25	1.48	6.95	706	368	13	754	608	1
C.	Rinnengraben												
14	Madonna	425	−900	207	24.35	2.09	15.41	1849	91	58	2371	2891	14
15	Sebastiani	345	−935	142	37.37	12.56	2.15	1032	700	18	699	393	9
16	Giuseppe	250	−955	63.8	27.02	27.29	9.77	1353	1013	25	629	2400	33
17	Zero	280	−880	0	37.33	13.31	0.81	850	670	18	477	1290	8
18	Giuseppe	270	−790	63.8	30.05	6.76	7.71	900	442	22	373	1495	9
D.	Giuseppe/1100												
19	Sebastiani	750	−450	142.5	48.75	7.99	0.81	385	871	14	267	174	21
20	Giuseppe	1080	−340	63.8	27.90	0.89	1.45	1325	1041	24	901	746	6
21	Zero	750	−340	0	29.31	3.71	6.94	1033	339	75	835	740	6
22	Zero	1170	−250	0.5	50.68	1.85	3.81	696	396	13	952	412	3
23	X Clara	1265	−180	−127	37.07	2.06	18.48	829	105	18	1371	775	12
24	XII Clara	1200	−180	−201.5	32.94	1.01	20.48	1171	69	20	711	2486	13

Table 1. *Continued*

I. Orebodies[a] n localization	x	y	z	Zn	Pb	Fe	Ge	Cd	Ga	Tl	As	Sb
E. Aloisi N												
25 Zero	780	−40	0	36.62	1.31	0.72	539	394	60	333	121	7
26 Zero	1240	0	0	55.42	2.99	2.98	821	320	12	1481	477	2
27 X Clara	1165	−10	−127	37.19	6.84	3.88	1387	297	49	994	2118	11
28 XII Clara	930	−20	−201.5	40.81	11.62	5.56	812	293	41	469	593	6
29 XII Clara	1530	20	−201.5	40.32	2.73	20.01	1070	110	17	2094	2789	10
30 XIII Clara	1120	−5	−240	33.13	3.65	0.55	1023	336	20	656	714	5
F. Fallbach												
31 Cinque Punte	1315	470	−4.95	38.48	4.25	3.97	842	370	17	1341	1287	23
32 X Clara	1240	550	−127	41.53	8.22	7.36	1121	504	16	1101	1787	6
G. Miner. E												
33 Luschari	220	425	55	22.06	21.44	2.76	985	1895	64	293	1675	153
34 Franc./Gius.	100	275	0	21.61	1.18	10.19	857	391	31	772	860	29
H. Aloisi S												
35 Zero	330	−25	0	40.66	2.12	0.82	488	378	39	363	241	14
36 V Lajer	535	−55	−74	42.17	1.34	3.65	592	924	16	696	1256	22
37 X Clara	135	−15	−127	38.76	4.76	0.17	481	869	17	277	569	6
38 X Clara	365	−50	−127	42.15	6.74	1.16	589	697	16	884	976	5
39 XII Clara	10	−10	−201.5	30.13	10.49	0.27	568	501	22	227	461	20
40 XII Clara	470	−80	−195	35.62	3.91	1.83	721	407	19	442	1886	3
41 XII Clara	475	−85	−207	37.81	4.93	3.61	644	568	18	609	616	36
42 XII Clara	510	−90	−207	51.09	2.51	2.72	356	386	13	452	269	3
43 XIII Clara	20	−30	−240	47.01	2.33	0.53	335	1095	14	264	275	5
44 XV Clara	10	−55	−321.4	41.88	6.06	2.96	716	495	16	521	421	3
45 XV Clara	−140	−30	−321.4	46.21	5.86	1.81	569	717	30	315	273	7
46 Rampa Pr.	−80	−70	−400	28.62	3.81	1.26	743	270	56	304	315	28
47 Rampa Pr.	−240	−110	−418	38.65	4.91	0.57	457	354	17	506	305	15
I. Aloisi S-A/B												
48 XV Clara	−285	240	−321.4	38.86	16.31	0.72	432	1884	52	240	1167	14
49 XV Clara	−250	60	−321.4	17.09	3.26	0.29	637	965	100	64	125	19
50 XVII Clara	−360	140	−400	44.56	10.53	0.22	560	1177	41	306	1489	9
51 XVII Clara	−200	40	−404	43.45	10.01	0.62	675	540	15	441	2054	5
52 XVII Clara	−130	10	−404	49.44	5.08	4.99	489	566	13	588	1351	9
53 XIX Clara	−425	80	−479	44.72	7.99	0.91	447	815	15	361	1670	2
54 XIX Clara	−570	110	−480	43.48	8.49	0.19	390	1491	45	208	193	17
J. Abendblatt												
55 Sebastiani	685	−635	142.5	38.17	12.85	1.76	677	814	17	539	417	34
56 Giuseppe	−65	−400	65.5	36.11	0.98	2.43	649	3048	48	81	1466	128
57 Giuseppe	630	−575	63.8	46.28	1.69	4.86	567	722	14	680	1813	13
58 Zero	670	−600	0	38.23	19.23	2.91	1093	365	17	707	634	6
59 Zero	340	−515	26	42.45	1.65	0.49	356	583	16	381	1782	4
60 Zero	210	−510	−18.5	17.78	10.24	0.42	1226	800	143	227	500	53
61 X Clara	−100	−445	−127	26.43	27.53	0.37	1495	820	25	270	1123	757
62 XIII Clara	80	−530	−240.7	38.82	25.88	0.81	674	916	17	188	603	35
K. Col.Principale												
63 VII Giovanni	900	−435	424	35.98	6.78	1.32	606	1240	19	594	1976	16
64 Sebastiani	375	−500	142.5	22.17	5.76	0.44	693	944	76	409	172	103
65 Giuseppe	260	−485	63.8	19.94	41.44	0.57	394	2163	34	134	363	406
66 Zero	220	−480	0	27.58	8.88	0.16	924	1387	68	356	211	6
67 V Lajer	90	−480	−59	55.86	0.44	0.06	279	2383	12	179	931	62
68 X Clara	25	−500	−127	34.09	1.57	0.47	403	586	20	227	1577	13
69 X Clara	−10	−250	−127	36.14	4.08	0.75	596	579	41	216	685	15
70 XIII Clara	−175	−380	−240.7	40.15	15.41	0.27	207	2826	17	93	148	210
L. Struggl-Udo												
71 IV Giovanni	410	−315	318	15.23	9.24	6.91	912	1101	158	1318	1201	133
72 IV Giovanni	430	−320	318	22.72	1.21	3.13	1689	198	133	2568	576	5
73 Giuseppe	90	−215	70	35.81	12.42	5.46	1169	270	19	951	1078	49
74 Giuseppe	75	−235	87	26.05	26.17	1.51	479	2313	26	136	186	1
75 XII Clara	−105	−205	−201.5	27.53	20.89	1.02	526	1701	73	128	407	238
76 XIII Clara	−140	−205	−240	24.02	15.34	2.27	765	3343	61	105	218	183
77 XV Clara	−135	−215	−321.4	48.01	0.81	0.44	312	2266	14	31	64	41
78 XVII Clara	−300	−255	−408.5	29.08	10.98	0.52	1022	1472	65	226	1015	93
79 XVII Clara	−180	−250	−408.5	43.78	1.42	1.75	386	935	46	225	1162	18

Table 1. *Continued*

I. Orebodies n localization	x	y	z	Zn	Pb	Fe	Ge	Cd	Ga	Tl	As	Sb
80 XIX Clara	−215	−250	−480	27.15	4.01	2.21	420	2157	59	52	799	30
81 XIX Clara	−280	−280	−480	23.45	36.54	1.28	750	1462	29	156	754	138
M. Abendschlag												
82 X Clara	−100	−455	−127	16.17	47.12	0.62	1149	2589	41	69	566	401
83 XII Clara	−720	−70	−201.5	31.47	10.77	1.55	507	2750	68	49	465	227
84 XII Clara	−430	−310	−201.5	30.38	22.95	0.42	795	1905	22	134	507	289
85 XV Clara	−200	−25	−345	46.53	11.71	0.17	497	2082	14	280	1303	26
86 XVII Clara	−655	−25	−408	26.84	8.25	0.26	630	2592	25	194	673	26
87 XVII Clara	−655	−70	−408	32.84	14.21	0.24	721	1536	20	354	1272	70
88 XVII Clara	−610	−70	−412	40.89	14.54	0.32	372	3496	16	97	666	134
89 XVII Clara	−600	−80	−408	50.82	2.36	0.11	334	1686	13	149	623	105
90 XVII Clara	−600	−80	−408	25.71	23.48	0.41	961	1410	55	148	876	29
91 XVII Clara	−495	−170	−415	23.17	13.61	0.67	550	2221	90	109	264	130
92 Rampa Pr.	−565	−120	−450.7	28.06	0.94	0.08	1078	995	98	392	318	23
93 XIX Clara	−465	−190	−480	35.82	6.81	0.39	388	1223	52	206	313	47
94 XIX Clara	−620	−70	−480	40.93	8.22	0.21	256	4367	46	175	220	173
N. V Allineam.												
95 Sond.1782	−595	−355	−474	12.42	9.99	0.61	924	1221	211	377	160	77
96 Sond.1761	−550	−425	−570	5.31	0.09	0.12	304	253	126	242	181	72
97 Sond.1760	−550	−435	−620	6.99	0.32	0.21	143	384	96	178	205	27
O. Vena W												
98 XV Clara	−315	−320	−353	32.16	8.45	0.23	223	4168	21	35	172	71
99 XIX Clara	−300	−285	−480	35.13	7.75	0.81	550	707	65	326	814	32
100 XIX Clara	−450	−300	−480	33.46	21.93	1.03	459	2931	20	176	268	233
P. IV Allineam.												
101 Rampa SW	−770	−300	−472.5	42.79	0.58	0.18	480	956	30	158	720	13
102 Rampa SW	−525	−305	−475	20.82	1.11	0.52	670	2723	132	203	188	87
Q. III Allineam.												
103 XIII Clara	−870	75	−242.7	31.12	5.92	0.42	673	2468	75	109	373	325
104 XVII Clara	−920	−5	−400	18.09	7.13	0.82	634	1257	141	56	245	103
105 Sond.1526	−755	−190	−400	13.21	1.13	0.99	498	737	153	87	416	35
106 Sond.1531	−755	−255	−440	11.95	0.15	0.17	56	1325	197	109	109	33
R. Allineam. SE												
107 XVII Clara	−850	110	−400	16.25	17.18	2.51	863	2209	103	161	1341	245
108 Sond.1542	−1205	300	−400	14.11	1.28	0.48	580	1256	143	136	217	68
109 Sond.1550	−1065	215	−470	11.19	9.55	0.97	210	1211	216	101	309	80
110 Sond.1551	−1205	215	−500	10.52	12.27	0.37	1142	1582	249	304	98	140
111 Sond.1543	−1065	220	−530	10.83	12.23	0.38	1115	991	142	64	348	110

[a] Abbreviations: I. labels for orebodies; n sample number; x,y,z sample coordinates.

of the calculated trace element contents of the individual orebodies are presented in Table 2.

For statistical computation (not discussed in detail in this report) a fixed value of 10 ppm was assigned to the samples with Ga contents (about 50% of samples) and Ge contents (two samples) below the detection limits (20 ppm).

4 Trace Element Associations and Distribution Patterns

Mineralogical and chemical data on sulphide phases (sphalerite, galena, pyrite-marcasite, unpubl. data of the Società Mineraria del Predil S.p.A.; Schroll 1981, 1983; Pimminger et al. 1985a,b), detailed microscope studies and statistical computation of trace elements (present authors) are best discussed by

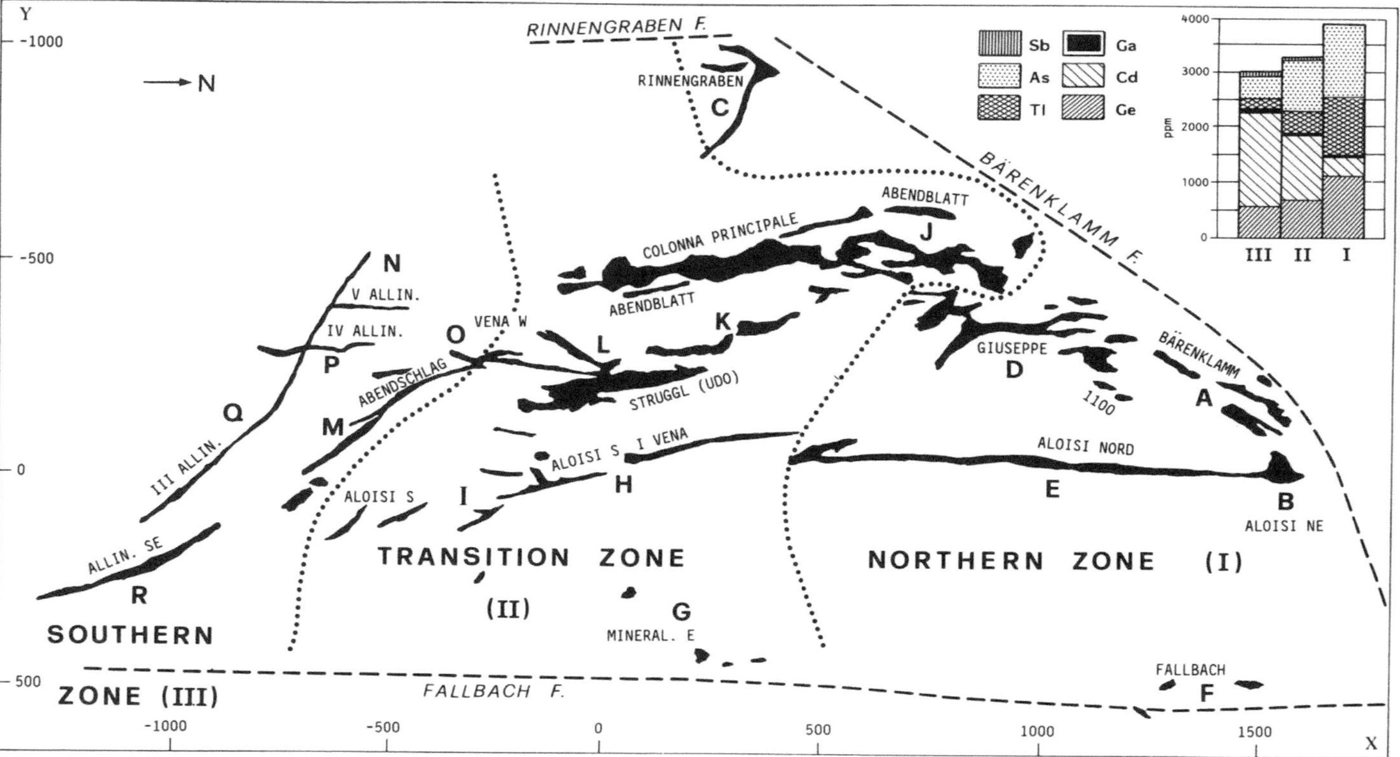

Fig. 5. Schematic mine plan of the Raibl deposit and distribution of the orebodies (labelled from *A* to *R*) and of the faults (*broken lines*) confining the deposit. The histogram emphasizes, from a geochemical point of view, three well- defined zones (*I-II-III* marked by *dotted lines*)

Table 2. Averages (m), ranges and standard deviations (sd) of the trace element contents of the individual orebodies, labelled from A to R (see Fig. 5)

I	Orebodies (n)[a]	Ge			Cd			Ga			Tl			As			Sb		
		m	range	sd	m	range	sd	m	range	sd	m	range	sd	m	range	sd	m	range	sd
A	Bärenklamm (7)	1593	817–2912	760	96	50–182	54	26	15–37	7	1197	376–2895	750	1378	200–2405	708	19	1–90	29
B	Aloisi NE (6)	970	706–1291	185	236	79–368	113	24	13–69	20	1171	754–1708	362	1467	415–3505	1133	6	1–12	4
C	Rinnengraben (5)	1197	850–1849	370	583	91–1013	306	28	18–58	15	910	373–2371	739	1694	393–2891	875	15	8–33	9
D	Giuseppe/1100 (6)	907	385–1325	312	470	69–1041	366	27	13–75	22	839	267–1371	327	889	174–2486	747	10	3–21	6
E	Aloisi N (6)	942	539–1387	263	292	110–394	88	33	12–60	18	1005	333–2094	162	1135	121–2789	969	7	2–11	3
F	Fallbach (2)	982	842–1121	437	437	370–504	67	17	16–17	1	1221	1101–1341	120	1537	1287–1787	250	14	6–23	9
G	Mineral. E (2)	921	857–985	64	1143	391–1895	752	48	31–64	17	532	293–772	240	1268	860–1675	408	91	29–153	62
H	Aloisi S (13)	558	335–743	126	724	378–924	245	23	13–56	12	451	227–884	185	605	241–1886	472	13	3–36	10
I	Aloisi S-A/B (7)	519	390–675	100	1063	540–1884	457	40	13–100	29	315	64–588	157	1150	125–2054	677	11	2–19	6
J	Abendblatt (8)	842	356–1495	361	1009	365–3048	788	37	14–143	41	384	88–707	219	1042	417–1813	545	129	4–757	240
K	C.na Principale (8)	513	207–924	221	1514	579–2826	794	36	12–76	23	276	93–594	156	758	148–1976	648	104	6–406	131
L	Struggl-Udo (11)	766	312–1689	394	1565	198–3343	892	62	14–158	44	536	31–2568	754	678	64–1201	395	84	1–238	75
M	Abendschlag (13)	634	256–1149	278	2219	995–4367	912	43	13–98	28	181	49–392	101	620	220–1303	337	129	23–401	112
N	V Allineamento (3)	457	143–924	337	619	253–1221	429	144	96–211	49	266	178–377	83	182	160–205	18	59	27–77	22
O	Vena W (3)	411	223–550	138	2602	707–4168	1432	35	20–65	21	179	35–326	119	418	172–814	283	112	32–233	87
P	IV Allineamento (2)	575	480–670	119	1840	956–2723	884	81	30–132	51	180	158–203	23	454	188–720	266	50	13–87	37
Q	III Allineamento (4)	465	56–673	245	1447	737–2468	632	142	75–197	44	90	56–109	22	286	109–416	120	124	33–325	119
R	Allineamento SE (5)	782	210–1142	351	1450	991–2209	424	171	103–249	54	153	64–304	82	463	98–1341	448	129	68–245	63

[a] Abbreviations: I labels for orebodies; (n) number of samples; m average; sd standard deviation.

distinguishing the following two groups of elements: the Ge-Cd-Ga and Tl-As-Sb groups.

4.1 Ge-Cd-Ga Group

This element group is related to the macro- and microscopically different sphalerite phases of the Raibl deposit. The single sulphide-phase analyses, carried out in 1960 by the Società Mineraria del Predil for an economic reassessment of lead and zinc ores, confirmed the total lack of Ge and Cd in galena and pyrite-marcasite. Despite detailed microscope studies, no Ge-, Cd- or Ga-bearing phases were found. Johan et al. (1983) described several Ge-rich minerals present as microscopic inclusions in sphalerites of a sulphide deposit from the Pyrenees, France. The existence of Ge microphases in sphalerite is also thought to be possible by Moh and Jager (1978). Pimminger et al. (1985a) excluded the presence of such Ge microphases in sphalerite of the Raibl deposit, but consider conceivable Ge-Tl-S phases due to the high positive correlation between Ge and Tl.

Regarding the mechanism of Ge substitution in sphalerite, Bernstein (1985) considers the substitution for both Zn and S, proposed by Malevskiy (1966), as reasonable, since sphalerite and Ge metal have the same lattice (diamond-type). This mechanism allows high Ge concentrations. Referring to the Ge-rich sphalerite of the Saint-Salvy deposit (Montagne Noire, France), Johan (1988) suggested a coupled substitution mechanism involving essentially the monovalent ions Cu and Ag, which are lacking in the Raibl sphalerites (Pimminger et al. 1985b) but which, on the other hand, are essentially of collomorphic nature and reach up to 3000 ppm of germanium. Johan (1988) concluded, however, that different substitution mechanisms may occur simultaneously in natural sphalerites. Similar mechanisms may exist (Wedepohl 1978) with regard to the high Cd concentrations in sphalerites (up to more than 4000 ppm in the Raibl deposit), in agreement with the usually pronounced negative correlation to Ge.

In the Raibl deposit the distribution of Ge, Cd and Ga (Figs. 5, 6a) shows (1) a regular decrease in the total concentration from south to north; (2) a characteristic distribution of these elements throughout the orebody complex, specifically an increase in Ge and a decrease in Cd and Ga from south to north; and (3) a strong antithetic distribution of Ge and Cd, the latter positively correlated to Ga (Table 3).

4.2 Tl-As-Sb Group

The mode of occurrence of Tl in sulphides might be explained by the presence of specific thallium mineral phases (Sobott et al. 1987) or Tl-bearing sulphosalts, i.e. the Pb-As-Sb-S phases, geocronite and jordanite, the latter with up to more than 1.5% Tl (Wedepohl 1974).

The high Tl contents of some Raibl ores were related to colloform sphalerites according to Hegemann (1960) and Dessau (1967), whereas Schroll

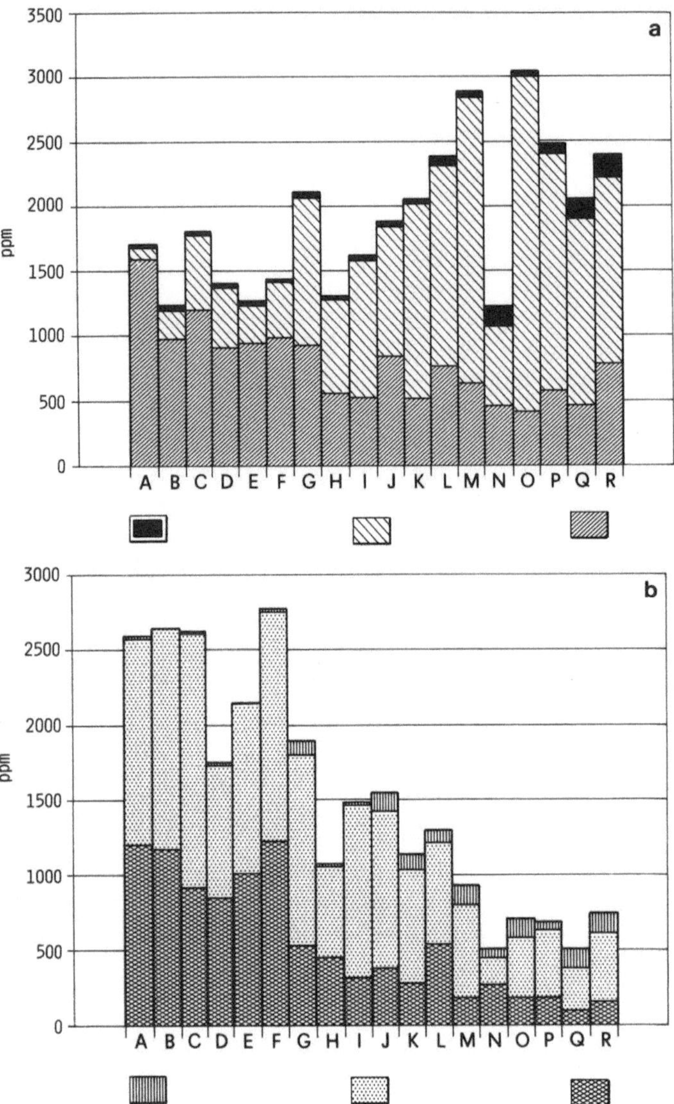

Fig. 6. Histograms defining the N-S trending of **a** the average Ge, Cd, Ga contents with respect to sphalerite and **b** the average Tl, As, Sb contents with respect to bulk sulphides (sphalerite + galena + pyrite-marcasite). For explanation of *A* to *R*, see Fig. 5

(1981, 1983) considered the special enrichments of Tl, As, Sb of the Alpine-type deposits as compounds of possible complex microphases. Later investigations with electron probe analysis on Bleiberg mineralizations (Pimminger et al. 1985a,b) confirmed such microphases, which could be found as presumably Tl-bearing jordanite inclusions in dolomite and sphalerite. Since 1966 (Venerandi) jordanite has been recognized in the Raibl deposit. Recent detailed microscope studies (unpubl.) confirmed the distribution of this Pb-As-S

Table 3. Correlation matrix of trace elements of the 111 ore samples (Table 1)

	Ge	Cd	Ga	Tl	As	Sb
Ge	1.000					
Cd	−0.3988	1.000				
Ga	−0.0128	0.1092	1.000			
Tl	0.4959	−0.5495	−0.1911	1.000		
As	0.4475	−0.3280	−0.3031	0.4480	1.000	
Sb	−0.0037	0.4754	0.1308	−0.3142	−0.1856	1.000

Significance probabilities (n = 111): 99% confidence r > 0.249
95% confidence r > 0.191

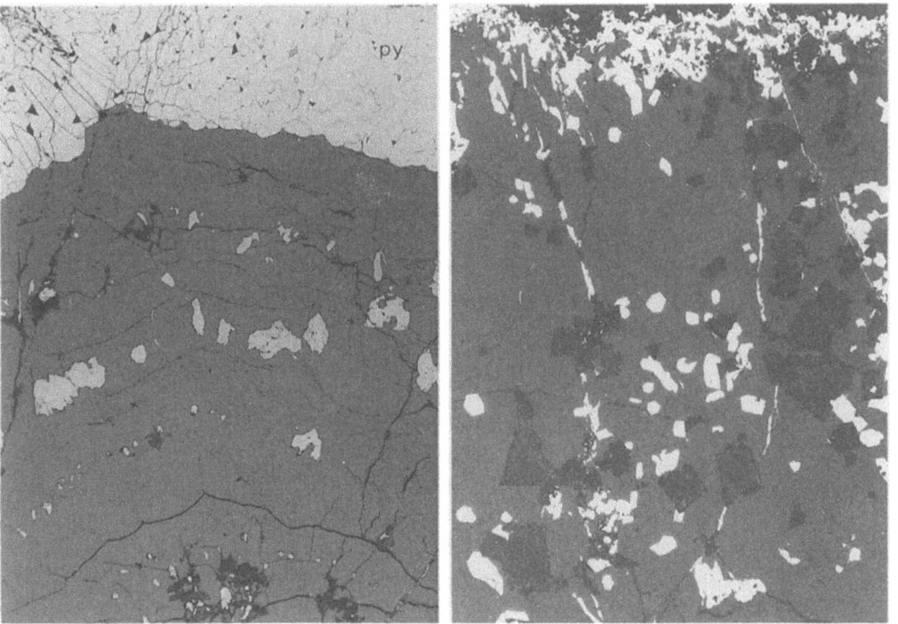

Fig. 7a,b. Aloisi NE – XIII Clara, Raibl. Polished section, 160× **a** Grape-like patches (aggregates of subparallel crystals) of jordanite (*white*) in and between the concentric bands of botryoidal sphalerite (*grey*). At the *top* an overgrowth of pyrite-marcasite (*py*). **b** Euhedral jordanite (*white*) in sphalerite (*grey*). *Black* is dolomite, at the *top* and as crystals replacing sphalerite

phase as small patches and fine veinlets particularly widespread in the ore paragenesis of the northern marginal zone of the deposit (Fig. 7a,b).

As far as the distribution of these trace elements in the deposit is concerned (Figs. 5, 6b), the following points are noteworthy: (1) the significant decrease southwards in the total concentration of Tl, As and Sb; (2) the substantial decrease southward in Tl and As and increase in the low Sb concentrations; (3) the high positive correlation between As and Tl, both being antithetic to Sb (Table 3).

5 Zoning Features of the Deposit

In the structurally confined area of the Raibl deposit (Figs. 3, 5) a strict relationship between geological evidence (Broglio Loriga et al. 1988; Doglioni 1988a; Omenetto and Brigo 1991; Omenetto and Brigo, in prep.), the sulphide ore paragenetic association (di Colbertaldo 1948; Brigo and Omenetto 1978) and trace element geochemistry could be established. The statistical interpretation of the varying distribution patterns of trace elements, based on the speculative grouping of the orebodies, allowed one to distinguish a lateral zoning, lacking any vertical variation. In Fig. 5 three zones are outlined and characterized by their global trace element amounts that are also listed in Table 4 together with other meaningful coefficients.

The main geological, paragenetic and chemical features of the three zones are described below.

(I) Northern Marginal Zone (Orebodies A to F). In this zone, north of the platform-basin margin, the main ore traps are (1) the talus breccias or debris-flow accumulations (for example, the orebodies Bärenklamm, Aloisi NE, Fallbach) which are related to an emersion surface and arranged near the N-S striking Rinnengraben and Fallbach faults (the western and the eastern borders of the deposit, respectively) and near the NE-SW striking arched Bärenklamm fault (the northwestern border of the deposit), and (2) some fracture-controlled veins, the most important being the N-S striking Aloisi N vein, which intersects nearly the whole thickness of the Dolomia Metallifera.

The ore paragenesis is dominated by botryoidal, red-brownish sphalerite and pyrite-marcasite, rare galena and Tl-rich jordanite. Bituminous dolomitic sediments with disseminated microgranular sphalerite, globular pyrite and some barite compose the matrix of the ore breccias.

This zone (Table 4) shows the highest average concentration of all the analyzed trace elements (3896 ppm), is characterized by the lowest average amount of Ge-Cd-Ga association (1447 ppm) and the highest average amount of Tl-As-Sb association (2419 ppm). Individually, the high average contents of Ge, As, Tl contrast with the lowest average contents of Cd, Sb and Ga. This zone also shows the highest Zn/Pb metal ratios with a mean value of about 10 and the lowest average metal grade of 4 wt% (Zn + Pb).

Concerning the ore-forming conditions, Ge-rich sphalerites are considered to be deposited from low-temperature fluids having a low to moderate sulphur activity (Bernstein 1985) and the Tl-enrichments (associated with As and Sb) are also indicators for low-temperature and low-pH environments which govern the deposition of botryoidal intergrowths of sphalerite, pyrite-marcasite and rare galena (Sobott et al. 1987).

Moreover, it is interesting to mention that the Ga/Ge and Ga/Zn ratios in the sphalerites of Raibl (Ga/Ge: −1.6 to −0.7; Ga/Zn: −2.4 to −4.1) are very similar to those measured by Möller et al. (1983) in the sphalerites of the Bleiberg mine (Ga/Ge: −1.5 to −0.5; Ga/Zn: −4.5 to −5.0).

(II) Transitional Zone (Orebodies G to L). This zone pertains to the platform domain underlying the northern deposits of the Calcare del Predil Formation,

Table 4. Average values of trace elements and some significant coefficients of zones I, II and III of the Raibl deposit[a]

Zones	I (n)	Ge	Cd	Ga	Tl	As	Sb	Σ TE	Ge + Cd + Ga	Tl + As + Sb	Zn/Pb	Zn + Pb
(I) Northern zone:	A..F (32)	1099	352	26	1057	1350	12	3896	1477	2419	10	4
(II) Transition zone:	G..L (49)	687	1169	41	416	917	72	3302	1897	1405	5	5
(III) Southern zone:	M..R (30)	554	1696	103	175	404	101	3031	2357	669	3	6

[a] Abbreviations: I labels for orebodies; (n) number of samples; TE = trace elements; Zn/Pb and Zn + Pb values derived from beneficiation data provided by SIMS.p.A.

the basal unit of the onlapping basinal wedge. Variation in thickness or lack of these basal sediments, relevant to morphological highs and lows of the platform and conceivably originated by synsedimentary tectonics, could be confirmed (Broglio Loriga et al. 1988). Ore traps are represented (Fig. 5) by great, lens-shaped solution-collapse breccias (Colonna Principale, Struggl-Udo) and by some N-S and NNW-SSE striking veins (Abendblatt, Aloisi S-I vena, Aloisi S-A/B).

The significant variation in ore paragenesis is due to the diminution of pyrite-marcasite, the appearance of macrobotryoidal yellow sphalerite, the increasing amounts of galena and the local concentrations of a white-grey microcrystalline sphalerite at the contact of platform carbonate and bituminous shales and dolostones (Calcare del Predil Formation). The composition of the rarer Tl-bearing Pb-As-(Sb)-S phase should also be varied, because of the sharp Sb increment of about 80%.

The total average concentration of all the trace elements (3302 ppm) decreases by 15%, while the total average amount of the Ge-Cd-Ga association (1897 ppm) increases by 28% and the total average amount of the Tl-As-Sb association (1405 ppm) decreases by 43%. The Zn/Pb metal ratio also decreases abruptly to a mean value of about 5, while the average metal grade increases to 5% (Table 4). In particular, from zone I to II both a sharp decrease in Ge, Tl and As and increase in Cd, Sb, Ga took place (Figs. 5, 6).

The logarithmic histogram of the Ge/Cd ratios of the individual orebodies (Fig. 8) displays sharply antithetic values emphasizing the southward transition to depositional environments which may be consistent with an increased oxidation potential (pH-Eh diagrams in Brookins 1988).

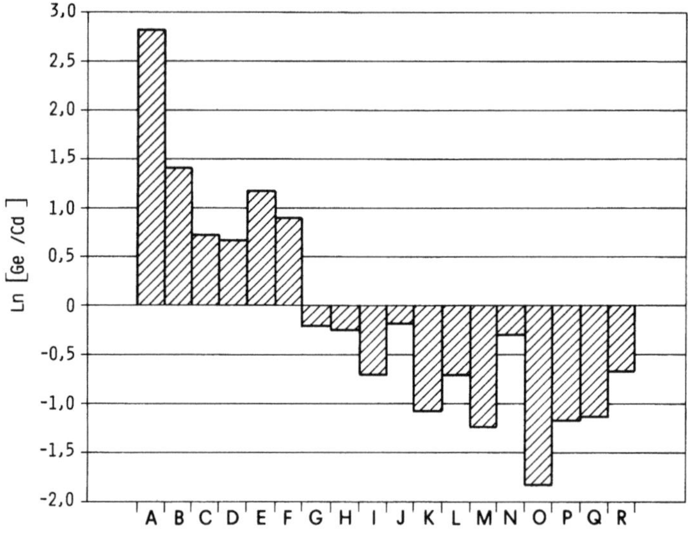

Fig. 8. Logarithmic histogram of the Ge/Cd ratio. The abrupt variation in the values from the northern (from *A* to *F*) to the southern zone (from *G* to *R*) is apparent; see also Fig. 5

(III) Southern Zone (Orebodies M to R). The present, small, southernmost zone of the mine affects the terminal, wedge-shaped portion of carbonate platform, closing at the down-lap margin to the Buchenstein Formation. The exclusively fracture-controlled orebodies are NW-SE striking veins (Abendschlag, Allineamento SE, III Allineamento) prevailing over N-S ones (Vena W, IV and V Allineamento).

Ore paragenesis consists mainly of coarse-grained yellow sphalerite and significant amounts of galena. White-grey microcrystalline sphalerite occurs frequently in the same manner as described for zone II. In spite of the very low Tl, As and the higher Sb contents, the element proportions suggest the presence of rare-Tl-bearing geocronite, which exists also (Lagny et al. 1973) in the coeval Salafossa deposit (Fig. 1).

A gradual decrease, at a rate of about 8%, in the total average concentration of trace elements (3031 ppm) characterizes this zone (Fig. 5, Table 4). The slight variation agrees with a rate of increase of 20% for the Ge-Cd-Ga association (2357 ppm). Individually, Ge, Tl and As decrease slightly, while an increase in Cd, Sb, Ga is apparent. The Zn/Pb metal ratio is represented by a further reduced mean value of about 3, the average metal content by further upgrading to 6%.

The strongly reduced Tl and As concentrations and the highest Cd concentration in sphalerite with a lower but nearly constant amount of Ge should indicate an increased pH environment (Sobott et al. 1987; Brookins 1988) in this sector of the deposit.

6 Concluding Remarks

This chapter contributes to the knowledge of the ore geochemistry in the Raibl carbonate-hosted lead-zinc deposit (eastern Southern Alps, Italy). Analysis of Ge, Cd, Ga, Tl, As, Sb on a statistically reliable population of Zn-Pb sulphide-rich samples has revealed a clearly zoned pattern (Figs. 5, 7) with sharply antithetic behaviour of Ge, Tl, As, enriched in the northern zone, in contrast to the Cd (Sb, Ga) contents, which show a conspicuous increase in the southern sector.

This zoning is tentatively explained in terms of concomitant variations in pH, Eh, degree of saturation, and the sulphur and oxygen fugacities during ore deposition, taking also into account the role of crystallization/recrystallization processes overprinting the primary colloform depositional fabrics of the sulphide phases.

The observed coincidence between the resulting geochemical zones and parts of the deposit having a distinctive palaeogeographic signature (passing from basin to platform throughout the slope facies, from south to north) and bearing different types of orebodies is likely not accidental and probably has a precise genetic significance. However, a number of connecting elements must still be investigated and elucidated before this factual correspondence at the deposit scale can be fit into a coherent genetic model. In particular, the sequence of tectonic phases active in the Raibl site from the Mid-Triassic to

Neogene epoch needs to be more carefully differentiated with respect to the timing and type of mineralizing processes.

Acknowledgements. We are very grateful to P. Omenetto for constructive discussions and important contributions. Thanks are also extended to M. Boni, A. Ferrario, L. Fontboté and F. Rodeghiero for thoughtful comments. The paper was significantly improved by consideration of the constructive criticism of the referees, for which we are very grateful. We appreciate the support received on many occasions by the mining company SIM S.p.A. and especially by G. Cerrato. Thanks also to E. Schroll who provided the trace element analyses carried out in the laboratories of the Bundesversuchs- und Forschungsanstalt – Arsenal – (Vienna, Austria). This research was supported by grants from Italian M.P.I. (40%).

References

Assereto R, Desio A, di Colbertaldo D, Passeri LD (1968) Note Illustrative della Carta Geologica d'Italia alla scala 1:100000, foglio Tarvisio. Serv Geol d'Ital, Poligrafica Ercolano Napoli, pp 1–70

Assereto R, Brigo L, Brusca C, Omenetto P, Zuffardi P (1976) Italian ore/mineral deposits related to emersion surfaces: a summary. Mineral Depos 11:170–179

Bechstädt T, Brandner R, Mostler H, Schmidt K (1978) Aborted rifting in the Triassic of the Eastern and Southern Alps. Neues Jahrb Geol Paläontol Abh 156:157–178

Bernstein LR (1985) Germanium geochemistry and mineralogy. Geochim Cosmochim Acta 49:2409–2422

Brandner R (1984) Meeresspiegelschwankungen und Tektonik in der Trias der NW-Tethys. Jahrb Geol Bundesanst 126:435–475

Brigo L, di Colbertaldo D (1972) Un nuovo orizzonte metallifero nel Paleozoico delle Alpi Orientali. In: Proc II Int Symp Min Dep Alps, Ljubljana, pp 109–124

Brigo L, Omenetto P (1978) The lead and zinc ores of the Raibl (Cave del Predil – northern Italy) zone: new metallogenic data. In: Zapfe H (ed) Scientific results of the Austrian projects of IGCP until 1976. Springer, Wien New York, pp 103–110

Brigo L, Omenetto P (1985) Lithogeochemical observations on some ore-bearing Triassic sequences of the Italian Southern Alps. Monograph Series on Mineral Deposits 25, Borntraeger, Berlin, pp 95–104

Brigo L, Kostelka L, Omenetto P, Schneider HJ, Schroll E, Schulz O, Štrucl I (1977) Comparative reflection on four alpine Pb-Zn deposits. In: Klemm DD, Schneider HJ (eds) Time and strata-bound ore deposits. Springer, Berlin Heidelberg New York, pp 273–293

Broglio Loriga C, Doglioni C, Neri C (1988) Studi stratigrafici e strutturali nell'area mineraria di Raibl. Società Italiana Miniere S.p.A., priv rep, 47 pp

Brookins DG (1988) Eh-pH diagrams for geochemistry. Springer, Berlin Heidelberg New York, 176 pp

Cerny I (1989) Die karbonatgebundenen Blei-Zink-Lagerstätten des alpinen und ausseralpinen Mesozoikums. Die Bedeutung ihrer Geologie, Stratigraphie und Faziesgebundenheit für Prospektion und Bewertung. Arch Lagerstättenforsch Geol Bundesanst Wien 11:5–125

Colbertaldo D di (1948) Il giacimento piombo-zincifero di Raibl in Friuli (Italia). 18th Int Geol Congr, London. Industrie Poligrafiche Longo & Zoppelli, Treviso, 149 pp

Crerar D, Wood S, Brantley S, Bocarsly A (1985) Chemical controls on solubility of ore-forming minerals in hydrothermal solutions. Can Mineral 23:333–352

D'Argenio B, Horvath F (1984) Some remarks on the deformation history of Adria, from the Mesozoic to the Tertiary. Ann Geophys 2:143–146

Dessau G (1967) Gli elementi minori nelle blende e nelle galene della miniera di Salafossa (S. Pietro e S. Stefano di Cadore, Alpi Orientali italiane). Confronti con i giacimenti del Bergamasco e di Raibl. Atti G St Geomin Agordo: 123–134

Doglioni C (1986) Tectonics of the Dolomites (Southern Alps, northern Italy). J Struct Geol 10:1–13

Doglioni C (1988a) Examples of strike-slip tectonics on platform-basin margins. Tectonophysics 156:293–302

Doglioni C (1988b) Note sull'evoluzione strutturale della zona di Raibl. In: Broglio Loriga C, Doglioni C, Neri C (eds) Studi stratigrafici e strutturali nell'area mineraria di Raibl. Società Italiana Miniere S.p.A., priv rep, 47 pp

Haq BL, Hardenbol J, Vail P (1987) Chronology of fluctuating sea levels since the Triassic. Science 235:1156–1167

Hegemann F (1960) Über extrusiv-sedimentäre Erzlagerstätten der Ostalpen, II Teil: Blei-Zink Lagerstätten. Erzmetall 12:122–127

Johan Z (1988) Indium and germanium in the structure of sphalerite: an example of coupled substitution with copper. Mineral Petrol 39:211–229

Johan Z, Oudin E, Picot P (1983) Analogues germanifères et gallifères des silicates et oxides dans les gisements de zinc des Pyrénées centrales, France: argutite et carboirite, deux nouvelles espèces minerales. Tschermaks Min Petrol Mitt 31:97–119

Köppel V, Schroll E (1988) Pb-isotope evidence for the origin of lead in strata-bound Pb-Zn deposits in Triassic carbonates of the Eastern and Southern Alps. Mineral Depos 23:96–103

Lagny P, Omenetto P, Ottemann J (1973) Géocronite dans le gisement plombo-zincifère de Salafossa (Alpes orientales italiennes). Neues Jahrb Min Monatsh 12:529–546

Lemoine M, Trümpy (1987) Pre-oceanic rifting in the Alps. Tectonophysics 133:305–320

Malevskiy AY (1966) Form of germanium in sphalerite. Earth Sci Sect 167:81–83

Massari F (1986) Some thoughts on the Permo-Triassic evolution of the South-Alpine area (Italy). Mem Soc Geol Ital 24:179–188

Moh GH, Jager A (1978) Phasengleichgewichte des Systems Ge-Pb-Zn-S in Relation zu Germaniumgehalten alpiner Pb-Zn-Lagerstätten. Verh Geol Bundesanst 3:437–440

Möller P, Dulski P, Schneider H-J (1983) Interpretation of Ga and Ge content in sphalerite from the Triassic Pb-Zn deposits of the Alps. In: Schneider H-J (ed) Mineral deposits of the Alps and of the Alpine epoch in Europe. Springer, Berlin Heidelberg New York, pp 213–222

Omenetto P, Brigo L (1991) Carbonate-hosted lead-zinc deposits in the framework of the Triassic Dinaric-East Alpine province. In: Verbano (ed) Zuffar' days, Symp Cagliari, Oct 10–15, 1988. Germignaga Varese, pp 61–71

Pimminger M, Grasserbauer M, Schroll E, Cerny I (1985a) Anwendung der Ionenstrahlmikroanalyse zur geochemischen Charakterisierung von Zinkblenden. Arch Lagerstättenforsch Geol Bundesanst Wien 6:209–214

Pimminger M, Grasserbauer M, Schroll E, Cerny I (1985b) Trace element distribution in sphalerites from Pb-Zn-ore occurrences of the Eastern Alps. Tschermaks Min Petrol Mitt 34:131–141

Sangster DF (1983) Mississippi Valley-type deposits: a geological mélange. In: Kirsvarsanyi G, Grant SK, Pratt WP, Koenig JW (eds) Proc Int Conf on Mississippi Valley-type lead-zinc deposits. University of Missouri, Rolla, Missouri, pp 7–19

Sangster DF (1990) Mississippi Valley-type and Sedex lead-zinc deposits: a comparative examination. Trans Inst Min Metall Sect B 99:B21–B42

Schroll E (1981) REM-Untersuchungen an Schalenblenden: ein Beitrag zur As- und Tl-Führung von Sphaleriten. Fortschr Mineral 59:178–179

Schroll E (1983) Geochemical characterization of the Bleiberg type and other carbonate-hosted lead-zinc mineralizations. In: Schneider H-J (ed) Mineral Deposits of the Alps and of the Alpine epoch in Europe. Springer, Berlin Heidelberg New York, pp 189–197

Schroll E (1984) Geochemical indicator parameters of lead-zinc ore deposits in carbonate rocks. In: Wauschkuhn A, Kluth C, Zimmermann RA (eds) Syngenesis and epigenesis in the formation of mineral deposits. Springer, Berlin Heidelberg New York, pp 294–305

Schroll E (1990) Die Metallprovinz der Ostalpen im Lichte der Geochemie. Geol Rundsch 79:479–493

Sobott RJ, Klaes R, Moh GH (1987) Thallium-containing mineral systems. Chemie Erde 47:195–218

Venerandi I (1966) Sulla presenza di jordanite nel giacimento di Raibl. Atti Symp Int Giac Min Alpi 1:243–250

Wedepohl KH (1974–1978) Handbook of geochemistry. Springer, Berlin Heidelberg New York

Wilson EN, Hardie LA, Phillips OM (1990) Dolomitization front geometry, fluid flow patterns, and the origin of massive dolomite: the Triassic Latemar buildup, northern Italy. Am J Sci 290:741–796

The Genesis of the Pennine Mineralization of Northern England and Its Relationship to Mineralization in Central Ireland

D.G. Jones, J.A. Plant and T.B. Colman[1]

Abstract

We propose a modified episodic basin dewatering model for the Pennine Pb-F-Ba-Zn mineralization involving the mixing of deep compaction-driven brines from Carboniferous basins with a shallow gravity flow of less saline waters. Deep flow was focused by convection in buried high heat production granites between major fluid expulsion events. The evidence favours a Permian date for the major mineralization.

The genesis of the major Irish Zn-Pb-Ba deposits is most widely attributed to a modified Sedex model involving Carboniferous seawater circulated to great depths by convection. Recent data, though, indicate high geothermal gradients and suggest that such deep convection cells were not necessary. Alternative, basin dewatering models have also been proposed. However, it is doubtful that a sufficient local thickness of basinal Carboniferous sediments was present at the time of mineralization, so these models may not be viable.

There are close similarities between the Dinantian basins of northern England and Ireland. There is thus a strong possibility of Irish-type mineralization in northern England. Also, studies of northern England may shed light on the genesis of Irish ore deposits. The Irish and Pennine deposits can be related to different phases of basin evolution. The Irish deposits formed during Early Carboniferous crustal extension and associated high geothermal gradients. In contrast, the Pennine deposits formed during a period of declining geothermal gradients following the deformation, uplift and erosion of the Carboniferous rocks.

1 Introduction

The mineralization of the South Pennine Orefield (SPO) and the mineral deposits of central Ireland both occur in Carboniferous (Dinantian) carbonate-dominated successions. The Northern Pennine Orefield (NPO) of the Askrigg and Alston blocks lies within a more varied succession of shales with numerous sandstone and limestone beds. They exhibit considerable differences in their morphology, mineralogy and time of formation which have given rise to numerous hypotheses for their genesis. Increasingly, the idea that the Pennine

[1] Minerals and Geochemical Surveys Division, British Geological Survey, Keyworth, Nottingham, NG12 5GG, UK

Spec. Publ. No. 10 Soc. Geol. Applied to Mineral Deposits
Fontboté/Boni (Eds.), Sediment-Hosted Zn-Pb Ores

mineralization represents a subset of the generalized Mississippi Valley type (MVT) mineralization (Dunham 1983) and that the major Irish deposits have affinities with the sedimentary exhalative (Sedex) class of deposits (e.g. Briskey 1986; Sangster 1990) has gained favour (see Tables 1 and 2). However, there still appears to be no consensus for the formation of the mineralization of

Table 1. Comparison of general MVT mineralization with Pennine-style (NPO: Northern Pennine Orefield - Alston Block; SPO: South Pennine Orefield)

MVT	Pennine
Host rocks	
Limestone often dolomitized	Limestone, locally sandstone and dolerite
Age of host rocks	
Cambrian-Jurassic Peaks in Cambro-Ordovician and Carboniferous	Carboniferous-Triassic, mainly Dinantian
Depositional environment of mineralization	
Replacement, solution breccias and cavities in lithified host rocks near edge of stable platform	Mainly veins in lithified host rocks in platform carbonates on positive blocks, some replacement deposits
Tectonic setting	
On margins of, and some distance from major basins. Fertile basins tend to be less homogeneous with basal sandstone aquifers	On margins of secondary basins within major basin. Depth of burial of portion of basin may be important to generate the observed fluid temperatures
Igneous rocks	
Mainly absent except in Illinois-Kentucky (Hicks Dome)	Caledonian granite cores to NPO. Basic sill (NPO) and interbedded volcanics (SPO) control some mineralization
Burial at time of mineralization	
Shallow < 2 km	Shallow < 2 km
Presumed timing of mineralization in relation to basin development	
Middle stage onward 10–40 Ma after initiation of basin	Middle stage > 50 Ma after initiation of basin
Tectonic regime during mineralization	
Uplift and compression - Upper Mississippi Valley and Illinois-Kentucky. Subsidence of basin and expulsion of brines (Pine Point)	Uplift and compression or renewed extension
Geothermal gradient	
Normal	Normal
Alteration of host rocks	
Dolomitization, dissolution	Silicification, dolomitization, dissolution
Elements in ores	
Zn $>$ Pb. Trace Ba F (Tri State Upper Mississippi Valley) Pb $\gg$ Zn. Trace Ba F (SE Missouri) Ba $>$ Zn, Pb $>$ F (central and SE Missouri) Ba, F, Zn $>$ Pb (central Tennessee and central Kentucky) F $>$ Zn $>$ Pb, Ba (Illinois-Kentucky)	F $>$ Pb $>$ Ba $\gg$ Zn (central NPO and SPO) Ba $>$ Pb $>$ F $>$ Zn (outer NPO)

Table 2. Comparison of the major stratiform massive sulphide Zn-Pb-(Ba) mineralization of central Ireland and general Sedex mineralization

Sedex	Central Ireland
Host rocks	
Deep-water marine clastics; starved euxinic basinal sediments; shallow water marine calcareous facies and evaporites	Shallow water limestone, associated with later Waulsortian (patch) reefs <200 m water depth
Age of host rocks	
Proterozoic-Tertiary; peaks in mid-Proterozoic and Devonian	Carboniferous; Courceyan
Depositional environment of mineralization	
Exhalative, syndiagenetic	Exhalative, syndiagenetic
Position of orefield in relation to basin	
At margin of, or within, basin especially within third-order basin	At basin margin close to Lower Palaeozoic basement
Relation of mineralization to faulting	
On, or adjacent to, major basin controlling fault	On major syndepositional fault or associated fault
Burial of host rocks at time of mineralization	
None, near-surface	None, near-surface
Igneous rocks	
No direct association but mafic volcanics commonly within basin; often below the mineralized horizon	Not associated. Some mafic volcanism after mineralization
Timing of mineralization in relation to basin development	
Early to middle stage	Early to middle stage, depending on whether Munster basin (Devonian) is considered
Tectonic regime during mineralization	
Extension	Extension
Geothermal gradient	
Increasing; enhanced	Increasing; enhanced
Elements in ores	
$Zn > Pb \gg Cu$ Ba can be major (Aberfeldy, Meggen).	$Zn > Pb, (Cu) > (Ba)$ Ba can be major (Ballynoe)

the two areas. In this chapter we investigate the Pennine mineralization in particular, and try to relate its genesis to that of the major central Ireland stratiform Zn-Pb deposits (e.g. Navan, Tynagh and Silvermines). Using basin analysis studies, we propose a simple overall synthesis of the formation of the mineral deposits of the two areas. We do not attempt to explain the genesis of the complete spectrum of base metal mineralization in central Ireland, but briefly consider some of the other styles of mineralization present later in this section.

The Pennine mineralization of northern England occurs in the Northern Pennine (Alston and Askrigg) and South Pennine orefields (Fig. 1). The mineral deposits occur principally as large numbers of long (up to 1 km),

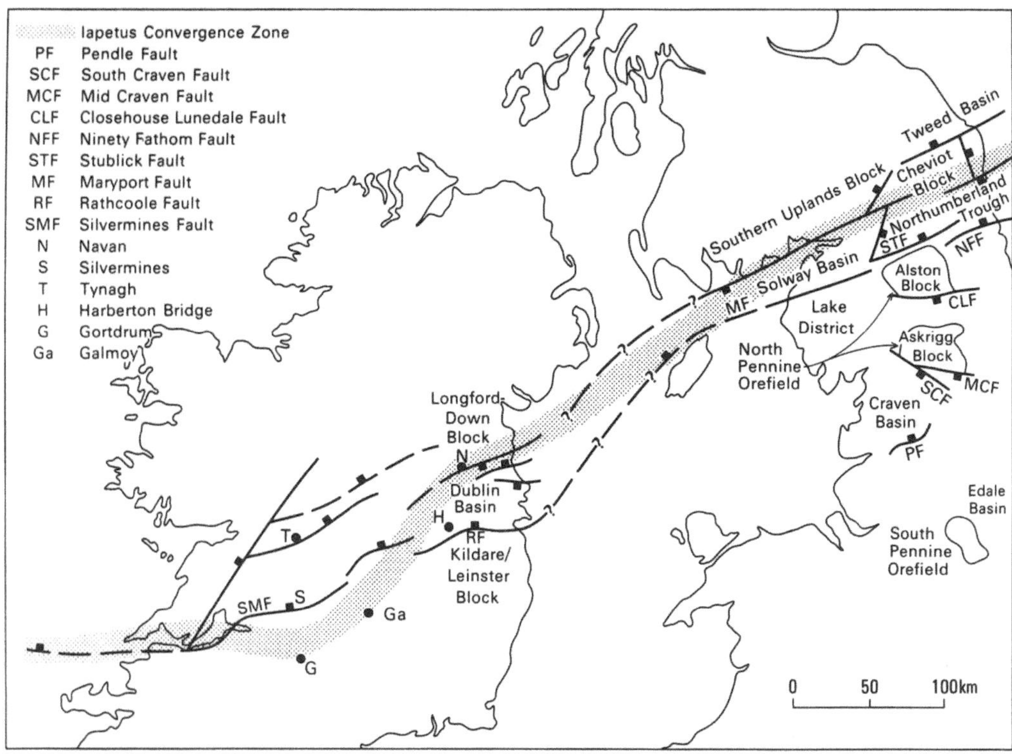

Fig. 1. Location of the Pennine orefields and major Irish base metal deposits in relation to inferred Dinantian synsedimentary faults and basins along the line of the Iapetus Convergence Zone. (After D.W. Holliday, D.J. Evans, J.D. Cornwell, M.C. Barr, A.J.W. McDonald, R.A. Chadwick in J.A. Plant and D.G. Jones unpubl. report 1991; Todd et al. 1991)

narrow (<20 m), subvertical quartz/carbonate/sulphate/fluorite/sulphide vein ore shoots (Dunham 1991). These are of limited vertical extent, being confined to a small number of massive, competent limestone and sandstone horizons of Asbian, Brigantian and Pendleian age and within a major Stephanian dolerite sill, the Whin Sill, in the Alston orefield. Semi-concordant replacement deposits also occur in the Alston and South Pennine orefields. The main ore minerals are fluorite and galena, with subsidiary, though locally important, baryte, witherite, calcite, sphalerite and chalcopyrite. Total production for the Pennine orefields is estimated to exceed 6 Mt lead, 6.5 Mt fluorite, 2 Mt baryte, 0.9 Mt witherite and 0.1 Mt zinc.

The genesis of the Pb, F, Ba (Zn) deposits of the Northern Pennine Orefield has been the subject of much debate which has intensified in recent years. A detailed review of the models proposed by various authors and of the characteristics of the orefields of the Askrigg and Alston blocks (Northern Pennine Orefield) and the South Pennine Orefield (SPO) is given in Plant and Jones (1989) and only a brief summary is included here.

The base metal mineralization of central Ireland has recently been reviewed by McArdle (1990). He divides the deposits into four classes:

1. Stratiform massive sulphide Zn-Pb-(Ba) deposits. These include the major Navan, Silvermines and Tynagh deposits.
2. Disseminated and breccia-hosted Zn-Pb deposits. These include the Harberton Bridge deposit.
3. Vein and disseminated Cu deposits such as Gortdrum and Mallow.
4. Vein baryte deposits.

The most significant economic deposits are those of the first class. They all occur in Courceyan shallow-water carbonates and are associated with major E-W to NE-SW trending faults. The mineralization consists of very fine-grained, honey-brown sphalerite, galena and pyrite in a very fine-grained carbonate matrix. A separate synchronous massive baryte lens formed at Silvermines. At Silvermines much of the mineralization is synsedimentary and of Sedex style; at Navan a mixture of syngenetic and syndiagenetic styles occur while at Tynagh the bulk of the deposit was formed during early diagenesis.

The breccia-hosted mineralization at Harberton Bridge consists of sphalerite, galena and pyrite in Courceyan to Arundian limestones (Emo 1986). Brecciation occurs through a 500-m section of carbonates. The mineralization is thought to have occurred during hydrothermal karstification (Emo 1986).

The Gortdrum (Cu-Ag-Hg) and Mallow (Cu-Ag) deposits consist of structurally controlled epigenetic mineralization in Courceyan rocks. At Gortdrum the mineralization is closely associated spatially with high level basic intrusions (Steed 1986).

The most comprehensive review of the Irish deposits is given in numerous papers within the volume edited by Andrew et al. (1986) and more recently by McArdle (1990).

2 Sources and Transport of Mineralizing Fluids

2.1 Pennine Mineralization

2.1.1 Sources

The classical studies of the ore zoning on the Alston Block in the Northern Pennine Orefield (e.g. Dunham 1934, 1948) suggested the presence of a buried granite which was thought to have been responsible for generating magmatic hydrothermal mineralization comparable to that associated with the Variscan granite batholith of southwest England. However, borehole evidence showed subsequently that the granite was Caledonian rather than Variscan and that it had been eroded prior to the deposition of the Carboniferous limestones which host much of the mineralization (Dunham et al. 1965). Thus, the zoning (Fig. 2) could not be explained simply as the result of hydrothermal mineralization associated with granite magmatism. Nevertheless, Smith (1974) demonstrated that fluorites from the Alston area contained large quantities of REE. These were shown to have chondrite-normalized patterns containing features (positive Eu anomalies and HREE enrichment) similar to those in plagioclase feldspar

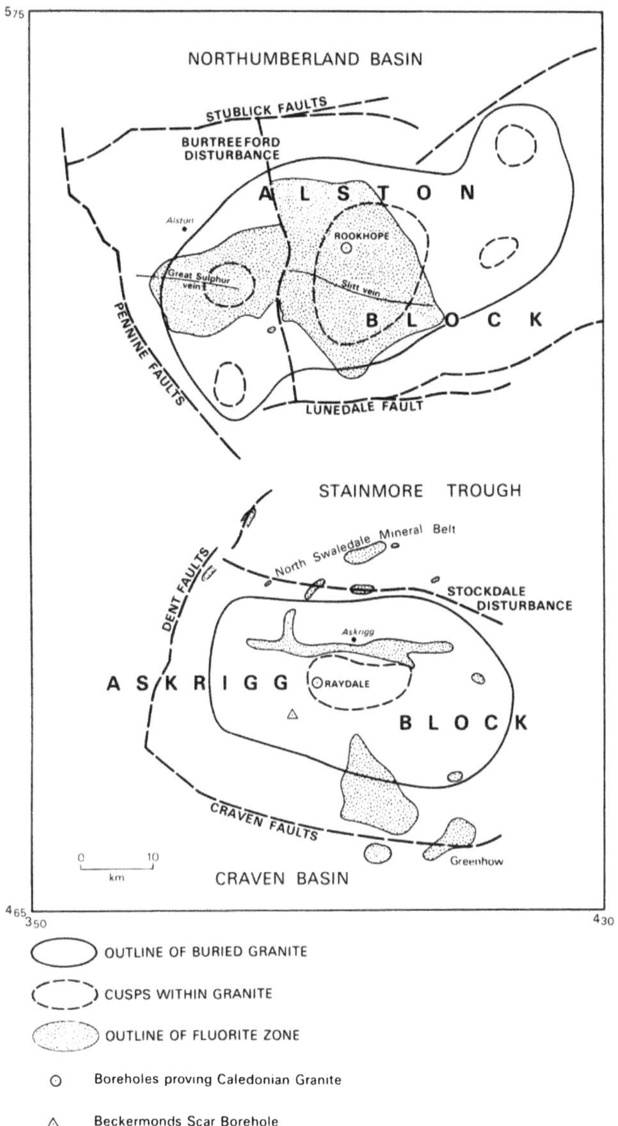

Fig. 2. Northern Pennine Orefield: relationship between buried granites, fault-bounded blocks and mineralization. The inner fluorite zone of the Alston Block is closely associated with cupolas (*cusps*) on the Weardale granite. It is surrounded by a lower temperature baryte zone. Zoning is much less well defined on the Askrigg Block

and heavy minerals in the underlying Weardale Granite (Shepherd et al. 1982). However, it is more likely that the positive Eu anomalies are due to relatively high temperature (>200–250 °C) fluid-rock interaction of acid chloride-rich brines (e.g. Bau and Möller 1992). Under these conditions positive Eu anomalies appear to be produced regardless of the rock type involved (Michard

1989). Maximum fluid temperatures for the Alston Block, as indicated by fluid inclusions, are greater than those for the S. Pennines, where positive Eu anomalies are absent.

More recently, emphasis has shifted to models involving a sedimentary basin source for the ore fluids involved in the formation of the Pennine ore deposits. Overall there is strong evidence that a major component of the ore-forming system was formational water derived from shales. Oxygen isotope data (enrichment in ^{18}O), δD values, the sulphide sulphur isotopic data and the composition of fluid inclusions (which are in the Na-Ca-Cl system) are in the range of those from the main stage of Mississippi Valley type (MVT) mineralization which are of undisputed oilfield brine affinity (Taylor 1974). Ford (1976) and Worley and Ford (1977) have proposed that the source of the fluids was Carboniferous basins beneath the North Sea but other authors, notably Solomon et al. (1971), Robinson and Ineson (1979) and Plant and Jones (1989) have suggested that onshore Carboniferous basins were the main source of the Pennine ore fluids. Jones and Plant (1989) presented geochemical evidence that derivation of ore fluids from basins such as the Edale Basin could explain the chemistry and Pb isotopic signatures of the orefields. Mass balance calculations of the ore-forming elements Pb, Zn, Ba and F and thermal modelling of the Edale basin confirmed that it was a potential source of ore fluids for the SPO (Plant and Jones 1989).

Recently, Halliday et al. (1990) have also argued, on the basis of a limited number of Sm-Nd and Sr isotopic determinations, for a Carboniferous sedimentary source of the Pennine ore fluids. They do not consider the granite to have contributed REE or alkaline earth elements to the ore fluid. However, their own εNd data for the northern Pennine fluorites could be explained by the interaction of Carboniferous basinal fluids with Weardale granite at about 280 Ma.

Halliday et al. (1990) argue that a Lower Carboniferous sedimentary source best fits the Nd and Sr data. However, this appears to be based on single shale and limestone analyses from the relatively thin succession on the Alston Block. Further work is clearly needed to confirm that such signatures are also typical of the basinal Carboniferous sequences most frequently invoked as the source of the mineralization.

Interaction of migrating basinal fluids with the buried granites of the Northern Pennine Orefield was incorporated into the models of Sawkins (1966), Solomon et al. (1971), Shepherd et al. (1982) and Brown et al. (1987). Brown et al. (1987) presented detailed geochemical and petrological evidence that the Weardale Granite had been hydrothermally altered and mineralized during its emplacement in Devonian times and that the hydrothermal system had been reactivated subsequently during Pennine ore formation: the high heat production and reactivation of the major fracture systems of the buried intrusion serving to focus the flow of ore fluids migrating from sedimentary basins. They argued that a contribution from granitic rocks to the ore fluid was indicated by the occurrence of minor amounts of a variety of Co, As, Sb, Bi, Mo, Sn and REE minerals in the deposits sited around the cupolas of the granite. Rankin and Graham (1988) interpreted Na:Li ratios in fluid inclusions in terms of Li enrichment of fluids by interaction with the granite. There is thus a consensus that the ore fluids were derived from Carboniferous basins and, with the exception of Halliday et al. (1990), that in the Alston Block the fluids interacted with buried Caledonian granite.

Sheppard and Langley (1984) argued on the basis of $\delta^{18}O$ and δD data and the concentration of major elements in modern Na-Ca-Cl brines of meteoric origin from collieries in northeast England that ore fluids could be generated rapidly by dissolution, exchange and/or filtration reactions between meteoric water, evaporites and clays. They also suggested that such fluids could have been available for flushing out to produce mineralization on more than one occasion. Some of the brines have high Br : Cl ratios consistent with reactions with organic matter (hydrocarbons), they are moderately acid ($pH < 6.1$) and they may be enriched in Ba or sulphate depending on the particular unit in the Westphalian from which they are derived. The fluids studied by Sheppard and Langley vary considerably, however, at the local (mine) scale, whereas the fluid inclusion data of, for example, Rogers (1977), Shepherd et al. (1982) and Atkinson (1983) indicate the relative homogeneity of the ore fluids in the individual Pennine orefields. The REE data on fluorites also suggest fluid homogenization throughout the source region of each orefield with infinite reservoir conditions of fluid generation. Hence, a model that involves extensive equilibration of fluids within sedimentary basins better fits the evidence than one that involves recharge and interaction of meteoric or other supergene water with sediments.

2.1.2 Transport

The main discussion now centres on the processes involved in the transport of the ore fluids and their timing, with convective, gravity and compaction-driven models favoured by different authors. Halliday et al. (1990), although describing the mineralization of the North Pennines as MVT, proposed a model closely similar to that originally put forward by Russell (1978) to explain the genesis of many of the Irish deposits; mineralization being produced by "anomalously deep penetration of the crust by fluids, concomitant with the production and reactivation of fracture permeability induced by unusual regional lithospheric tension". Moreover, they suggest that the most likely age for the fluorite mineralization of the Alston Block is 200 Ma (i.e. Late Triassic/Early Jurassic).

A simple convective model does not account for the Pb, F dominance of the orefields, a feature which cannot be explained using simple solubility relationships in brine systems (full discussion in Plant and Jones 1989). Deep geological/hydrogeological evidence also argues against lithospheric tension oriented N-S or NNW-SSE and capable of generating the E-W or ENE-WSW trending vein systems of the Pennine orefields at 200 Ma and suggests that fluid expulsion from Carboniferous basinal sediments was highly unlikely at this time (J.A. Barker, R.A. Downing, D.W. Holliday, R. Kitching in J.A. Plant and D.G. Jones, unpubl. BGS report 1991). At about 200 Ma deep lateral flow could have been directed towards the Pennine High. However, the data of Cathles and Smith (1983) suggest that fluid expulsion from basins occurs over a few tens of millions of years, increasing to a maximum at about 40 million years then decreasing. Their model implies that major mineralization at 200 Ma, at least 100 million years after the *end* of the Carboniferous, would have been highly unlikely. Part of the difficulty in applying the findings of Halliday et al. (1990) may be the dating of the deposits using Rb-Sr isochrons in fluid inclusions collected from vein systems overlying the Weardale high heat production (HHP) granite, since this may have remained at a high temperature for a longer period of time than in the orefield generally. Fluid inclusions may thus record the time at which re-equilibration of Rb-Sr systems ended in an anomalously hot area of the orefield rather than the age of the Pennine mineralization generally. Convective models, particularly those involving deep penetration of the crust during lithospheric extension, are, therefore, considered inappropriate for the Pennine mineralization.

The stratigraphic and structural evolution of the Pennine region indicates that the uplift associated with the movements at the end of the Carboniferous would have created a shallow gravity flow system that existed from the Stephanian to Lower Permian times (J.A. Barker, R.A. Downing, D.W. Holliday, R. Kitching in J.A. Plant and D.G. Jones, unpubl. BGS report 1991). The flow directions would have been towards the inland basins that existed in Early Permian times extending from the uplifted areas in the Pennines, where the mineralization is now found, towards the source rocks in the basinal areas. Hence, it is difficult to apply a gravity model of fluid movement for the Pennine orefields since the ore fluids would be driven away from the orefields towards the eastern sea-board of Britain.

The suggestion of Dunham (1983) that the Pennine mineralization represents a subtype of the Mississippi Valley type of ore deposits provides a starting point for further consideration of the ore-forming processes and their timing (see Table 1). Such deposits are thought generally to form from basinal brines as proposed initially by White (1958). This is consistent with evidence outlined earlier that the Pennine ore fluids were derived from Carboniferous shale basins with large-scale equilibration of fluids and their source rocks prior to mineralization.

Fluids capable of transporting metals may move in sedimentary basins in response to such factors as thermal gradients and variations in topography, gravity and deformation (Hanor 1979; Anderson and Macqueen 1982) but the general model applied to MVT deposits is that proposed originally by Jackson and Beales (1967) for the Devonian Pine Point deposit of the Canadian North-West Territories. The formation of this deposit is suggested to have resulted from dewatering of basin facies shales with lateral migration into karstified limestone on adjacent emergent blocks. Dewatering occurred initially during the compaction of the basin sediments by lithostatic pressure. Complex phase transformations may have occurred as the clay-dominated sediments lost their pore water and as the temperature and lithostatic pressure rose with increasing depth of burial. The fluids, charged with metals in solution, moved up-dip towards the basin margins and deposition of the metals and other ore-forming elements occurred following incursion of the fluids into permeable areas of adjacent platform carbonates. This general model for MVT mineralization and the modified model of episodic dewatering (e.g. Cathles and Smith 1983) emphasize compaction as the principal process involved in the mobilization of

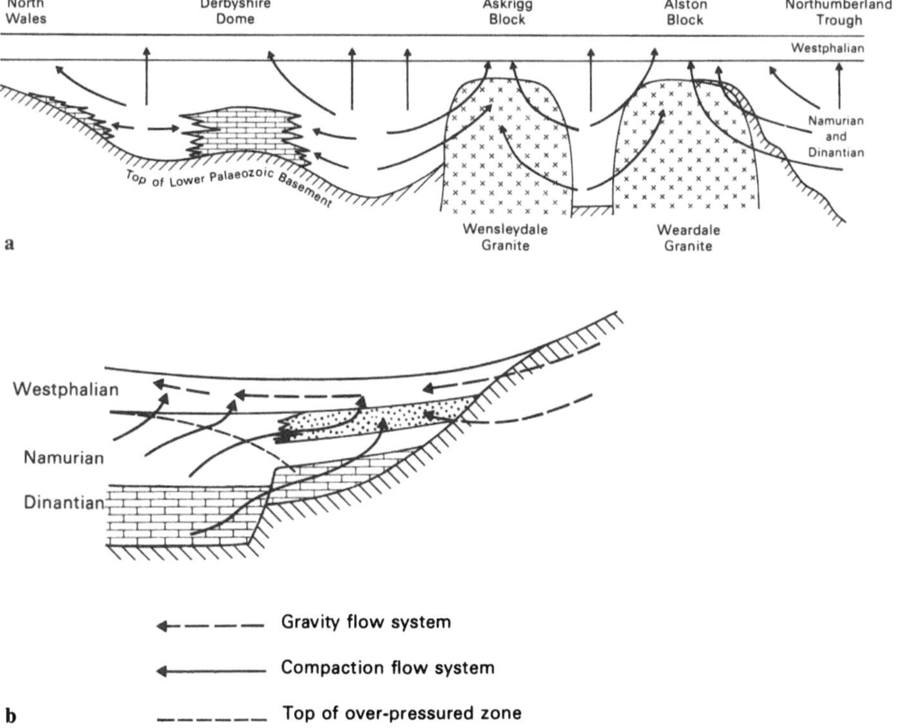

Fig. 3. a Directions of groundwater flow from Dinantian-Namurian basins in northern England under the influence of tectonic compressive stresses in Late Carboniferous to Early Permian times. **b** Diagrammatic section of groundwater flow paths in the Pennines during Late Carboniferous/ Early Permian times showing interception of compaction flow by gravity flow. (After J.A. Barker, R.A. Downing, D.W. Holliday, R. Kitching in J.A. Plant and D.G. Jones unpubl. report 1991)

ore fluids. In the case of the episodic dewatering model, dewatering is considered to reach a peak about 40 Ma into the history of the basin, after deep burial of the basinal sediments.

The basic basin dewatering model for MVT orebodies formed the basis for the models of Solomon et al. (1971), Worley and Ford (1977) and Robinson and Ineson (1979) for the Pennine ore deposits as discussed earlier. In order to explain features such as the difference in the age of the mineralization relative to the age of the basins, the fracture-bound nature of the mineralization and the dominance of Pb and F in the ore assemblage, Plant and Jones (1989) argued for a modification of this model. This involved fluid release by seismic pumping related to Late Carboniferous or Early Permian tectonism with dewatering of Visean-Namurian shale basins and fluid reservoirs that had become overpressured as a result of Namurian-Westphalian sedimentation (Fig. 3a). In the Pennine orefields (as in those of central Kentucky and Illinois-Kentucky in North America) there is an association between Pb dominance in the sulphide assemblage, a high fluorite content in the orefields generally and the fracture control of orebodies (Dunham 1983). Plant and Jones (1989) suggested that fracturing released highly evolved basinal fluids in which the activity of major species was controlled by diagenetic phase transformations rather than by solubility relations. Hence, regional tectonism may be one of the most important factors in the genesis of Pb-F-dominant MVT deposits generally. According to this model mineralized veins are formed as a result of the incursion of moderately acid, highly saline ($NaCl$-$CaCl_2$-H_2O) brines generated under infinite reservoir conditions and carrying hydrocarbons, Pb, Zn, Ba and F into fracture systems in limestone sequences. Sulphide precipitation was suggested to occur as a result of reduction of sulphate in limestone formational fluids by hydrocarbons in the brines. Fluorite was deposited as a result of the release of calcium by acid neutralization reactions associated with sulphide precipitation. Baryte was precipitated by mixing of the Ba-enriched brine with SO_4^{2-} above and around the zone of sulphide deposition.

In the North Pennines the buried high heat production Weardale and Wensleydale granites were suggested by Brown et al. (1987) and Plant and Jones (1989) to have locally focused fluids into hydrothermal convection cells, resulting in the observed spatial association between mineral zonation and the subcrop of buried granites (Fig. 3a). The relationship is particularly clear in the case of the Weardale Granite, in which hydrothermal alteration/re-equilibriation occurred initially during emplacement of the intrusion in the Lower Devonian and subsequently as a result of Pennine ore deposition. Temperature maxima coincide with areas of intense mineralization in the fluorite zone, centred over the negative gravity anomaly associated with the granite (Fig. 2), with temperature minima in the baryte zone around the margin of the buried intrusion (see Dunham 1983 for a summary of fluid inclusion temperatures). Elsewhere the role of low-density chemically evolved granites in the basement was suggested to increase the buoyancy and rigidity of crust locally and to propagate fracture systems from basement to cover (Plant and Jones 1989).

Such a model accounts for the common surface geological-geochemical features of the Pennine orefields, including the mineral assemblage with $Pb + F > Ba + Zn$ and the structural control of mineral veins, whereby permeability was related principally to movement of fracture systems rather than karstification or near-surface groundwater movements. It also accounts for differences between the North and South Pennine Orefields: for example, the increased temperature and K/Na ratios indicated by fluid inclusion studies in the Alston Block compared to the Askrigg and South Pennine orefields.

Recent BGS data (J.A. Plant and D.G. Jones, unpubl. BGS report 1991) have led to some modification of the Plant and Jones (1989) model. The new results are most consistent with a combination of a shallow gravity flow system with oxygenated sulphate-rich waters and a deeper compaction-driven flow of more saline fluids from Viséan-Namurian basins (Fig. 3b). Palaeogeographic reconstructions for the Early Permian suggest that the uplift at the end of the Carboniferous would have created a gravity flow system that continued into Lower Permian times (J.A. Barker, R.A. Downing, D.W. Holliday, R. Kitching in J.A. Plant and D.G. Jones unpubl. BGS report, 1991). This system would have been relatively shallow as northern England had been reduced to a peneplain. In the basinal areas formed in Carboniferous times sediments were probably still being compacted and may have been overpressured. This is a situation that exists today in the Pannonian Basin in central Europe (Ottlick et al. 1981) and it is recognized that well-sealed reservoirs may remain overpressured after uplift and erosion have removed the overlying sediments that caused the original overpressure.

Compressive tectonic forces in Late Carboniferous times would have increased the fluid pressures, originally due to compaction, until hydraulic fracturing took place. The development of stress was probably cyclic, building up slowly until the rocks fractured, releasing the fluids rapidly. The conduit would then reseal as the stress declined. This periodic fracturing of the overpressured seal would have allowed the rapid movement of fluids upwards and towards the margins of the Carboniferous depocentres where suitable host rocks for mineral deposition occurred. The shallow gravity systems would have prevented the rise of the deep-seated brines to the surface but mixing of the two waters is likely to have occurred at depth.

This conceptual model is similar to that proposed to explain Pb-Zn mineralization elsewhere; a number of authors (for example, Sharp 1978; Leach and Rowan 1986; Clendenin and Duane 1990; Bethke and Marshak 1990; Symons and Sangster 1991) have proposed that MVT Pb-Zn and Ba mineralization in northern Arkansas and southern Missouri was the result of the sudden release, in discrete pulses, of northward migrating overpressured brines from the Arkoma Basin to the south during the Westphalian to Permian, Ouachita orogeny. Other models involving simple basin compaction do not fit with the observed features of the mineralization. For example, Cathles and Smith (1983) pointed out that fluid migration rates have to be much greater than those produced simply by compaction if mineralization temperatures of 100–200 °C were to be maintained from source area to deposition zone. Bethke (1985) showed that fluid expulsion in non-overpressured basins approximates to a steady-state process and heat transfer is by conduction so that normal geothermal gradients exist at basin margins. He concluded that this is unlikely to cause mineralization and that sudden dewatering of overpressured strata, with rapid transfer of hot brines, is a requirement.

Such a model is also consistent with the isotopic evidence for mixed seawater and freshwater sources for mineralization in the South Pennine Orefield presented by Robinson and Ineson (1979). Convection within the Weardale and Wensleydale granites appears also to have been a feasible process superimposed on the compactional system, although of much lesser importance during major expulsions of basinal fluids.

A recent study encompassing new Pb isotopic data for galenas and for a range of potential Pb sources around the South Pennine area (Lower Palaeozoic shales, Viséan-Namurian shales and Old Red and Namurian sandstones) is summarized by Jones et al. (1991). The data are most consistent with a Carboniferous basin source of Pb with extraction of Pb at around 245 Ma. The mineralization in the northern Pennines is known to mostly post-date the Whin Sill which was emplaced at around 295 Ma. It appears likely that the bulk of the mineralization was formed between these two dates since, as discussed earlier, the evidence in favour of a 200 Ma date is not compelling. A model involving overpressured fluids, generated during late Westphalian to Early

Permian times still seems most appropriate but, instead of one major episode of dewatering, an episodic (multiple) dewatering model is considered more appropriate. This fits best with the hydrogeological arguments above and accords well with the paragenetic evidence from the orefields which demonstrates clearly a complex history of vein infilling.

2.2 Irish Mineralization

The Irish deposits are not spatially related to any outcropping granite though probable concealed granites occur close to Navan (Kentstown pluton) and Silvermines (Nenagh pluton). These plutons, inferred from the gravity data, are possibly of Caledonian age (Phillips and Sevastopulo 1986). If this age is correct, they may have exerted a structural control on the formation of the deposits but they could not have had a primary role in fluid formation.

The copper-dominant deposits of the Limerick and Tipperary areas may be genetically related to the Chadian-Asbian Limerick basic volcanics. The Gortdrum Cu-Hg-U (Steed 1986), Mallow Cu-Ag and Ballyvergin Cu deposits are all epigenetic vein, replacement and stockwork deposits in Courceyan limestones with local structural control of the mineralization. Formation is ascribed to hydrothermal fluids derived variously from volcanic sources, deep circulation in the underlying Lower Palaeozoic rocks and from the Devonian basin to the south.

Russell (1978, 1983, 1986) proposed, in the case of the major Zn-Pb deposits, that mineralizing fluids were generated from Carboniferous seawater which penetrated, and reacted with, the underlying Caledonian basement. He suggested that convection of pore fluids occurred in continental crust as a result of increased permeability during crustal extension, but when the geothermal gradient was normal (i.e. about 30 °C/km). Russell postulated that the convection cells responsible for the Irish deposits became progressively larger (up to 20 km radius) and more deeply excavating (up to 10 km max. depth) with time. Mills et al. (1987) presented Pb isotopic data for the Navan mine in support of the Russell model. Lead in the stacked ore lenses is progressively less radiogenic with decreasing age, which is interpreted as reflecting the increase in the depth to which mineralizing fluids penetrated with time. Sulphur isotope data for the Navan, Tynagh and Silvermines deposits (e.g. Boast et al. 1981) suggest mixed, deep-seated and seawater sources for the stratiform sulphides with the baryte being derived from Late Carboniferous seawater.

LeHuray et al. (1987), in a widespread study of the lead isotopes of the Irish deposits, also concluded that deep hydrothermal convection was required, not only to account for the observed lead isotope ratios, but also because they suggest that derivation from contemporaneous basins is incompatible with their known dimensions and especially their thin Courceyan successions. Dixon et al. (1990) presented further evidence from a study of whole rock lead isotopes which supported the hypothesis that the lead in the deposits was derived from the underlying Lower Palaeozoic basement.

An alternative proposal was made by Lydon (1986) who considered that the ore fluids involved in the formation of the Irish deposits were generated by the dewatering of Devonian-Early Carboniferous basins. The fluids reacted with the clastic basin infill at depths of 1–2 km to produce metalliferous brines. A cover of Early Carboniferous shales provided thermal insulation to enable the brines to reach temperatures exceeding 200 °C under an enhanced geothermal gradient (which may have reached 75 °C/km in places, Strogen et al. 1990). Periodic extension of the basins permitted release of the metalliferous fluids via listric faults into the Carboniferous sediments. A similar model was proposed by Williams and Brown (1986). The thin Courceyan succession at Navan, which is only 30 m above the Lower Palaeozoic basement, thickens rapidly to the south with the Trim borehole, 20 km south of Navan, proving at least 800 m thickness of Courceyan rocks (Sheridan 1972). The early clastic basin infill is probably derived from eroded Caledonian rocks to the north and could therefore have a similar Pb isotope composition. However, the Carboniferous basins were probably not sufficiently developed at the time of mineralization to have been a major source of fluids (McArdle 1990). Therefore, the Russell (1978, 1983, 1986) convection model perhaps best explains the current information but the evidence of an enhanced geothermal gradient (Strogen et al. 1990) suggests that such deep convection cells were not necessary to raise the fluids to the appropriate temperatures.

3 English Dinantian Basins: A Possible Setting for "Irish-Style" Deposits

There are close similarities in the tectonic setting and sedimentary history of Dinantian basins in Ireland and northern England. These suggest a strong likelihood of "Irish-type" mineralization in northern England. They also imply that studies of northern England may shed light on the genesis of the Irish deposits.

In the case of the central Ireland deposits the importance of NE-SW to ENE-WSW lineaments has been noted by numerous authors and summarized by Hitzman and Large (1986) and supports strongly previous suggestions that they are inherited from the Caledonian basement (Phillips et al. 1988). The area is characterized by a particularly well-developed set of NE-SW lineaments which correspond generally with the margins of the gravity lows recognized by Brown and Williams (1985) and with parts of the inferred Dinantian syndepositional fault system shown in Fig. 1. The Navan deposit lies close to the intersection of the NE-SW lineaments and a N-S lineament which can be recognized on images of the processed gravity data. (J.D. Cornwell, pers. comm.).

The projection of the main geophysical lineaments identified in Ireland across the Irish Sea (Max et al. 1983) suggests that many of the structures continue into northern England (Fig. 1). The Virginia and Graignammanagh-Wicklow lineaments, which bound the main zone in which mineral deposits occur in Ireland, can be equated broadly with a lineament in the Southern Uplands and the South Solway lineament respectively in Britain. The Navan-Silvermines lineament in the centre of the area lies close to the probable trace of the Iapetus Convergence Zone as proposed previously by Phillips et al. (1976). This separates Caledonian rocks of the Avalonian and Laurentian continents which, as widely accepted, became juxtaposed following closure of the Iapetus Ocean. The precise trace of the Iapetus suture has recently been reviewed by Todd et al. (1991) who suggest a number of changes to its course through central Ireland.

Unfortunately, no seismic reflection data are available for Ireland, but in northern England seismic data from the Northumberland Basin, which lies over the probable continuation of the Iapetus Convergence Zone in Britain, and the Craven Basin to the south of the Askrigg Block, which had a Carboniferous tectonic and sedimentary history more closely similar to that of Ireland, provide detailed information on the Lower Dinantian setting in which Irish-style deposits are thought to have formed. This period was dominated by crustal extension associated with the syn-rift deposition of sedimentary sequences which are predominantly siliciclastic in the NE (the Northumberland Basin) and become progressively more carbonate-rich southwestward across NW England and Ireland. Gardiner and MacCarthy (1981) have suggested that the Devonian-Carboniferous basins of southern Ireland probably formed by rifting in response to extension. A direct association between rifting and metallogeny has been proposed by Deeny (1987). In the case of the Northumberland Basin a lithospheric extension factor of 1.30 is considered most consistent with the new cross-sectional reconstructions prepared using seismic data and with the relatively low maturity levels of organic matter in the basin away from igneous rocks (Kimbell et al. 1989). The higher extension factors of 1.7 proposed previously by Leeder and McMahon (1988), and especially those of approximately 8 suggested by Haszeldine (1989) are considered unrealistic.

The Northumberland Trough and its westerly continuation, the Solway Basin, lie within the hanging wall block of a proposed late Caledonian shear zone along the trace of the Iapetus Convergence Zone and it suggested that the basin formed as a direct response to extensional reactivation of the shear zone in Carboniferous times (Chadwick and Holliday 1991). The tectonic setting of the Northumberland Basin thus appears to be closely similar to that of central Ireland; the Maryport-Stublick-Ninety Fathom and Rathcoole-Silvermines faults (Fig. 2) representing major, synsedimentary basin bounding, fault systems comprised of a number of individual fault segments, generally arranged en echelon.

In the case of the Craven Basin the estimates of extension (D.J. Evans, unpubl. data) are similar to the values calculated for the Northumberland Trough (Kimbell et al. 1989) but lower than those proposed previously by Leeder (1982), Dewey (1982) and Haszeldine (1989). Overall, the Craven basin is an extensive, composite asymmetrical graben or trough, bounded to the N, NW and SE by major NE-SW to E-W trending synsedimentary faults (Riley 1991). These include the Mid-Craven and Pendle faults. Within the basin a complex series of tilted fault blocks define intrabasinal highs and lows which actively controlled deposition during the Dinantian (Arthurton et al. 1988).

The tectonic setting of the Lower Dinantian, the period during which most of the Irish-style deposits formed, thus approximates closely the waxing or rift phase of the McKenzie (1978) model for basin development as proposed previously by Leeder (1982) and Plant and Jones (1989). The bunching of isotherms and synsedimentary faults at basin margins, as indicated by the model, suggests that these are most likely to the locus of ore deposition. Such a model is consistent with the occurrence of Irish-style ore deposits in Ireland along the trace of major NE-SW lineaments interpreted here as major basinal synsedimentary fault systems bounding gravity lows. New evidence that the geothermal gradient in central Ireland was greater than 75 °C/km (Strogen et al. 1990) and the presence of Chadian-Brigantian alkaline-tholeiitic basic volcanics southeast of Limerick and at Edenderry southwest of Dublin is also most consistent with the extensional rifting phase of basin development according to the McKenzie model.

4 Conclusions

Overall, basin analysis studies based on seismic and other geophysical and geological data combined with geochemical and isotopic studies of ore minerals and likely source rocks offers a good basis for understanding ore deposits associated with sedimentary basins. The Irish and Pennine-style deposits can then be related to different phases in basin-forming cycles which affected NW Europe from Late Palaeozoic to Mesozoic times. Hence, the major Sedex Irish-style deposits were formed in geothermal systems over zones of high heat flow and tectonism in the crust. They are related to the waxing phase of a regional Lower Carboniferous regime of crustal extension and basin formation associated with the rise of hot asthenosphere beneath the crust, which was characterized by contemporaneous high geothermal gradients and syndepositional faulting along reactivated basement fault systems and subsequent alkaline basaltic magmatism (Table 3).

In contrast Pennine-style ore deposits were formed following a period characterized by declining geothermal gradients (contemporaneous magmatism becoming tholeiitic by the Early Permian and, finally, ceasing) and by regional subsidence of the crust with overpressuring of basins as a result of the deposition of thick Westphalian sequences; they were deposited from fluids similar to oilfield brines expelled from overpressured basins by regional tectonism probably in Lower Permian times (Table 3).

Pennine-style ore deposits are thus considered to be a special case of Mississippi Valley type ore deposits whereby evolved fluids from overpressured basins were injected into opening regional fracture systems during tectonism probably during the time interval between 290 and 245 Ma. The most appropriate model for Pennine-style ore deposits appears to be a modification of the episodic dewatering model of Cathles and Smith (1983), mineralization forming by the mixing of deeper compaction-driven fluids from Carboniferous basins with a shallow, less saline gravity-driven flow. Stable isotope data (e.g. Solomon et al. 1971; Robinson and Ineson 1979) indicate such mixed fluid sources and hydrogeological evidence argues for the interception of a deeper basin-derived flow by a shallow gravity-driven flow arising from upland areas projecting from the Permian peneplain. Release of the overpressured fluids largely post-dated

Table 3. Simplified chronology of events relating to the Pennine and Irish mineralization (see text for more detailed discussion)

Late Devonian–Early Courceyan
- Crustal extension
- Basin initiation
- Syndepositional faults develop especially on reactivated Caledonian structures

Late Courceyan–Chadian
- Widespread carbonate sedimentation
- Active syndepositional faulting with associated syndepositional to syndiagenetic stratiform Zn-Pb-Ba mineralization in Ireland. High geothermal gradient

Chadian–Arundian
- Continuing carbonate sedimentation and growth faulting with onlap onto a successively wider area

Holkerian–Early Namurian
- Growth faulting declining in importance
- Carbonates give way to shales
- Local alkaline basaltic volcanism (Limerick, S. Pennines)

Namurian–Westphalian
- Crustal downwarp, growth faulting of lesser importance
- Thick clastic sedimentation
- Basaltic volcanism (more tholeiitic) in English East Midlands
- Overpressuring of basinal sequences
- Generally declining geothermal gradients

Late Westphalian–Permian
- Variscan orogeny
- Injection of Whin Sill (Alston Block); approx 5000 km^2 dolerite sill
- Dewatering of overpressured basins with lateral and upward fluid movement to give Pennine mineralization (local focusing of fluids by high heat production granites)

Triassic–Present
- Uplift and erosion

the emplacement of the Whin Sill and could have been related mainly to a later episode of crustal extension during the Permian.

Fluid flow from the basins appears, on the basis of minor Co, As, Sb, Bi, Mo, Sn and REE minerals and Na:Li ratios in inclusion fluids, to have been focused through high heat production granites in the northern Pennines (especially the Alston Block), giving rise to the close spatial association of the mineralization with cupolas on the underlying granite.

Acknowledgements. This paper stems from work funded partly by the UK Department of Trade and Industry and the Mineral Industry Research Organisation which is gratefully acknowledged. We would like to thank all our colleagues on this project and in particular Dr. J.A. Barker, Dr. M.C. Barr, Dr. J.D. Cornwell, Dr. R.A. Downing Dr. D.J. Evans, Dr. D.W. Holliday, Dr. R. Kitching, Dr. A.J.W. McDonald and Dr. I.G. Swainbank. The text has benefitted from internal reviews by Dr. D.J. Evans, Dr. H.W. Haslam and Dr. D.W. Holliday. Valuable comments were made by P. McArdle, Prof. L. Fontboté and an anonymous referee.

References

Anderson GM, Macqueen RW (1982) Ore deposits models-6. Mississippi Valley type lead-zinc deposits. Geosci Can 9:108–117

Andrew CJ, Crowe RWA, Finlay S, Pennell WM, Pyne JF (eds) (1986) Geology and genesis of mineral deposits in Ireland. Irish Assoc Econ Geol, Dublin, 711 pp
Arthurton RS, Johnson EW, Mundy DJC (1988) Geology of the country around Settle. Mem Br Geol Surv, Sheet 60, 147 pp
Atkinson PA (1983) Fluid inclusion and geochemical investigation of the fluorite deposits of the Southern Pennine Orefield. Thesis, University of Leicester
Bau M, Möller P (1992) Rare earth element fractionation in metamorphogenic hydrothermal calcite, magnesite and siderite. Mineral Petrol 45:231–246
Bethke CM (1985) A numerical model of compaction-driven groundwater flow and heat transfer and its application to the palaeohydrology of intracratonic sedimentary basins. J Geophys Res 90:6817–6828
Bethke CM, Marshak S (1990) Brine migrations across North America – the plate tectonics of groundwater. Annu Rev Earth Planet Sci 18:287–315
Boast AM, Swainbank IG, Coleman ML, Halls C (1981) Lead isotope variations in the Tynagh, Silvermines and Navan base metal deposits, Ireland. Trans Inst Min Metall Sect B Appl Earth Sci 90:B115–B119
Briskey JA (1986) Descriptive models of sedimentary exhalative Zn-Pb. In: Cox DP, Singer DA (eds) Mineral deposit models. Bull US Geol Surv 1693, pp 211–212
Brown C, Williams B (1985) A gravity and magnetic interpretation of the structure of the Irish Midlands and its relation to ore genesis. J Geol Soc (Lond) 142:1059–1075
Brown GC, Ixer RA, Plant JA, Webb PC (1987) Geochemistry of granites beneath the North Pennines and their role in mineralisation. Trans Inst Min Metall Section B Appl Earth Sci 96:B65–B76
Cathles LM, Smith AT (1983) Thermal constraints on the formation of Mississippi Valley-type lead-zinc deposits and their implications for episodic basin dewatering and deposit genesis. Econ Geol 78:983–1002
Chadwick RA, Holliday DW (1991) Deep crustal structure and Carboniferous basin development within the Iapetus Convergence Zone, northern England. J Geol Soc (Lond) 148:41–53
Clendenin CW, Duane MJ (1990) Focused fluid flow and Ozark Mississippi Valley-type deposits. Geology 18:116–119
Deeny DE (1987) Central Irish geology/metallogeny: a Lower Carboniferous rifting-related exhalative catastrophy? Mineral Depos 22:116–123
Dewey JF (1982) Plate tectonics and the evolution of the British Isles. J Geol Soc (Lond) 139:371–412
Dixon PR, LeHuray AP, Rye DM (1990) Basement geology and tectonic evolution of Ireland as deduced from Pb isotopes. J Geol Soc (Lond) 147:121–132
Dunham KC (1934) Genesis of the North Pennine ore deposits. Q J Geol Soc (Lond) 90:689–720
Dunham KC (1948) Geology of the Northern Pennine Orefield, vol 1. Tyne to Stainmore. Econ Mem Geol Surv GB, 357 pp
Dunham KC (1983) Ore genesis in the English Pennines: a fluoritic subtype. In: Kisvarsanyi G, Grant SK, Pratt WP, Koenig JW (eds) International conference on Mississippi Valley type lead-zinc deposits, University of Missouri. Rolla Press, Rolla, pp 86–112
Dunham KC (1991) Geology of the Northern Pennine Orefield, vol 1. Tyne to Stainmore, 2nd edn. Econ Mem Brit Geol Surv, 299 pp
Dunham KC, Dunham AC, Hodge BL, Johnson GAL (1965) Granite beneath Visean sediments with mineralisation at Rookhope, north Pennines. Q J Geol Soc (Lond) 121:383–417
Emo GT (1986) Some considerations regarding the styles of mineralisation at Harberton Bridge, County Kildare. In: Andrew CJ, Crowe RWA, Finlay S, Pennell WM, Pyne JF (eds) Geology and genesis of mineral deposits in Ireland. Irish Assoc Econ Geol, Dublin, pp 461–469
Ford TD (1976) The ores of the South Pennines and Mendip Hills, England – a comparative study. In: Wolf KH (ed) Handbook of strata-bound and stratiform ore deposits. II. Regional studies and specific deposits, vol 5. Regional studies. Elsevier, Amsterdam, pp 161–195
Gardiner PRR, MacCarthy IAJ (1981) The Late Paleozoic evolution of sourthern Ireland in the context of tectonic basins and their transatlantic significance. In: Kerr JW, Fergusson AJ (eds) Geology of the North Atlantic borderlands. Can Soc Petrol Geol Mem 7:683–725
Halliday AN, Shepherd TJ, Dickin AP, Chesley JT (1990) Sm-Nd evidence for the age and origin of a Mississippi Valley type ore deposit. Nature 344:54–56

Hanor JS (1979) The sedimentary genesis of hydrothermal fluids. In: Barnes HL (ed) Geochemistry of hydrothermal ore deposits. Wiley, New Yrok, pp 137–172

Haszeldine RS (1989) Evidence against crustal stretching, north-south tension and Hercynian collision, forming British Carboniferous basins. In: Arthurton RS, Gutteridge P, Nolan SC (eds) The role of tectonics in Devonian and Carboniferous sedimentation in the British Isles. Occas Publ Yorks Geol Soc 6:25–33

Hitzman MW, Large D (1986) A review and classification of the Irish carbonate-hosted base metals deposits. In: Andrew CJ, Crowe RWA, Finlay S, Pennell WM, Pyne J (eds) Geology and genesis of mineral deposits in Ireland. Irish Assoc Econ Geol, Dublin, pp 217–238

Jackson SA, Beales FW (1967) An aspect of sedimentary basin evolution: the concentration of Mississippi Valley-type ores during the late stages of diagenesis. Bull Can Pet Geol 15:393–433

Jones DG, Plant JA (1989) Geochemistry of shales. In: Plant JA, Jones DG (eds) Metallogenic models and exploration criteria for buried carbonate-hosted ore deposits – a multidisciplinary study in eastern England. Brit Geol Surv, Keyworth, Inst Min Metall, London, pp 65–94

Jones DG, Plant JA, Colman TB, Swainbank IG (1991) New evidence for Visean-Namurian shales as the source of the Pennine mineralisation of England. In: Pagel M, Leroy JL (eds) Source, transport and deposition of metals. Proc 25 years SGA Anniversary Meet, Nancy, 30 Aug–3 Sept 1991. Balkema, Rotterdam, pp 309–312

Kimbell GS, Chadwick RA, Holliday DW, Werngren OC (1989) The structure and evolution of the Northumberland Trough from new seismic reflection data and its bearing on modes of continental extension. J Geol Soc (Lond) 146:775–787

Leach DL, Rowan EL (1986) Genetic link between Ouachita foldbelt tectonism and the Mississippi Valley-type lead-zinc deposits of the Ozarks. Geology 14:931–935

Leeder MR (1982) Upper Palaeozoic basins of the British Isles – Caledonide inheritance versus Hercynian plate margin processes. J Geol Soc (Lond) 139:479–491

Leeder MR, McMahon AH (1988) Upper Carboniferous (Silesian) basin subsidence in northern Britain. In: Besley BM, Kelling G (eds) Sedimentation in a synorogenic basin – the Upper Carboniferous of northwest Europe. Blackie, Glasgow, pp 43–52

LeHuray AP, Caulfield JB, Rye DM, Dixon PR (1987) Basement controls on sediment-hosted Zn-Pb deposits: a Pb isotope study of Carboniferous mineralisation in central Ireland. Econ Geol 82:1695–1709

Lydon JW (1986) Models for the generation of metalliferous hydrothermal systems within sedimentary rocks and their applicability to the Irish Carboniferous Zn-Pb deposits. In: Andrew CJ, Crowe RWA, Finlay S, Pennell WM, Pyne J (eds) Geology and genesis of mineral deposits in Ireland. Irish Assoc Econ Geol, Dublin, pp 555–577

Max MD, Ryan PD, Inamdar DD (1983) A magnetic deep structural geology interpretation of Ireland. Tectonics 2:431–451

McArdle P (1990) A review of carbonate-hosted base metal-baryte deposits in the Lower Carboniferous rocks of Ireland. Chron Rech Min 500:3–29

McKenzie DP (1978) Some remarks on the development of sedimentary basins. Earth Planet Sci Lett 40:25–32

Michard A (1989) Rare earth element systematics in hydrothermal fluids. Geochim Cosmochim Acta 53:745–750

Mills H, Halliday AN, Ashton JH, Anderson IK, Russell M (1987) Origin of a giant orebody at Navan, Ireland. Nature 327:223–226

Ottlick P, Galfi J, Horvath F, Stegena L (1981) The low enthalpy geothermal resource of the Pannonian Basin, Hungary. In: Rybach L, Muffler LJP (eds) Geothermal systems: principles and case histories. Wiley, New York, pp 221–245

Phillips WEA, Sevastopulo GD (1986) The stratigraphic setting and structural setting of Irish mineral deposits. In: Andrew CJ, Crowe RWA, Finlay S, Pennell WM, Pyne J (eds) Geology and genesis of mineral deposits in Ireland. Irish Assoc Econ Geol, Dublin, pp 1–30

Phillips WEA, Stillman CJ, Murphy T (1976) A Caledonian plate tectonic model. J Geol Soc (Lond) 132:779–609

Phillips WEA, Rowlands A, Coller DW, Carter JS, Vaughan A (1988) Structural studies and multidata correlation of mineralization in central Ireland. In: Boissonnas J, Omenetto P (eds) Mineral deposits within the European Community. Springer, Berlin Heidelberg New York, pp 353–377

Plant JA, Jones DG (eds) (1989) Metallogenic models and exploration criteria for buried carbonate-hosted ore deposits – a multidisciplinary study in eastern England. Brit Geol Surv, Keyworth, Inst Min Metall, London, 161 pp

Plant JA, Jones DG, Brown GC, Colman TB, Cornwell JD, Smith K, Smith NJP, Walker ASD, Webb PC (1988) Metallogenic models and exploration criteria for buried carbonate-hosted mineral deposits: results of a multidisciplinary study in eastern England. In: Boissonas J, Omenetto P (eds) Mineral deposits within the European Community. Springer, Berlin Heidelberg New York, pp 321–352

Rankin AH, Graham MJ (1988) Na, K and Li contents of mineralizing fluids in the Northern Pennine Orefield, England and their genetic significance. Trans Inst Min Metall Sect B Appl Earth Sci 97:B99–B107

Riley NJ (1991) Stratigraphy of the Worston Shale Group, Dinantian, Craven Basin, NW England. Proc Yorks Geol Soc 48:163–187

Robinson BW, Ineson PR (1979) Sulphur, oxygen and carbon isotope investigations of lead-zinc-baryte-fluorite-calcite mineralization, Derbyshire, England. Trans Inst Min Metall Sect B Appl Earth Sci 88:B107–B117

Rogers PJ (1977) Fluid inclusion studies in fluorite from the Derbyshire orefield. Trans Inst Min Metall Sect B Appl Earth Sci 86:B128–B132

Russell MJ (1978) Downward-excavating hydrothermal cells and Irish-type ore deposits: importance of an underlying thick Caledonian prism. Trans Inst Min Metall Sect Appl Earth Sci B87: B168–B171

Russell MJ (1983) Major sediment-hosted exhalative zinc + lead deposits: formation from hydrothermal convection cells that deepen during crustal extension. In: Sangster D (ed) Sediment-hosted stratiform lead-zinc deposits. Min Assoc Can, Short Course Handbook 8, pp 251–282

Russell MJ (1986) Extension and convection: a genetic model for the Irish Carboniferous base metal and baryte deposits. In: Andrew CJ, Crowe RWA, Finlay S, Pennell WM, Pyne JF (eds) Geology and genesis of mineral deposits in Ireland. Irish Assoc Econ Geol, Dublin, pp 545–554

Sangster DF (1990) Mississippi Valley-type and sedex lead-zinc deposits: a comparative examination. Trans Inst Min Metall Sect B Appl Earth Sci 99:B21–B42

Sawkins FJ (1966) Ore genesis in the North Pennine orefield in the light of fluid inclusion studies. Econ Geol 61:385–401

Sharp JM (1978) Energy and momentum transports model of Ouachita Basin and its possible impact on formation of economic mineral deposits. Econ Geol 73:1057–1068

Shepherd TJ, Darbyshire DPF, Moore GR, Greenwood DA (1982) Rare earth element and isotope geochemistry of the North Pennine ore deposits. In: Gites filoniens Pb Zn F Ba de basse temperature du domain varisque d'Europe et d'Afrique du Nord. Symposium Orleans. Bull Bur Rech Geol Min Sect II (2-3-4):371–377

Sheppard SMF, Langley KM (1984) Origin of saline formation waters in northeast England: application of stable isotopes. Trans Inst Min Metall Sect B Appl Earth Sci 93:B195–B201

Sheridan DJR (1972) The stratigraphy of the Trim No. 1 well, Co. Meath and its relationship to Lower Carboniferous outcrop in east-central Ireland. Bull Geol Surv (Ireland) 1:311–344

Smith FW (1974) Yttrium content of fluorite as a guide to vein intersections in partially developed fluorspar ore bodies. Trans Am Inst Min Eng 256:95–96

Solomon M, Rafter TA, Dunham KC (1971) Sulphur and oxygen isotope studies in the northern Pennines in relation to ore genesis. Trans Inst Min Metall Sect B Appl Earth Sci 80:B259–275

Steed GM (1986) The geology and genesis of the Gortdrum Cu-Ag-Hg orebody. In: Andrew CJ, Crowe RWA, Finlay S, Pennell WM, Pyne JF (eds) Geology and genesis of mineral deposits in Ireland. Irish Assoc Econ Geol, Dublin, pp 481–499

Strogen P, Jones GLl, Somerville ID (1990) Stratigraphy and sedimentology of Lower Carboniferous (Dinantian) boreholes from west Co. Meath, Ireland. Geol J 25:103–137

Symons DTA, Sangster DF (1991) Paleomagnetic age of the central Missouri baryte deposits and its genetic implications. Econ Geol 86:1–12

Taylor HP Jr (1974) The application of oxygen and hydrogen isotope studies to problems of hydrothermal alteration and ore deposition. Econ Geol 69:843–883

Todd SP, Murphy FC, Kennan PS (1991) On the trace of the Iapetus suture in Ireland and Britain. J Geol Soc (Lond) 148:869–880

White DE (1958) Liquid of inclusions in sulfides from Tri-State (Missouri-Kansas-Oklahoma) is probably connate in origin. Bull Geol Soc Am 69:1660 (Abstr)

Williams B, Brown C (1986) A model for the genesis of mineral deposits in Ireland. In: Andrew CJ, Crowe RWA, Finlay S, Pennell WM, Pyne JF (eds) Geology and genesis of mineral deposits in Ireland. Irish Assoc Econ Geol, Dublin, pp 579–590

Worley NE, Ford TD (1977) Mississippi Valley type orefields in Britain. Bull Peak Dist Mines Hist Soc 6:201–208

Genesis of Sphalerite Rhythmites from the Upper-Silesian Zinc-Lead Deposits – A Discussion

M. Sass-Gustkiewicz and K. Mochnacka

Abstract

Based on the data collected from the Pomorzany Zn-Pb mine, the authors present a formation model of "rhythmically banded sphalerite ores" ("rhythmites") known from the Mississippi Valley-type (MVT) Zn-Pb deposits. The origin of these ores is controversial and the discussion reflects the differences in syngenetic versus epigenetic interpretation. Meso-, macro- and microscopic observations lead to the conclusion that the ores are of metasomatic origin and resulted from two processes: (1) replacement contemporaneous with dissolution; (2) cavity filling. The genetic model explains the transition from primary depositional rhythmicity (abab) to a superimposed, non-depositional, bipolar one (abcdcba).

1 Introduction

The so-called "rhythmically banded sphalerite ores" (or "rhythmites") are well known from various MVT localities including the Upper Silesian Zn-Pb deposits. The origin of this specific type of sulphide accumulation is still a matter of discussion. Two contrasting genetic concepts have been developed by Bogacz et al. (1973) and Fontboté and Amstutz (1982). The difference between the concepts is as controversial as the syngenetic versus epigenetic interpretation. However, attention should be paid to both ideas since these have been based on the same dataset derived from the same mine (Trzebionka).

Observations presented in this chapter originate from the Pomorzany mine. Rhythmites found here appear to be identical with those from the Trzebionka mine (for detailed descriptions, see Bogacz et al. 1973; Mochnacka and Sass-Gustkiewicz 1981; Fontboté and Amstutz 1982). The authors attempt to interpret these structures in terms of the criteria applied by advocates of the two contrasting genetic concepts.

The main features of rhythmites and their two genetic interpretations given by Bogacz et al. (1973) and by Fontboté and Amstutz (1982) are listed and cited in Table 1. The authors will comment on them in the same order using the same headings.

Institute of Geology and Mineral Deposits, University of Mining and Metallurgy, al. Mickiewicza 30, 30-059 Krakow, Poland

Spec. Publ. No. 10 Soc. Geol. Applied to Mineral Deposits
Fontboté/Boni (Eds.), Sediment-Hosted Zn-Pb Ores

Table 1. Selected features of sphalerite rhythmites and their genetic interpretation

Bogacz et al. (1973)	Fontboté and Amstutz (1982)
Host rocks	
"the dolomite is a crystalline neosome and shows metasomatic cross-cutting contacts against the surrounding carbonates".	"ores occur in carbonates, mostly dolomites, formed in tidal flat environments".
Contact relationships	
"the contacts of ores with enclosing dolomite are cross-cutting showing somewhat jagged appearance".	"great geometric congruence of the ore textures on all scales with the enclosing rock . . ."
Succession of geometric elements in rhythmites (see Table 2)	
Metasomatic features	
"metasomatic replacement of cavity walls has contributed to the formation of ores"	"the lack of evidence of replacement must also be pointed out"
Adherence	
"the banded ores reflect the pattern of primary stratification"	"the rhythmic banding is not 'inherited' from preexisting depositional rhythmicity"
Origin	
"rhythmites are a result of dissolution of the OBD, deposition of sulphides in open cavities and metasomatic replacement of the cavity walls"	"diagenetic crystallization rhythmites are a result of differentiation by fractional crystallization during diagenesis"

2 Host Rocks

The host rock for the Upper Silesian Zn-Pb ores is the ore-bearing dolomite (OBD). The origin of the OBD is interpreted either as metasomatic by Bogacz et al. (1973) or as diagenetic (tidal-flat) by Fontboté and Amstutz 1982).

Based on the observations on all the scales (meso-, macro- and microscopic), the authors agree with the opinion of Bogacz et al. (1973) and interpret the OBD as a neosome in relation to the pre-existing, enclosing carbonate rocks (limestones and early diagenetic dolomites) of tidal-flat origin. For a detailed discussion on the origin of the OBD, see Bogacz et al. (1975).

3 Contact Relationships

Contacts between the rhythmites and the hosting OBD are perceived in completely different ways: cross-cutting (Bogacz et al. 1973) or geometrically congruent (Fontboté and Amstutz 1982).

The contacts show, in fact, both features perfectly observable on the scale of a single orebody. For tabular bodies contacts tend to follow the bedding planes at a distance of several tens of metres but they can change abruptly into cross-cutting contacts if the thickness of an orebody swells rapidly 10–20 times, forming the nest. If the hosting OBD is laminated, the contacts are jagged (see Bogacz et al. 1973, Pl. XXXII, Fig. 1). If the OBD is spotted, the contacts are irregular and, finally, if the host rock is structureless, the contacts are gradational.

3 Succession of Geometric Elements in Rhythmites

Banding in sphalerite rhythmites originates from the repetition of various geometric elements easily distinguishable in and comparable between the orebodies. Mineralogy of the elements appears to be identical (Table 2), but both the time relationships between the elements and the genetic interpretation of the processes involved are different for the two groups of authors. Bogacz et al. (1973) claim that the elements "bear a record of three processes: 1 – cavity making; 2 – cavity filling; 3 – metasomatic replacement". Furthermore, "these processes were essentially contemporaneous the replacement following by the short margin the formation of cavities". Finally, "the formation of cavities and cavity filling always precede the replacement and formation of replacement grains". Such an interpretation differs essentially from that of Fontboté and Amstutz (1982; see Table 2). This controversy is a main moot point in the whole discussion on the origin of rhythmites.

In the authors' opinion, based here on microscopic observations, the processes involved in the formation of rhythmites follow the concept of Bogacz et al. (1973), but the succession of the formation processes of geometric elements corresponds to that of Fontboté and Amstutz (1982).

Metasomatic transformation of limestones into the sphalerite ores has been described in detail by Mochnacka and Sass-Gustkiewicz (1981). The general idea has been applied here to explain the origin of rhythmically banded sphalerite ores. Figure 1A illustrates the selected steps of transformation (Nos. 1–10) as observed in both transmitted and reflected light.

The OBD has originated from the dolomitization of limestones (1). The first metasomatic process was the formation of dolomitic limestone composed of calcite and dolomite I (2) with the local recrystallization of the latter.

Table 2. Comparison of geometric elements distinguished in sphalerite rhythmites by [1], Bogacz et al. (1973); [2], Fontboté and Amstutz (1982) (with genuine names); and [3], present authors (cf. Figs. 1 and 2)

Essentials of ore rhythm [1]	Generations of diagenetic crystallization rhythmite [2]	Geomeric elements of bipolar rhythmicity [3]
1 Solution cavities	III Empty central space	d Empty space
2 Sphalerite Encrustations	II c Pyrite and marcasite b Subhedral sphalerite a Schalenblende	c Marcasite ZnS IIIb crystalline ZnS IIa collomorph
3 Sphalerite Replacement rims	I Aggregates of sphalerite and inclusions of dolomite	b Zns II replacement rims ZnS I disseminated
4 Relics of unmineralized dolomite		a Relics of unmineralized dolomite

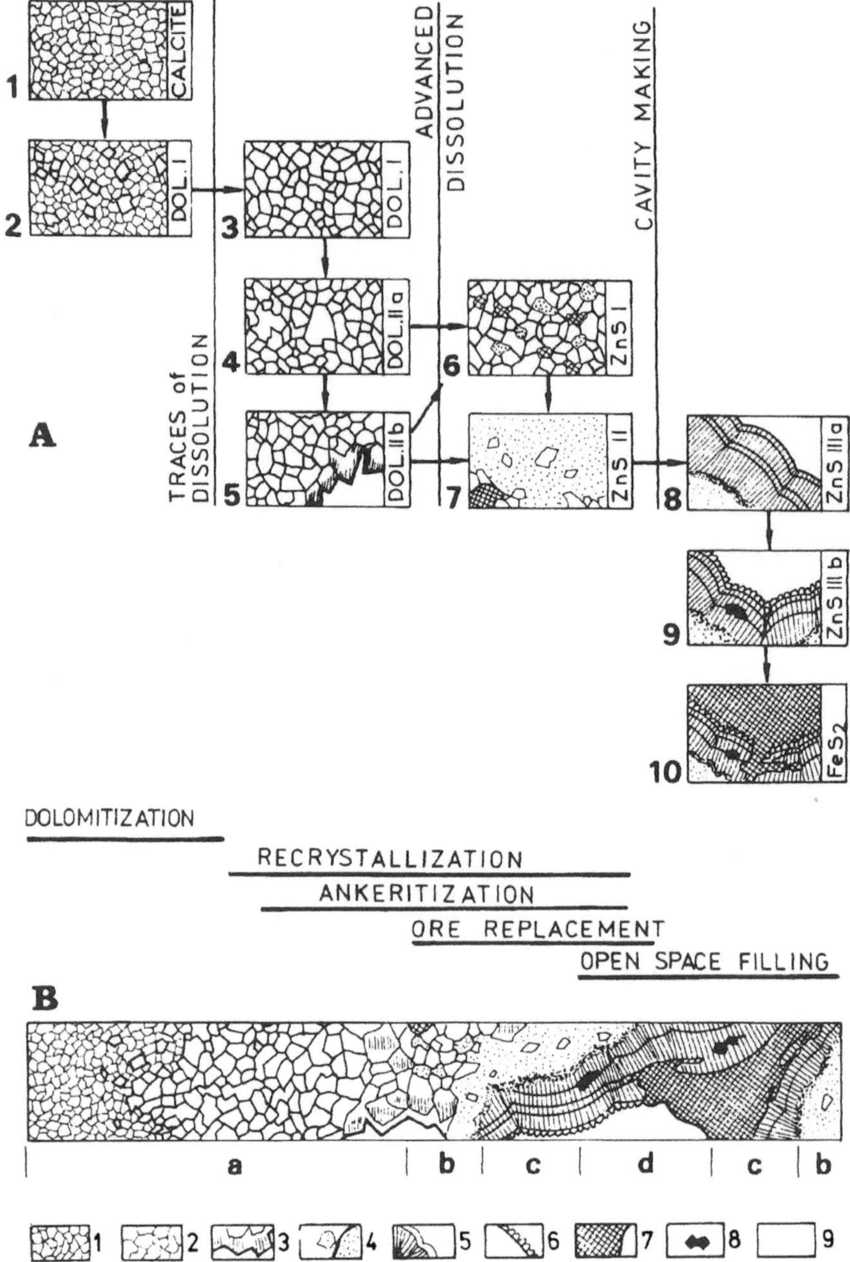

Fig. 1A,B. Development of processes involved in sphalerite rhythmite formation. **A** Schematic drawings of selected steps (1–10) in rhythmite formation (for explanations see text) (modified after Mochnacka and Sass-Gustkiewicz 1981); **B** schematic drawings of sphalerite rhythmite and their geometric elements *a*, *b*, *c*, *d*. *1* Calcite; *2* dolomite; *3* ankeritic zones in dolomite crystals; *4* disseminated crystals and aggregates of sphalerite; *5* collomorph sphalerite; *6* crystalline sphalerite; *7* iron sulphides; *8* galena; *9* empty voids

Progressive dolomitization led to the formation of pure dolostone (dolomite I) (3). Recrystallization of dolomite I (4) resulted in coarse-grained, idiotopic or partly hipidiotopic dolostone (dolomite IIa). Dolomitization was associated with the dissolution and formation of vugs. These vugs have been partly filled with the idiotopic dolomite rhombs of concentric structure (dolomite IIb) with outer zones composed of ankerite (5).

Formation of dolomite II (a and b) was accompanied by the appearance of disseminated sphalerite I of metasomatic origin (6). Massive ZnS (sphalerite II) with relics of dolomite rhombs originate from the aggregation of disseminated crystals (7). Sphalerite II occurs in an already completely recrystallized dolomite. The idiotopic/hipidiotopic rhombs of dolomite show typical concentric zonation with an outer ankeritic zone – a feature especially well developed in rhombs of dolomite protruding into the empty voids (5). The increasing number of voids is an effect of a more advanced dissolution which accompanies the formation of metasomatic sphalerite ores.

Contacts between ZnS I and II are gradational. The number and size of sphalerite crystals increase towards the empty voids. The final effects are massive sphalerite II replacement rims around the walls of the voids.

A further step in the mineralizing processes is the infilling of empty voids with ZnS III. This generation can be divided into two subgenerations: IIIa and IIIb. The ZnS IIIa is collomorph and forms alternating, light honey and dark brown layers (8). The ZnS IIIb is crystalline, light yellow and covers the surface of the ZnS IIIa (9).

Marcasite, pyrite and galena appear in various stages of metasomatic replacement of the OBD. Marcasite is the most common mineral and may replace sphalerite III (10) and its older generations and may fill the empty voids. The occurrence of these sulphides will not be described in detail as they do not affect the discussion on rhythmites.

The filling of empty voids formed during metasomatic replacement of the OBD could proceed any time after the completion of this process, i.e. also during the formation of a karst deposit.

4 Metasomatic Features

The genetic concepts either prove (Bogacz et al. 1973) or refute (Fontboté and Amstutz 1982) the involvement of metasomatic replacement in the formation of rhythmites.

Summarizing the results of microscopic observations so far, the authors suggest that:

1. Rhythmites originate from the sequence of metasomatic processes: dolomitization, recrystallization, ankeritization and replacement of the OBD by sphalerite (see also Arne and Kissin 1989);
2. The sequence of these processes is the same in all the studied samples;
3. Replacement may cease at any moment, leaving the full spectrum of metasomatites: dolomitic limestones, dolostones, ankeritic dolostones, dolostones with disseminated sphalerite, dolostones with sphalerite aggregates and massive sphalerite ores.

5 Adherence

Both contrasting concepts include adherence as a crucial argument in the genetic discussion (Table 1). It appears, however, that congruency versus lack of congruency between the ore rhythms and the sedimentary rhythms is understood differently by both groups of authors. Bogacz et al. (1973) claim that these structures are inherited because "the lack of real congruency between ore rhythms and sedimentary rhythms testifies strongly against the syngenetic origin of the ores following the pattern of sedimentary structures", and "it is only the general pattern of ores that is partly congruent with the general trend of sedimentary interfaces". In contrast, Fontboté and Amstutz (1983, p. 332) believe that the rhythmicity of banded ores cannot be inherited from primary structures because they do not repeat the rhythmicity of "Sedimentary rhythmic structures".

Comments on this problem are based on the microscopic studies summarized in Fig. 1.

Geometrical elements of rhythmites (a, b, c and d in Fig. 1B) are related to the succeeding metasomatic processes observed under the microscope (Fig. 1A). Thick horizontal lines illustrate the range of specific processes.

Traces of dissolution observed under the microscope allow one to distinguish the steps in this process (Fig. 1A, vertical lines). The following features are evident:

1. Each step of metasomatic replacement is accompanied by dissolution of carbonates;
2. Dissolution voids are observed exclusively in samples which show evidence of replacement;
3. Intensity of dissolution is directly proportional to the intensity of replacement.

These features prove that close genetic relationships exist between metasomatic replacement and dissolution of the OBD.

Dissolution processes easily develop in carbonate rocks which are readily dissolved by even slightly acid solutions. Deposition of sulphides in the Upper Silesian ores appears to result from the mixing of cold, meteoric, oxidizing and hot, ascending, metal- and sulphur-bearing solutions. Consequent oxidation of the latter is accompanied by the release of substantial amounts of H^+ ions which contributes to the dissolution reaction (Barnes 1983).

It should be emphasized here that the sphalerite rhythmites are an integral but younger part of a sulphide karst deposit. Dissolution of host rocks proceeded, without any question, contemporaneously with the deposition of sulphides, which points to an identical geochemical regime governing the whole ore-forming process.

Taking into account the replacement-dissolution relationships (Fig. 1A) and the microscopic structure of geometric elements (Fig. 1B), the genetic model is proposed for the origin of rhythmites. The model attempts also to explain how the rhythmites follow the general bedding pattern (i.e. the sedimentary structures) but do not repeat the sedimentary rhythms. In other words, it seems important to understand how the primary "abab" depositional rhythmicity can be transformed into the new, superimposed, bipolar "abcdcba" rhythmicity (Fig. 2).

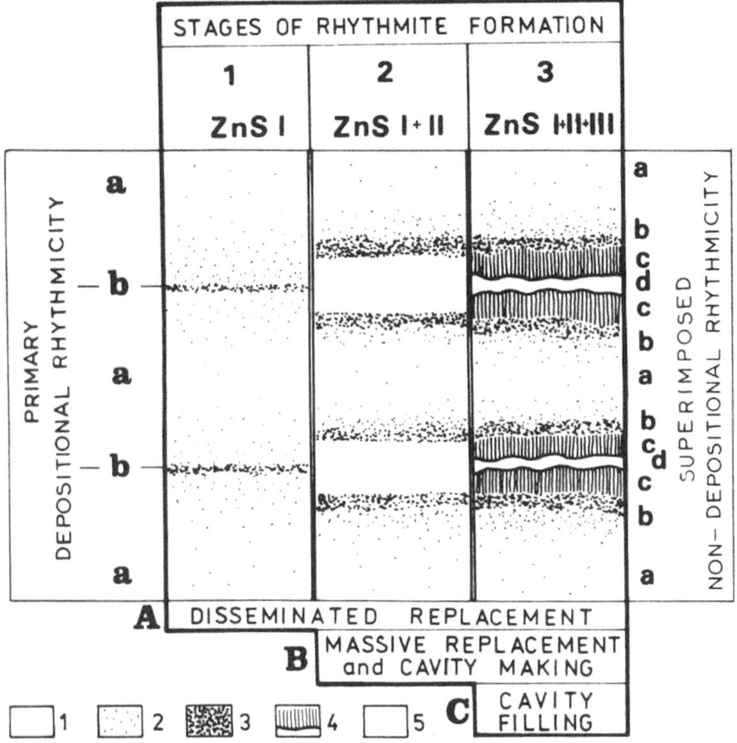

Fig. 2. Stages of rhythmite formation (*1*, *2*, *3*) related to the mineralizing processes (*A*, *B*, *C*) and the transformation of primary rhythmicity (*abab*) into the bipolar one (*abcdcba*). *1* Relics of unmineralized OBD; *2* disseminated sphalerite I; *3* replacement rims of sphalerite II; *4* encrustations of sphalerite III; *5* empty voids

Three stages of rhythmite formation are distinguished (Fig. 2, stages 1, 2, 3) based on the dominating mineralizing processes: disseminated replacement, massive replacement (both closely connected with cavity making) and cavity filling (A, B, C in Fig. 2).

In the first stage disseminated replacement follows mainly the sedimentary discontinuities (b). This process is accompanied by the formation of early, microscopic dissolution voids. As the reaction proceeds, these geometrically oriented open spaces may facilitate the transfer of mineralizing solutions in a preferable direction.

Further progress of replacement (stage 2) causes the enlargement and confluence of small voids into the larger (macroscopic) sheet cavities enclosed between the two parallel surfaces. Continuously supplied mineralizing solutions penetrate into the pore space of the OBD along the cavity walls and spread into the unaltered parts causing contemporaneous mineralization (sphalerite) and dissolution (dolomite). Therefore, the enlargement of the sheet cavities affects also the dolostone enclosing the already formed disseminated sphalerite crystals. These crystals are released from the hosting rock and removed by the solution flowing through the cavities. The process proceeds until the dolomite crystals are completely leached out and/or replaced by sphalerite II in the zone adjacent to the walls of the cavities. Such a mechanism is responsible for sheet-cavity making and for the formation of their massive replacement rims (Fig. 2, stage 2). Metasomatic replacement of the OBD and the formation of a disseminated sphalerite halo develop until the massive replacement rims "seal" the walls of the cavities and preclude further penetration of solutions into the pore space.

In the final stage 3 (Fig. 2) the newly formed empty voids are filled with sphalerite III.

Development of replacement and dissolution causes the transformation of primary depositional rhythmicity (abab) into the superimposed, non-depositional one (abcdcba) (Fig. 2). The latter is typically bipolar because the replacement process develops symmetrically from the primary, "starting" surface "b".

If the primary, depositional rhythmicity is closely spaced and if the metasomatic processes last for a sufficiently long time, the mechanism described above may cause almost complete replacement of dolostone.

6 Origin

The two contrasting and conceivable genetic concepts relate the origin of sphalerite rhythmites either to the dissolution of the OBD, deposition of sulphides in open spaces and metasomatic replacement of OBD by ZnS (Bogacz et al. 1973) or to diagenetic crystallization resulting from differential crystallization fractionation (Fontboté and Amstutz 1982).

The authors confirm, in general, the opinion of Bogacz et al. (1973). The only but essential difference is the succession of the processes involved in the formation of rhythmites. The authors regard metasomatic replacement and dissolution to be genetically closely related, contemporaneous and always preceding the cavity filling. Therefore, the succession includes two processes: (1) replacement always connected with dissolution and (2) cavity filling, instead of: (1) dissolution (cavity making); (2) cavity filling; (3) replacement, as reported by Bogacz et al. (1973).

The considerations presented above seem to be valid also in the case of cavernous and irregularly banded sphalerite ores which reveal the same type of rhythmicity. Irregularly banded sphalerite ores occur in the close vicinity of the inherited rhythmites within the same orebodies but are incongruent with sedimentary structures and, thus, could not be inherited from the palaeosome. This problem has been touched on by Bogacz et al. (1973) but has not been solved as yet. The genetic model for these structures will be proposed in a separate paper.

Acknowledgements. A syllabus of this paper was presented during the IAS Symposium, "Sedimentology related to mineral deposits", held in Beijing, China, in 1988. Thanks are due to Professor G.C. Amstutz, who kindly suggested the publication of its extended version. The authors also appreciate the discussion and language corrections from Dr. W. Mayer.

References

Arne DC, Kissin SA (1989) The significance of "diagenetic crystallization rhythmites" at Nanisivik Pb-Zn-Ag deposit, Baffin Island, Canada. Mineral Depos 24:230–232

Barnes HL (1983) Ore-depositing reactions in Mississippi Valley type deposits. In: Kisvarsanyi G (ed) Proc Vol Int Conf on M-V-T Pb-Zn deposits. Univ Missouri-Rolla, Rolla, Missouri, pp 77–85

Bogacz K, Dzulynski S, Haranczyk C, Sobczynski P (1973) Sphalerite ores reflecting the pattern of primary stratification in the Triassic of the Cracow-Silesian region. Ann Soc Geol Pol 43:285–300

Bogacz K, Dzulynski S, Haranczyk C, Sobczynski P (1975) Origin of the ore-bearing dolomite in the Triassic of the Cracow-Silesian Pb-Zn district. Ann Soc Geol Pol 45:139–155

Fontboté L, Amstutz GC (1982) Observations on ore rhythmites of the Trzebionka Mine, Upper Silesian-Cracow region, Poland. In: Amstutz GC, El Goresy A, Frenzel G, Kluth C, Moh G, Wauschkuhn A, Zimmermann R (eds) Ore genesis, the state of the art. Springer, Berlin Heidelberg New York, pp 83–91

Fontboté L, Amstutz GC (1983) Diagenetic crystallization rhythmites in Mississippi Valley type ore deposits. In: Kisvarsanyi G, Grant SK, Pratt WP, Koenig JW (eds) Proc Vol Int Conf on MVT Pb-Zn deposits. Univ Missouri-Rolla, Rolla, Missouri, pp 328–337

Mochnacka K, Sass-Gustkiewicz M (1981) The metasomatic zinc deposits of the Pomorzany mine (Cracow-Silesian ore district, Poland). Ann Soc Geol Pol 51:133–151

Note of the Editors

This contribution discusses the interpretation given by Fontboté and Amstutz (1982) to the rhythmic ore textures of the Upper Silesian zinc-lead deposits described also by Bogacz et al. (1973). In particular, it is criticized that clear evidence for replacement and metasomatism was disregarded. These criticisms are justified, as already noted in Fontboté and Gorzawski (1990), when describing similar rhythmic ore textures of the San Vicente MVT deposit in Peru where it is outlined that the rhythmic textures are the result of the reaction of an influxing brine with the host rock under considerable burial. The question of the creation of the conspicuous rhythmicity of the banded ores in Upper Silesia and elsewhere is discussed in Fontboté (1993). In contrast to Sass-Gustkiewicz and Mochnacka (this chap.), who propose that the banding is essentially inherited from primary depositional rhythmicity, Fontboté (1993) prefers to see these fabrics as the result of a self-organization process, i.e. the spontanous transition of a nonpatterned state to a patterned state without the intervention of a patterned external cause, under conditions far from equilibrium. The regular spacing of the bands is interpreted as a result of a kinetic feedback between areas of dissolution. Initial hetereogeneities, for instance, primary bedding or structural fracturing, may influence but not fully determine the textural evolution leading to the rhythmic banding. This interpretation is largely compatible with the model presented by Sass-Gustkiewicz and Mochnacka (L.F.).

Fontboté L (1993) Self-organization fabrics in carbonate-hosted ore deposits: the example of diagenetic crystallization rhythmites (DCRs). In: Fenoll Hach-Ali P, Torres-Ruiz J, Gervilla F (eds) Current research in geology applied to ore deposits. Proceedings of the Second Biennial SGA Meeting, Granada, Spain, Sept. 9–13 1993, pp 11–14

Fontboté L, Gorzawski H (1990) Genesis of the Mississippi Valley-type Zn-Pb deposit of San Vicente, central Peru: geological and isotopic (Sr, O, C, S) evidences. Econ Geol 85:1402–1437

Geochemometrical Studies Applied to the Pb-Zn Deposit Bleiberg/Austria

E. Schroll[1], H. Kürzl[2] and O. Weinzierl[3]

Abstract

The concept of geochemometry, based on the combination of chemical and isotope data, aims at the development of ore deposit models and genetic classifications in an objective, quantitative way.

In a case study geochemical information on the Pb-Zn deposit Bleiberg/Austria is analyzed by using statistical-graphical methods studying single-element features as well as multivariate patterns in an exploratory, unconventional way. The results show that it is possible to use these techniques to characterize typical ore units with different genetic implications of the deposit investigated.

First attempts have been made to develop a high dimensional, general geochemical model of carbonate-hosted Pb-Zn deposits by testing the multivariate statistical approach according to inherent interelement relationships and individual element characteristics.

1 Introduction

Petrogeochemical analysis replaced chemical bulk analysis by partial analysis, including even trace elements. Classification of magmatic rocks and identification of geotectonic domains based on geochemical analysis have become already standard methods in petrology.

Mineral and ore deposits can be evaluated as specific cases of the genesis of rocks. The diversity and affinity of deposits, the differentiation within them as well as the often observable chemical reactions with the host rock have made the application of geochemical methods a quite unequal and difficult exercise. In general, not the ore itself but single mineral phases are more suitable for investigation.

This type of research, based on the idea of the "guide elements" (Goldschmidt 1924) and the corresponding quantitative analysis of the mineral phases, finally led to the concept as defined by the senior author (Schroll

[1] Bundesversuchs- und Forschungsanstalt Arsenal Postfach 8 A-1031 Vienna, Austria
[2] LMS Umweltsysteme Gesellschaft mbH, Franz-Josef-Straße 6 A-8700 Leoben, Austria
[3] Institut für Geodatenerfassung und Geosystemanalyse Montanuniversität Leoben, Roseggerstraße 15, A-8700 Leoben, Austria

Spec. Publ. No. 10 Soc. Geol. Applied to Mineral Deposits
Fontboté/Boni (Eds.), Sediment-Hosted Zn-Pb Ores

1976, 1979, 1984, 1985a,b, 1990) to include statistically not only chemical concentrations but also data from isotope analyses in the sense of a "geochemometrical" interpretation.

The constantly growing amount of analytical measurements, generated by modern, computerized techniques, needs a new approach in data reduction and interpretation. The development of methods in statistical and graphic analysis in combination with powerful new computerized assessment systems has prepared the ground for a new attempt to bring the idea of "geochemometry" into realization.

The investigation discussed in this chapter, which forms part of a larger research program, is limited to the geochemometric presentation of the Pb-Zn deposit Bleiberg (also named Bad Bleiberg or Bleiberg/Kreuth) in Carinthia, Austria.

2 Concepts of Geochemometry

The term "geochemometry" proposed by Schroll (Schroll and Caglayan 1986; Schroll 1990) describes a methodology for "surveying" geological bodies by using techniques of analytical geochemistry combined with suitable mathematics to develop models in an objective, exploratory way. The combination of chemical concentrations, isotope data and other measurable parameters provides the essential, basic information for geochemical characterization of geological bodies and ore deposits.

All these parameters are dependent on the physico-chemical conditions and dynamic systems prevailing in the environment of the geological source rocks and locations of deposition. It can be inferred that the geochemical data derived from these milieus include condensed information about the genetic history of the geological bodies investigated.

The dimension of time, important for all geochemical processes, can be included in the relevant data model by adding isotope data and derived ratios insofar as they can be connected to radioactive decay processes or to seawater conditions, as in the case of stable isotopes.

The obtained data system represents a complex network of relations. Exact monitoring of these interobject dependencies, however, does not seem to be a realistic task because of various difficulties in data generation and multivariate object characterization. It must be considered, for example, that the sample material has to be destroyed before examination and the supply of samples can vary to a certain extent. Additionally, the amount of quantitative data generated in this way cannot give any additional information concerning the texture or structure of the mineral paragenesis of the ore investigated.

Since representative in situ measurements cannot be obtained easily by the applied analytical techniques, the effect of average samples cannot be eliminated. It must also be considered that the sample taken should represent the underlying geological process sufficiently and that the number of samples cover natural variances. Therefore, mineralogical and petrographical investigations are supplementary approaches employed to characterize the sample geochemically.

The ideal result of such investigations would be an accurate estimation and complete understanding of the average composition of the orebody in terms of the sum of ore and gangue minerals according to their major, minor and trace element contents. In contrast to the more homogeneous rock formations, this is rarely possible with regard to orebodies. More often, a rough estimation of ore grades, element ratios or element concentrations can already be considered a success.

Therefore, the procedure seems to be more successful when working with trace element concentrations. In the examination of the Pb-Zn ore types, sphalerite and galena are the most suitable indicator minerals. They can be investigated together with gangue minerals like the carbonate minerals calcite and dolomite including their relation to the carbonatic host rocks, as well as the fluorite and sulphate minerals like anhydrite and baryte.

Isotope data strongly support the interpretation of other chemical data and also allow the detection of hidden genetic features. In this project, data derived from S-isotope measurements of sulphides and sulphates as well as C- and O-isotope measurements in carbonates and the Pb isotopes of Pb ores have been added to the chemical dataset.

However, even with these data the limits have not yet been reached. Sr isotopes in rocks and ores, Pb isotopes of the host rock or the H-isotopes in fluid inclusion and other data from fluid inclusion measurements could represent further possibilities to distinguish the geochemical parameters of an ore deposit.

3 Statistical-Graphical Methods

Modelling geological processes, e.g. the genetic development of ore deposits according to geochemical data, is a challenging task and requires new and unconventional approaches. In dealing with complex, natural phenomena the sampling procedure and the generation of analytical data are crucial factors for a reliable interpretation. Experience shows, however, that sampling and analytical variability can still be high and that they are largely uncontrolled. These manifold errors introduced under such circumstances add some uncertainty to the data, which in turn can create outliers and discontinuities in the distribution. Therefore, the analysis and interpretation of this data type may be considered as "soft data" in contrast to "hard data", which are derived from a well-defined environment where parameters are clearly defined and any change is carefully controlled.

Application of statistics to geochemical datasets has been in line with the technical developments in analytical chemistry and data processing systems for the past 20 years, when routine multi-element measurements started on a larger scale. A comprehensive overview of statistical methods applied mainly to regional reconnaissance data (e.g. stream sediment surveys) is documented in Howarth (1983).

A new way of interpreting such data types came into use with the introduction of "exploratory data analysis" (EDA) as developed by Tukey (1977). This technique reveals a variety of properties best suited for dealing with inconsistent, heterogeneous datasets and exhibits further conceptional advantages over classical statistical models by introducing an outlier-resistant, robust calculation of distribution parameters and correlations (Huber 1981). In many case studies and projects EDA has been successfully applied for the interpretation of geochemical data (Kürzl 1988a,b; Kürzl and O'Connor 1988; Dutter 1987).

In this chapter, however, this concept of EDA with its graphical-analytical techniques and multivariate, robust statistical models is applied probably for the first time to this type of geochemical data for an ore deposit in such a comprehensive and consequent way. The discussed case study presents a few examples of the analytical capabilities offered by EDA. It represents a new dynamic procedure to explore data structures and multivariate relations to find further unknown properties of the deposits analyzed. It starts by examining the behaviour of single variables by graphic techniques like box plots, density traces or one-dimensional scatter plots.

The condensed, graphic description of an empirical distribution of geochemical variables by the box plot is well suited to make comparisons between different groups like ore units or individual deposits.

A variety of multivariate symbols can be applied also in an exploratory way to characterize multivariate relations by generating patterns of the individual objects, e.g. deposits or ore units.

Different multivariate statistical techniques are used for geochemical classification. For the most useful ones robust techniques have been introduced which lead to stabler and more reliable results without eliminating the datasets of outliers or inconsistencies. The principal component analysis (PCA) with the biplot as its graphic display (Gabriel 1971) and correspondence analysis (Howarth 1983) proved to be most efficient for interpreting multivariate geochemical deposit data. Although quite similar in statistical concept, the two methods complement one another. Correspondence analysis represents the differences in shape (i.e. differences in the scale of the geochemical variables are removed), while the PCA provides the size and shape of the data vectors. Further detailed descriptions and references concerning the applied multivariate data analysis techniques can be found in Barnett (1981). To assist further detailed interpretation (e.g. discovery of natural groups) supplementary categorical data (e.g. different ore units of Bleiberg or different ore deposits) are encoded and represented by different symbols in the graphics.

All these multivariate statistical techniques need complete, stable data matrices, which contain about three times more objects than variables in the system in order to produce reliable and stable results. For the Bleiberg deposit this prerequisite has been met in most cases. However, for multivariate comparison with other deposits far more data should be available, thus allowing a detailed analysis of the system to deduce subtile genetic differences in the element concentrations in time and space.

4 Geology of the Bleiberg Ore Deposit

The Pb-Zn deposit of Bleiberg is located in the Mesozoic of the Gailtal Alps west of the town Villach in Carinthia/Austria (13°40′ E long. and 46°40′ N Lat). Bleiberg is the largest known Pb-Zn deposit in Austria. However, larger deposits of the same type exist in Raibl/Italy (Cave di Predil) and in Meziča (Miess)/Slovenia (Brigo et al. 1977). Most of the other base metal mineralizations of the same age – about 200 are known in the Eastern Alps – are characterized by minor metal enrichments. The latest presentation of the geology and genesis of the Bleiberg deposit is given by Cerny (1989b) and Zeeh and Bechstädt (this Vol.).

Up until now the Pb-Zn deposit of Bleiberg has produced about 3 million t of metal (Pb + Zn). A total metal content of up to 5 million t can be inferred (Cerny 1989a). The deposit itself extends 12 km in strike and 800 m have been explored in depth by underground mine workings. However, the extent of the whole deposit and its definite limits and borders to the host rock have not yet been explored in detail.

The distribution of the ore minerals Ba and F in the Bleiberg deposit is shown in Fig. 1. The ratio of Pb/Zn exhibits a wide range, from 1 to 20. For the total deposit a ratio of 1 to 1.5 has been calculated. This fact may be of special genetic importance with regard to the origin of the metals in that it may indicate a source of sedimentary clastic rocks.

Employing mineralogical investigations, about 50 different mineral phases could be detected. Two-thirds of them are secondary minerals of the oxidation zone. The most important primary ore minerals are galena, sphalerite and iron sulphides (marcasite and pyrite). Molybdenite is rare and can be found independently some distance away from the main Pb-Zn orebodies. Gangue is represented by calcite, dolomite, baryte (with galena) and fluorite (with sphalerite). Wulfenite is the most interesting mineral of the oxidation zone. Vanadium minerals are rare.

Traces of molybdenum are detectable in bituminous carbonate rocks, especially of the Cardita strata, in concentrations of 1 to 10 ppm Mo. The same concentration range of Mo is found in the ore minerals.

Figure 1 shows the stratigraphical situation of the five geochemical statistical groups related to the ore units presented in this study. The sample groups of the Kalkscholle (Erlach district) are genetically inhomogeneous and consist mainly of massive zinc ore but they also contain some breccia ores.

5 Geochemical Database

For statistical analysis and interpretation a dataset of nearly 4000 single data for the Bleiberg deposit was available during the last years. Together with further data derived from similar deposits, about 20 000 data could be gathered for the whole research project. Due to the fact that the geochemical investigation of the Bleiberg deposit covers a time period of about 40 years, the dataset does not correspond ideally to the requirements necessary for reliable data analysis and interpretation, mainly with regard to homogeneity.

Chemical data of 146 samples of sphalerite with about 10–20 single variables (depending on the analytical methods used) as well as 57 samples of galena with about 10–15 single variables were collected. The samples are mostly hand specimens. The mineral separation was done by handpicking or by subsequent enrichment by flotation in the laboratory. Some samples were

Table 1. Average values (in ppm) and variances of trace element distributions in sphalerite and galena (N = number of samples) of the Bleiberg deposit

Element	Sphalerite[a]					Galena				
	N	Md	Mn	rMn	rS	N	Md	Mn	rMn	rS
Ag	64	8	11	10	10	57	1.5	3.6	2.6	2.5
As	145	174	295	210	185	57	90	220	185	205
Cd[b]	145	1800	1910	1860	820	54	11	20	14	10
Cu	71	12	18	16	14	57	0.25	0.5	0.4	0.4
Fe	146	3450	4550	4240	3530	–	–	–	–	–
Ga	146	11	18	12	10	–	–	–	–	–
Ge[b]	146	200	335	270	230	57	0.1	0.6	0.3	0.2
In	146	0.1	2	0.1	0	57	0.1	0.1	0.3	0
Mn	145	24	32	27	15	57	0.3	1.6	1	1.2
Mo	139	0.8	4.1	1.5	1.8	57	0.1	1.6	0.4	0.5
Ni	139	3.4	4.6	3.9	3.2	–	–	–	–	–
Sb	145	1	3.7	1.8	1.4	57	35	76	59	64
Sn	63	0.1	0.5	0.2	0.2	57	0.1	0.3	0.2	0
Tl	146	59	97	78	66	57	1.7	3.4	2.1	1.6
V	139	1.1	2.2	1.9	1.8	57	0.1	0.2	0.1	0

[a] Md = median; Mn = arithmetic mean; rMn = robustly calculated mean; rS = standard deviation, also robustly calculated.

[b] The data of Cd and Ge can be compared with the real results of mining. For the period 1980 up to 1991 the following arithmetic averages were calculated: Cd 1910 ppm and Ge 174 ppm (Cerny and Schroll 1991). The arithmetic means (Mn) show excellent correspondence, while the median value (Md) of Ge is only comparable in the order of magnitude, since there is a larger variance in the distribution of Ge (see Fig. 2) and overrepresentation of samples of germanium-rich colloform sphalerite.

taken directly from the runoff concentrates of the mill. The corresponding statistics of the trace elements in sphalerite and galena are presented in Table 1.

The samples were taken at different times and by different persons. One part of the sample collection comes from detailed sampling carried out by Prof. Dr. O. Schulz and his team (University of Innsbruck). The chemical analyses were completed under the auspices of the IUGS-IGCP research program No. 6, 1983 to 1985 (Kostelka et al. 1986) and the raw mineral research program of the Austrian government ending 1989 (Cerny and Schroll 1991).

The analyses were carried out in the Geotechnical Institute BVFA Arsenal (Vienna) mainly by a specially developed method of optical emission spectroscopy with a carbon-arc excitation, which is especially suited for trace elements like Ge, Ga, In and Tl, which exhibit excellent detection limits.

Complementarily and in part with some redundancy, other methods were also applied, e.g. atomic emission spectroscopy with (AES), inductively coupled plasma excitation (ICP) or wavelength-dispersive X-ray fluorescence analysis (WD-XRF). In addition, (ICP-AES) mass spectroscopic data of four sphalerite samples were included (courtesy of Prof. Dr. Möller, Hahn-Meitner Institute, Berlin).

With respect to sulphur isotope data, 291 sulphide, 42 sulphate, 25 carbon and 25 oxygen isotope as well as 27 lead isotope measurements were integrated into the database. The sulphide data show a wide range in variation, from $\delta^{34}S$ −40‰ to −1‰ (CDT), according to mineral phase and ore types. Sulphates show a variation of 9 to 28‰. The mean value of baryte is around 15‰, for the other sulphates around 18‰. Grey anhydrite in concordant layers exhibits 15‰, which is the same isotope composition of seawater sulphate from the Triassic time period (Schroll et al. 1983 and unpubl. data).

The composition of carbon and oxygen in the host rock (e.g. Wettersteinkalk) is around $\delta^{13}C$ 2.6‰ and $\delta^{18}O$ −7.5‰ (PDB). Hydrothermal processes during diagenesis have only a very small influence on the C-isotope composition by changing slightly in the direction of the lighter C-isotope. Oxygen reaches maximum values of up to −16.5‰ (PDB) in gangue minerals or in late diagenetic carbonate formations of the swell facies (Kappel and Schroll 1982 and unpubl. data).

The lead isotopes do not show any significant changes within the whole ore deposit. The lead of the ore, however, is not identical to the lead of the host rock. The only exception is the lead of galena in the geodes from the first "Cardita schist". These geodes, which contain mainly iron sulphides, are interpreted as products of lateral secretion (Köppel and Schroll 1988).

The temperature range of the ore mineralization is between 100 to 200 °C. the leaching temperature of the metals from the source rocks is estimated at 200 °C based on the Ge/Ga ratio (Möller 1985). The rock sequence carrying the ore deposit has undergone a heating process by tectonic superimposition in the range of 100 to 150 °C (Kappel and Schroll 1982). Investigations of fluid inclusions show homogenization temperatures up to 200 °C, confirming the above-mentioned temperature range (Zeeh and Bechstädt, this Vol.).

6 Data Analysis and Geochemical Interpretation

The distribution of trace elements like Cd, Ga, Ge, Mn, Tl and Mn in sphalerite from the whole Bleiberg deposit is shown in Fig. 2. Without knowing the genetic differences in the Bleiberg ore units, at least three different populations can be distinguished from the distribution patterns. A similar situation can be seen by comparing the $\delta^{34}S$ density distributions in sphalerites (Fig. 3).

This rough subdivision of the ore deposit into statistically different units, which are also genetically determined, reveals significant differences. The distribution shown in Figs. 2 and 3 indicates that several genetically different units of ore must exist. By using the exploratory data analysis technique and a box-plot comparison as in Fig. 4, the Erzkalk ore units and the Kalkscholle show similarities, indicating a genetic relationship. The same observation could be made for Maxer Baenke and Cardita, when only the trace element Ge is used for classification. By using Cd, the form of the box plot and mean levels change, but it must also be noted that the number of data also changes significantly between the ore units. In the bottom row of the box-plot comparison in Fig. 4 the $\delta^{34}S$ values are displayed graphically.

The differences between the ore units are much more pronounced and the similarities between Erzkalk and Kalkscholle have disappeared. The geochemical characterization of the whole Bleiberg deposit by trace elements in

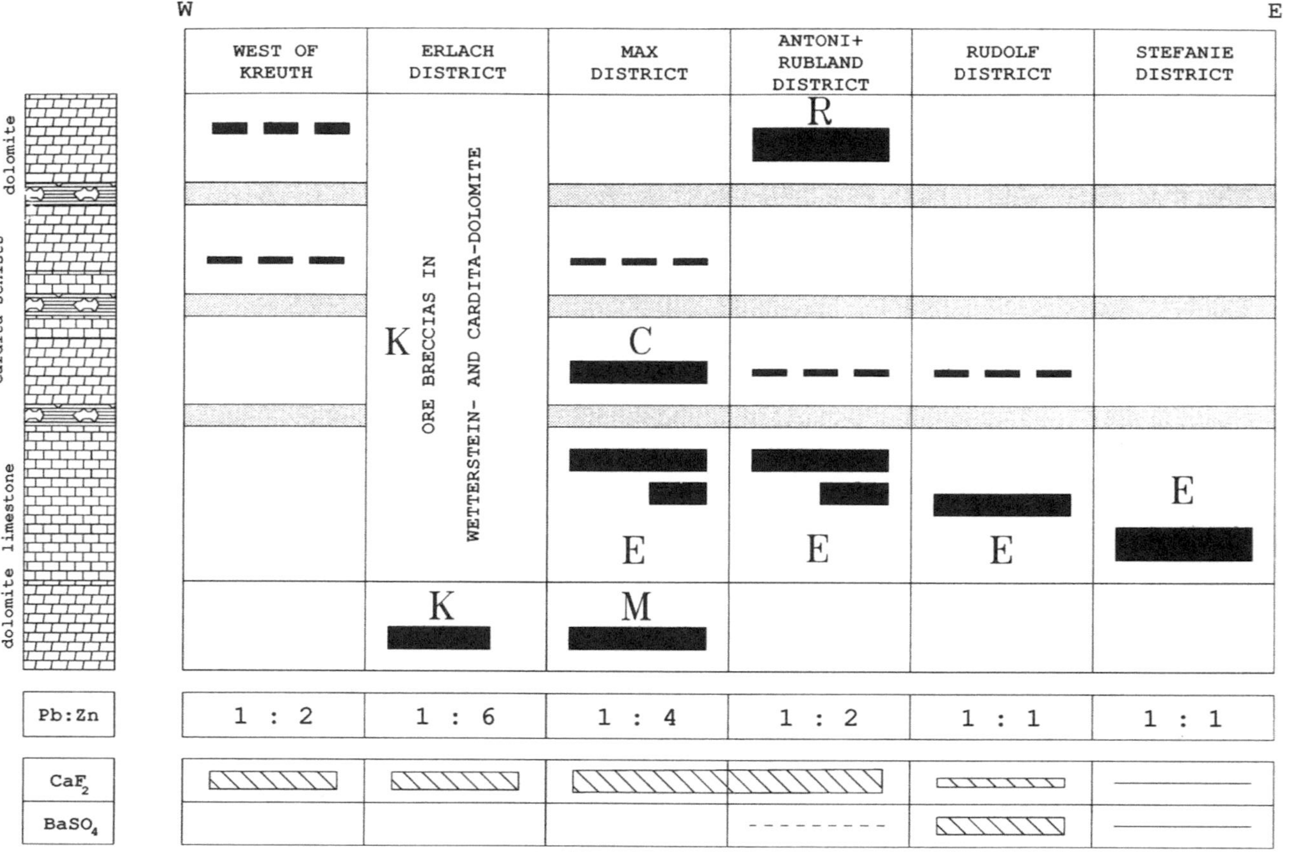

Fig. 1. The spatial distribution of the elements Pb, Zn (Ba, F) within the depositional units of the Bleiberg deposit (after Cerny, 1989a). *C* Cardita; *E* Erzkalk; *K* Kalkscholle; *M* Maxer Bänke; *R* Rubland. The intensity of the ore mineralization is indicated by the thickness of the *black bars* and reflects the statigraphical situation of the ore types associated with the geochemical data used in this work

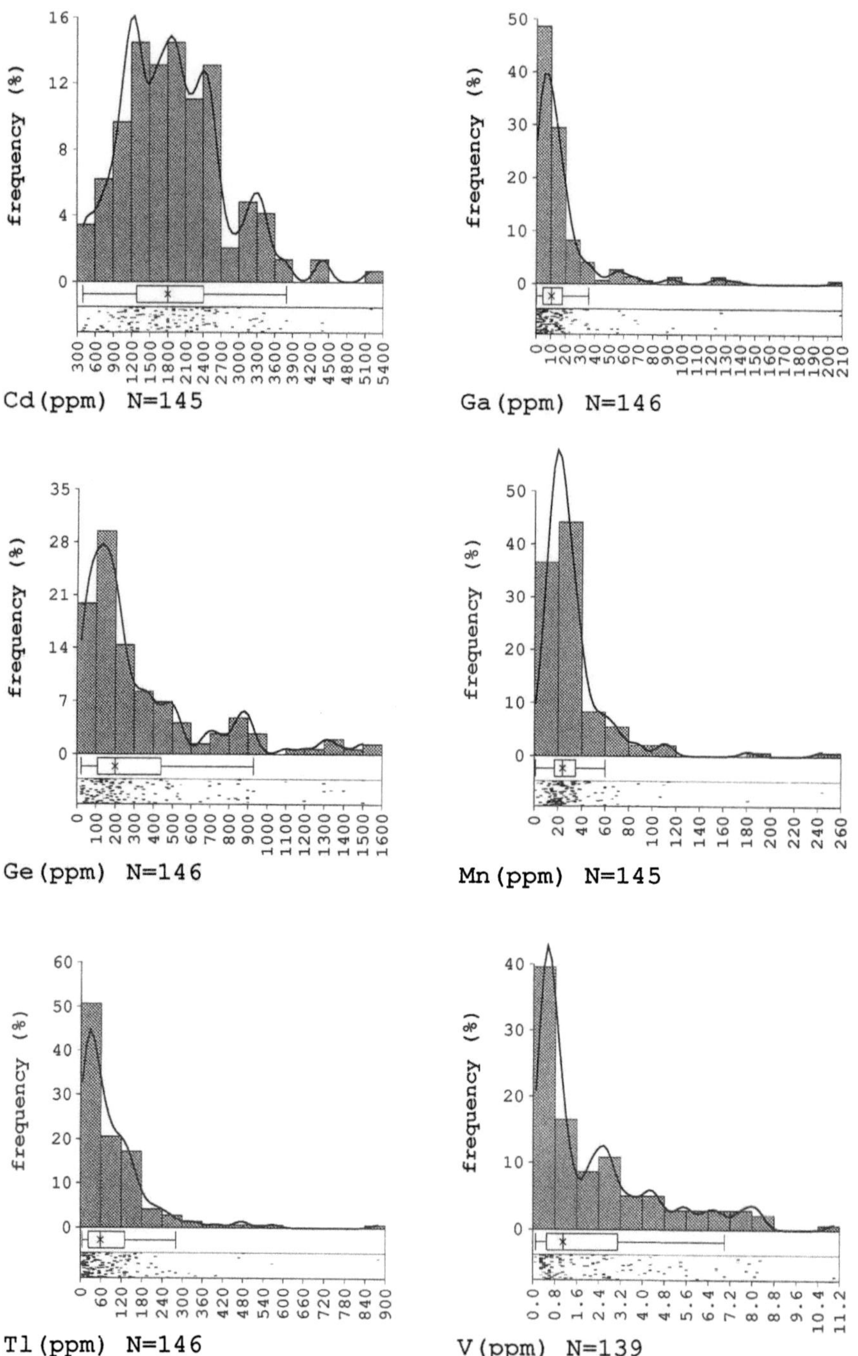

Fig. 2. Characterization of the density distribution of selected trace elements from sphalerites of the Bleiberg deposits by using graphic techniques of EDA

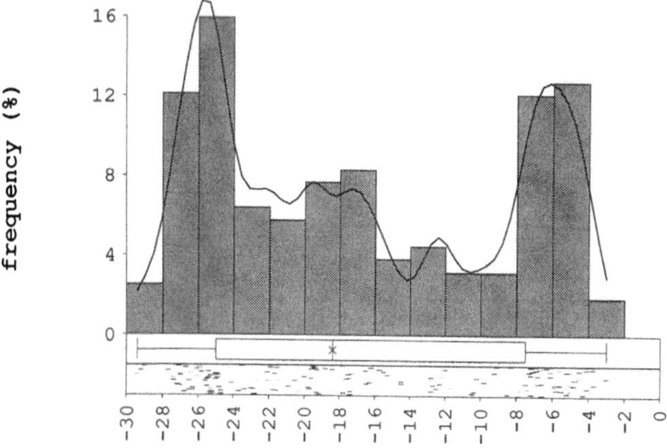

Fig. 3. Density diagram of the $\delta^{34}S$ values of sphalerites of the Bleiberg deposit using a similar presentation as in Fig. 2

sphalerite and galena is documented in Table 1. In Table 1 the arithmetic mean, the classical estimator of central tendency, can be compared with "central values" calculated with robust methods by weighting outliers and uncontrolled data influences. Therefore, the robust calculated mean is generally lower than the arithmetic one. The robust calculated variance representing the standard deviation has also been added in order to make a more reliable comparison of element variability between the two ore minerals and other similar ore deposits.

Quite remarkable are the values of Cu and Ag, which indicate Ag in sphalerite and the depletion of Cu. This observation may be of some importance genetically.

A summary of the bivariate behaviour of trace elements in sphalerite is presented in Fig. 5 in form of a "draftman's display" (Tukey and Tukey 1981). In this presentation the correlation matrix is depicted graphically and nonlinear behaviour as well as sample groupings, which cannot be easily traced by using numerical analysis only, can also be detected.

A straightforward way to graphically classify objects with multivariate data is by using multivariate symbols. An example for this technique is seen in Fig. 6. The use of a face as a multivariate symbol is much more expressive (Fig. 7). The affinity of Erzkalk and Kalkscholle and the differences between Maxer Baenke and Cardita are remarkable. The face of "Bleiberg avg." represents the average of the four units investigated.

When data for the same trace elements, derived from sphalerite and galena, are available, then the multivariate comparison of trace element behaviour within and between two mineral phases is possible. Principal component analysis (PCA) is also quite useful for this purpose and the results, also calculated by robust methods, are displayed in Fig. 8. However, the number of

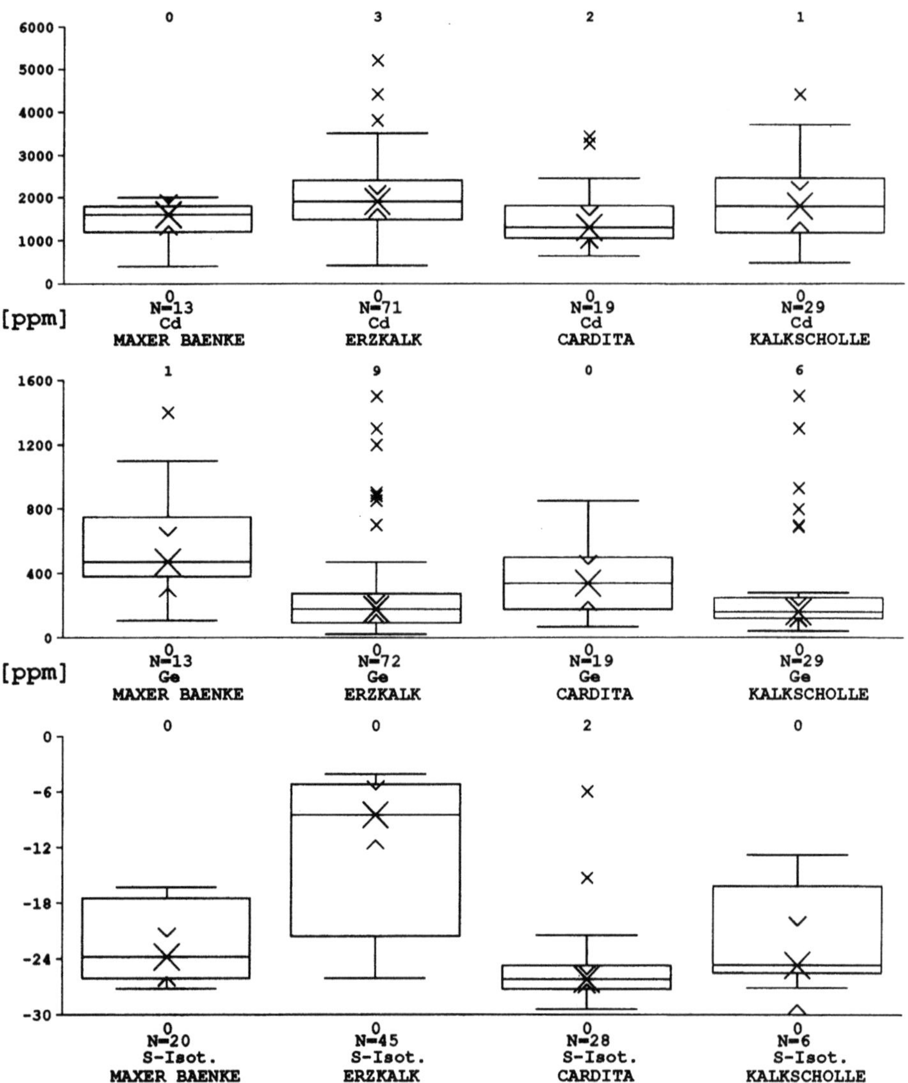

Fig. 4. Box-plot comparison of selected trace elements and isotope data from sphalerite of the different ore units within the Bleiberg deposit

common trace elements is limited to four, e.g. As, Ge, Mn and Tl. In addition, S-isotope data for both minerals are included. The results of the first three eigenvectors in terms of variable loadings can be found on the right-hand side of Fig. 8.

Figure 9 shows a PCA with nine trace elements and S-isotope data of sphalerite preclassified according to five ore units of the Bleiberg deposit. The corresponding loadings of the first three eigenvectors are shown on the right-hand side of Fig. 9.

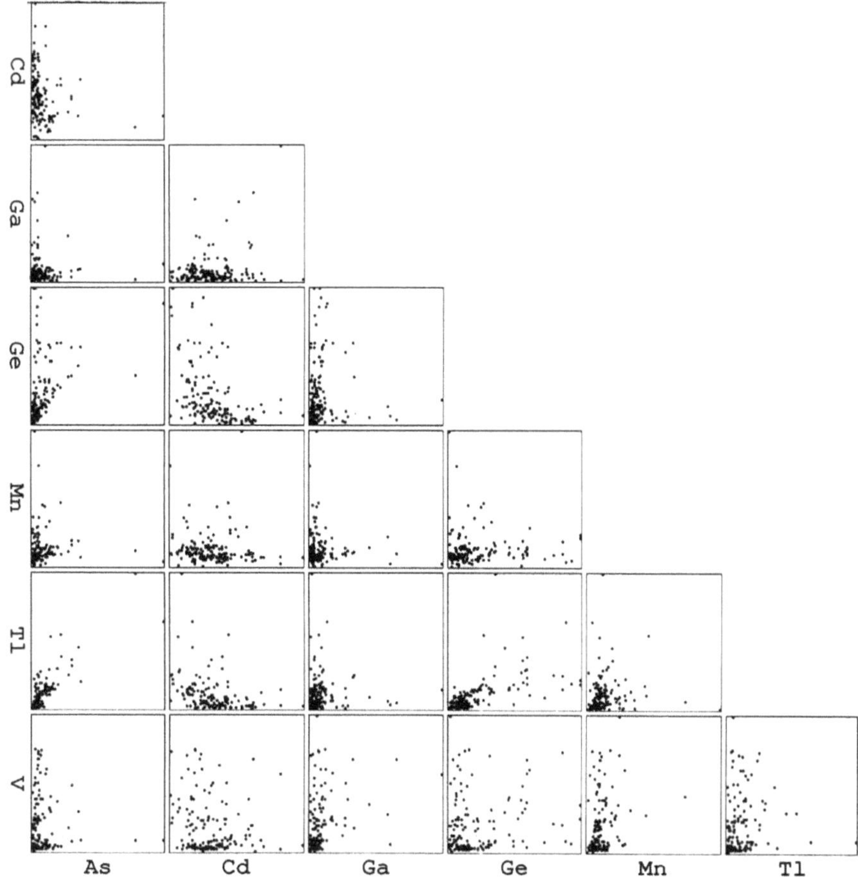

Fig. 5. A draftman's display of selected trace elements (N = 157) in sphalerite of the Bleiberg deposit for graphic evaluation of bivariate element behaviour. The interelement relations, e.g. the positive correlation of As and Tl or Ge and Tl as well as the negative trend with Ge and Cd or the independence of Mn and Cd, can be directly observed

The elements representing overall variance are As, Cd, Ge, Ni, Tl, including the S-isotopes, and can be found in the first eigenvector. Elements having a high discriminating power and included in the second eigenvector are As, Ga, Mo, Ni and V. The results do not give a homogeneous grouping, however, major trends and tendencies of the individual ore units can be observed. The strong correlation between Cd and S-isotopes influences the separation of samples significantly, showing a strong negative correlation with the other element group. The high correlation of As, Ni and Tl is also enhanced.

One important aim of the present investigation is the geochemical classification of the Bleiberg deposit in relation to other carbonate-hosted deposits. Several multivariate methods have been used and tested to obtain reliable and stable results. Correspondence analysis works according to a similar concept as PCA, but it removes the influence of absolute element values and identifies the

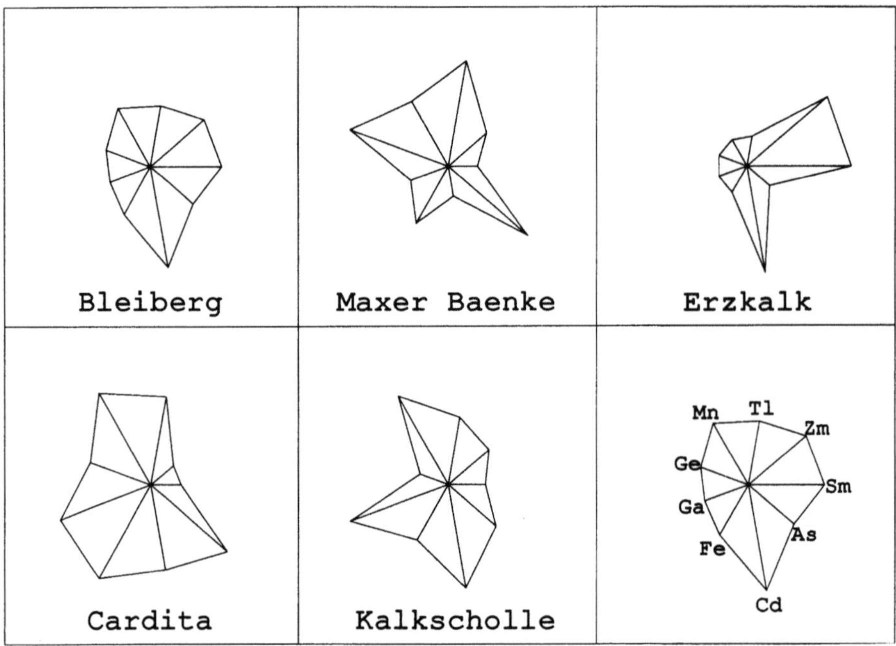

Fig. 6. By using a star as a multivariate symbol, 9 geochemical parameters exhibit the difference between the four major ore units of the Bleiberg deposit by forming different graphic patterns. For the whole Bleiberg deposit the shape is quite homogeneous due to the averaging effect which can be observed when the different units are displayed together. The pronounced change in patterns between the main ore units is a significant hint to a distinct geological evolution within the deposit

relationship of element associations in the geochemical characterization of ore deposits.

The results of such an analysis are presented in Fig. 10. For the multivariate analysis 11 variables were used (trace elements like As, Cd, Ga, Fe, Mn, Tl in sphalerite, S-isotope values of sulphides, sphalerite and sulphates and the μ_2-values of the Pb-isotope composition). Based on these variables, a homogeneous dataset for 25 deposits could be obtained as input matrix for the correspondence analyses. The results projected on the plane of the first and second axis are shown in Fig. 10.

In the Karawanke mountain range the mineralization of Windisch-Bleiberg corresponds to the Erzkalk type, while in Hochobir the predominance of Ga observed in the deposit is quite unusual (see also Cerny et al. 1982).

Using other datasets, it could be possible to show the position of further important and genetically interesting carbonate-hosted ore deposits, e.g. Les Malines (France), which would be located around the Bleiberg type, while the MVT deposits of North America would plot in a distinctly different area, indicating their own genetic, evolutionary processes (Schroll et al. 1991).

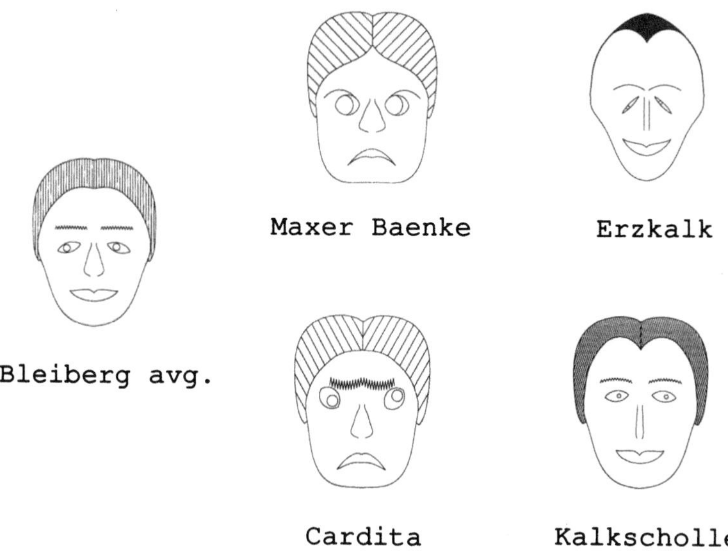

Explanation

Asrm -	eye size	- Asrm
Germ -	pupil size	- Germ
Gars -	position of pupil	- Gars
Ferm -	eye slant	- Ferm
Fers -	horiz. pos. of eye	- Fers
Garm -	vert. pos. of eye	- Garm
Sbrs -	curvature of eyebrow	- Sbrs
Sbrm -	density of eyebrow	- Sbrm
Mnrs -	horiz. pos. of eyebrow	- Mnrs
Inrm -	vert. pos. of eyebrow	- Inrm
Mnrm -	upper hair line	- Mnrm
Gers -	lower hair line	- Gers
Tlrm -	face line	- Tlrm
Cdrm -	darkness of hair	- Cdrm
Asrs -	hair shading slant	- Asrs
Tlrs -	nose line	- Tlrs
Cdrs -	size of mouth	- Cdrs
Cdrm -	curvature of mouth	- Cdrm

Fig. 7. Faces have been used to represent multivariate data in an efficient way in the form of condensed patterns (Chernoff 1973). The five faces are established by using 18 statistical parameters derived from 10 geochemical variables of sphalerite (As, Cd, Fe, Ga, Ge, In, Mn, Sb, Tl, V). These parameters are assigned to 18 components of the face, e.g. eye size, pupil size, nose line, etc. For example, the mean values of Ge in the different ore units define the pupil size, while the eye size is related to the mean values of As. The face as a multivariate symbol, representing high dimensional data in a very compact form, should be evaluated more as an overall impression applied to general comparisons

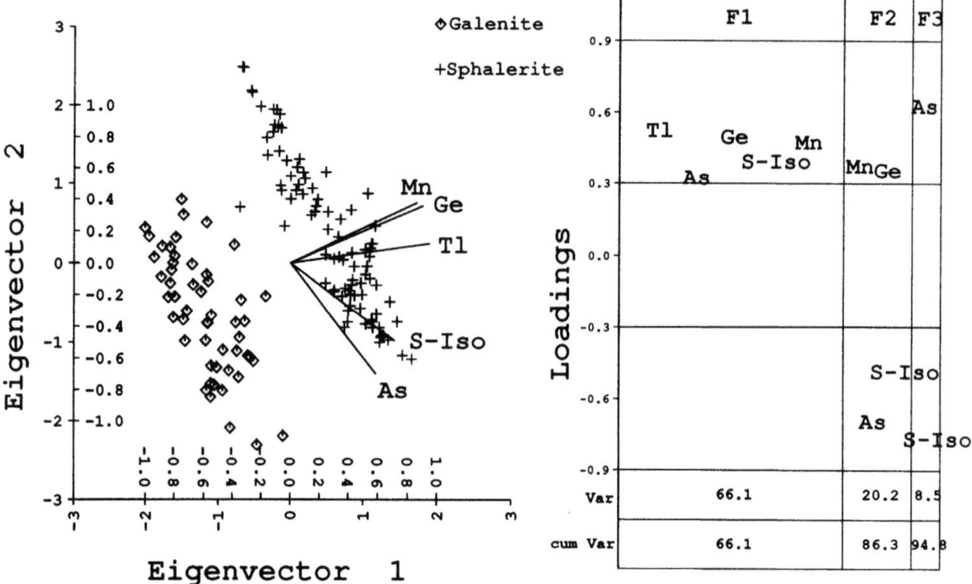

Fig. 8. ZnS and PbS trace elements and S-isotope data of the Bleiberg deposit in a biplot based on a principal component analysis. The loadings of the first three eigenvectors are displayed on the right. The biplot, representing the plane of the first and second eigenvector, shows a distinct grouping of the two sample types in relation to certain variable associations of the two mineral phases. Sphalerite is generally enriched in the trace elements in comparison to the galena data which exhibit element deficiency. Within both groups As and the S-isotope ratios appear as an intersample discriminator, thus indicating the same geochemical trend. This corresponds to the known differences between the element ratios As/Tl with 1 to 10 in sphalerite and 100 to 1000 in galena

7 Conclusion

The approach of this geochemometrical study using multivariate geochemical discrimination of ore deposits provides a technique that can deal with and analyze complex systems such as those existing in the selected example of the carbonate-hosted Pb-Zn deposit of Bleiberg. The availability of data of chemical concentrations of minor and trace elements in a single mineral phase made it possible to characterize the major ore types of this ore deposit. The graphic presentations of the geochemical data clearly show their variation and affinities. The combination of chemical and isotope data support further the determined position of the Bleiberg deposit in the Alpine Pb-Zn mineralizations and in the group of sediment-hosted Pb-Zn deposits in general.

This approach shows new ways of characterizing and classifying ore deposits by using geochemical data only. The only restrictions to comprehensive and complete geochemical deposit modelling are the lack of data even from

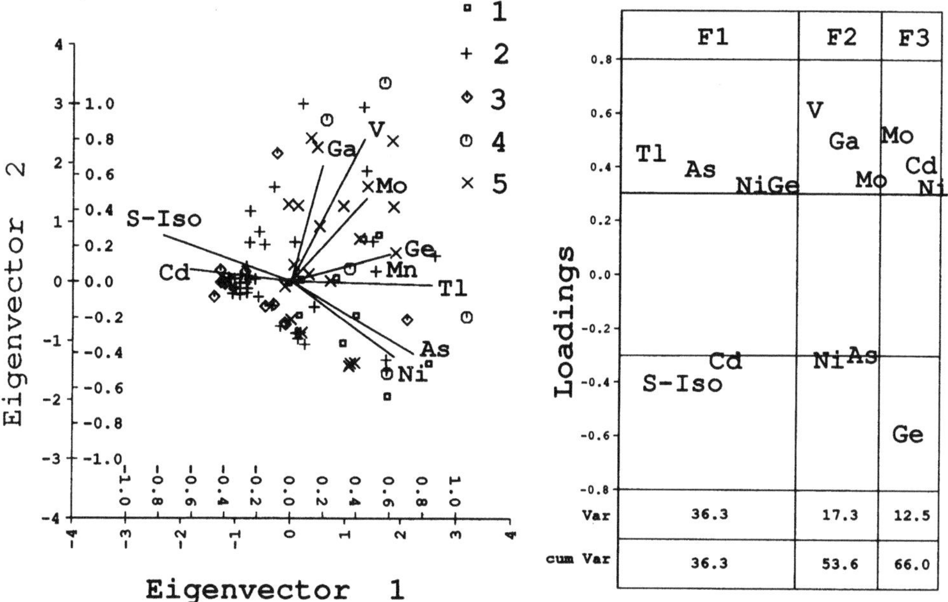

Fig. 9. ZnS trace element and isotope data of the Bleiberg deposit in a biplot. The corresponding loadings of the first three eigenvectors are displayed on the *right*. *1* Maxer Baenke; *2* Erzkalk; *3* Rubland; *4* Cardita; *5* Kalkscholle. Clusters of the Erzkalk samples are dominated by Cd and S-isotopes with deficiencies of other trace elements, although a few samples are scattered over the whole plot. The samples of the ore subunit Rubland (see Fig. 1) follow this trend quite clearly. The few samples representative of the Cardita units also show a high variance, but they are also strongly influenced by the trace element association. Kalkscholle ore clusters around the multivariate mean showing with a tendency to the trace element association of Ga-V-Mo. A few samples are enriched in Ni-As which form a separate subgroup. Similar to the Maxer Baenke ore, samples are situated around the Ni-As association, revealing a variance dominated nearly exclusively by these two elements

well-explored and well-known deposits and the heterogeneous databases available to date with regard to reliability and comparability.

The techniques used for graphic analysis and multivariate interpretation are a suitable way to deal with high dimensional data, and models can be investigated and established graphically in an exploratory manner. The number of ore deposits to be included in a multivariate analysis system are practically unlimited and around 40 geochemical variables can be easily managed.

Acknowledgements. The results presented in this study are part of a major project funded by the Austrian Scientific Research Foundation under project No. P6798-GEO, "Geochemometry of Sediment-Hosted Pb-Zn Deposits" (Schroll et al. 1991). Further support was granted by the Joanneum Research Association, Graz, the research organization of the province of Styria. Special thanks are due to Prof. J. Wolfbauer, Director of the Institute of Environmental system Analysis, Joanneum Research Association, for this steady support and his permission to use their data processing systems and software packages. Appreciation is also extended to Dr. Bäck from Logistic-Management Service for support in the preparation of this work and to Mrs. S. Tiefenbrunner for typing the manuscript.

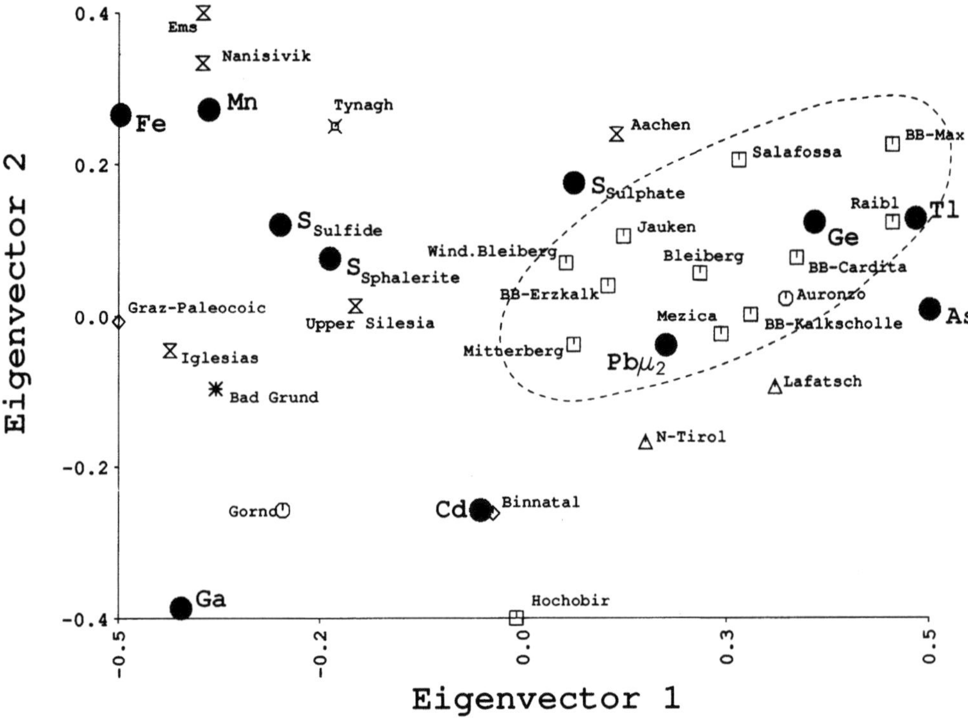

Fig. 10. By using the results of a correspondence analysis the observations (deposits) and variables (trace elements of ZnS and isotope data) are displayed in the plane of the first and second eigenvectors. The classification model is based on association patterns of the variables included in the analysis. The Bleiberg deposit with its ore units is mainly characterized by the association of Ge-Tl-As. The hatched line indicates high within-deposit variation between the extremes of Maxer Baenke and the Erzkalk. In this area Meziča, Raibl, Salafossa and the Anisian deposits of Auronzo (Italy) can also be found. Meziča indicates a strong affinity to Bleiberg, which is quite remarkable. In the Karawank mountain range the Windisch Bleiberg deposit corresponds to the Erzkalk type, while Hochobir, dominated by Ga, is atypical (see also Cerny et al. 1982). Among the 25 classified ore deposits, those dominated by the Fe-Mn association as well as by Ga and Cd are distinctly discriminated and separated from the Bleiberg group

References

Barnett V (ed) (1981) Interpreting multivaviate data. Wiley, Chichester, 374 pp

Brigo L, Kostelka L, Omenetto P, Schneider H-J, Schroll E, Schulz O, Strucl I (1977) Comparative reflections on four Alpine Pb-Zn deposits. In: Klemm D, Schneider H-J (eds) Time and stratabound ore deposits. Springer, Berlin Heidelberg New York pp 273–291

Cerny I (1989a) Die karbonatgebundenen Blei-Zink-Lagerstätten des alpinen und außeralpinen Mesozoikums. Arch Lagerstättenforsch Geol Bundesanst Wien 11:5–125

Cerny I (1989b) Current prospecting strategy for carbonate-hosted Pb-Zn mineralizations at Bleiberg Kreuth (Austria). Econ Geol 84:1430–1435

Cerny I, Schroll E (1991) Forschungsbericht zu Projekt ÜLG 13: Erfassung heimischer Vorräte an hochtechnologisch interessanten Spezialmetallen. Geol Bundesanstalt Wien, 80 pp (unpubl)

Cerny I, Scherer J, Schroll E (1982) Blei-Zink-Verteilungsmodelle in stilliegenden Blei-Zink-Revieren der Karawanken. Arch Lagerstättenforsch Geol Bundesanst Wien 2:15–22

Chernoff H (1973) The use of faces to represent points in k-dimensional space graphically. J Am Statist Assoc 68:361–368
Dutter R (1987) Robust statistical methods applied in the analysis of geochemical variables. In: Sendler W (ed) Contributions to stochastics. Physica Verlag, Heidelberg, pp 89–100
Gabriel KR (1971) The biplot graphic display of matrices with applications to principal components analysis. Biometrika 58:453–467
Goldschmidt VM (1924) Geochemische Verteilungsgesetze der Elemente II, Beziehungen zwischen den geochemischen Verteilungsgesetzen und dem Bau der Atome. Videnska Skr I Mat-Naturvidensk Kl (Kristiania) 4:1–37
Howarth RJ (ed) (1983) Statistics and data analysis in geochemical prospecting. In: Govett GJS (ed) (1983) Handbook of exploration geochemistry, vol 2. Elsevier, New York, 437 pp
Huber PJ (1981) Robust statistics. Wiley, New York, 308 pp
Kappel E, Schroll E (1982) Ablauf und Bildungstemperatur der Blei-Zink Vererzungen von Bleiberg-Kreuth, Kärnten. Carinthia II (Klagenfurt) 172/92:49–62
Köppel V, Schroll E (1988) Pb-isotope evidence for the origin of lead in strata-bound Pb-Zn deposits in Triassic carbonates of the Eastern and Southern Alps. Mineral Depos 23:96–103
Kostelka L, Cerny I, Schroll E (1986) Coordination of diagnostic features in ore occurrences of base metals in dolomite and limestone. Final Rep IGCP-Proj 6, Schriftenreihe der Erdwiss Kommission, Österr Akad Wiss Wien, Bd 8, pp 283–298
Kürzl H (1988a) Exploratory data analysis: recent advances for the interpretation of geochemical data. J Geochem Explor 30:309–322
Kürzl H (1988b) Graphical displays of multivariate geochemical data on scatter plots and maps as an aid to detailed interpretation. In: McCarn D (ed) Geological data integration techniques. Proc Technical Commitee Meeting organized by the International Atomic Energy Agency, 1986, IAEA-Tec Doc-472, Vienna, pp 245–272
Kürzl H, O'Connor PJ (1988) Multivariate patterns derived from graphical analysis of regional geochemical data as an efficient guide to mineralization. In: MacDonald DR, Mills KA (eds) Prospecting in areas of glaciated terrain 1988. The Canadian Institute of Mining and Metallurgy, Halifax, pp 469–484
Möller P (1985) Development and application of the Ga/Ge-geothermometer for sphalerite from sediment hosted deposits. In: Germann K (ed) Geochemical aspects of ore formation in recent and fossil sedimentary enviroments. Monograph Series on Mineral Deposits, Borntraeger, Berlin, 1–14
Schroll E (1976) Analytische Geochemie, Bd 2. Enke, Stuttgart, 374 pp
Schroll E (1979) Progress in the knowledge of indicator elements. In: Ahrens LH (ed) Origin and distribution of the elements. Pergamon, Oxford, pp 213–216
Schroll E (1984) Geochemical indicator parameters of lead zinc ore deposits in carbonate rock. In: Wauschkuhn A, Kluth C, Zimmermann RA (eds) Syngenesis and epigenesis in the formation of mineral deposits. Springer, Berlin Heidelberg New York, pp 294–305
Schroll E (1985a) From the guide element to the geochemical classification. In: Germann K (ed) Geochemical aspects of ore formation in recent and fossil sedimentary environments. Monograph Series of Mineral Deposits, Borntraeger, Berlin, pp 1–14
Schroll E (1985b) Geochemische Parameter der Blei-Zink-Vererzung in Karbonatgesteinen und anderen Sedimenten. Arch Lagerstättenforsch Geol Bundesanst Wien 6, pp 167–178
Schroll E (1990) Die Metallprovinz der Ostalpen im Lichte der Geochemie. Geol Rundsch 79(2):479–493
Schroll E, Caglayan H (1986) The Pb-Zn deposit of Keban (SE Taurus Mts.) and its position in the Aegean – Tauridian metallogenetic province. In: Petraschek WE, Jankovič S (eds) Geotectonic evaluation and metallogeny of the Mediterranean Arcs and Western Asia. Österr Akad Wiss Schriftenreihe Evolwiss Komm Wien 8:75–86
Schroll E, Schulz O, Pak E (1983) Sulfur isotope distribution in the Pb-Zn-deposit Bleiberg (Carinthia, Austria). Mineral Depos 18:17–25
Schroll E, Kürzl H, Weinzierl O (1991) Geochemometrie (multivariate Charakterisierung und Klassifizierung) von sedimentgebundenen Pb-Zn Vererzungen auf der Basis geochemisch-geologischer Meßdaten und Fakten. Endbericht FWF-P-6798-Geo. (Forschungsfond zur Förderung der wissenschaftlichen Forschung Wien, 6 Bände: davon Textband, 240 pp
Tukey JW (1977) Exploratory data analysis. Addison-Wesley, Reading, MA, 506 pp
Tukey PA, Tukey IW (1981) Graphical display of data sets in 3 or more dimensions. In: Barnett V (ed) Interpreting multivariate data. Wiley, Chichester, pp 189–274

Mississippi Valley-Type, Sedex, and Iron Deposits in Lower Cretaceous Rocks of the Basque-Cantabrian Basin, Northern Spain

F. Velasco, J.M. Herrero, P.P. Gil, L. Alvarez and I. Yusta

Abstract

Economic and subeconomic carbonate-hosted Zn-Pb and Fe deposits occur in the Basque-Cantabrian region at two distinct stratigraphic horizons in carbonate rocks of Lower Cretaceous age. The lower Aptian (Gargasian) horizon contains the economically most important base metal and iron ore deposits. Subordinate ore occurrences are found in a second horizon of upper Aptian to lower Albian age. The ore deposits are found predominantly at platform margins. Synsedimentary block tectonics controls the palaeogeography and appears to be an important factor for mineralization, mainly creating palaeohighs and providing feeder channels.

Ore occurs in both horizons mainly as (1) massive sulphide lenses (Sedex deposits), and (2) carbonate replacement orebodies, in part stratiform at the deposit scale (MVT and iron deposits). Main ore minerals are sphalerite and galena, which are accompanied by subordinate pyrite, chalcopyrite, fluorite and barite. Dolomite, calcite, quartz, as well as siderite and ankerite are frequent gangue minerals. Siderite and ankerite constitute the main ore minerals in the ore deposits in the central part of the basin (e.g. Gallarta). In several districts a late generation of cross-cutting veins with similar paragenesis as in the stratabound deposits is recognized.

Most ore minerals have formed in middle to late diagenetic stages. Fluid inclusions of these stages display homogenization temperatures ranging between 120 and 230 °C and moderate to high salinities. Stable (C, O and S) and Pb-isotope data are consistent with ore deposition from hot brines having directly or indirectly leached Palaeozoic materials. The formation of both MVT (e.g. Reocín) and Sedex (e.g. Troya) deposits, as well as that of the iron deposits is seen as the differentiated response to the discharge of hot saline brines during basin evolution, according to local geological characteristics.

1 Introduction

The Basque-Cantabrian region, located in northern Spain between the Basque Palaeozoic massifs and the Asturian massif (Fig. 1), hosts several ore deposits

Departamento de Mineralogía y Petrología, Universidad del País Vasco, Apdo. 644, E-48080 Bilbao, Spain

Spec. Publ. No. 10 Soc. Geol. Applied to Mineral Deposits
Fontboté/Boni (Eds.), Sediment-Hosted Zn-Pb Ores

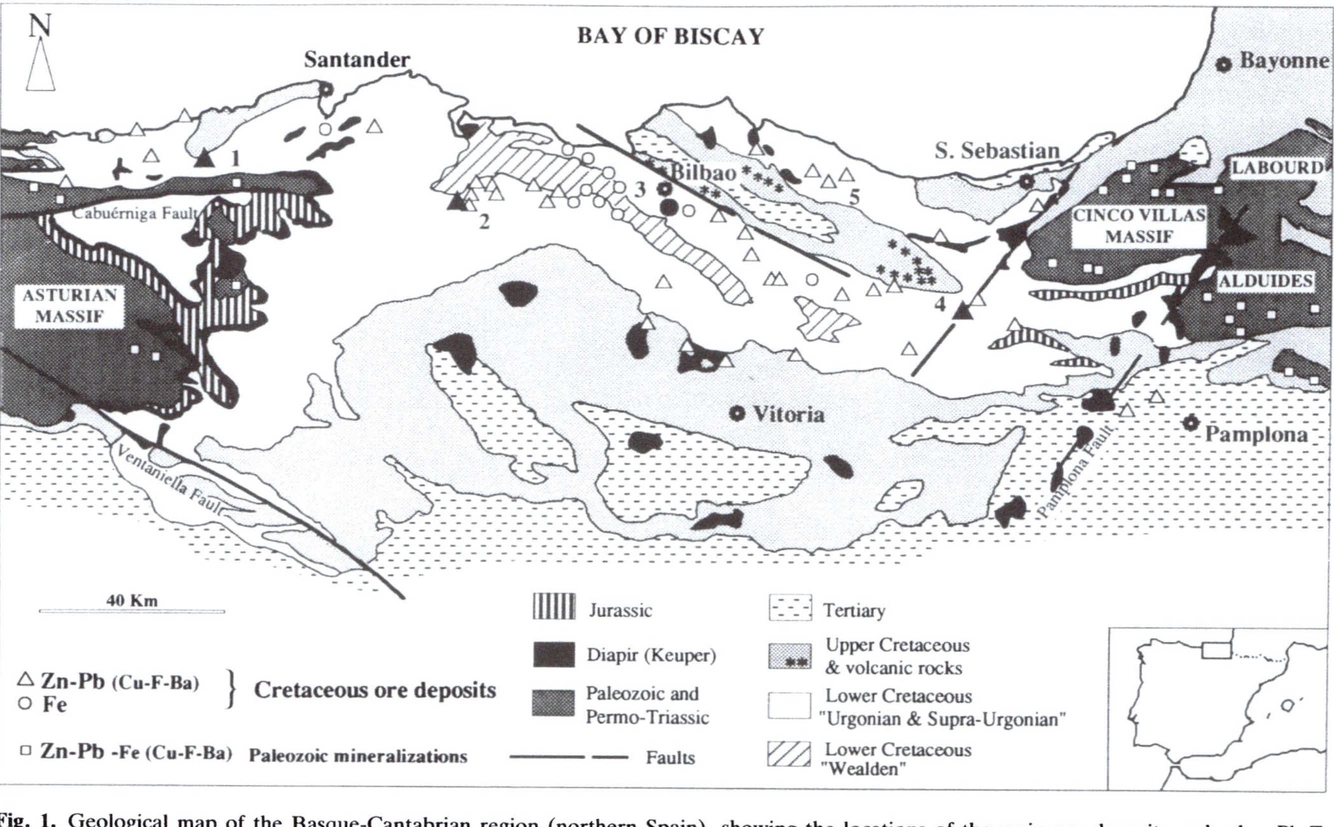

Fig. 1. Geological map of the Basque-Cantabrian region (northern Spain), showing the locations of the main ore deposits and other Pb-Zn occurrences. *1* Santander district (Reocín mine, Udías, La Florida, Orconera, Novales); *2* Western Biscay district (Txomin, Anselma, Matienzo, La Peña, El Mazo, Pozalagua, Siete-Puertas, Campo Fresco, La Rasa); *3* Bilbao district (Gallarta mine, Dícido, Galdames, Hoyo-Covarón, Alén, Bilbao, La Arboleda); *4* Guipúzcoa district (Troya mine, Legorreta, Katabera, Aralar); *5* northern Biscay district (Berriatua, Markina, Aulestia)

associated to Cretaceous carbonate rocks, some of which are still being exploited. The orebodies consist of variable concentrations of Zn, Pb, Cu, as well as, Fe, F and Ba minerals. Zinc, lead and iron are the main metals recovered. The most productive Pb-Zn ore deposits have been Reocín, Troya, Legorreta, and smaller deposits in the western Biscay district, whereas for Fe they have been Gallarta, Bilbao, Galdames, and Dícido. In addition, dozens of small ore occurrences are known, making this region a true metallogenic province.

The iron mining industry already mentioned by Plinius in his *Natural History* has constituted the greatest mining wealth in the region. The mineral was first worked in artisan ironworks and later in blast furnaces. Base metal sulphides and, occasionally, fluorite and barite ores were only treated from the middle of the nineteenth century; firstly, for lead and silver, then for zinc; and between the 1940 and 1975 for fluorite. Between 1909 and 1990, the district of Santander produced nearly 40 Mt of 8% combinated zinc, and lead, mainly from the Reocín orebody.

Presently, only Reocín (AZSA) and Troya (EXMINESA) mines are exploited. The former had in 1989 estimated reserves of around 30 million tons (Mt) of sulphides with a grade of 10.8% Zn and 1.5% Pb, whereas Troya reserves lie slightly below 4 Mt with grades of 11% Zn, 1% Pb and 0.2% Cu. Of subordinate importance was the Matienzo mine, western Biscay, which before exploitation was completely stopped in 1975, produced a mineral with an ore grade of 3% Zn, 2.4% Pb (with 220 g/t of Ag) and 5% F. One decade later, the RCAM/CEMINSA stopped the activity in the San José mine of Legorreta (Guipúzcoa), where the exploited mineral presented a grade of 10% Zn and 3% Pb. Recently, several mining companies have developed exploration programs for base metals but with few results up to now. From a stratigraphic and genetic point of view, numerous ore deposits of siderite, situated in the central sector of the basin, may be correlated with the base metal orebodies. Reserves of approximately 90 Mt (~35% Fe, 9% SiO_2) have been estimated in the Bilbao district. Taking into account the extracted mineral, the total volume of these deposits may be calculated at around 370 Mt.

Metallogenic studies carried out in this region (Herrero et al. 1982; Gil et al. 1984; Herrero and Velasco 1987, 1988, 1989; Herrero et al. 1987, 1988; Herrero 1989; Fernández-Martínez 1989; Gil et al. 1990; Velasco et al. 1990; Gil 1991; Seebold et al. 1992) point out the existence of different types of ores: (1) MVT deposits; (2) Sedex deposits; (3) carbonate-hosted iron deposits; and (4) lead-zinc and iron deposits resulting from epigenetic fracture filling of vein type which occur at several stratigraphic levels. They show a wide variation in the nature of the relations between mineralization and host rock, but parageneses and episodes of deposition can be compared to a certain extent.

The studies mentioned above support genetic hypotheses based on the circulation of basinal brines, variably enriched in metals, able to replace and deposit its content within and on the platform carbonate sediments. The ore formation took place probably during and after the thermal subsidence stage, and the location of deposition is controlled by tensional fractures. The differences observed between individual mining districts and orebodies must be the result of the special geological characteristics of the source areas, the specific mechanisms of ore deposition and the lithological units affected during migration of the hydrothermal fluids.

2 Geology

The Basque-Cantabrian basin represents the western end of the Pyrenean system (Fig. 1). It comprises a thick sequence of Mesozoic and Tertiary materials, among which those of Cretaceous age are dominant. This sedimentary basin is located between the Cinco Villas and Alduides Palaeozoic massifs to the east, Sierra de la Demanda to the south, and the Asturian massif to the west. It was first developed as an intracratonic basin interconnected with other European ones, which deposition of Triassic materials and relatively thin Jurassic sedimentary sequences. Subsequently, as result of the North Atlantic opening , an extensive and complex stage of rifting took place (Bay of Biscay) which determined thick accumulations of detritic sediments followed by platform carbonates with great lateral continuity, locally considerably thick. The total thickness in the central part of the basin exceeds 14000 m. The extensive bibliography about the evolution of this basin includes Rat (1959, 1983, 1988), Rat and Pascal (1979), Ramírez del Pozo (1971), García-Mondějar (1979, 1990a,b), Carcía-Mondějar et al. (1985), Pascal (1982, 1985), and Fernández-Mendiola (1986).

2.1 Triassic-Jurassic

The Basque-Cantabrian basin started to differentiate after the end of the Hercynian orogeny following three tectonic directions (Capote 1983): (1) a EW fracture system visible in the Cabuérniga fault and in the Leiza fault, the latter located in the southern part of the Cinco Villas massif; (2) NW-SE system, illustrated by the Ventaniella and Bilbao faults; (3) SW-NE, represented by the Pamplona fault. Under clearly distensive conditions, argillaceous, arenaceous, and evaporitic sequences typical of Keuper continental sediments were deposited. Concurrently with this crustal extension, lavas, sills, and dykes of basaltic composition and tholeiitic affinity were emplaced (Béziat et al. 1991). Jurassic basinal carbonates follow (Fig. 1, Table 1).

2.2 Lower Cretaceous

Cretaceous sedimentation starts mainly with sandstones and conglomerates deposited in braided and meandering rivers and distal fluvial fans, often showing cross-stratification. Main lithologies are quartz-arenites, subarkoses and lutites (lower Aptian: Kimmeridgian-Barremian and Bedoulian, according to Pujalte 1977 and García-Garmilla 1987). At the top of the transition zone, patches of limestones and some interbedded lignite beds occur.

Pascal (1985) and García-Mondějar (1979, 1990a) describe four Aptian-Albian biosedimentary systems or depositional sequences (S_1 to S_4, in Fig. 2)

Table 1. Generalized pre-Upper Cretaceous stratigraphic column, Basque-Cantabrian Basin (northern Spain)

					Lithology	Environment	Reference
	Cenomanian		U10[a]		Sandstone, turbidites, marl, mudstone	Supra-Urgonian facies Ranging from shallow siliciclastic platform to deep flysch. Local carbonate platforms	Rat (1959, 1983) Rossy (1988)
Lower cretaceous	Albian	Upper	S4	U8,9	Massive limestone (U7,9) Marl, shale (U8), local chert beds. Carbonatic sandstone (U6)	Urgonian facies Open marine platform (requienid rudists, foraminifera, algae, corals), sometimes restricted. Shelf margins with mud mounds, knoll reefs or sand-tide flat islands Slope with slumps, olistoliths and turbiditic sandstones Basin intraplatform with pelagic fauna (*Tritaxia*) and varying influx of quartz silt and argillaceous material Ore deposits preferentially at top of shelf margins	Pascal (1982, 1985) Rat and Pascal (1979) García-Mondéjar (1979, 1990a,b) García-Mondéjar et al. (1985) Fernández-Mendiola (1986) Rat (1983, 1988) Hines (1985) Herrero and Velasco (1989)
		Middle	S3	U6,7			
		Lower					
	Aptian	Clansayesian	S2	U4,5	Marl and shale in U4. Limestone and calcarenitic limestone in U5		
		Gargasian		U2,3	Shale and sandstone more or less carbonatic in U2; rudistid limestone in U3		
		Bedoulian	S1				
	Barremian		U1		Shale, sandstone and minor conglomerate. Lignite beds.	Purbeck and Wealden facies. Fluvial terrigenous sediments (channel, floodplain, alluvial fan)	Pujalte (1977) García-Garmilla (1987)
	Neocomian						
Basement	Jurassic				Carbonate and sandstone, clay, evaporites and diabases	Marine and continental	Rat (1959)
	Triassic						
	Carboniferous		Palaeozoic massifs		Shale and limestones	Marine (turbidites and platform) carbonates	Pesquera (1985) Pesquera and Velasco (1989)

[a] S = Depositional sequences; U = sedimentary units.

controlled by synsedimentary tectonics. They consist of terrigenous sediments followed by carbonate formations, mainly rudist limestones, which represent the most characteristic outcrops of the region. This Aptian-Albian sequence is called the "Urgonian complex" (Rat 1959) and is subdivided into the sedimentary units U_1 to U_9 (Table 1 and Fig. 2). Each sequence (S_1 to S_4) starts with detritic units. Their depositional characteristics pass from a transgressive to a regressive regimen, with the local presence of unconformities showing erosional surfaces, including palaeochannels and palaeokarsts. These unconformities have a great interest from a metallogenic point of view (see below). The thickness of Lower Cretaceous sediments in the Basque-Cantabrian basin varies greatly at a regional scale. A complete Aptian-Albian section in the Bilbao area reaches about 4000 m.

At the top of the last sequence (S_4), in the central and southern part of the basin, the sedimentation was predominantly detritic with a deposit of sandstone and clays of a fluvial to marine origin ("Supraurgonian complex", Rat 1959). In the northern part of the region, sedimentation was predominantly flyschoid (Rat 1959, 1983). In this area, interlayered alkaline volcanic flows, dated as Upper Albian-Cenomanian, occur (Azambre and Rossy 1976; Montigny et al. 1986; Rossy 1988).

It is interesting to emphasize the preference of mineralization for the carbonate platform margins. Abrupt borders are located in high-energy environments colonized by corals and algae where high porosity predominates. Occasionally, erosional shelf margins occur, characterized by debris, slope carbonate facies and fall of blocks towards the talus. Gradual margins are developed in both low- and high-energy environments. In the latter case, high porosity calcarenite bars were deposited. The presence of mud mounds and micritic mud reefs that locally reach a great vertical thickness (>70 m) at the platform margins is quite frequent. In talus zones, slump beds, olistoliths, breccias, and megabreccias are abundant. Lower margin slopes are characterized by marls and mudstones bearing siliceous sponge spicules as well as nodules and layers of chert. In the finest facies of terrigenous sediments siderite nodules occur. The basin facies are characterized by alternating carbonate shales, marls, and marly limestones with sponges and *Tritaxia*. Clay, silt-sized quartz, framboidal pyrite disseminations, and organic matter, in variable proportions, appear also.

2.3 Tectonics

Synsedimentary block tectonics controls the development of the above described Urgonian sedimentary environment. The main regional trends are extensional faults with NW-SE and NE-SW directions. These fault systems have been related to the left strike-slip movements of the Iberian and European plates (Rat 1988; García-Mondéjar 1990b). Subordinated east-west faults also occur. Some of these directions correspond to reactivation of old Hercynian accidents. A series of N 20° faults related to the direction (N 90°–N 120°) of the transform fault system (Riaza-Molina 1984), located in the Bay of Biscay, existed as normal faults during the lateral motion of the Iberian

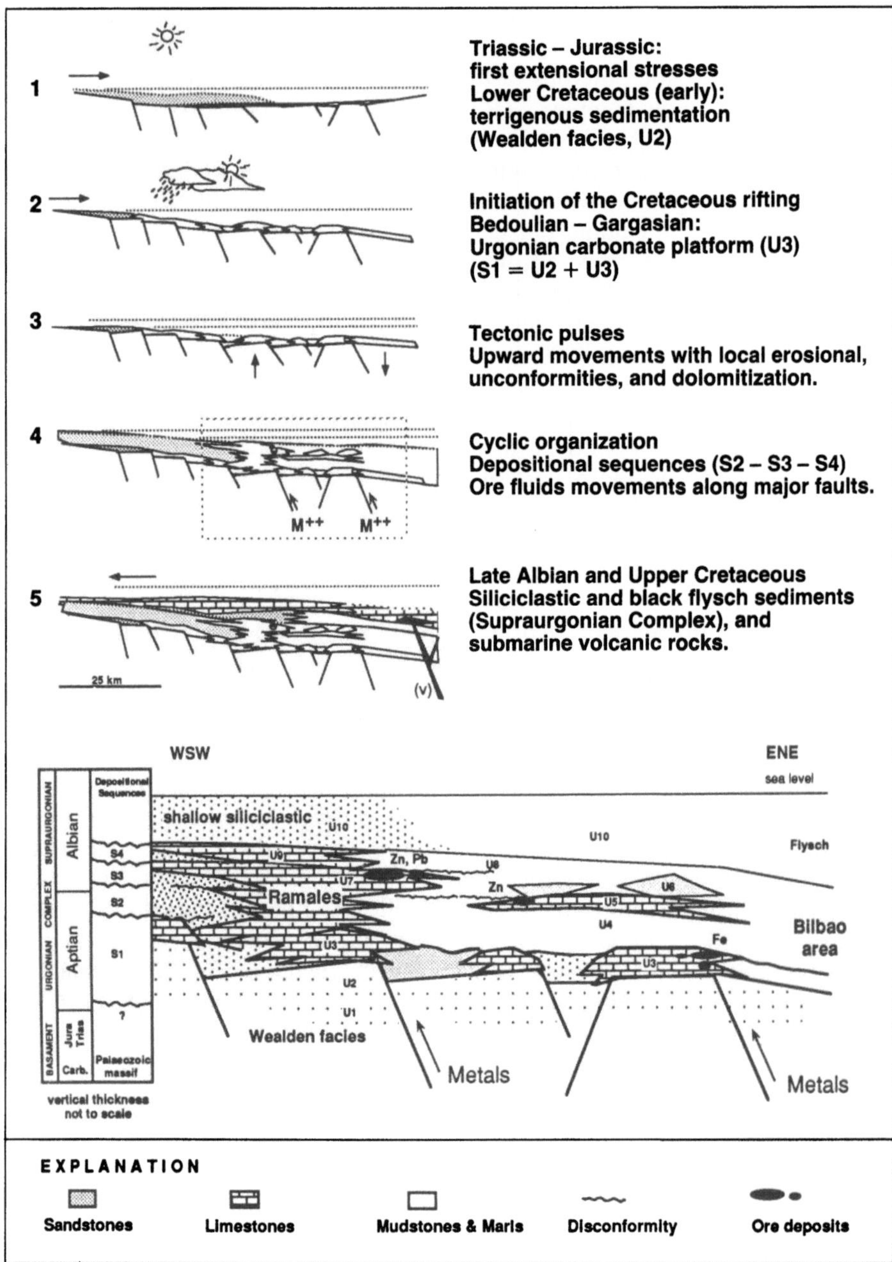

Fig. 2. Diagrammatic evolution of the Basque-Cantabrian basin according to Rat (1983) and García-Mondéjar et al. (1985). The depositional sequences related to tectonic pulses begin with terrigenous materials and terminate with carbonate platform. The *inset* is a schematic cross section in the middle part of the basin showing the relation between lithofacies assemblages and orebody locations

and European plates. Movements along these faults took place during Lower Cretaceous controlling the local development of the sedimentary basin. From the study of unconformities and erosional surfaces García-Mondéjar (1990a) has established more than six tectonic pulses strongly influencing the Cretaceous sedimentation. Among these, the early-lower Albian pulse (transition from S_2 to S_3) was the most important.

Furthermore, halokinetic activity of Keuper sediments (clays and evaporites) is also noted as an important factor. These materials formed upwellings on the seafloor, even piercing the Cretaceous sequence as diapirs. These processes were especially active during Upper Cretaceous. The formation of salt-induced highs favoured the construction of reef patches (García-Mondéjar 1979; Vadala et al. 1981; Hines 1985).

During Tertiary times, tectonic activity related to the main Alpine compressional phase (Upper Eocene and Oligocene) produced a series of folds in the sedimentary cover and the uplift of the Hercynian basement. In the western zone, folds are quite gentle, whereas in the central and eastern areas compressive tectonics produces major east-west-trending asymmetrical reverse faults. In these areas fold axes are nearly parallel to the general east-west North Pyrenean thrust, whereas to the west they tend to curve progressively towards the north, forming the Basque Arc (Fig. 1). At a later stage a series of distensive faults developed; these fractures are parallel as well as perpendicular to the folding axes, some of them are mineralized with Zn-Pb-F-Cu-Fe ("Alpine veins").

3 Lithofacies and Diagenesis of Host Rocks

The investigated ore deposits are hosted in carbonate formations that belong to the transition between the S_1–S_2, S_2–S_3 and S_3–S_4 depositional sequences (Fig. 2). As discussed above, the Urgonian carbonate complex is characterized by shelf cyclicity, with rapid prograding sedimentation ranging from shallow siliciclastic deposits to platform carbonates, up to euxinic basins influenced by fine argillaceous and siliceous filling. The most typical platform carbonates are constituted by requienid rudist limestones comprising molluscan wackstones, packstones and sometimes floatstones and micrite mounds. Large foraminifera (*Orbitolina*) allow one to date these rocks into the Aptian-Albian substage. Red algae, corals, bryozoan and sponges are also abundant. The S_2–S_3 transition is marked, in some places, by platform edge sands (grainstones) strongly cemented (blocky, syntaxial cements), indicating freshwater phreatic environments as the result of emersion processes. This kind of cementation is also observed in platform limestones, where diagenesis starts in the shallow-water marine zone and ends in the freshwater phreatic zone (Herrero and Velasco 1987; Herrero 1989). Calcitic cementation is locally followed by dolomitization or ankeritization, and subordinately silicification and sideritization. The organic silica particles are dissolved at basic pH and reprecipitated as cristobalite-tridymite (CT-opal), chalcedony and quartz in chert beds at the talus slope.

The precipitation of some authigenic minerals (framboidal pyrite, sphalerite, bipyramidal quartz), the transformation of clay and illite-smectite mixed layers, and the high reflectivity of organic matter (R_{max} values of vitrinite around 4–6%) denote progressive evolution of diagenesis under anoxic conditions. Paragenetic studies indicate that it is in an intermediate stage outlined by a progressive diminution of the illite-smectite mixed layered and kaolinite, counterbalanced by an enrichment in illite (Kübler index <0.52°, according to Arostegui et al. 1991) when siderite and base metal ore deposits start to form at the top and at the margins of the carbonate units.

4 Ore Types

The ore deposits in the Basque-Cantabrian province have been classified into four different types, taking into account their characteristics and the stratigraphic situation of the host rocks (Table 2): (1) stratabound orebodies in micritic limestones (corresponding to the U_3 unit of a lower Aptian age); (2) stratabound orebodies in calcarenites from U_5 and U_7 in the upper Aptian-lower Albian transit; (3) vein fillings hosted in carbonate and detritic series of Lower Cretaceous; (4) a fourth subordinate type of ore is related to Triassic rocks in the exposed Upper Cretaceous diapirs. The spatial relationship between mineralizations and the U_3, U_5 and U_7 units reveals the stratabound character of the ores and a clear sedimentologic control (Herrero and Velasco 1989).

According to the style of mineralization and the mode of occurrence, ore deposits may be subdivided into the following subtypes or groups, all of them occurring in platform limestones:

1. Massive sulphides and siderite;
2. Disseminations within the host carbonates;
3. Lenses congruent with stratification; and
4. Veins clearly discordant with the sedimentary host units.

Among the mentioned types there is a continuum with all transitions and complications in morphology (Fig. 3 and Table 2). Taking into account the exploited commodities, two large groups of deposits may be distinguished: (1) Zn-Pb ± (Fe-F-Ba) ore deposits and (2) Fe deposits. While the second group is almost exclusively constituted by siderite and ankerite, the first group presents several ore assemblages, with sphalerite, galena, pyrite, fluorite, chalcopyrite, barite, ankerite, and siderite as the main characteristic minerals.

The types with stratiform morphologies are those which presently have greater economic interest. The following examples illustrate some of them:

Troya Mine (Guipúzcoa). It is the best example of a stratiform orebody, hosted in Lower Cretaceous (Gargasian) carbonates. It is approximately 1500 m long in a N-S direction, between 50 and 250 m wide, and averages 5 m in thickness (Fernández-Martinez 1989). Its southern half is located in the carbonate U_3 unit (Fig. 3), whereas its northern part intersects the isopachs of the carbonate shelf, around 30–40 m below its top. At the bottom of the sulphide bed the carbonate shelf appears variably silicified and replaced by massive, coarse-grained siderite. Troya is very rich in pyrite and is the only deposit

Table 2. Characteristics of the main ore deposits in the Lower Cretaceous of the Basque-Cantabrian Basin (northern Spain)

Type	Morphology	Features	Ore elements[a]	Mineralogy[b]	Principal localities	Age[c]
Vein	Fracture filling, veinlet and stringer	Cemented fractures and brecciated rocks in active Tertiary faults, N165° hosted at U10, U7 and U3, respectively	Zn – Pb ± Cu Zn – Pb – F Fe	sl + gal + py + cp ± sulf sl + fl + gal + cp + dol sid + py + cp	Barambio, Galdakao, Axpe, Villareal, Ollerías, Antzuola, Atxondo Matienzo, La Peña, El Mazo Galdames, Sopuerta	Lower Cretaceous
Diapir	Massive, stock and disseminated ores; fissure veins	Minor dolomitization and silicification Disseminated ores (replacement and brecciated rocks)	Zn – Pb ± (Fe)	sl + gal + ba ± py	Jugo, Orduña	Cretaceous
Stratabound	Stratiform and massive	Sideritization and ankeritization (U5)	Fe	sid + ank (=goeth)	Dícido, Hoyo-Covarón, Setares, Orconera	Aptian – Albian (Clansayesian Lower Albian)
	Disseminated, massive	Dolomitization (U5)	Zn – Pb ± (F-Ba)	sl + fl + ba + gal	Anselma, Mañaria, Pozalagua	
	Stratiform, dissemination, palaeochannels	Dolomitization and silicification (U5 and U7) DCR. Palaeokarstic	Zn – Pb ± (Fe)	sl + gal + ank + py(mc)	Txomin, Legorreta, La Rasa, Campo Fresco, Siete Puertas, Lemona, Asaldita, Berriatua, etc.	
	Massive to stratiform	Replacement (U3) DCR and breccia	Fe	sid + ank	Gallarta, Bilbao, Alén, Federico, La Arboleda	Aptian (Bedoulian Gargasian)
	Stratiform to massive	Sedimentary deposit Recrystallization and replacement Banded and brecciated ores	Zn – Pb ± (Ba)	sl + gal + py + mc ± (ba)	Troya, Reocín, Novales, Udías, La Florida, Katabera	
	Unknown (tabular?)	Silicification Replacement	Zn – Pb – Cu	sl + gal + (aspy + sulf)	Aralar (Astonalde)	

[a] Ore elements are listed in order of decreasing economic importance (most important are underlined; minor but characteristic elements are enclosed in parentheses).
[b] Abbreviations: sid = siderite, py = pyrite, mc = marcasite, cp = chalcopyrite, sl = sphalerite, gal = galena, aspy = arsenopyrite, sulf = sulfosalts, fl = fluorite, dol = dolomite, ank = ankerite, ba = barite.
[c] Age of the host rocks.

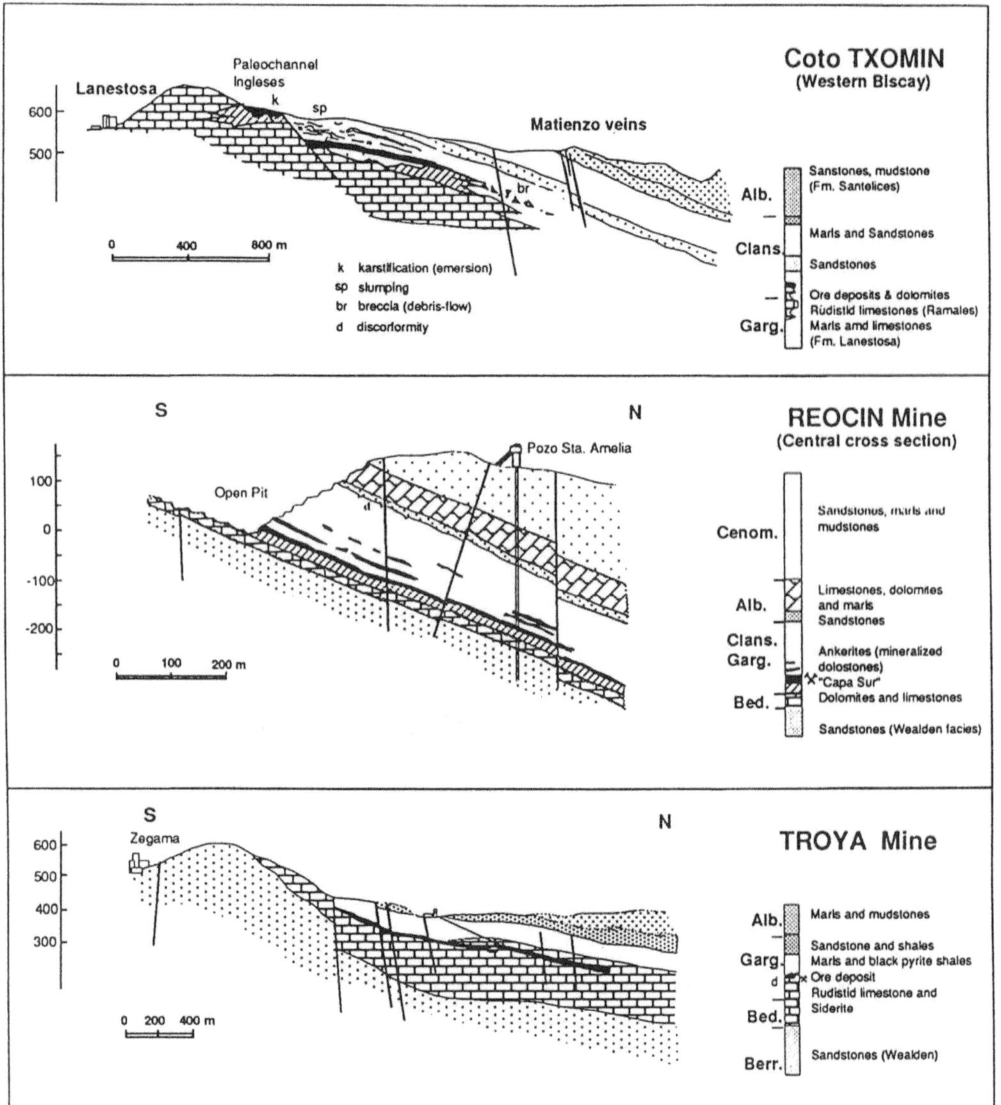

Fig. 3. Highly schematic cross section through the Txomin, Reocín and Troya deposits, showing stratiform morphology. A diagrammatic representation of the host rock stratigraphy appears on the *right*. *Berr.* Berriasian; *Bed.* Bedoulian; *Garg.* Gargasian; *Clans.* Clansayesian; *Alb.* Albian; *Cenom.* Cenomanian. Vertical scale in metres

of the region which can be described as a massive sulphide. The mineral assemblages have been described by Velasco et al. (1990). Their most representative are: (1) "pyritic facies" (py + mc ± sl ± cp) and (2) "zinc-bearing facies" (sl + py + mc + cp + ga + sid + ank + ba). In addition, a sideritic facies (sid ± ba + ank + py + sl + qtz) and a siliceous facies (qtz + sl + py) are invariably located at the base of the mineralization, with wider development near faults. Its morphological, textural and geochemical characteristics (see below; Velasco et al. 1990) suggest that Troya is a Sedex ore deposit.

Coto Txomin (Western Biscay). This occurrence is representative of numerous small deposits of the Carranza-Lanestosa area, associated with the facies change between the carbonate platform margin and the slope talus (Herrero 1989). In general, they exhibit stratiform geometry (Fig. 3) conforming to the U_7 top. Variable thickness ranges from 20–40 cm up to 1.5 m. The host limestones (U_7) preserve the features of a partialy emergent margin (Herrero and Velasco 1989). Their strong dolomitization is accompanied by the occurrence of excellent examples of diagenetic crystallization rhythmites (DCR in the sense of Fontboté and Gorzawski 1991) and diagenetic breccias. Associated with these metasomatic changes, sphalerite and galena, and scarce pyrite assemblages occur in the form of ore lenses.

Reocín Mine (Cantabria). It is located in the SE limb of the Santillana syncline, interbedded at the bottom of an iron-rich dolomite unit about 80 m thick (Fig. 3). the Reocín mine is the most important ore deposit in the Basque-Cantabrian basin. It is hosted by strongly ankeritized carbonate rocks of Gargasian-Clansayesian age lying over an 8-m-thick dolomitic bed of Gargasian age. It is a typical MVT deposit (Monseur 1967; Monseur and Pel 1972; Vadala et al. 1981; Bustillo 1985; Barbanson 1987; Seebold et al. 1992). Its morphology is variable. At the west side it is stratiform, with an extent greater than 1800 m ("Capa sur"), it has a thickness ranging between 1 and 1.5 m, and is very rich in massive or banded sphalerite, while at the east side ("Zona Barrendera"), the mineralization appears highly discordant as small and dispersed lenses and disseminations, along the iron-rich dolomite (even up to 100 m of thickness), voids and breccia fillings in the ankerite rocks. Between these two areas ores occur as lenses of variable extent and grades, in part superposed in a vertical section. Mineralogy is relatively simple: (1) sphalerite with rhythmic ("schalenblende") and botroidal morphologies, disseminated in the cavities of the host ankerite as well as filling fissures; (2) skeletal and xenomorphic galena grains replacing sphalerite and carbonates; (3) ankerite as granoblastic relicts; (4) marcasite as automorphic, bladed crystals growing in druses or as fibro-radial masses; (5) pyrite and "melnikovite-pyrite", very scarce as isolated crystals or in the proximity of fractures; (6) white sparry dolomite with curved faces followed by marcasite; and (7) white calcite, in several geodes.

5 Ore Mineralogy

Most of the mineralogical and geochemical studies have been established on the ore deposits exploited presently or in the recent past, e.g. Reocín, Troya Legorreta, Txomin. In addition, a complementary sampling has been carried out in old mine dumps and abandoned underground galleries of, in part, hardly accessible smaller ore deposits and occurrences.

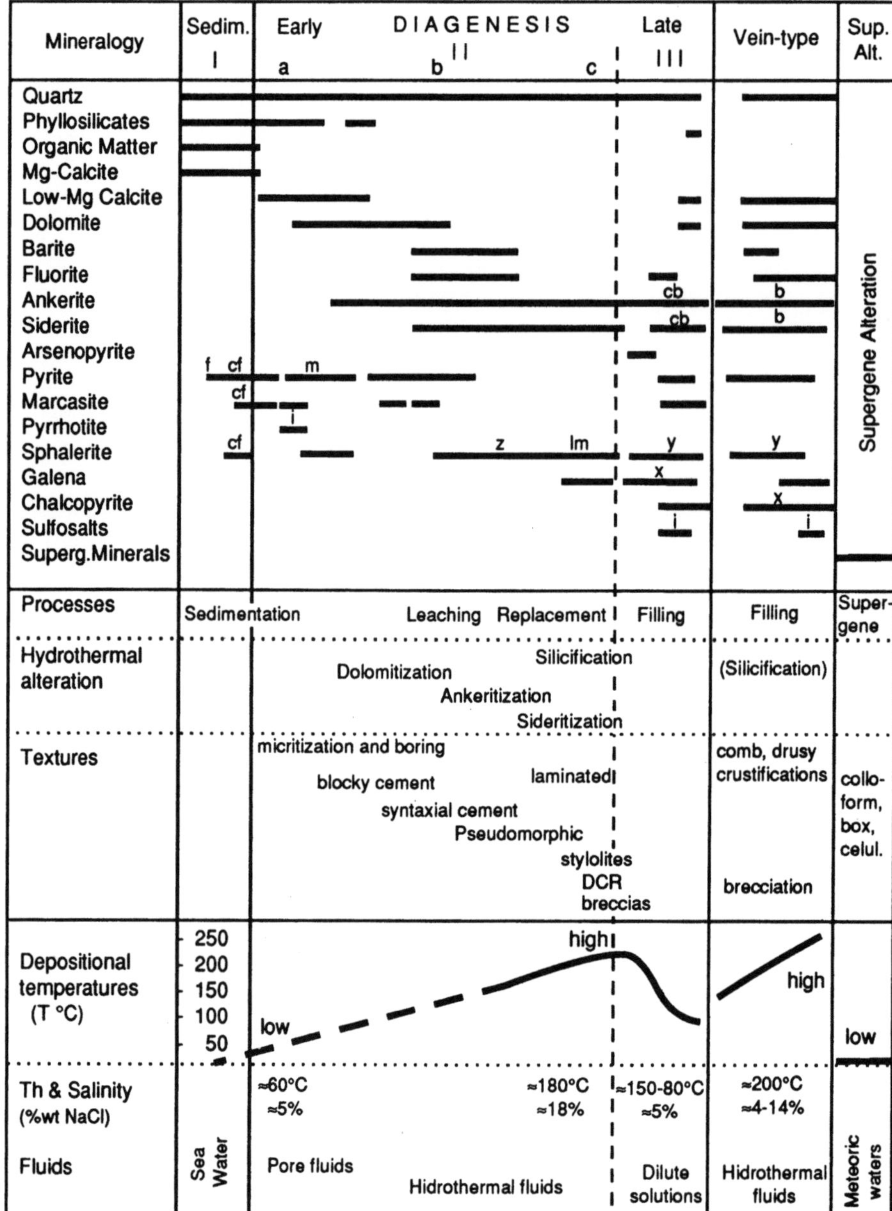

Fig. 4. Generalized paragenetic relationships of minerals from the main Pb-Zn and Fe deposits of the Basque-Cantabrian province. *f* Framboidal; *m* massive; *lm* well-laminated layers; *cf* collomorphic (only in Troya); *cb* collapse breccia; *z* zonated crystals; *i* inclusions; *x* xenomorphic void filling; *y* pale yellow; *DCR* diagenetic crystallization rhythmites; *I* sedimentation and early diagenesis; *II* and *III* burial diagenesis episodes; *a*, *b*, *c* subgenerations or distinct depositional phases

From the study of the main ore deposits (Vadala et al. 1981; Barbanson 1987; Herrero 1989; Rohou 1989; Velasco et al. 1990; Gil 1991) a simple mineralogy common to most ore deposits may be established. Without considering small differences among the ore deposits studied, a great similarity between the different paragenetic sequences is observed. These can be summarized in a generalized mineral paragenesis diagram (Fig. 4 and Table 3). Ore sedimentation (stage I) has only been recognized in Sedex-type mineralizations, metasomatic replacements from the beginning of diagenesis (stage II) and massive ore deposition with extensive replacements contemporary with the generalized dissolution of carbonates in the mineralization zones (stage III), and fillings of later cavities which represent the last stage of the main mineralization. Finally, as result of hydrothermal remobilization in the post-tectonic stage, new mineralizing episodes took place, producing cross-cutting vein filling with paragenesis similar to that already considered and crystals with a greater grain size.

Massive pyrite of the first stage (I) is abundant in Troya. Stages II and III control the most frequent paragenetic associations in all deposits, and present the maximum economic interest. The main sulphides represented are sphalerite, pyrite and/or marcasite, siderite and/or ankerite, galena and chalcopyrite. Some sulphosalts have been detected (tetrahedrite, boulangerite, bournonite, ullmannite), nearly always as tiny inclusions in galena (Recueto, El Mazo, Siete-Puertas, Balón, Aralar and Legorreta). Barite and fluorite are also found in a certain abundance, while arsenopyrite, very scarce, appears only in copper-rich parageneses from Sierra de Aralar. The distribution of iron sulphides in the mineral assemblages of the basin is rather heterogeneous: they are scarce in the central and western sector of the basin (Bilbao, Arcentales-Trucios, Reocín), and become abundant in the deposits from the eastern area (Legorreta, Troya, Aralar, Katabera). Calcite, ankerite (mainly in Reocín, Troya, Gallarta) or dolomite (mainly in Legorreta, Txomin, Anselma) and quartz are always present among the gangue minerals.

6 Lithogeochemistry

Lithogeochemical investigations carried out on the Lower Cretaceous shale and carbonate facies reveal Zn, Pb, Cu, (Mn) contents slightly lower than published average values for similar facies (Table 4). Both Wealden shales from the basin centre and those from the Arcentales region and the basin facies corresponding to the Aptian-Albian transition (Herrero 1989; Yusta et al. 1989; Gil 1991) show base metal contents below 50% of typical shales (Turekian and Wedepohl 1961; Vinogradov 1962; Dejonghe 1985). The only exceptions are found as anomalies in certain places which can be correlated with ore occurrences located in the Urgonian units. FeO and MgO contents of basal terrigenous formations are quite close to the values mentioned in the bibliography. Investigations presently being carried out try to establish whether the relatively low values of certain trace elements could be the result of leaching processes.

Table 3. Characteristics of the main diagenetic minerals from ore deposits

Mineral	Stage	Ore textures and occurrence
Sphalerite ZnS	I	Rare colloform aggregates and irregular to anhedral masses.
	IIa	Aggregates of fine to coarse subhedral to anhedral crystals, often zoned (2–10% FeS), colloform and well-laminated layers.
	IIb	Disseminated, massive and well-laminated layers (2–10% FeS). Oscillatory zoning.
	IIc	Pale yellow coarsely grained and xenomorphic void filling (<1% FeS), associated with calcite and white sparry dolomite.
Pyrite FeS_2	I	Framboidal pyrite and colloform aggregates. Syneresis textures (or contraction cracks).
	IIa	Fine-grained euhedral (cubes) occurring as massive assemblages; heterogeneous recrystallization. Pyrrhotite as relict diminutive inclusions.
	IIb	Euhedral crystals and replacement of marcasite (pseudomorphic textures).
Marcasite FeS_2	I	Several subgenerations: colloform aggregates, euhedral to needle crystals associated with framboidal pyrite.
	IIb	Euhedral crystals locally associated; some types are highly anisotropic.
	III	Geode-like and radiating arrangement of relatively large euhedral crystals.
Pyrrhotite $Fe_{1-x}Fe$	IIa	Very rare inclusions into massive pyrite.
Chalcopyrite $FeCuS_2$	IIc	Anhedral crystals associated with galena and pyrite. Very fine crystals replacing sphalerite ("watermelon" textures).
	III	Anhedral and massive filling of veinlet and void.
Galena PbS	IIc	Scarce cubic or dendritic growth crystals replacing carbonates (generation II) and ore layers.
	III	Xenomorphic void and veinlet filling.
Barite $BaSO_4$	II	Authigenic rosettes and individual large euhedral crystals. Sometimes dissolved and replaced or crustified by sphalerite.
Fluorite CaF_2	II	Cubic and subeuhedral crystals associated with barite, sphalerite and dolomite.
	III	Massive and coarsely grained aggregates filling veinlets.
Carbonates (Ca, Mg, Fe) $(CO_3)_2$	II	DCR, colapse breccia, cockade, disseminated dolomite, ankerite and siderite.
	IIb	In DCR: fine to medium crystalline anhedral carbonates (dark bands).
	IIc	Medium to coarse-grained subhedral iron-bearing carbonates, arranged in a bipolar pattern (white bands).
	III	Baroque dolomite and siderite and other geode-like white sparry carbonates.
		Xenomorphous filling composed of calcite or empty central spaces.

The distribution of trace elements in the carbonates is strongly controlled by diagenesis, with a great depletion in the mSr/Ca relation of the shallow facies. The Mn/Fe ratio decreases from the platform (0.117) towards the basin facies (0.010). However, a constant Mn/Fe weight ratio of 0.04 is observed in

Table 4. Abundance of selected elements in Lower Cretaceous rocks from the central area of the Basque-Cantabrian Basin

	n	Zn	Pb	Cu	Sr	FeO	MgO	Mn/Fe
Shale average (1)[a]	–	80–91	20–24	17–57	20–200	2.3–5.8	1.0–2.7	–
Shale U1 (2)	46	29	12	13	95	5.13	1.21	0.04
Shale U2 (3,4)	71	23	13	10	213	3.20	0.87	0.01
Marl U8 (5)	45	29	10	4	820	1.61	0.74	0.01
Limestone average (1)	–	6–24	4–18	1–12	100–990	0.1–2.7	–	–
Limestone S1 (3,4)	29	16	7	7	456	0.53	0.42	0.05
Limestone S2 (4,5)	98	11	3	2	264	0.08	0.60	0.12
Dolostone (5)	23	120	11	0	43	1.18	16.76	0.04

[a] Data from: (1) Connor and Shacklette (1975); (2) Yusta et al. (1989); (3) Gil (1991); (4) Herrero (1989); (5) Yusta et al. (1991). n = analyzed samples. FeO and MgO in percent, other values in ppm.

dolostones and dolomitized limestones hosting the ore deposits (Yusta et al. 1991).

7 Fluid Inclusions

Primary fluid inclusion data have been obtained from gangue minerals (barite, quartz, calcite, and celestite) of the main and late stages as well as from transparent ore minerals of the late stages (sphalerite, fluorite and siderite). Both stratabound and vein-type ores have been considered.

The most studied inclusions are biphasic liquid gas, with filling grades ranging from 0.95 to 0.97, and only some of them are triphasic liquid-solid vapour, including halite as daughter minerals and calcite-nahcolite as captive crystals. In some cases, liquid hydrocarbon phases and/or bitumen as immiscible globules have been recognized, mainly in baritic deposits associated with some Pb-Zn mineralizations close to Troya (Zerain). Figure 5 summarizes the most important data obtained from the microthermometric analysis of more than 1000 fluid inclusions (complete results in Alvarez et al., 1993). The studied minerals in most cases correspond to the main stage of ore deposition (e.g. sphalerite and fluorite from Anselma mine, sphalerite, and siderite from Troya mine, siderite, calcite, and celestite from Gallarta, fluorite from Matienzo, and sphalerite within septarians of Barambio occurrences).

Main stage fluid inclusions of Pb-Zn ± (F + Ba) stratabound mineralizations display homogenization temperatures in the range of 190–230 °C and salinities vary between 8 and 21 eqiv.wt% NaCl. A dominantly chloride complex composition is assumed (Alvarez et al., 1993). Fluid inclusions of later stages show lower homogenization temperatures and salinities. Mixing within colder, less saline fluids can be envisaged.

Siderite fluid inclusions from the Bilbao iron district show, for the main stage, homogenization temperatures of 190–230 °C and final melting temperatures from −9 to −15 °C, corresponding to chloride-rich solutions. Later stages

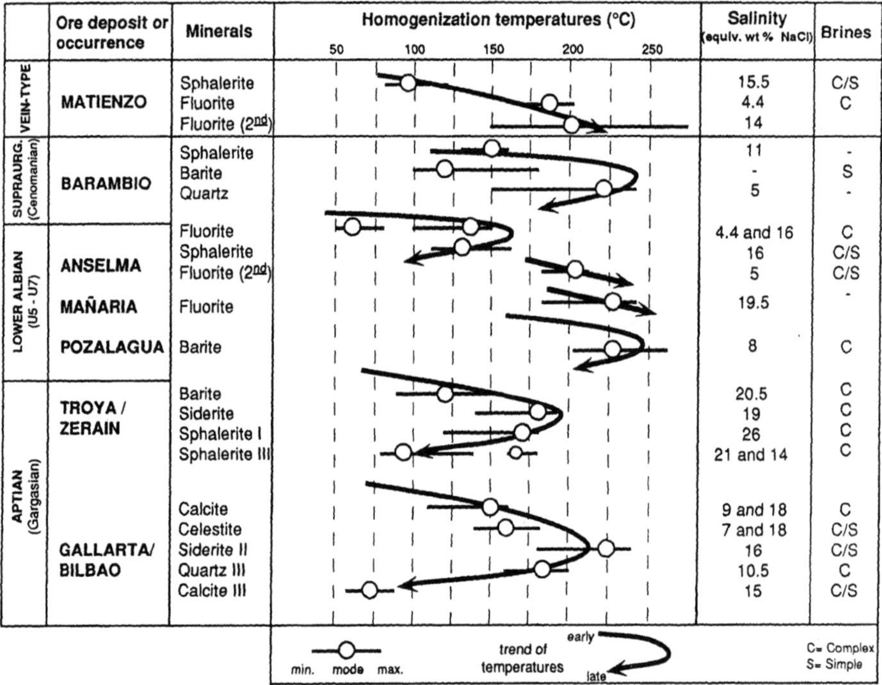

Fig. 5. Microthermometric results (°C) obtained on quartz, carbonates, barite, fluorite and sphalerite from different ore deposit types. *Arrows* denote temperature paths of hydrothermal solutions. Analytical results are presented in Alvarez et al. (1993)

show a progressive decrease in temperature, whereas the salinity remains quite constant. On the other hand, fluid inclusions in celestite and calcite from stromatactic cavities in the host rocks indicate that the carbonate sediments have undergone temperatures of 120 to 180 °C during their diagenetic history. Salinities range between 7 and 18 eqiv.wt% NaCl.

The global histograms of final melting temperatures and the correlation of these data with their corresponding salinities clearly suggest the similarity of the different brines and the existence of a fluid mixture during the mineralizing processes. Replacement textures observed in many ores and the evidence of dissolution-precipitation reactions over the earliest phases, describing successive pulsations, are coherent with the idea of fluid mixture processes. These characteristic have also been evidenced in other MVT ore deposits (Cathles and Smith 1983). The mineralizing processes took place at temperatures above 100–120 °C with averages near 180 °C. Although temperature ranges for MVT models (mainly 80–150 °C) may reach even 250 °C (Touray 1989), our values can be found within the highest temperature examples occurring in the Upper Mississippi Valley district.

The geothermometric results are consistent with vitrinite reflectance measurements on carbon-rich Lower Cretaceous sediments. High thermal gradients in the most subsiding central part of the basin with maximum reflectance values from 5 to 6.5% have been obtained. They could indicate burial temperatures of 250 °C (Bone and Russell 1988). These data confirm that in the Lower Cretaceous sedimentary rocks, especially in the basal parts of the sequence, anchizone conditions have been reached in certain areas (Arostegui et al. 1991).

8 Sulphur Isotopes

The analytical data summarized in this study correspond to sulphur isotopes of the main diagenetic stage (sphalerite, galena, pyrite and barite) and of vein-type paragenesis (sphalerite and galena) from Reocín, Troya and the western Biscay district (Fig. 6; complete results in Velasco et al., in prep.). The results of Troya and Reocín reflect a relatively homogeneous isotopic composition both for early and late diagenetic sulphides with a tight distribution for sphalerite of around 7‰. The western Biscay district that represents mostly data from vein-type minerals displays a different isotopic behaviour with a scatter between positive and negative values, although most data are included in a range between 0 and 7‰. The values in $\delta^{34}S$ of sulphates (Troya, Pozalagua and Anselma) also display relative homogeneity (25 and 30‰).

The similarity between the values obtained suggests that the sulphides had a common sulphur source. The prevailing positive values suggest the possibility of a deep homogeneous hydrothermal source. Inorganic reduction of sulphates could provide such a source.

Temperatures calculated with fractionation equations of barite-galena and sphalerite-galena (Ohmoto and Lasaga 1982) yield the following results: two barite-galena paragenetic pairs from Anselma and Pozalagua indicate temperatures of around 190 and 240–250 °C; six sphalerite-galena pairs from Anselma, 140–210 °C. For the Troya mine the temperatures obtained with four sphalerite-galena pairs are 183 and 197 °C. In Reocín the temperatures lie in the range 100–180 °C. Finally, for the post-tectonic vein-type orebodies hosted in the Supraurgonian complex, the values of sphalerite-galena fractionation show temperatures close to 145–190 °C (Matienzo) and 290 °C (La Peña) or values unreasonably high, suggesting sulphur isotopes disequilibrium.

Although geothermometric data show a certain scattering, it can be noted that most values lie within the 150–200 °C range, i.e. they are consistent with the fluid inclusion data. The sulphur geothermometry results must be accepted with great reserve since paragenetic equilibrium is not in all cases texturally clear.

9 Oxygen and Carbon Isotopes

Figure 7 (analytical results are contained in Velasco et al., in prep.) presents oxygen and carbon isotope ratios of 50 carbonate samples from the siderite deposits of Gallarta, the Troya mine and MVT occurrences west of Bilbao. The following minerals and generations are considered: (1) host rock calcite; (2) stage II sparry dolomite; (3) stages IIb and IIc sparry ankerite; (4) stages IIb and IIc sparry siderite; (5) stage III calcite and dolomite. Despite the fact

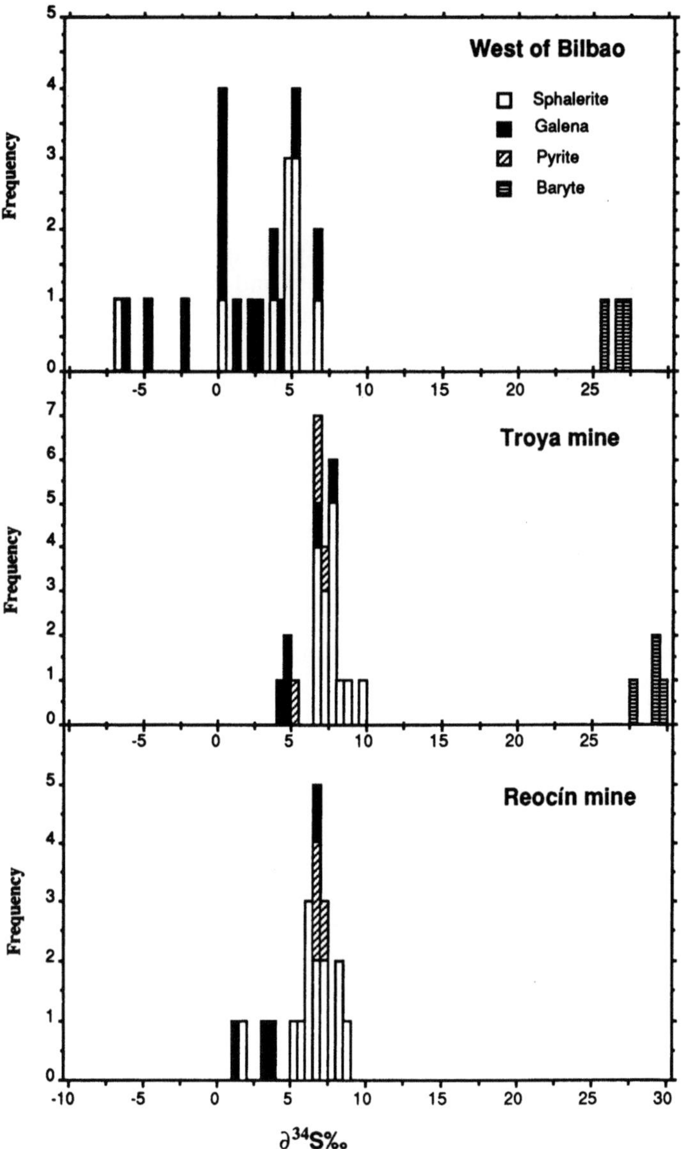

Fig. 6. Sulphur isotopic composition of ore-stage sulphides at western Biscay district (Txomin, Anselma, Pozalagua, Matienzo), Troya mine, and Reocín mine. Analytical results are presented in Velasco et al. (in prep.)

that samples are from different districts, certain common trends can be recognized. The sparry crystals of siderite from IIb and IIc generations, very abundant in some massive or banded mineralizations (Gallarta and Troya), display light average values of $\delta^{13}C$ (−1.5‰) and relatively light values of $\delta^{18}O$ (between 17.5% and 23.6‰). The ankerites and dolomites associated with

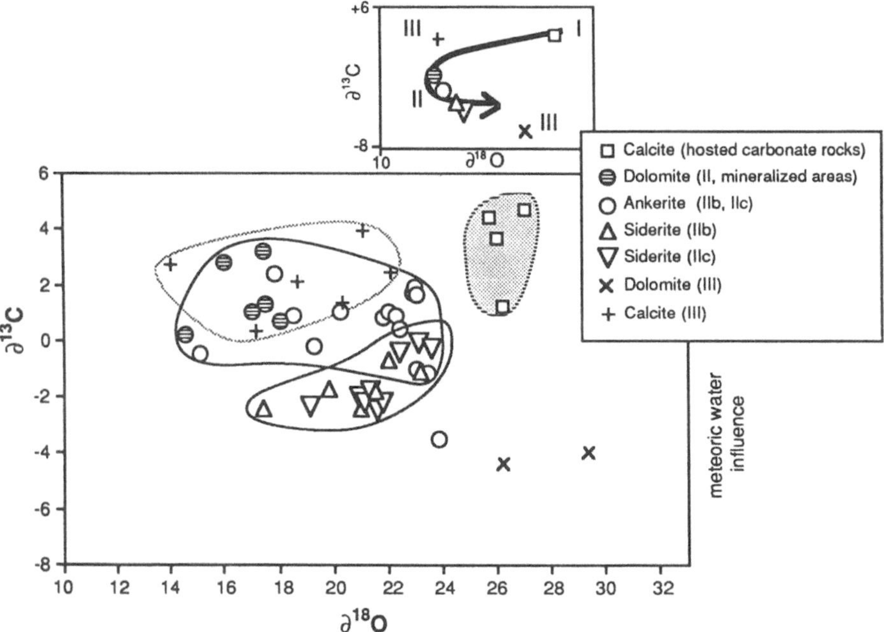

Fig. 7. Variation of $\delta^{13}C$ vs. $\delta^{18}O$ isotopes for calcites, dolomites, ankerites and siderites from host carbonates and mineralizations (western Biscay district, Gallarta and Troya). The path represents the coupled O-C depletion. Analytical results are presented in Velasco et al. (in prep.)

lead-zinc mineralizations are also characterized by light oxygen isotopes (18.5 and 21‰), whereas iron-rich carbonates present heavier values. The corresponding carbon isotopes show variable values, systematically lighter, as the fluids precipitated minerals increasingly rich in iron.

Comparing the values of neoformed carbonates with those of their precursors (Urgonian limestones), a clear depletion in ^{12}C and ^{18}O of the fluid solutions during late diagenesis is observed. This evolution could be explained as the result of a temperature increase in the connate water during compaction, or due to the incorporation of light carbon, liberated during diagenetic changes.

10 Lead Isotopes

Lead isotope analyses of 18 galena samples from stratabound and vein-type ore deposits and smaller occurrences are plotted in Fig. 8. As discussed in Velasco et al. (in prep.) the data suggest that a significant part of the lead has been scavenged from pre-Mesozoic rocks, either by brines directly migrating through the basement (in the sense of Russell 1983), or indirectly by leaching of eroded and redeposited basement materials. It can be noted that the Lower Cretaceous galenas are enriched in 206Pb and 208Pb with respect to those of Palaeozoic Sedex deposits, and these with respect to the Cambrian(?) Sierra de

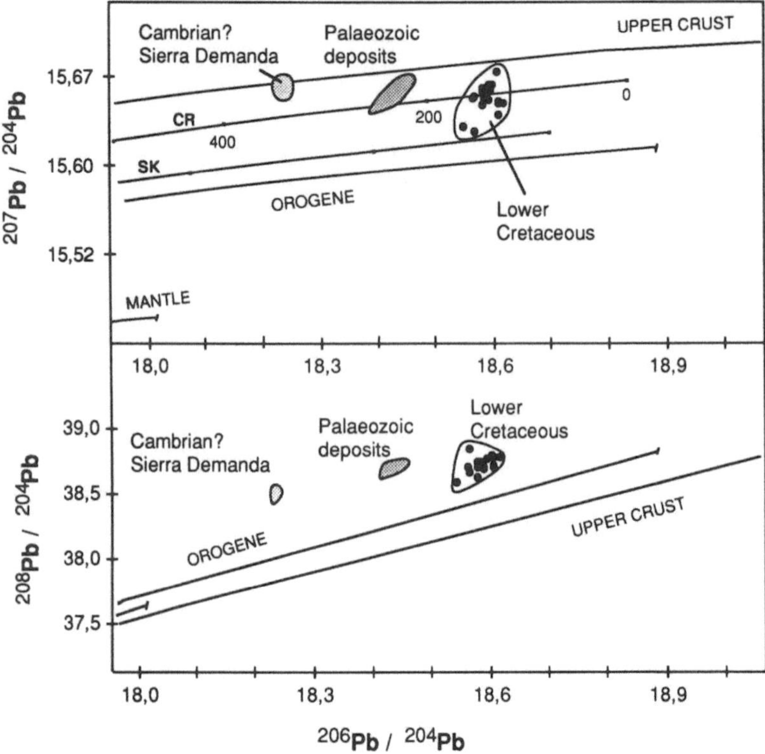

Fig. 8. Lead isotope plots of galena-lead from the Pb-Zn occurrences in relation to the orogene evolution curves from the plumbotectonics model II of Zartman and Doe (1981). These data points are compared with other galena-lead (*shaded areas*) from Carboniferous Sedex deposits (Cinco Villas massif, western Pyrenees) and the Cambrian(?) Sierra de la Demanda. Note the similarity of the lead isotope ratios of the stratabound Lower Cretaceous galenas. *SK* and *CR* are lead isotope evolution curves of Stacey and Kramers (1975) and Cumming and Richards (1975). Analytical results are presented in Velasco et al. (in prep.)

la Demanda ore deposits. Derivation from similar sources at different ages could explain these preliminary data.

11 Hydrothermal Fluids and Genetic Model: Discussion

The ore deposits of the Basque-Cantabrian zone are located near the margins of the platforms, supporting the idea of a strong tectonic control. These fractures have probably favoured the circulation of solutions, acting as a feeder of ascending hydrothermal fluids. The particular hydrology of each locality as well as the, in part, structurally controlled sedimentological and palaeogeographic features determine the development of different morphological types of ore deposits (replacement, open-space filling, disseminated, bedded, massive). The frequent spatial association of ore deposits of different morphologies in a single mineralized assemblage is consistent with this hypothesis. The nature of carbonate sediment and the porosity determine the ability to accept replacement mineralizations.

On the basis of textural and mineralogical studies a multi-stage diagenetic origin can be envisaged, responsible for the replacement phenomena and the massive precipitation of sulphides, especially during the latest stages. Dissolution of carbonate host rocks, indicating an acidic medium, is shown by different replacement textures including pseudomorphs. Successive removals and re-equilibrium implied important volume changes, accompanied by a strong increase in porosity. In these environments, diagenetic crystallization rhythmites and collapse breccias develop, cemented with white sparry carbonates. In places, this mineralizing environment was probably located at the seafloor surface, where the conditions were very close to those typical of exhalative sedimentary phenomena, as is the case in the southern part of the Troya mine.

Wall rocks next to the orebodies locally exhibit minor responses to the action of ore fluids. Along synsedimentary faults the rocks show an increase in grade of alteration assemblages, including dolomite, iron carbonates and/or silicification. These hydrothermal alterations do not extend more than a few metres away from the ore contacts. Data of carbon and oxygen isotopes seem to corroborate that an important part of the hydrothermal alteration, associated with the mineralizing processes, took place in environments that evolved towards increasing temperatures, as the burial diagenesis advanced.

Fluid inclusion data, morphology, mineralogy, textures of the minerals and the geochemical data are arguments in favour of a hydrothermal origin with circulation at a moderate temperature (120–230 °C) and with medium-high salinity arising from sediment dewatering during burial stages. The complexity of the hydrothermal systems suggests mixing of two chemically distinct brines: a surficial meteoric water and a deeper and saturated basinal brine. Deeper saline hot basinal brines, able to leach and transport the metals from the underlying basal units, circulated in zones where 250 °C was locally exceeded. These temperatures are consistent with anchizone paragenesis in the central part of the basin (Arostegui et al. 1991). The other fluids had a moderate temperature, were less saline and oxidizing, and had a possible marine meteoric origin (meteoric-phreatic pore solutions).

The geothermometric measures, which indicate that 200 °C was exceeded in the basin during mineralization, correspond to depths of 4000 m. If these calculations are correct and if we assign a normal residence time to the sediments and do not generalize the existence of anomalous geothermal environments, the approximate age for the mineralizing event would correspond to the last upper Aptian (Clansayesian) to lower Albian, contemporary with the most intense pulsation of the rifting and most anomalous geothermal environment.

Lead isotope analyses indicate that the metals were leached from underlying rocks (basement and detrital rocks derived from it as the U_1 and U_2 Wealden units). This is mainly supported by the relative homogeneity of the lead isotope ratios and the clearly more radiogenic values than Palaeozoic Sedex deposits described by Pesquera et al. (1985) and Pesquera and Velasco (1989).

Metal base sulphide and iron mineralizations in the Basque-Cantabrian basin are in most cases epigenetic, developed after deposition of the host carbonates, but some syngenetic ores are deposited at the seafloor surface (e.g. Troya). Even if they show a great diversity in morphologies and mineral parageneses, both types of ores formed as a result of the same type of discharge of ore solutions during Lower Cretaceous. Their diversities reflect the local geological setting and the variations in chemical composition and source of the ore solutions.

Acknowledgements. The authors wish to acknowledge the suggestions and critical observations made by Profs. L. Fontboté and M. Boni as well as their invitation to contribute to this book. We are thankful for the permission to use the facilities offered by mining companies (EXMINESA, AZSA and AGRUMINSA) and access to open pits and underground workings. This investigation was financed by the Research Projects 130.310-050/87, 130.310-078/88 and 130.310-115/89 of the Universidad del País Vasco/Euskal Herriko Unibertsitatea.

References

Alvarez L, Velasco F, Gil PP (1993) Nature of mineralizing fluids in the siderite deposits of Bilbao district, northern Spain. Bol Soc Esp Mineral 16:101–113

Arostegui J, Zuluaga MC, Velasco F, Ortega-Huertas M, Nieto F (1991) Clay mineralogy and illite-smectite distribution in the central Basque-Cantabrian basin (Iberian Peninsula). A tentative interpretation. Clay Minerals 26:535–548

Azambre B, Rossy M (1976) Le magmatisme alcalin d'âge crétacé dans les Pyrénées occidentales et l'Arc basque; ses relations avec le métamorphisme et la tectonique. Bull Soc Géol Fr 18:1725–1728

Barbanson L (1987) Les minéralisations Zn, Pb, Ba, Hg, Cu de socle et couverture carbonatés de la Province de Santander (nord de l'Espagne). These, Université de Orléans, 292 pp

Béziat D, Joron JL, Monchoux P, Treuil M, Walgenwitz F (1991) Geodynamic implications of geochemical data for the Pyrenean ophites (Spain-France). Chem Geol 89:243–262

Bone Y, Russell NJ (1988) Correlation of vitrinite reflectivity with fluid inclusion microthermometry: assessment of the technique in the cooper/Eromanga basins, South Australia. Aust J Earth Sci 35:567–570

Bustillo M (1985) Contribución al conocimiento de las mineralizaciones Pb-Zn del tipo Reocín en el sector Oeste de Cantabria. Estud Geol 41:127–138

Capote R (1983) La evolución tardihercínica. In: Comba JA (ed) Geología de España. Libro Jubilar J.M. Rios. IGME Madrid 2:17–36

Cathles LM, Smith AT (1983) Thermal constraints on the formation of Mississippi Valley-type lead-zinc deposits and their implications for episodic basin dewatering and deposit genesis. Econ Geol 78:983–1002

Connor JJ, Shacklette HT (1975) Background geochemistry of some rocks, soils, plants, and vegetables in the conterminous United States. US Geol Surv Prof Pap 574-F, 168 pp

Cumming GL, Richards JR (1975) Ore lead isotope ratios in a continuously changing Earth. Earth Planet Sci Lett 28:155–171

Dejonghe L (1985) Contribution à l'étude métallogenique du sinclinorium de Verviers (Bélgique). PhD Thesis D, Univ Pierre et Marie Curie, Paris 6, 389 pp

Fernández-Martinez J (1989) Criterios sedimentológicos en la definicion de un área de interés metalogenético en torno a Mina Troya (Guipúzcoa). In: Robles S, García Mondejar J, Garrote A (eds) XII Congr Espanol de Sedimentologia. Leioa, Bilbao, pp 319–329

Fernández-Mendiola PA (1986) El Complejo Urgoniano en el sector oriental del Anticlinorio de Bilbao. Kobie Ser CN 16:7–184

Fontboté L, Gorzawski H (1990) Genesis of the Mississippi Valley-type Zn-Pb deposits of San Vicente, central Peru: geological and isotopic (Sr, O, C, S) evidences. Econ Geol 85(5): 1402–1437

García-Garmilla F (1987) Las formaciones terrígenas del 'Wealdense' y del Aptiense Inferior en los anticlinales de Bilbao y Ventoso (Vizcaya, Cantabria): Estratigrafía y sedimentación. Thesis, Univ del País Vasco, 340 pp

García-Mondéjar J (1979) El Complejo Urgoniano del sur de Santander. PhD Thesis, Univ Bilbao. University Microfilms International, Ann Arbor, 673 pp

García-Mondéjar J (1990a) The Aptian-Albian carbonate episode of the Basque-Cantabrian Basin (northern Spain): general characteristics, controls and evolution. In: Tucker M (ed) Carbonate platform and basin sedimentary systems. International Association of Sedimentologists (IAS), Spec Publ 9, pp 257–290

García-Mondéjar J (1990b) Strike-slip subsidence of the Basque-Cantabrian Basin of northern Spain and its relationship to Aptian-Albian opening of Bay of Biscay. In: Tankard AJ, Balwill HD (eds) Extensional tectonics and stratigraphy of the North Atlantic margins. AAPG Mem 46:395–409

García-Mondéjar J, Hines FM, Pujalte V, Reading HG (1985) Sedimentation and tectonics in the Western Basque Cantabrian area (northern Spain) during Cretaceous and Tertiary times. In: Mila MD, Rosell J (eds) 6th IAS European regional meeting, excursion guidebook, Lleida, 9, pp 309–392

Gil PP (1991) Las mineralizaciones de hierro en el Anticlinal de Bilbao: Mineralogía, Geoquímica y Metalogenia. PhD Thesis, Univ País Vasco, 343 pp

Gil PP, Martinez R, Velasco F (1984) Ritmitas diagenéticas en las mineralizaciones de hierro de Bilbao. In: Calvo JP, Martin C (eds) I Congr Español de Geología, II, Segovia, pp 491–499

Gil PP, Crespo J, Velasco F, Herrero JM, Casares MA (1990) Mineralizaciones de hierro asociadas a calcarenitas de edad albiense, Mina de Dícido (Cantabria). Bol Geol Min 101:76–88

Herrero JM (1989) Las mineralizaciones de Zn, Pb, F en el sector occidental de Vizcaya: Mineralogía, Geoquímica y Metalogenia. PhD Thesis, Univ País Vasco, 285 pp

Herrero JM, Velasco F (1987) A diagenetic model with mixing of basinal and meteoric waters for the Legorreta Zn-Pb stratabound ore deposit in the Basque-Cantabrian region (northern Spain). In: Sassi S (ed) 8th IAS Regional Meeting of Sedimentology, Abstr, Tunis, pp 264–265

Herrero JM, Velasco F (1988) Tipología de los yacimientos de Fe y Pb-Zn-F (Ba) de la cuenca cretácica vasco-cantábrica. Bol Soc Esp Mineral 11:176–178

Herrero JM, Velasco F (1989) Control sedimentológico de las mineralizaciones de Zn-Pb-F en el sector occidental de Vizcaya. XII Congreso Español de Sedimentología. Leioa, Bilbao, pp 307–318

Herrero JM, Velasco F, Fortuné JP (1982) Estudio preliminar sobre las mineralizaciones de hierro y plomo-cinc-fluor en ambiente carbonatado del oeste de Vizcaya. Bol Soc Esp Mineral 5:183–190

Herrero JM, Velasco F, Fano H (1987) Dolomitización, silicificación y concentración de sulfuros de Zn, Py y Fe en calcarenitas urgonianas (Mina de Legorreta, Guipúzcoa). Bol Geol Min 88(4):516–526

Herrero JM, Perez-Alvarez M, Touray JC, Velasco F (1988) Late thermal events in the Mesozoic Basque basin: evidence from secondary fluid inclusions in Mississippi-Valley type fluorite occurrences. Bull Minéral 111:413–420

Hines FM (1985) Sedimentation and tectonics in north-west Santander. In: Mila MD, Rosell J (eds) 6th IAS European regional meeting, Lleida, excursion guidebook, 9, pp 371–392

Monseur G (1967) Synthèse des connaissances actuelles sur le gisement stratiforme de Reocín. In: Brown JS (ed) Genesis of stratiform lead-zinc-barite-fluorite deposits (Mississippi Valley type deposits). Econ Geol Monogr 3:278–293

Monseur G, Pel J (1972) Reef environment and stratiform ore deposits (essay of a synthesis of the relationship between them). In: Amstutz GC, Bernard AJ (eds) Ores in sediments. Springer, Berlin Heidelberg New York, pp 195–207

Montigny R, Azambre B, Rossy M, Thuizat R (1986) K-Ar study of Cretaceous magmatism and metamorphism in the Pyrenees: age and length of rotation of the Iberian Peninsula. Tectonophysics 129:257–273

Ohmoto H, Lasaga AC (1982) Kinetics of reactions between aqueous sulfates and sulfides in hydrothermal systems. Geochim Cosmochim Acta 46:1727–1745

Pascal A (1982) Evolution des systémes biosédimentaires urgoniens en Espagne du Nord. Neues Jahrb Geol Paläontol Abh 165:77–86

Pascal A (1985) Les systémes biosédimentaires urgoniens (Aptien-Albien) sur la marge nord-ibérique. Mém Géol Univ Dijon 10:569 pp

Pesquera A (1985) Contribución a la Mineralogía, Petrología y Metalogenia del Macizo Paleozoico de Cinco Villas (Pirineos Vascos). PhD Thesis, Univ del País Vasco, 579 pp

Pesquera A, Velasco F (1989) The Ardíturri Pb-Zn-F-Ba deposits (Cinco Villas massif, Basque Pyrenees). A deformed and metamorphosed stratiform deposit. Mineral Depos 24:199–209
Pesquera A, Velasco F, Fortuné JP, Tollon F (1985) Les mineralisations de type exhalatif sedimentaire "Sedex" a Pb-Zn-F-Ba du district d'Arditurri, Guipuzcoa (Espagne). C R Acad Sci París 300:463–468
Pujalte V (1977) El Complejo Purbeck-Weald de Santander. Estratigrafía y Sedimentación. Thesis, Univ de Bilbao, 202 pp
Ramirez del Pozo J (1971) Bioestratigrafía y Microfacies del Jurásico y Cretácico del Norte de España (Region Cantábrica). Mem IGME 78:357 pp
Rat P (1959) Les pays Crétacés Basco-Cantabriques (Espagne). Thesis, Publ de l'Université de Dijon, XVIII, 525 pp
Rat P (1983) Les Régions Basco-cantabriques et Nord-Ibériques. In: Vue sur le Crétacé Basco-cantabrique et nord-ibérique. Une marge et son arrière-pays, ses environnements sédimentaires. Mém Géol Univ Dijon 9:1–24
Rat P (1988) The Basque-Cantabrian basin between the Iberian and European plates: some facts but still many problems. Rev Soc Geol Esp 1:327–348
Rat P, Pascal A (1979) De l'étage aux systémes bio-sedimentaires urgoniens. Géobios Mem Espec 4:385–399
Riaza-Molina C (1984) De la importancia de las fallas N20 en la creación del Golfo de Vizcaya. Estudio de detalle del Arco Santanderino. In: Calvo JP, Martin C (eds) I Congr Español de Geología, Segovia, III, pp 265–278
Rohou P (1989) Evolution tectonosédimentaire et métallogenique de la region Cantabrique au sud de Santander: relations entre phases tectoniques, paléogeographie et minéralisations Zn-Pb et Fe. Thesis, Univ d'Orleans, 409 pp
Rossy M (1988) Contribution a l'etude du magmatisme mesozoique du domaine pyreneen. I- Le Trias dans l'ensemble du domaine. II- Le Cretacé dans les Provinces Basques d'Espagne. PhD Thesis, Univ Franche-Comté, Besançon, 368 pp
Russell MJ (1983) Major sediment-hosted exhalative zinc+lead deposit: formation from hydrothermal cells that deepen during crustal extension. In: Sangster DF (ed) Sediments-hosted stratiform lead-zinc deposits. Short Course, Mineral Assoc Canada, Victoria, pp 251–282
Seebold I, Fernández G, Reinoso JR, Alonso JA, Escayo MA, Gómez F (1992) Yacimientos estratoligados de blenda, galena y marcasita en dolomías. Mina de Reocín (Cantabria). In: García Guinea J, Martinez-Frias J (eds) Recursos Minerales de España. CSIC Textos Universitarios, pp 947–967
Stacey JS, Kramers JD (1975) Approximation of terrestrial lead isotope evolution by a two-stage model. Earth Planet Sci Lett 26:207–221
Touray JC (1989) Etudes d'inclusions fluides et modélization sur la génèse des gîtes du type "Mississippi Valley". Chron Rech Min 495:21–30
Turekian KK, Wedepohl KH (1961) Distribution of the elements in some major units of the earth's crust. Geol Soc Am Bull 72:175–195
Vadala P, Touray JC, García-Iglesias J, Ruiz F (1981) Nouvelles données sur le gisement de Reocín (Santander, Espagne). Chron Rech Min 462:43–59
Velasco F, Fernández JM, Herrero JM, Fano H (1990) Mineralogía y texturas del yaciminento de sulfuros de Mutiloa: Mina Troya, Guipúzcoa. Bol Soc Esp Mineral 13-1:53–54
Velasco F, Pesquera A, Herrero JM (in press) Lead isotope compositions of galenas in carbonate-hosted deposits of the Basque-Cantabrian region, northern Spain. Mineral Depos (in preparation)
Vinogradov AP (1962) Sredniye soderzhaniya khimischeskikh elementov glavnykh tipakh izverzhennykh gornykh porod zemnoi kory. Geokhimiya 7:560–561
Yusta I, Aróstegui J, García-Garmilla F, Herrero JM, Velasco F (1989) Sedimentation and diagenesis in the Villaro Wealden (Lower Cretaceous, Basque-Cantabrian region, northern Spain). In: Kázmér M (ed) 10th IAS Regional Meeting on Sedimentology, Hungarian Geol Inst Budapest, pp 259–260
Yusta I, Herrero JM, Velasco F (1991) Pautas de distribución de Sr/Ca, Mg/Ca y Mn/Fe en carbonatos urgonianos del oeste de Vizcaya (España). Geogaceta 10:100–103
Zartman RE, Doe BR (1981) Plumbotectonics. The model. Tectonophysics 75:135–162

Carbonate-Hosted Pb-Zn Mineralization at Bleiberg-Kreuth (Austria): Compilation of Data and New Aspects

S. Zeeh and T. Bechstädt

Abstract

The lead-zinc deposit of Bleiberg-Kreuth is hosted by carbonate rocks of Middle to Upper Triassic (Ladinian/Carnian) age. There is a clear connection between palaeogeography (e.g. palaeotopographic highs within a lagoon) and ore emplacement. The sequence of carbonate cementation of the host rock gives good insight into the diagenetic processes and allows the relative dating of ore formation. Fluid inclusions from fluorite and carbonate cements indicate initial temperatures during ore emplacement from 80 up to 200 °C, followed by a temperature increase up to 250 °C during late burial diagenesis. Hydrothermal transport of metals and ore formation within a medium to deep burial environment are assumed.

1 Introduction

The lead-zinc deposit of Bleiberg-Kreuth is situated in the easternmost part of the Gailtal Alps (Fig. 1) within a thick, mainly carbonate sequence of Triassic rocks. Out of a total of approximately 3000 m of sedimentary rocks in the Bleiberg area, approximately 350 m are ore-bearing (Fig. 2). The Pb-Zn ores, which are free of silver and copper, are mainly accompanied by fluorite, rarely barite and anhydrite, as well as many other accessory minerals (Schroll 1953, 1984a; Schroll et al., this Vol.).

The Bleiberg-Kreuth deposit forms part of a metallogenic province with several stratabound lead-zinc deposits (e.g. Mežica, Raibl). These deposits are situated on both sides of the Periadriatic suture, and are located within shallow-water carbonates of Middle to Upper Triassic (Ladinian/Carnian) age. They are often combined under the term "Bleiberg" or "Alpine type" (Maucher and Schneider 1967; Sangster 1976; Brigo et al. 1977), which were thought to be different from the Mississippi Valley-type deposits (MVT). This distinction was mainly based on the following features:

1. The apparent occurrence of external sedimentary ore textures (Schulz 1964; Maucher and Schneider 1967);

Geologisch-Paläontologisches Institut der Universität Heidelberg, Im Neuenheimer Feld 234, 69120 Heidelberg, Germany

Spec. Publ. No. 10 Soc. Geol. Applied to Mineral Deposits
Fontboté/Boni (Eds.), Sediment-Hosted Zn-Pb Ores

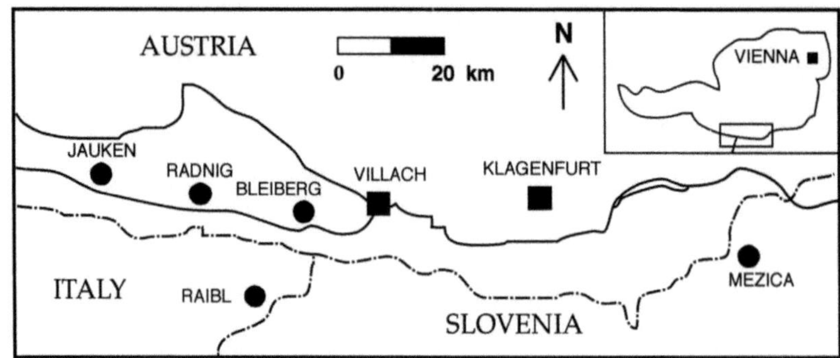

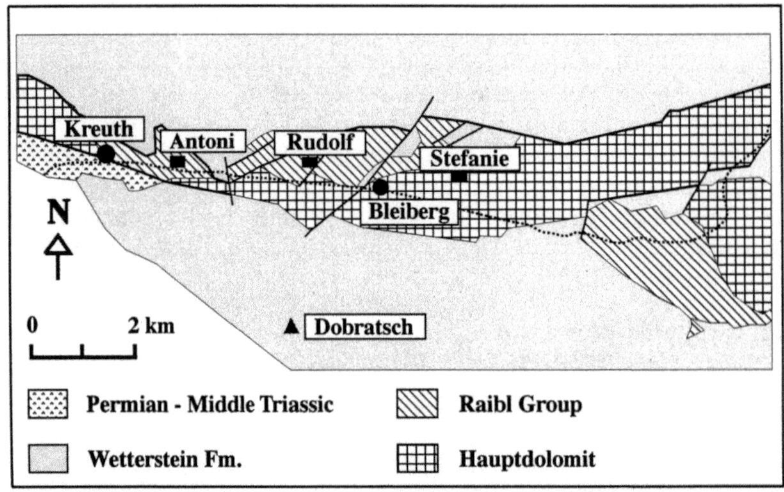

Fig. 1. A Location map of the studied area. **B** Schematic geological map of the Bleiberg area. (Redrawn after Holler 1977)

2. An assumed association with volcanism, being of the same age as the ore-hosting sediments (uppermost Wetterstein and Raibl Fms.) (Maucher and Schneider 1967);
3. The relatively low radiogenic isotopic composition of the ore lead ("B-type") compared to the future model ages of the "J-type".

In our opinion, these parameters cannot be used as arguments to distinguish between MVT and Bleiberg types because:

1. The apparently external sedimentary ore textures often represent internal sediments. In other cases, they can be interpreted as metasomatic features (mimetic replacement);
2. There is no evidence of volcanic activity in the Drau Range or in the Northern Calcareous Alps (Bechstädt 1975a) during deposition of the host rocks (Carnian);

3. The isotopic composition of the ore lead indicates that the crystalline basement is the main source, whereby overlying rocks might have contributed to a minor degree (Köppel and Schroll 1985). This interpretation is comparable to the assumed sources of other MVT deposits.

This chapter presents geochemical, sedimentological and diagenetic data, which give new information on the genesis of these deposits. The data suggest, in accordance with Sangster (1986), that the "Bleiberg-type" deposits actually belong to the "normal" Mississippi Valley type.

2 Stratigraphy and Facies of the Triassic in the Area of the Bleiberg-Kreuth Deposit

At the base of the Triassic sequence (Fig. 2), on top of the metamorphic basement (Gailtal crystalline) and the Carboniferous of Nötsch, Permian/Scythian sandstones occur (Gröden Formation and Formation of the Alpine Buntsandstein; Krainer 1989). These clastic sedimentary rocks are followed by gypsum-bearing red and green marly and sandy sedimentary rocks of the Scythian stage (Werfen Formation; Krainer 1989). These are overlain by marly limestones of the Anisian "Alpine Muschelkalk", representing shallow-water marine sediments, followed by the Anisian "Zwischendolomit", which locally carries small stratabound Pb-Zn mineralizations. These are limited to the northern rim of the Gailtal Alps (Cerny 1989b).

The overlying carbonate platform of the Wetterstein Fm. shows different lithologies. In most of the mining area it is mainly represented by dolomitic rocks. A cross section perpendicular to the platform margin exhibits different facies realms (Ott 1967; Sarnthein 1967; Colins and Nachtmann 1978; Henrich and Zankl 1986; Zeeh 1990). In the area of Bleiberg-Kreuth a S to N transect of the uppermost Wetterstein Fm. shows the following environments (Fig. 3):

- Reef facies of the Dobratsch mountain, which is characterized by corals, sponges (Sphinctozoa) and the microproblematicum *Tubiphytes obscurus*.
- Back reef facies consisting of fine reef debris.
- Near-reef lagoonal to tidal flat facies showing alternations of subtidal (with the Dasycladaceae *Teutloporella herculea*) and inter- to supratidal sediments.
- Far-reef lagoonal facies, dominated by subtidal carbonates and the dasycladacean algae, *Poikiloporella duplicata*, which also occur in the subtidal sediments of the inner platform further north.
- Adjacent "Bleiberg facies" consisting of cyclic sequences of alternating subtidal and intertidal to supratidal carbonates (Bechstädt 1973, 1975b). Emersion layers are frequently found intercalated. These are characterized by microkarst as well as calcretes, greenish marls and lithoclastic-wackestones/floatstones with black pebbles. Synsedimentary or early diagenetic evaporites, including grey anhydrite and barite (Schulz 1968; Schroll and Wedepohl 1972), are sometimes finely dispersed within the carbonate rocks. The Bleiberg facies represents a cyclically exposed palaeotopographic high, which

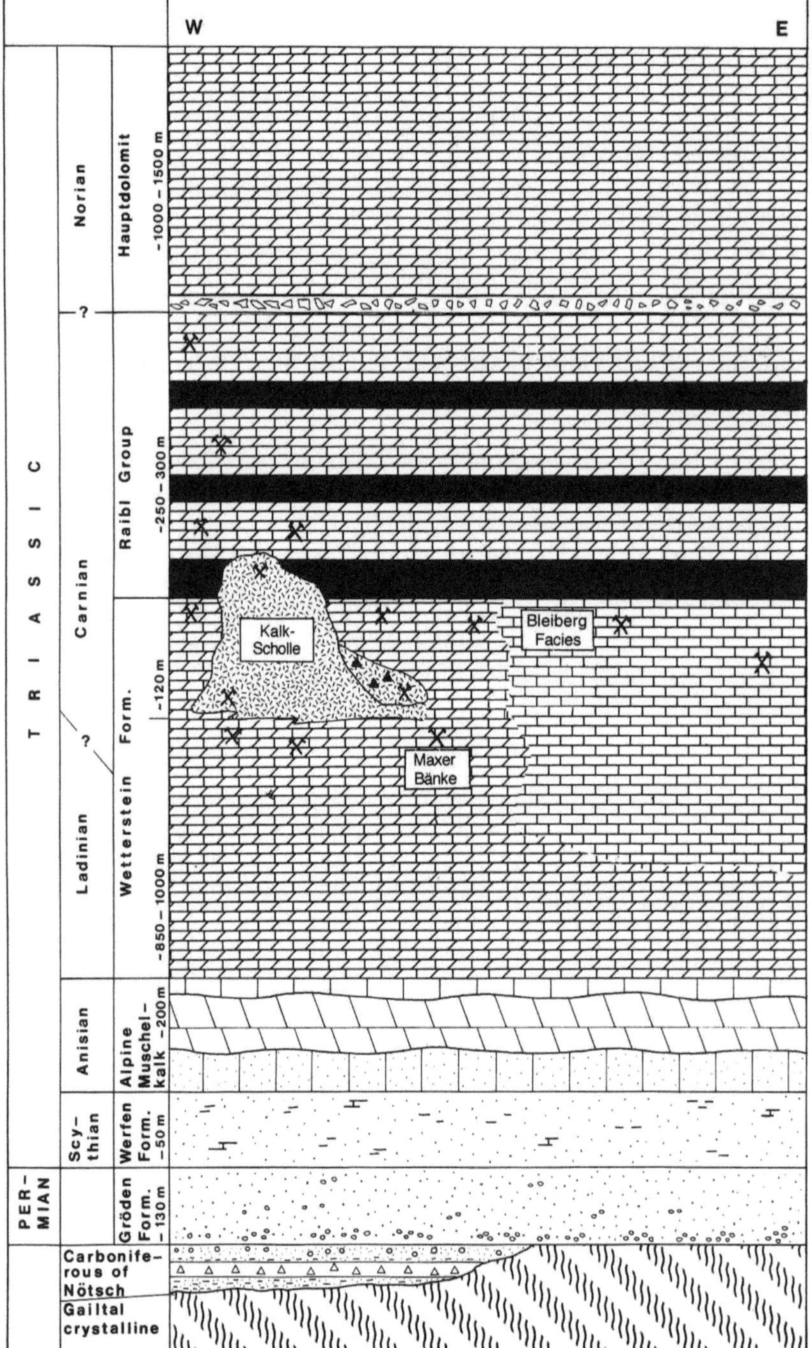

Fig. 2. Stratigraphy in the area of the Bleiberg-Kreuth lead-zinc mine

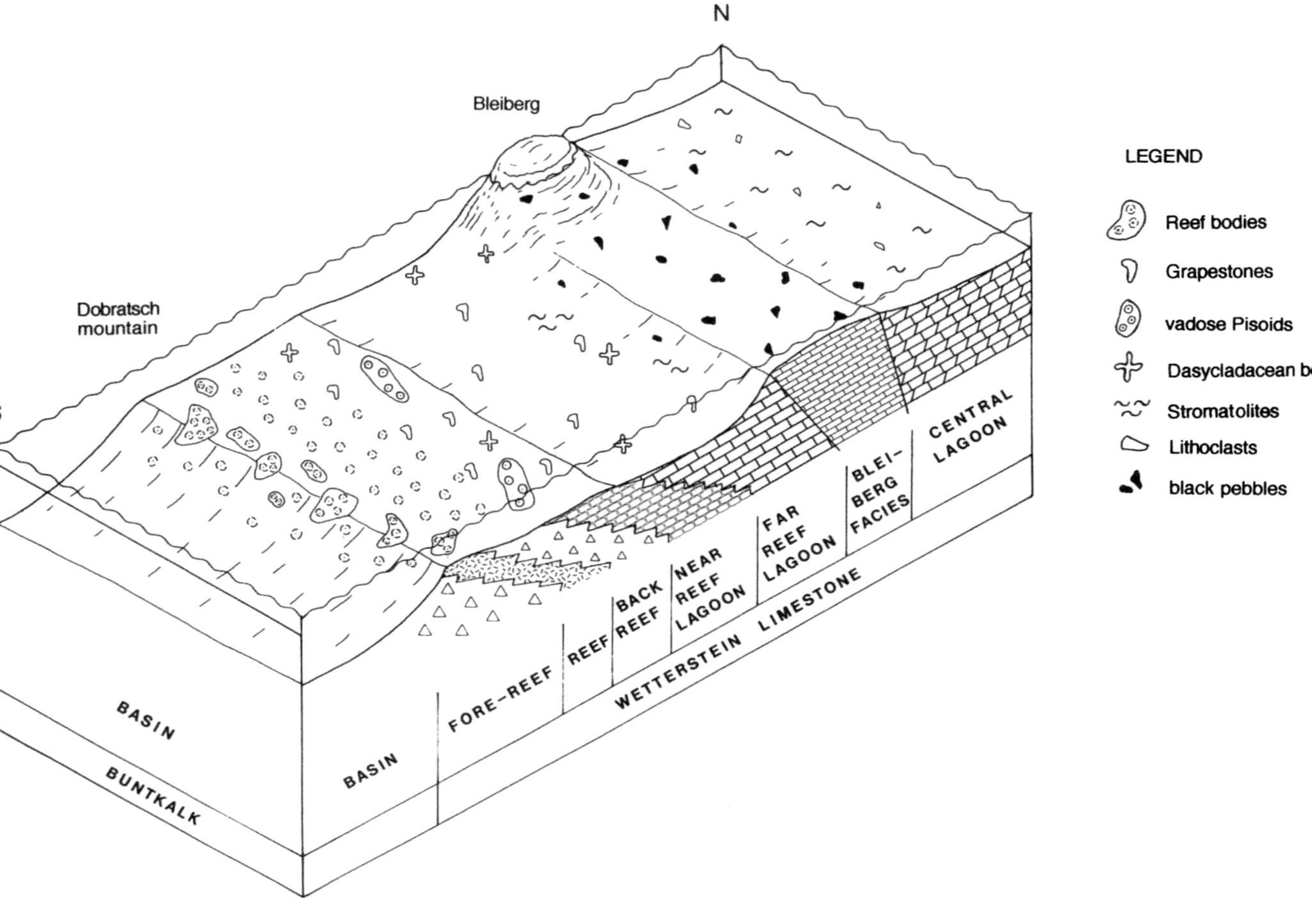

Fig. 3. Facies zones of the uppermost Wetterstein Fm.

is parallel to the platform margin for several kilometres. In the vicinity, other palaeotopographic highs ("Kalkscholle", "Riedhartscholle" and "Josefischolle", all possibly horst blocks) occur, which were subaerially exposed most of the time, deeply karstified, and affected by massive dolomitization (Bechstädt 1975c; Cerny 1989a,b). The Bleiberg facies thins towards these highs (Kostelka 1971; Bechstädt 1975c), which consist of sub- to intertidal carbonates of Ladinian/lower Carnian age. All of these areas, which were emerged for longer or shorter periods of time (Bleiberg facies, Kalk, Riedhart- and Josefischolle), form the major lead-zinc bearing rock types of the upper Wetterstein Fm.

- Central lagoonal to tidal flat facies, which are located further to the north and show even less faunal diversity. The interfingered subtidal and inter- to supratidal carbonates are interpreted as a product of a restricted sedimentary environment, located behind areas of small topographic relief.

The Wetterstein Fm. is followed by the Raibl Group, which consists of three, 10- to 40-m-thick terrigenous clastic and 30- to 70-m-thick carbonate-evaporite units. These are interpreted as alternating carbonate and clastic third-order cycles caused by sea-level fluctuations (Hagemeister 1988; Bechstädt and Schweizer 1991). Only the three carbonate units of the Raibl Group are ore-bearing; this applies especially to the lowermost carbonate interval with occasionally rich stratabound ores in the western part of the Bleiberg deposit.

The Raibl Group, in turn, is overlain by non-ore-bearing Norian dolomites (Hauptdolomit), which exceed 1000 m in thickness. A breccia horizon, occurring at the base of the Hauptdolomit, separates the third Raibl carbonate sequence from the Hauptdolomit.

These carbonate sequences are closely linked to the development of the Tethys. Tectonism, due to rifting (Bechstädt et al. 1978), subduction with back-arc rifting (Castellarin et al. 1980, 1988; Bosellini et al. 1982) or, as more recently suggested, transpression and transtension (Brandner 1984; Doglioni 1984, 1987; Blendinger 1985), disrupted a stable carbonate platform in upper Anisian to Ladinian time. Reefs were said to start their growth on elevated horst blocks. These models only consider the influence of tectonic activity, which is more prominent in the Southern Alps. In the Northern Calcareous Alps the development of the Wetterstein platform was apparently initiated by a sea-level rise during upper Anisian/lower Ladinian time (Rüffer and Bechstädt 1991), followed by extensional tectonics. Due to this combination of factors, basinal areas (e.g. Reifling and Partnach Fms.) and carbonate platforms (e.g. Wetterstein Fm.) formed, which prograded during Ladinian/Carnian time.

The carbonates of the Raibl Group and the Hauptdolomit were deposited on a stable shelf, which was influenced by sea-level changes. Extensional phases from Upper Triassic to Lower Cretaceous times (cf. Winterer and Bosellini 1981; Eberli 1988; Ratschbacher et al. 1989) are connected with the opening of the Penninic Ocean, which caused a general deepening of the area under consideration.

Table 1. Occurrence, sedimentological parameters and mineral assemblage of the ores at Bleiberg-Kreuth

	Maxer Bänke	Bleiberg facies	Josefi-, Riedhart-, Kalkscholle	Cardita orebody
Stratigraphic position	Ladinian/Carnian (170–360 m below 1st Raibl shale)	Carnian (0–60 m below 1st Raibl shale)	Ladinian/Carnian (0–200 m below 1st Raibl shale)	Carnian (1st, 2nd, 3rd Raibl dolomites)
Sedimentological parameters	Cyclic sedimentation of sub-, inter-, supratidal carbonates ("green marls")	Cyclic sedimentation of sub-, inter-, supratidal carbonates and emersion layers ("green marls", black pebbles, calcretes)	Sub-, inter-, supratidal carbonates, brecciated and strongly dolomitized and recrystallized	Cyclic sedimentation of sub-, inter-, supratidal carbonates
Occurrence of Pb and Zn ores	Stratabound	Stratabound and discordant (fissures, veins)	Disseminated and massive ores, ore clasts	Stratabound
Zn : Pb ratio	10:1	4:1 (W part of the Bleiberg mine) 2:1–1:1 (E part of the Bleiberg mine)	6:1–10:1	1:1–8:1
Main paragenesis	ZnS, PbS, FeS_2	ZnS, PbS, FeS_2, CaF_2, $BaSO_4$, $CaSO_4$	ZnS, PbS, FeS_2, (CaF_2, $BaSO_4$)	ZnS, PbS, FeS_2, ($BaSO_4$, CaF_2)

3 Pb-Zn Ores at Bleiberg-Kreuth

Mining activities in the Bleiberg area have been documented since 1333 (Holler 1953). In 1867 the Bleiberger Bergwerksunion (BBU) was organized, which still manages mining and ore beneficiation. The extension of the ore deposit is reflected by more than 1200 km of galleries, while mining at present is restricted to only the western part of the deposit (Antoni, Kreuth; Fig. 1B).

A total amount of more than 3 million t of Pb and Zn metal has been worked during the last 700 years; about 2 million t of metal are still available (Cerny 1989b). Metal content varies from 1% Zn to more than 30% Zn in massive sphalerite bodies. Production at present is directed towards ores containing more than 7% Zn (Cerny 1989a).

The Pb-Zn ores within the deposit of Bleiberg-Kreuth are concentrated in different units, described in the following (Table 1).

Maxer Bänke. The so-called Maxer Bänke of the Wetterstein Fm. occurs about 180–370 m below the first Raibl shale and consists of a thick cyclic sequence of shallow marine carbonates, marly limestones and so-called green marls (greyish carbonates with a clay content of up to 10%; Hagenguth 1984). The green marls, formerly interpreted to be of tuffaceous origin, represent the product of terrigenous weathering (Hagenguth 1984). Concordant and discordant orebodies occur within the Maxer Bänke, consisting mainly of sphalerite, galena and pyrite or marcasite. The average metal content of this sequence is about 0.4% Pb and 2.8% Zn with a total of 60 000 t of Pb and Zn.

Bleiberg Facies. Mining at Bleiberg-Kreuth was restricted for more than six centuries (Holler 1953) to the metal-rich stratabound mineralizations of the uppermost Wetterstein Fm., e.g. Bleiberg facies. Nine of the emersion layers within the 60-m-thick Bleiberg facies form lithostratigraphic markers (Holler 1936, 1960) that facilitate exploration and mapping of Alpine structures.

The average metal content of this sequence is about 1% Pb and 5% Zn and approximately 3 million t of Zn and Pb metal have been produced since mining started. The ores occur within oval-shaped bodies arranged subparallel to the bedding and within discordant fissures and veins. Sphalerite, galena and pyrite are the main ore components. Fluorite and blue-coloured anhydrite are the main accessory minerals in the western part of the mine, while barite prevails in the eastern part. Various carbonate cements occur within the orebodies, post- and/or pre-dating ore formation.

Kalk-, Riedhart- and Josefischolle. In 1951, a so far unknown type of mineralization was discovered in the western part of the mine. This was the so-called Kalkscholle, a stock-shaped orebody of about 2 million m^3 hosted by dolomitic rocks (Fig. 4). This is in contrast to the previously known cross sections of pipe-shaped and concordant orebodies of Kreuth and vein-shaped discordant mineralizations of Bleiberg, which did not exceed more than 60 to 80 m^2. Zinc-rich ores occur as networks and within breccias, and in some areas as coarse masses of sphalerite (Cerny 1989a,b). The average ore content of this orebody was 5% (0.5% Pb, 4.5% Zn).

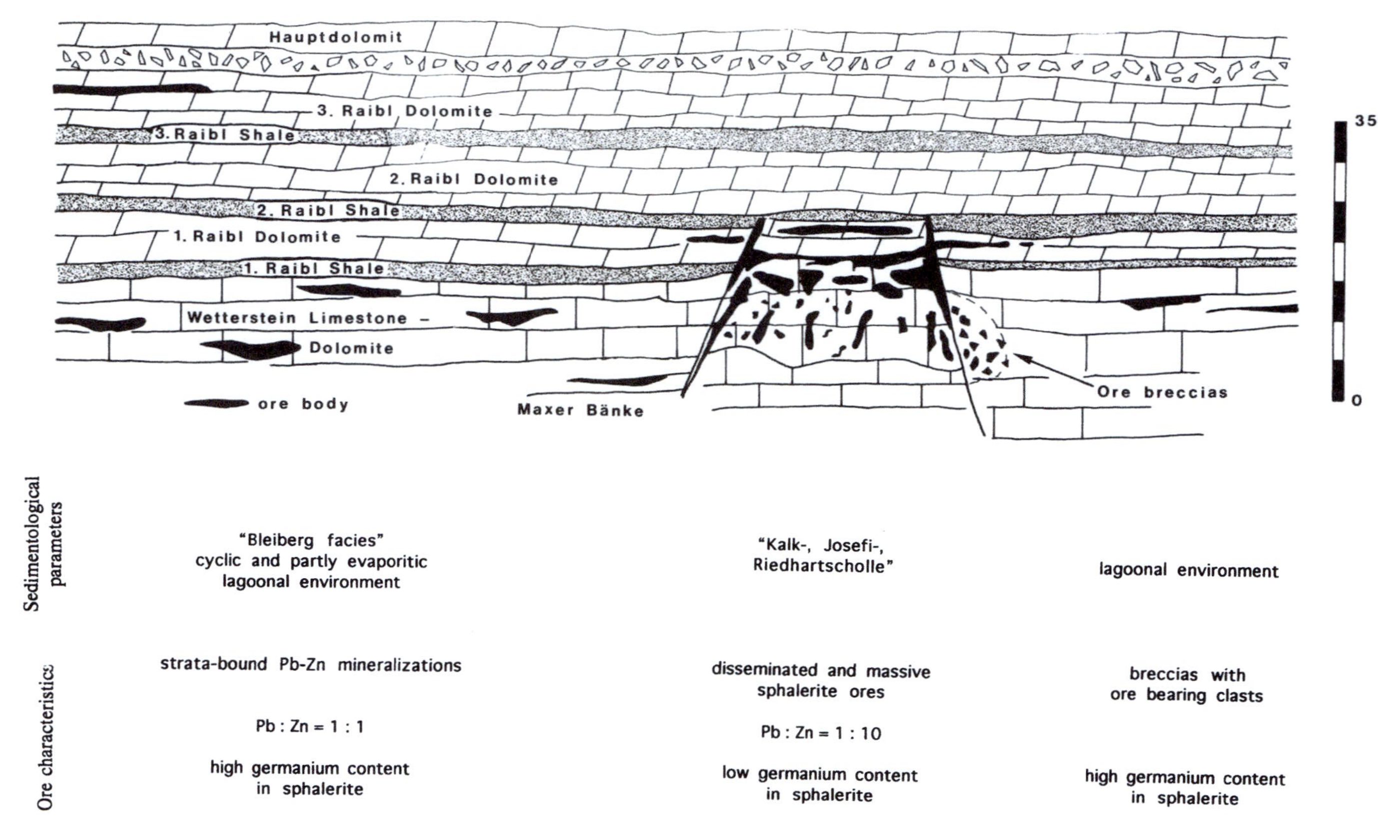

Fig. 4. Schematic palaeogeographic reconstruction of the lead-zinc bearing sequence at Bleiberg-Kreuth. (After Cerny 1989a)

The Kalkscholle orebody (and the Carditascholle, see below) formed the basis of an increase in production to 0.5 million t/year and for the introduction of more highly mechanized mining methods. Further sedimentological and geochemical investigations resulted in the discovery of two further orebodies of this type (Josefischolle, Riedhartscholle). According to present data, these three orebodies contain 8 million t of ore, with a metal content (Zn, Pb) of about 6%. The Zn/Pb ratios range from 6:1 to 10:1.

As a result of intensified underground exploration, an additional new ore type, consisting of ore-bearing clasts within a breccia, has been recently discovered south of the Riedhartscholle (Fig. 4). The clasts were derived from the Riedhartscholle and the overlying Raibl Fm. (no clasts from the Hauptdolomit could be found). The terrigenous matrix of the breccias has the same lithology as that of the Raibl shales (Cerny 1989a).

The contact of these ore-bearing breccias to the surrounding rocks is unclear because of extremely complicated Alpine tectonic structures within and around the orebodies. The genesis of the breccias is therefore still open to discussion. They have been explained as synsedimentary rock slides (Cerny 1989a,b) as well as internal breccias (Bechstädt 1975c). The critical sedimentary contacts to the surrounding rocks are, in most cases, obliterated by Alpine tectonics. It is also nuclear whether the ore-bearing clasts represent resedimented ores, or whether mineralization affected the clasts after formation of the breccia.

Cardita Orebody. Facies studies have shown that mineralization in the Raibl Group was dependent on the palaeomorphology of the underlying facies. Carbonates of the first Raibl dolomite, situated above the palaeotopographic highs of the Riedhart-, Josefi- and Kalkscholle, are reduced in thickness and carry massive stratabound mineralizations. Carbonates of the Raibl Group, however, which are situated above the Bleiberg facies, are only locally mineralized (Bechstädt 1975c, 1979; Cerny and Kostelka 1987).

As mentioned above, the lowermost carbonate unit (first Raibl dolomite) of the Raibl (Cardita) Group is of high economic significance. The Cardita orebody is situated to the west of the mining area and consists of massive mineralizations of 1 to 5 m thickness, which are arranged subparallel to the bedding. The metal content is about 10% Zn and Pb with a Zn/Pb ratio of 1:6.

4 Carbonate Cementation Sequences and Ore Formation

Diagenetic studies have begun recently within the uppermost Wetterstein Fm. of the Drau Range (Zeeh 1990) and the western part of the Northern Calcareous Alps (Maul 1991). They show a complex succession of cementation.

The extent of early diagenetic to shallow burial cementation depends on the local influence of the vadose and the meteoric regimes (Zeeh et al. 1988). The various cement and dolomitization types which result from these diagenetic stages (Fig. 5) occur in the Wetterstein Fm. of the Drau Range (Zeeh 1990) and in different parts of the Northern Calcareous Alps (Henrich and Zankl 1986; Maul 1991). Cements with a meniscus outline (sensu Richter 1976)

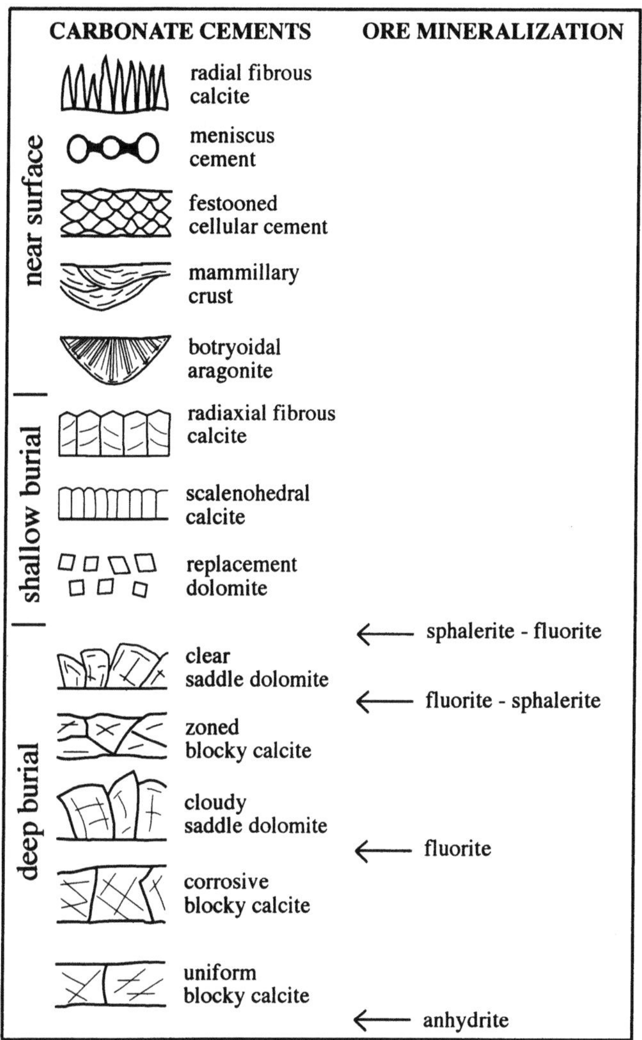

Fig. 5. Cement stratigraphy of the uppermost Wetterstein Fm. and the occurrence of sphalerite, fluorite and anhydrite with respect to the late diagenetic cements

represent the meteoric-vadose environment, while festooned cellular cements, mammillary crusts, "botryoidal aragonites" (now transformed to low magnesium calcite), as well as fibrous calcites are more typical of the marine vadose/phreatic environment. Widespread radiaxial-fibrous calcites, scalenohedral calcites (dogtooth cements) and idiomorphic pore-filling dolomites ("replacement dolomite" sensu Henrich and Zankl 1986) are most likely of marine origin. Maul (1991) suggested that the latter two were formed by meteoric waters and/or in a mixing zone. In this case the freshwater lens would have been extremely large. This is considered unlikely, because only short and

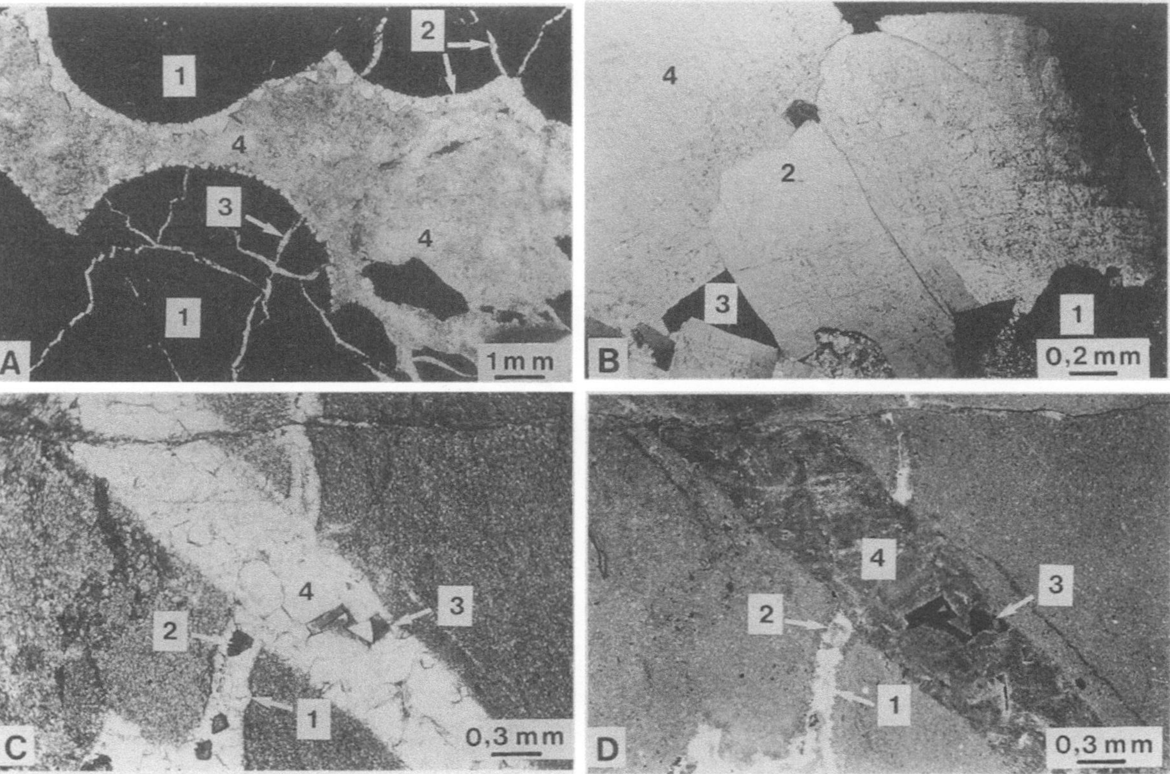

Fig. 6A–D. Carbonate cements and ore minerals. **A** Colloform sphalerite (*1*) cross-cut by fissures containing fluorite (*2*) and clear saddle dolomite (*3*). The cavity within the sphalerite crust is filled with fluorite (*2*) and clear saddle dolomite, which is replaced by cloudy saddle dolomite (*4*). Cardita orebody; transmitted light. **B** Sphalerite (*1*) is followed by cloudy saddle dolomite (*2*), fluorite (*3*) and uniform blocky calcite (*4*). Bleiberg facies; crossed nicols. **C** A fissure, containing clear saddle dolomite (*1*) and sphalerite (*2*) is cross-cut by a fissure filled with fluorite (*3*), which is replaced by cloudy saddle dolomite (*4*). Josefi-Scholle; transmitted light. **D** Same section as **C**, under cathodoluminescence, the clear saddle dolomite luminesces yellow and the cloudy saddle dolomite red

locally restricted emersion phases influenced parts of the Wetterstein platforms. The shallow burial diagenesis was terminated by the scalenohedral calcite and the "replacement dolomite".

Progressively deeper burial cementation exhibits a repeated succession of dolomitic and calcitic cements (Fig. 5) in the Wetterstein Fm. and in the Raibl Group. These cements were distinguished by petrography and optical cathodoluminescence. Deeper burial cementation began with a first generation of saddle dolomite ("clear saddle dolomite"), represented by relatively small crystals, appearing clear in transmitted light. The following calcitic blocky cement ("zoned blocky calcite") revealed by cathodoluminescence a zonation from non-luminescing to red, orange, yellow and orange luminescing zones. Dedolomitization of the earlier clear saddle dolomite is associated with the occurrence of this zoned blocky calcite. Orange-yellow luminescing microfractures (only visible by cathodoluminescence; Fig. 7B) are connected with the occurrence of the respective luminescence zones of the zoned blocky calcite. A second generation of saddle dolomite ("cloudy saddle dolomite") consists of relatively large crystals (Fig. 6B) with a mostly cloudy appearance in thin sections. The following two generations of calcitic blocky cements show crystal sizes up to several millimetres. Both are well distinguished by cathodoluminescence. Corrosion of the first generations of cements is a typical feature of a predominant non-luminescing or dull luminescing blocky calcite, sometimes with a small red and yellow luminescing outer rim ("corrosive blocky calcite"), while the following uniform red-orange luminescing blocky calcite ("uniform blocky calcite") often replaces former cements (Fig. 6B).

Corrosive processes were also responsible for the formation of cavities, which were subsequently filled by internal sediments. These sediments consist of crystal relics and insoluble material, some of which were mistaken by earlier authors (e.g. Schulz 1964) as external products.

The sequence of cements described above enables the relative dating of the various ore-forming processes. Initial results have given some insight into the complex history of mineralization (Zeeh 1990) involving galena, sphalerite, and fluorite. Some examples are described below.

Within the Cardita orebody (first Raibl dolomite horizon) sphalerite and galena are cross-cut by fissures bearing fluorite. Fluorite occurs also in cavities within sphalerite crusts. A later system of fissures with clear saddle dolomite cross-cut the sphalerite as well as the fluorite. Clear saddle dolomite and large crystals of cloudy saddle dolomite fill the remaining pores the cavities (Fig. 6A). This sequence indicates a precipitation of galena, sphalerite and fluorite before the clear saddle dolomite.

In samples from the Josefischolle, sphalerite and fluorite occur in a first generation of fissures together with clear saddle dolomite. A cross-cutting generation of fissures contains a second type of fluorite, which is replaced by cloudy saddle dolomite (Fig. 6C,D). Within pores, sphalerite was precipitated after a dolomitized scalenohedral cement and before clear saddle dolomite. Zoned blocky calcite and cloudy saddle dolomite filled the remaining pores. Samples from a core, drilled outside the mining area, show that the above mentioned second generation of fluorite might have been formed before the zoned blocky calcite.

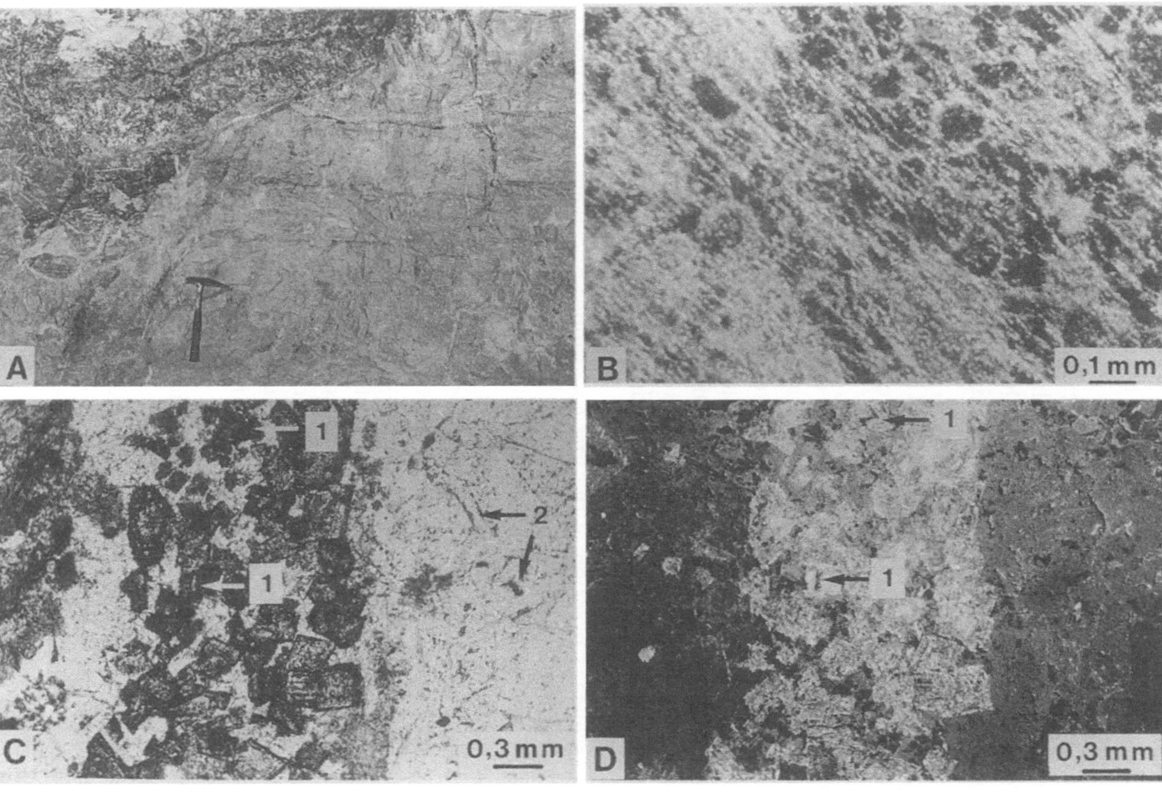

Fig. 7A–D. Ore formation and carbonate cements. **A** Part of the "Konradi" orebody with galena, sphalerite and pyrite (cf. Fig. 8, profile B). Note the discordant contact of the orebody to the surrounding rock, indicating ore precipitation within cavities. **B** Peloid-wackestone cross-cut by yellow luminescing microfractures. Sample from the vicinity of the Konradi orebody; cathodoluminescence. **C** Clear saddle dolomite partially replaced by fluorite (*1*). Fluorite also replaces cloudy saddle dolomite; relics of this dolomite are sometimes visible (*2*). Sample from a lead-zinc occurrence near Radnig; transmitted light. **D** Same section as **C**, dolomitic relics luminesce red; fluorite luminesce dark-blue; cathodoluminescence

A rare third generation of fluorite (as seen in the Radnig area) was developed after the formation of cloudy saddle dolomite (Fig. 6B) and sometimes replaces all preceding cements as well as the sediment (Fig. 7C,D). Corrosive and uniform blocky calcite post-dates this generation of fluorite. Additionally, in the Jauken mine (50 km westward of Bleiberg-Kreuth), a second generation of sphalerite and galena occurs after clear saddle dolomite.

These examples show a threefold repeated precipitation of fluorite, which has been described earlier (e.g. Schneider 1954; Schulz 1968). The REE distributions of these three fluorite stages show a significant decrease in HREE from the first to the third stage (Hein 1986). According to Möller (1983), this mirrors mobilization and recrystallization processes: the first generation of fluorite would represent the primary phase, while the second and third generations represent mobilization phases.

The thus far mentioned two generations of sphalerite (before and after clear saddle dolomite) can only be dated approximately in relation to the clear saddle dolomite. The connection to the other carbonate cements and to fluorite remains unclear.

According to the present results, the different generations of galena, sphalerite and fluorite follow the "replacement dolomite" and are post-dated by uniform blocky calcite. Only blue-coloured anhydrite, which accompanies the ore minerals in parts of the mine of Bleiberg-Kreuth, was precipitated after the uniform blocky calcite. A late-stage precipitation of this type of anhydrite was already proposed by Schroll (1953) and Kappel and Schroll (1982).

While carbonate cements of near-surface to shallow burial diagenesis are often preserved, they never post-date the formation of ore minerals. This does not support a synsedimentary exhalative origin for the Pb-Zn mineralization, as assumed by many authors (e.g. Schulz 1959, 1964; Hein 1986).

5 Geochemical Criteria and Fluid Inclusions of the Ores and the Host Rock

Studies on the geochemistry of the ores and the host rock started in the late 1960s. Analysis of the main and trace elements of whole rock samples by atomic absorption spectrometry (AAS) is a standardized process for exploration within the Bleiberg-Kreuth mine. There are abundant geochemical data and publications (e.g. Cardiche-Loarte and Schroll 1973; Cerny 1983; Krainer 1985; Hein 1986; Hagemeister 1988; Zeeh 1990). In contrast, fluid inclusion studies have started only recently. An outline of the geochemical and fluid inclusion results are given below.

5.1 Main and Trace Element Studies

Within the ore-bearing strata (Wetterstein Fm. and Raibl Group) the proportion of dolomite and limestone changes with dolomite prevailing within the lower Wetterstein Fm. In the eastern part of the mine, the uppermost Wetterstein Fm. is mainly built up by limestone, in the western part mainly by

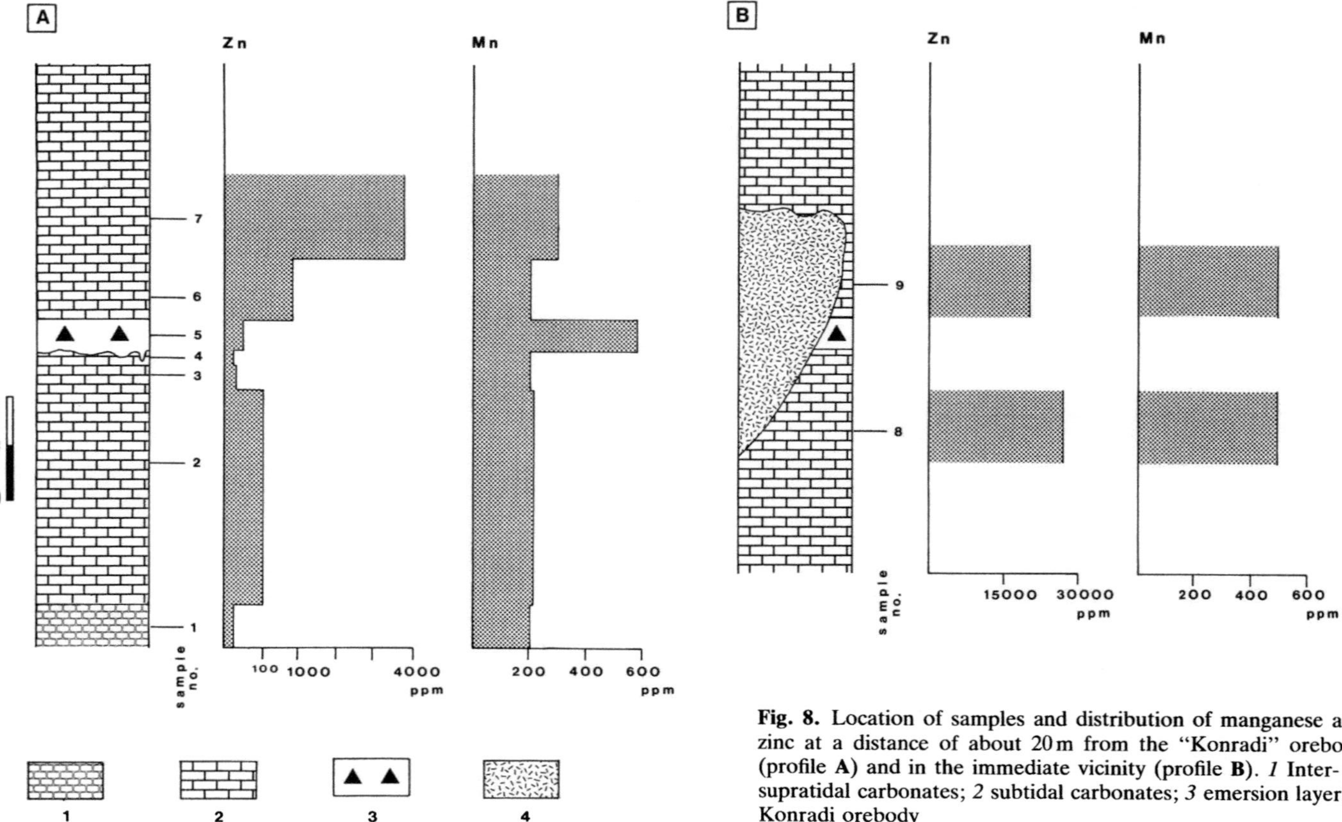

Fig. 8. Location of samples and distribution of manganese and zinc at a distance of about 20 m from the "Konradi" orebody (profile **A**) and in the immediate vicinity (profile **B**). *1* Inter- to supratidal carbonates; *2* subtidal carbonates; *3* emersion layer; *4* Konradi orebody

dolomite. Within the carbonate sequences of the Raibl Group the distribution of dolomite and limestone changes irregularly (Cerny 1983; Hagemeister 1986, 1988). Most of the lead-zinc ores are bound to dolomitic host rocks (Cerny 1983).

Pb-Zn orebodies are often characterized by increased contents in Mn, Fe, quartz, and Zn. Such haloes are of great importance in the exploration of lead-zinc ores (Cerny 1989a,b). Samples of subtidal limestone from the immediate vicinity of the "Konradi" orebody situated at the 11th floor Revier Stefanie (Fig. 7A) show an average Mn content of about 490 ppm (measured by AAS) and a Zn content of >20 000 ppm. Radiaxial fibrous calcites contain up to 860 ppm Mn, while these cements usually do not contain more than 110 ppm Mn. The whole rock (i.e. sediment and near-surface to shallow burial cements) shows by cathodoluminescence an overall yellow-orange luminescence.

Subtidal limestone 20 m away from this orebody (Fig. 8) contains about 230 ppm Mn; the Zn content varies from 19 to 3750 ppm. Radiaxial fibrous calcites show here an Mn content of 130 ppm. By cathodoluminescence yellow luminescing microfractures are visible, which are related to the yellow luminescing zone of the zoned blocky calcite. Closer to the orebody the quantity of yellow luminescing microfractures increases up to the overall yellow-orange luminescence of the whole rock described above.

Sphalerite is generally poor in Mn (Pimminger et al. 1985), and is therefore not responsible for the halo. Moreover, the data from the Konradi orebody indicate that the Mn halo is connected with the formation of the zoned blocky calcite (see Sect. 4) and is therefore a deep burial product. The late diagenetic incorporation of Mn into stable carbonate phases, however, is not well understood at the moment. Whether the Mn haloes are related to primary mineralization or remobilization during diagenesis is also an open question.

Within the Bleiberg mine the Ge contents of some sphalerite types vary. Sphalerites from the Bleiberg facies contain much more Ge (up to 600 ppm) than sphalerites from the Kalk-, Riedhart- and Josefischolle (max. 100 ppm). This may be caused by mobilization processes influencing the latter orebodies. In contrast, ore-bearing clasts derived from the Riedhartscholle contain higher Ge contents. This could be due to the siliciclastic matrix of the ore breccias, which might have prevented recrystallization and mobilization of the ores (Cerny 1989a,b).

5.2 Results of Preliminary Fluid Inclusion

This work presents the first fluid inclusion data of carbonate cements and fluorites from Bleiberg-Kreuth. Primary fluid inclusions are quite rare and could be found only in deep burial cements; from each of the measurable fluid inclusion-bearing crystals, about 10–15 measurements were made. A short synthesis of the data is given below.

Fluorites from the Cardita orebody, which were formed before the clear saddle dolomite, contain 10–18 μm, primary fluid inclusions with homogenization temperatures (Th) ranging from 170 to 195 °C. The melting temperatures of ice (Tm: −18 to −15 °C; Te: −38 to −34 °C) indicate a salinity of 18 to

21 wt% equiv. NaCl and a predominance of NaCl within the fluid. Measurements of secondary fluid inclusions yielded Th in the range of 110 to 130 °C.

Primary fluid inclusions have also been found in fluorites from an orebody within the Wetterstein Fm. They presumably represent the third generation of fluorite, formed after zoned blocky calcite. They show Th from 120 to 142 °C with a salinity of the fluid at about 22 to 24.7 wt% equiv. NaCl (Tm: −24 to −20.4 °C) and a predominance of $CaCl_2$ (Te: −52 to −49 °C).

Further measurements of primary fluid inclusions in fluorites (from the same locality), which occur in fissures and could not be exactly dated with respect to the carbonate cementation sequence, display Th from 84 to 120 °C. A few fluid inclusions homogenize at temperatures up to 144 °C, but they seem to be influenced by necking-down processes. The melting temperatures of ice indicate a low salinity (Tm: −6.3 to −3.8 °C) and prevailing NaCl (Te: −23 to −20.5 °C).

These fluid inclusion data from fluorites indicate precipitation or mobilization processes at a temperature range of 84 to 195 °C. This broad range of temperature reflects the different stages of fluorite precipitation during diagenesis. The fluid inclusion data of carbonate cements (Zeeh 1990) from the Wetterstein Fm. support this range of formation temperatures. Furthermore, a drop in Th during diagenesis is indicated by the cloudy saddle dolomite, followed by an increase to the highest measured temperatures with the formation of the corrosive and the uniform blocky calcite (cf. Table 2).

The present number of samples is relatively small and effects, such as "stretching" (Larson et al. 1973; Bodnar and Bethke 1984), leading to determination errors of the measured temperatures, cannot be completely excluded. A maximum temperature of up to 250 °C during burial (uncorrected for overburden pressure), as indicated by the fluid inclusion data, is in accordance with the assumptions made by Niedermayr et al. (1984), and with fluid inclusion studies from the Northern Calcareous Alps (Maul et al. 1989).

In the area of investigation, there is a strong discrepancy between the maximum temperatures deduced from fluid inclusions (up to 250 °C) and from fluorescence of organic matter, vitrinite reflectance and illite crystallinity, evaluated as 100 °C (Kappel and Schroll 1982). This discrepancy also exists in parts of the Northern Calcareous Alps (cf. Maul 1991; Petschick 1989) and

Table 2. Homogenization and melting temperatures of the deeper burial carbonate cements of the Wetterstein Fm. of the Drau Range

Carbonate cement (No. of samples)	Th (°C)	Tm (°C)	Te (°C)
Clear saddle (1)	90–101	−15 to −12	−52
dolomite (3[a])	155–182[a]	−9 to −2.7[a]	−37 to −26[a]
	155–184	−15 to −13	−55 to −52
Zoned blocky calcite (1)	128–162	−5 to −3	−30 to −28
Cloudy saddle dolomite (2)	210–252	−5 to −3	−23 to −20
Corrosive blocky calcite (1)	200–250	−25 to −21	−47 to −45
Uniform blocky calcite (1)			

[a] Data from Maul (1991), Northern Calcareous Alps.

might be caused by the duration of heating, different overburden pressures, and different thermal conductivities of the rocks. In other parts of the Northern Calcareous Alps, the maximum recorded temperature (Kralik et al. 1987; Petschick 1989) by vitrinite reflectance and illite crystallinity is about 300 °C. K/Ar isotopic age determinations indicate that this maximum temperature might be an Early Cretaceous thermal event and not a Tertiary "Alpine" event (Kralik et al. 1987). These maximum temperatures are close to the maximum Th of fluid inclusions from uniform blocky calcite (200–250 °C) and therefore the formation of the uniform blocky calcite during Early Cretaceous times was assumed (Zeeh and Bechstädt 1989). Furthermore, the decrease in temperature following the formation of the zoned blocky calcite might indicate more than one thermal event. Based on these data and assumptions, a relationship between carbonate cementation, ore formation and extensional phases, causing pulses of ascending hydrothermal fluids, was proposed. The hydrothermal fluids possibly did not heat up the whole rock pile, but affected the areas of mineralization (Zeeh and Bechstädt 1989).

5.3 Sulphur Isotopes

Sulphur-isotope studies (e.g. Schroll and Wedepohl 1972; Schroll and Pak 1983; Schroll et al. 1983), summarized in Fig. 9, show a broad range of mostly negative $\delta^{34}S$ values for the sulphides and positive values for the sulphates. The above mentioned authors assumed Triassic seawater as the source of sulphur, due to the similar isotopic composition of the measured sulphates and Triassic seawater. The differing isotopic composition of the sulphides was explained by bacterial reduction (Schroll and Wedepohl 1972; Schroll et al. 1983).

The sulphur isotope composition of the sulphides from the Raibl Group and Maxer Bänke differs from the sulphides of the Bleiberg facies (Fig. 9). Sulphides from the latter are about 25‰ lighter than sulphates, while sulphides from the Maxer Bänke and the Raibl Group are 40 to 50‰ lighter. These differences were explained by Kappel and Schroll (1982), Schroll et al. (1983) and Schroll (1984b) by influences of the depositional environment and of later diagenetic and redepositional processes. However, this interpretation is doubtful. It is based on assumed, original, significant differences in the organic matter content of the "less bituminous" carbonate rocks of the Bleiberg facies to the "more bitumimous" carbonate rocks of the Maxer Bänke and the Raibl Group. According to Schwarcz and Burnie (1973), sulphur isotopes of sedimentary sulphides are generally lighter than corresponding seawater sulphates. In clastic (non-carbonatic), deep euxinic basins, the isotope composition is about 50‰ lighter. In a clastic shallow marine, brackish or freshwater environment (e.g. tidal lagoons, estuaries) the difference is up to 25‰. However, facies analysis (cf. Hagenguth 1984; Hagemeister 1986; Zeeh 1990) clearly indicates shallow marine conditions and certainly no euxinic environments during deposition of the Wetterstein and Raibl carbonates. The differences in organic matter content are, most likely, due to later migration into porous dolomites.

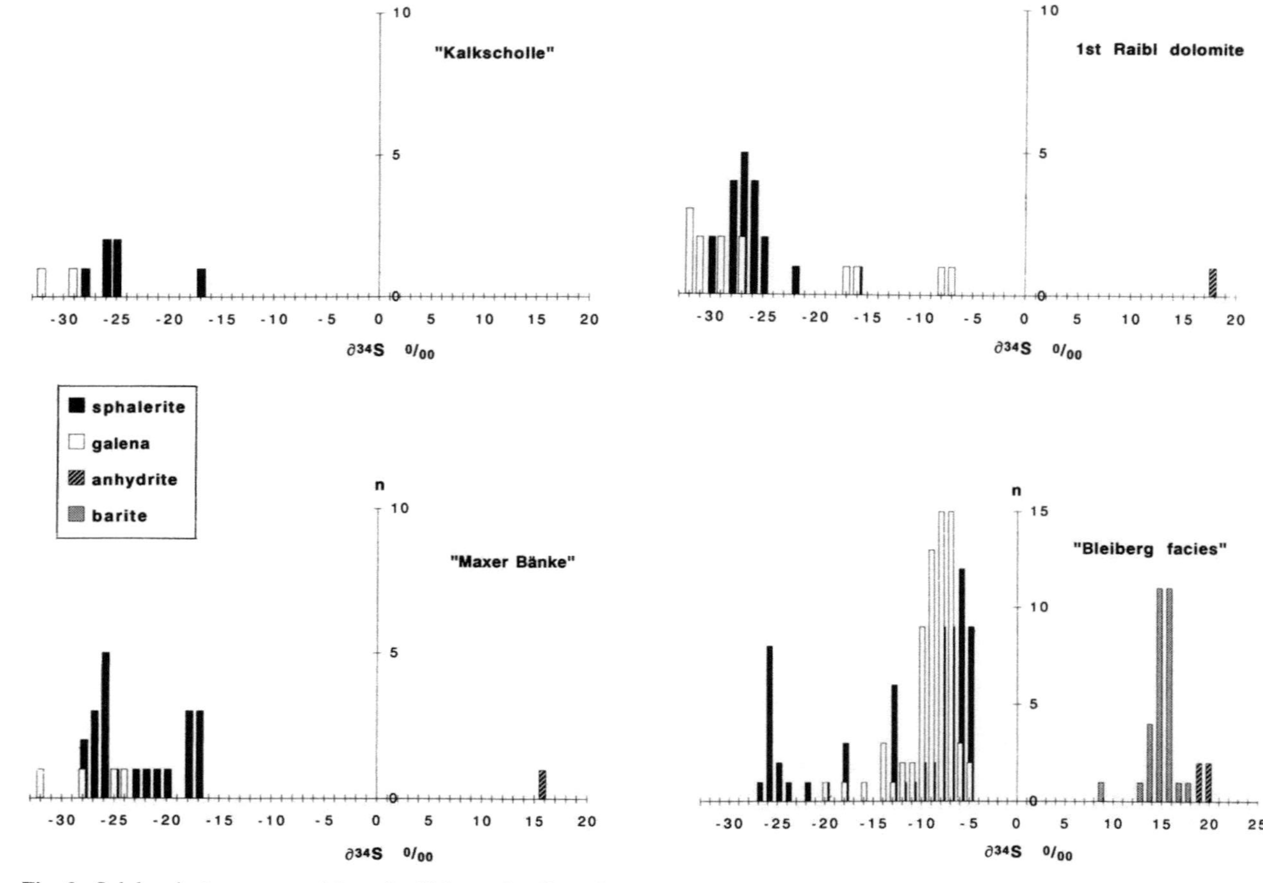

Fig. 9. Sulphur isotope composition of sulfides and sulfates from the Maxer Bänke, Kalkscholle, Bleiberg facies and the first Raibl dolomite. (Data from Schroll et al. 1983)

5.4 Lead Isotopes

The $^{207}Pb/^{206}Pb$ isotopic ratio of the Bleiberg-Kreuth ore lead shows moderately radiogenic values ("B-type") with a Carboniferous model age. Lead-isotope studies (Köppel and Schroll 1985, 1988) also show a different isotopic composition of ore lead and trace lead of the host rock. Isotopic similarity between feldspar lead from the crystalline basement and ore lead indicates a derivation of metals from Palaeozoic metasediments, with a possible minor amount of lead from Permian sandstones and Triassic volcanic rocks (Köppel and Schroll 1985, 1988).

5.5 Carbon and Oxygen Isotopes

The C- and O-isotope studies (Zeeh 1990; Zeeh et al., in prep.) show slightly lighter oxygen values in sediment and early diagenetic cements ($\delta^{13}C$: 2.16 to 3.02‰, $\delta^{18}O$: −7.35 to −4.16‰ PDB) compared to calcite precipitated from Triassic seawater ($\delta^{18}O \approx -2.75$‰: Frisia-Bruni et al. 1989). This shift to lighter values might be due to modification during diagenesis. The deeper burial carbonate cements exhibit characteristic light oxygen values of burial diagenesis ($\delta^{13}C$: 0.95 to 3.79‰, $\delta^{18}O$: −17.56 to −5.38‰ PDB). It should be stressed that no indication of meteoric influence during burial diagenesis is given by the C- and O-isotope values.

6 Conclusions

The genesis of the Bleiberg deposit, and of similar lead-zinc deposits in the Eastern Alps, has been interpreted in a variety of ways. The post-tectonic (Tertiary), magmatic-hydrothermal hypothesis for the ore emplacement (Tornquist 1927; Petrascheck 1931) has been the accepted model for a long time. Schneider (1954), Taupitz (1954) and Schulz (1959, 1964) recognized the pre-tectonic development of the ores and assumed external precipitation. This model is mainly based on the rare, small scaled and laterally discontinuous occurrences of ore-bearing sedimentary rocks. The authors assumed a transport of the metals by low temperature, hydrothermal fluids, precipitating ores within small depressions of the marine environment (Sedex type of emplacement, according to present-day nomenclature). Fabrics characteristic of ore emplacement within cavities were reported by Siegl (1956). Bechstädt (1973) showed that the shallow lagoonal character of the sediments was interrupted by several emersion phases and proposed (Bechstädt 1975a,b) an ore enrichment by karstic processes, following the "karst model" of Bernard (1973). Bechstädt (1979), and later, Bechstädt and Döhler-Hirner (1983), assumed that the metals were derived from an uplifted region in the north, and were enriched within instable minerals. Further concentration occurred during later diagenesis within karst cavities. Dzulynski and Sass-Gustkiewicz (1977) proposed a model of intense hydrotherms, leaching the carbonate host rock ("hydrothermal karst"), and precipitating metals within the newly formed cavities in contact with phreatic (marine or meteoric) waters. Köppel and Schroll (1985) assumed that descending surface waters during Lower to Middle Triassic penetrated the Palaeozoic metasediments, were heated up and mobilized the metals. The hydrothermal brines ascended and ores were precipitated externally as well as internally in Wetterstein and Raibl times.

All these theories reflect more or less the state of knowledge at the respective times. According to the accumulated information to date, some theories now seem to be unlikely. Of particular importance are the Pb-isotope data (Köppel and Schroll 1985) which show a distinct difference between ore lead and trace lead within the host rock. This excludes all theories based on mobilization of ore lead from the host rock. Considering additionally the extremely shallow marine depositional environment, the hypotheses assuming an external sedimentary emplacement of the non-disseminated ores are highly speculative (cf. also Finlow-Bates and Tischler 1983). As a result of the shallow water depths, an external ore precipitation would only be possible within small, pond-like depressions under anoxic conditions. These are not present; the ore-filled "pools, craters and pockets" (Schneider 1964) are, in fact, fillings of internal cavities (see above). The observations summarized in the present work show clearly that the Bleiberg ores are distinctly different from Sedex deposits and clearly result from epigenetic, mainly MVT, mineralization.

Sulphur-isotope studies have shown that the composition of the sulphate-sulphur is in concordance with Triassic seawater composition (Schroll et al. 1983). The theory of derivation of sulphide-sulphur from bacterial reduction of Triassic sulphate-sulphur (Kappel and Schroll 1982) is mainly based on the assumed external ore emplacement and low temperatures during diagenesis (100–110 °C). Fluid inclusion data, however, indicate higher temperatures and possibly hydrothermal pulses (see above). Therefore, other processes should be considered, such as transport of the sulphide-sulphur by the ore-forming fluids (Williams and Rye 1974; Ohmoto and Rye 1979), leaching of pre-existing sulphides (Ohmoto and Rye 1979) or inorganic sulphate reduction (Anderson 1991).

The following data should be considered in modelling the Bleiberg-Kreuth Pb-Zn deposit:

1. The ore lead is derived mainly from the Palaeozoic basement;
2. The occurrence of lead-zinc ores is largely connected with palaeogeographic structures (i.e. palaeotopographic highs within the Wetterstein platforms);
3. Relative dating of the ore emplacement by the sequence of carbonate cementation reveals that the earliest ores were precipitated after the formation of shallow burial cements;
4. Fluid inclusion studies indicate temperatures between 80 and 200 °C during ore formation.

Based on these facts and considering the Triassic to Cretaceous tectonic evolution of the Eastern Alps, the following model is proposed which, in part, incorporates the hypotheses given by Köppel and Schroll (1985, 1988).

Extensional tectonics, starting in the Triassic and continuing into the Jurassic, opened pathways for waters to descend into the underlying sediments (where they were probably enriched in chlorides) and into the basement rocks. Metals were leached from the metasediments of the crystalline basement and possibly from Permian and Triassic rocks. Subsequently, hydrothermal fluids ascended along pre-existing faults into the carbonates of the Wetterstein Fm. and Raibl Group, which had been buried at this time to medium/deeper depths. The internal emplacement of the first generation of ores was followed by precipitation of carbonate cements (clear saddle dolomite). Renewed ore supply and/or ore mobilization and migration at higher temperatures (up to 200 °C) caused the emplacement of late-stage ores (after clear saddle dolomite). Assuming a Cretaceous age for the uniform blocky calcite (which predates the ore formation), these thermal events and mineralization phases might be correlated, in a very hypotheticaly way, with extensional tectonic phases from Upper Triassic to Early Cretaceous (Zeeh and Bechstädt 1989).

The relationship of the lead-zinc ores with palaeogeographic structures (i.e. palaeotopographic highs) can be explained by the higher primary porosity (due to the early meteoric influence), favouring the fluid flow rate, in contrast to other facies. Another fact might be the occurrence of sedimentary sulphates (grey-coloured anhydrite, barite) and sulphides (disseminated pyrite within black pebbles) within the Bleiberg facies, giving a direct source of sulphur.

Another point to be considered is the concentration of the ore deposits within the uppermost part of the Wetterstein Fm., which is overlain by the first Raibl shale. The sealing effect of this shale to ascending solutions (Bechstädt 1979) is demonstrated by the only local occurrence of Pb-Zn ores within the Raibl Group. Subordinate ores within the Raibl Group occur only where Raibl and Wetterstein carbonates are in direct contact due to early or late tectonism.

Acknowledgements. We are indebted to Dr. R. Rieken (Göttingen) for the fluid inclusion measurements, to Prof. J. McKenzie and Dr. P. Kindle (Zurich) for C- and O-isotope measurements and to Prof. V. Köppel (Zurich), Prof. E. Schroll (Vienna) and Dr. I. Cerny (Bleiberg) for stimulating discussions. Financial support was provided by the Deutsche Forschungsgemeinschaft.

References

Anderson GM (1991) Organic maturation and ore precipitation in southeast Missouri. Econ Geol 86:909–926

Bechstädt T (1973) Zyklotheme im hangenden Wettersteinkalk von Bleiberg-Kreuth (Kärnten/Österreich). Veröff Univ Innsbruck (Festschr Heißel) 86:25–55

Bechstädt T (1975a) Lead-zinc ores dependent on cyclic sedimentation. Mineral Depos 10:234–248

Bechstädt T (1975b) Zyklische Sedimentation im erzführenden Wettersteinkalk von Bleiberg-Kreuth (Kärnten, Österreich). Neues Jahrb Geol Paläontol Abh 149:73–95

Bechstädt T (1975c) Sedimentologie und Diagenese des Wettersteinkalkes von Bleiberg-Kreuth. Ein Hinweis zur Genese der Blei-Zink-Erze. Berg- Hüttenm Mh 120:466–471

Bechstädt T (1979) The lead-zinc deposits of Bleiberg-Kreuth (Carinthia, Austria): palinspastic situation, paleogeography and ore mineralization. Verh Geol B-A 1978/3:221–235

Bechstädt T, Döhler-Hirner B (1983) Lead-zinc deposits of Bleiberg-Kreuth. In: Scholle P, Bebout DG, Moore CH (eds) Carbonate depositional environments. Am Assoc Petrol Geol Mem 33:55–63

Bechstädt T, Schweizer T (1991) The carbonate-clastic cycles of the East-Alpine Raibl Group: result of third-order sea-level fluctuations in the Carnian. Sediment Geol 70:241–270

Bechstädt T, Brandner R, Mostler H, Schmidt K (1978) Aborted rifting in the Triassic of the Eastern and Southern Alps. Neues Jahrb Geol Paläontol Abh 156:157–178

Bernard AJ (1973) Metallogenic processes of intra-karstic sedimentation. In: Amstutzt GC, Bernard AJ (eds) Ores in sediments. Springer, Berlin Heidelberg New York, pp 43–57

Blendinger W (1985) Middle Triassic strike-slip tectonics and igneous activity of the Dolomites (Southern Alps). Tectonophysics 113:105–121
Bodnar RJ, Bethke PM (1984) Systematics of stretching of fluid inclusions: fluorite and sphalerite at 1 atmosphere confining pressure. Econ Geol 79:141–161
Bosellini A, Castellarin A, Doglioni C, Guy F, Perric MC, Rossi PL, Simboli G, Sommavilla E (1982) Geologia della conca di arabba e dei rilievi circostanti. In: Castellarin A, Vai GB (eds) Guida alla geologia del Sudalpino centro-orientale. Soc Geol Ital Guide Geol Reg. Mem Soc Geol Ital Suppl C 24:243–254
Brandner R (1984) Meeresspiegelschwankungen und Tektonik in der Trias der NW-Tethys. Jahrb Geol B-A 126:435–475
Brigo L, Kostelka L, Omenetto P, Schneider H-J, Schroll E, Schulz O, Strucl I (1977) Comparative reflections on four alpine Pb-Zn deposits. In: Klemm DD, Schneider H-J (eds) Time and strata bound ore deposits. Springer, Berlin Heidelberg New York, pp 273–293
Cardiche-Loarte L, Schroll E (1973) Die Verteilung und Korrelation einiger Elemente in einem Erzkalkprofil der Bleiberger Fazies (Bleiberg/Kärnten-Rudolfschacht). Tschermaks Min Petrol Mitt 20:59–70
Castellarin A, Lucchini F, Rossi PL, Simboli G, Bosellini A, Sommavilla E (1980) Middle Triassic magmatism in Southern Alps: a geodynamic model. Riv Ital Paleontol Stratigr 85:1111–1124
Castellarin A, Lucchini F, Rossi PL, Selli L, Simboli G (1988) The Middle Triassic magmatic-tectonic arc development in the Southern Alps. Tectonophysics 146:79–89
Cerny I (1983) Pb-Zn-Erzmobilisationen in Dolomitgesteinen der Draukalkalpen (Kärnten, Österreich). In: Petrascheck WE (ed) Ore mobilization in the Alps and in SE-Europe. Österr Akad Wiss Erdwiss Komm 6:31–38
Cerny I (1989a) Current prospecting strategy for carbonate-hosted Pb-Zn mineralizations at Bleiberg-Kreuth (Austria). Econ Geol 84:1430–1435
Cerny I (1989b) Die karbonatgebundenen Blei-Zink-Lagerstätten des alpinen und außeralpinen Mesozoikums. Die Bedeutung ihrer Geologie, Stratigraphie und Faziesgebundenheit für Prospektion und Bewertung. Arch Lagerstätten forsch Geol Bundesanst Wien 11:5–125
Cerny I, Kostelka L (1987) The development of the geological groundwork as basis for ore prospecting at Bleiberg-Kreuth, Austria. In: Jancovic S (ed) Mineral deposits of the Tethian Eurasian metallogenetic belt between the Alps and the Pamirs (selected examples). UNESCO-IGCP Proj 169, Belgrad, pp 62–68
Colins E, Nachtmann W (1978) Geologische Karte der Villacher Alpe (Dobratsch), Kärnten. Mitt Ges Geol Bergbaustud Österr 25:1–10
Doglioni C (1984) Triassic diapiric structures in the central Dolomites (northern Italy). Eclogae Geol Helv 77:261–285
Doglioni C (1987) Tectonics of the Dolomites (Southern Alps, northern Italy). J Struct Geol 9:181–193
Dzulynski S, Sass-Gustkiewicz M (1977) Comments on the genesis of the Eastern-Alpine Zn-Pb deposits. Mineral Depos 12:219–233
Eberli GP (1988) The evolution of the southern continental margin of the Jurassic Tethys ocean as recorded in the Allgäu Formation of the Austroalpine nappes of Graubünden (Switzerland). Eclogae Geol Helv 81:175–214
Finlow-Bates T, Tischler SE (1983) Controls on Alpidic mineralization styles. In: Schneider H-J (ed) Mineral deposits of the Alps and of the Alpine epoch in Europe. Springer, Berlin Heidelberg New York, pp 7–18
Frisia-Bruni S, Jadoul F, Weissert H (1989) Evinosponges in the Triassic Esino Limestone (Southern Alps): documentation of early lithification and late diagenetic overprint. Sedimentology 36:685–699
Hagemeister A (1986) Zyklische Sedimentation auf einer stabilen Karbonatplattform: die Raibler Schichten (Karn) des Drauzuges (Österreich). Thesis, Albert-Ludwigs University, Freiburg
Hagemeister A (1988) Zyklische Sedimentation auf einer stabilen Karbonatplattform: die Raibler Schichten (Karn) des Drauzuges/Kärnten (Österreich). Facies 18:83–122
Hagenguth G (1984) Geochemische und fazielle Untersuchungen an den Maxerbänken im Pb-Zn-Bergbau von Bleiberg-Kreuth. Mitt Ges Geol Bergbaustud Österr S-H 1
Hein UF (1986) Zur Geochemie des Fluors im Nebengestein und Spurenelementfraktionierung in Fluoriten der kalkalpinen Pb-Zn-Lagerstätten. Berl Geowiss Abh A 81

Henrich R, Zankl H (1986) Diagenesis of Upper Triassic Wetterstein reefs of the Bavarian Alps. In: Schroeder JH, Purser BH (eds) Reef diagenesis. Springer, Berlin Heidelberg New York, pp 245–268

Holler H (1936) Die Tektonik der Bleiberger Lagerstätte. Carinthia II Spec Publ 7:1–82

Holler H (1953) Der Blei-Zinkerzbergbau Bleiberg, seine Entwicklung, Geologie und Tektonik. Carinthia II 143:35–46

Holler H (1960) Zur Stratigraphie des Ladin im östlichen Drauzug und in den Nordkarawanken. Carinthia II 70:63–75

Holler H (1977) Geologisch-tektonische Aufnahmen westlich der Bleiberger Lagerstätte. Carinthia II Sh 33

Kappel F, Schroll E (1982) Ablauf und Bildungstemperatur der Blei-Zink-Vererzung von Bleiberg-Kreuth/Kärnten. Carinthia II 172/92:49–62

Köppel V, Schroll E (1985) Herkunft des Pb der triassischen Pb-Zn-Vererzungen in den Ost- und Südalpen, Resultate bleiisotopengeochemischer Untersuchungen. Arch Lagerstätten forsch Geol Bundesanst Wien 6:215–222

Köppel V, Schroll E (1988) Pb-isotope evidence for the origin of lead in strata-bound Pb-Zn deposits in Triassic carbonates of the Eastern and Southern Alps. Mineral Depos 23:96–103

Kostelka (1971) Beiträge zur Geologie der Bleiberger Vererzung und ihrer Umgebung. Carinthia II Sh (Festschr Kahler) 28:283–289

Krainer K (1985) Beitrag zur Mikrofazies, Geochemie und Paläogeographie der Raibler Schichten der östlichen Gailtaler Alpen (Raum Bleiberg-Rubland) und des Karwendel (Raum Lafatsch/Tirol). Arch Lagerstätten forsch Geol Bundesanst Wien 6:129–142

Krainer K (1989) Zum gegenwärtigen Stand der Permoskythforschung im Drauzug. Carinthia II 179/99:371–382

Kralik M, Krumm H, Schramm JM (1987) Low grade and very low grade metamorphism in the Northern Calcareous Alps and in the Greywacke Zone: illite – crystallinity data and isotopic ages. In: Flügel HW, Faupl P (eds) Geodynamics of the Eastern Alps. Deuticke, Wien, pp 164–178

Krumm H (1984) Anchimetamorphose im Anis und Ladin (Trias) der Nördlichen Kalkalpen zwischen Arlberg und Kaisergebirge – ihre Verbreitung und deren baugeschichtliche Bedeutung. Geol Rundsch 73:223–257

Larson LT, Miller JD, Nadeau JE, Roedder E (1973) Two sources of error in low temperature inclusion homogenization determination, and corrections on published temperatures for the east Tennessee and Laisvall deposits. Econ Geol 68:113–116

Maucher A, Schneider H-J (1967) The Alpine lead-zinc ores. Econ Geol Monogr 3:71–89

Maul B (1991) Vom vadosen Bereich zur Anchimetamorphose. Diagenese des Oberen Wettersteinkalks der westlichen Nördlichen Kalkalpen. Thesis, Albert-Ludwigs University, Freiburg

Maul B, Schweizer T, Zeeh S, Bechstädt T (1989) Zementation und Korrosion im Wettersteinkalk und in den Arlbergschichten (Ladin/Cordevol, westliche Ostalpen). Geol Paläontol Mitt Innsbruck 16:169–172

Möller P (1983) Lanthanoids as a geochemical probe and problems in lanthanoid geochemistry. Distribution and behaviour of lanthanoids in non-magmatic phases. In: Sinha SP (ed) Systematics and the properties of the lanthanides. NATO ASI Ser C Math Phys Sci 109:561–616

Niedermayr G, Mullis J, Niedermayr E, Schramm J-M (1984) Zur Anchimetamorphose permoskythischer Sedimentgesteine im westlichen Drauzug, Kärnten – Osttirol (Österreich). Geol Rundsch 73:207–221

Ohmoto H, Rye RO (1979) Isotopes of sulfur and carbon. In: Barnes HL (ed) Geochemistry of hydrothermal ore deposits. Wiley, New York, pp 509–567

Ott E (1967) Segmentierte Kalkschwämme (Sphinctozoa) aus der alpinen Mitteltrias und ihre Bedeutung als Riffbildner im Wettersteinkalk. Bayer Akad Wiss Math-Naturwiss Kl Abh NF 131

Petrascheck WE (1931) Die mechanischen Gesetzmäßigkeiten der Bruchtektonik in Bleiberg (Kärnten). Zentralbl Min B 1931:477–483

Petschick R (1989) Zur Wärmegeschichte im Kalkalpin Bayerns und Nordtirols (Inkohlung und Illit-Kristallinität). Frankf Geowiss Arb Ser C 10

Pimminger M, Grasserbauer M, Schroll E, Cerny I (1985) Trace element distribution in sphalerites from Pb-Zn ore occurrences of the Eastern Alps. Tschermaks Min Petrol Mitt 34:131–141

Ratschbacher L, Frisch W, Neubauer F, Schmid SM, Neugebauer J (1989) Extension in compressional orogenic belts: the Eastern Alps. Geology 17:404–407

Richter DK (1976) Gravitativer Meniskuszement in einem holozänen Oolith bei Neapolis (Süd-Peloponnes, Griechenland). Neues Jahrb Geol Paläontol Abh 151:192–223

Rüffer T, Bechstädt T (1991) Eustatische Kontrolle auf die Frühphase der Wettersteinkalkentwicklung (Nördliche Kalkalpen, Westtirol). Sediment 91 Kurzfassungen der Beiträge Senckenberg-Am-Meer Ber 91/2:102–104

Sangster DF (1976) Carbonate-hosted lead-zinc deposits. In: Wolf KH (ed) Handbook of strata-bound and stratiform ore deposits, vol 6. Elsevier, Amsterdam, pp 447–456

Sangster DF (1986) Age of mineralization in Mississippi Valley-type (MVT) deposits: a critical requirement for genetic modelling. In: Andrew CJ, Crowe RWA, Finlay S, Pennell WM, Pyne JF (eds) Geology and genesis of mineral deposits in Ireland. Irish Assoc Econ Geol, Dublin, pp 625–634

Sarnthein M (1967) Versuch einer Rekonstruktion der mitteltriadischen Paläogeographie um Innsbruck, Österr Geol Rundsch 55:116–127

Schneider H-J (1954) Die sedimentäre Bildung von Flußspat im Oberen Wettersteinkalk der nördlichen Kalkalpen. Bayer Akad Wiss Math-Naturwiss Kl Abh NF 66:1–37

Schroll E (1953) Mineralparagenese und Mineralisation der Bleiberg-Kreuther Blei-Zink-Lagerstätte. Carinthia II 143:47–55

Schroll E (1984a) Mineralisation der Blei-Zink-Lagerstätte Bleiberg-Kreuth (Kärnten). Aufschluss 35:339–350

Schroll E (1984b) Geochemical indicator of lead-zinc ore deposits in carbonate rocks. In: Wauschkuhn A, Kluth C, Zimmermann RA (eds) Syngenesis and epigenesis in the formation of mineral deposits. Springer, Berlin Heidelberg New York, pp 294–305

Schroll E, Pak E (1983) Sulfur isotope investigations of ore mineralizations of the Eastern Alps. In: Schneider H-J (ed) Mineral deposits of the Alps and the Alpine epoch in Europe. Springer, Berlin Heidelberg New York, pp 169–175

Schroll E, Wedepohl KH (1972) Schwefelisotopenuntersuchungen an einigen Sulfid- und Sulfatmineralen der Blei-Zink-Erzlagerstätte Bleiberg/Kreuth, Kärnten. Tschermaks Min Petrol Mitt 17:286–290

Schroll E, Schulz O, Pak E (1983) Sulfur isotope distribution in the Pb-Zn deposit Bleiberg (Carinthia, Austria). Mineral Depos 18:17–25

Schulz O (1959) Beispiele für die synsedimentäre Vererzungen und paradiagenetische Formungen im älteren Wettersteindolomit von Bleiberg-Kreuth. Berg- Hüttenm Mh 105:1–11

Schulz O (1964) Lead-zinc deposits in the Calcareous Alps as an example of submarine-hydrothermal formation of mineral deposits. Dev Sedimentol 2:47–52

Schulz O (1968) Die synsedimentäre Mineralparagenese im oberen Wettersteinkalk der Pb-Zn-Lagerstätte Bleiberg-Kreuth (Kärnten). Tschermaks Min Petrol Mitt 12:230–289

Schwarcz HP, Burnie SW (1973) Influence of sedimentary environments on sulfur isotope ratios in clastic rocks: a review. Mineral Depos 8:264–277

Siegl W (1956) Zur Vererzung der Blei-Zink-Lagerstätte von Bleiberg. Berg- Hüttenm Mh 101: 108–119

Taupitz KC (1954) Die Blei-, Zink- und Schwefelerzlagerstätten der nördlichen Kalkalpen westlich der Loisach. Thesis, Bergakademie Clausthal

Tornquist A (1927) Die Blei-Zinkerzlagerstätte von Bleiberg-Kreuth in Kärnten. Springer, Wien

Williams N, Rye DM (1974) Alternative interpretation of sulphur isotope ratios in the McArthur lead-zinc-silver deposit. Nature 247:535–537

Winterer EL, Bosellini A (1981) Subsidence and sedimentation on Jurassic passive continental margin, Southern Alps, Italy. Am Assoc Petrol Geol Bull 65:394–421

Zeeh S (1990) Fazies und Diagenese des obersten Wettersteinkalkes der Gailtaler Alpen. Freiburger Geowiss Beitr 1:210+VI

Zeeh S, Bechstädt T (1989) Die Blei-Zink-Erze im Wettersteinkalk der Ostalpen: Eine jurassisch-kretazische Vererzung. Geol Paläontol Mitt Innsbruck 16:133–134

Zeeh S, Bechstädt T, Maul B (1988) Fazies und frühe Diagenese im obersten Wettersteinkalk (Drauzug, Nördliche Kalkalpen). Bochumer Geol Geotech Arb 29:245–247

Australian, Chinese, and North African Deposits

Australian Sediment-Hosted Zinc-Lead-Silver Deposits: Recent Developments and Ideas

P.J. Legge[1] and I.B. Lambert[2]

Abstract

Australia is particularly well endowed with large stratiform sediment-hosted Zn-Pb-Ag deposits of Middle Proterozoic age, the main examples being *Broken Hill*, *Mount Isa*, *Hilton*, *McArthur River* and *Century*. These deposits formed in intracratonic rifts during a remarkably brief mineralizing event around 1690–1670 Ma, and their genesis appears to have been linked, at least indirectly, to igneous events. The mineralization in each comprises syngenetic and/or diagenetic components. Broken Hill can be distinguished from the other deposits, having formed in a rift characterized by more abundant bimodal volcanism and lower S availability, which was later subjected to high grades of regional metamorphism.

It appears that two features, which evolved in the Early to Middle Proterozoic, paved the way for the formation of major stratiform sediment-hosted Zn-Pb-Ag deposits. These were the enrichment of Pb in the upper crust through cycles of K-rich felsic igneous activity and the accumulation of thick piles of sediments, including evaporites, in intracratonic rifts. Compared with volcanogenic Cu-Zn deposits which are common in Late Archaean and Phanerozoic terranes, the sediment-hosted Zn-Pb-Ag deposits apparently formed at lower temperatures and from more saline fluids.

In the Palaeozoic of eastern Australia, there are no significant stratiform sediment-hosted Zn-Pb-Ag deposits, but there are important Zn-Pb-rich volcanogenic massive sulphide deposits including *Rosebery*, *Que River*, *Hellyer*, *Benambra*, *Woodlawn* and *Thalanga*. These occur in predominantly felsic volcanic sequences, and some of them pass laterally into carbonaceous shale lenses. The ores formed at and near the sea bed in fault controlled sub-basins within generally shallow water basins, from hydrothermal fluids which included major components of seawater. There are also significant discordant sediment-hosted deposits at *Elura*, *CSA* and *The Peak*, in Siluro-Devonian flysch facies above a speculated rift sequence in the Cobar Basin formed over a basement detachment ramp. These appear to have been emplaced from metamorphic fluids, during a period of transfer/wrench faulting, possibly accompanied by granite intrusion. It is postulated that the metals were dissolved largely from

[1] North Broken Hill Peko Limited 476 St. Kilda Road Melbourne, Vic, 3004 Australia
[2] Elisian Resources 2 Bonwick Place Garran, ACT, 2605 Australia

Spec. Publ. No. 10 Soc. Geol. Applied to Mineral Deposits
Fontboté/Boni (Eds.), Sediment-Hosted Zn-Pb Ores

the underlying trough sequence, with Late Silurian volcanogenic mineralization being the major source.

In Western Australia, several deposits of Middle Palaeozoic, carbonate-hosted Zn-Pb display typical characteristics of Mississippi Valley type deposits, including *Cadjebut*, *Blendevale*, and *Admiral Bay* in the Canning Basin, and *Sorby Hills* in the neighbouring Bonaparte Gulf Basin. They formed epigenetically from highly saline, hydrocarbon-bearing basin brines, which dissolved metals from the deep basin strata and basement, and were expelled mainly along fault zones, to form both space-fill and replacement deposits.

1 Introduction

This chapter concentrates on the geological settings and genesis of the major Australian Proterozoic sediment-hosted Zn-Pb-Ag deposits and prospects of northern and southern Australia, emphasizing recent developments and concepts. It also briefly reviews the Palaeozoic volcanogenic massive sulphide deposits and the discordant sediment-hosted Zn-Pb-Ag-(Cu-Au) mineralization of eastern Australia and the carbonate-hosted Zn-Pb deposits of northwestern Australia. Further information on the Australian deposits can be obtained from the papers cited and the comprehensive volume edited by Hughes (1990).

Size and grade data for the Zn-Pb-Ag mines and principal prospects are included in Table 1, and basement geology is shown on the location maps (Figs. 1, 2 and 3). Figure 4 presents a schematic cross section relating time and tectonic events to mineralization. Reference to the new international nomenclature of geological time for the Proterozoic is included.

The oldest Archaean rocks in Australia occur in the Pilbara Block where a greenstone and granite terrain (3600–3000 Ma) contains several small volcanic hosted massive sulphide occurrences: although none are currently in production, exploration is active. Several significant deposits of volcanic hosted Cu-Zn occur in the younger Yilgarn Block (2800–2600 Ma) including *Golden Grove (Scuddles and Gossan Hill)* and *Teutonic Bore*. These deposits are referred to in Table 1 and Fig. 1 but are not discussed further in this chapter.

Widespread tectonic activity affected Australia during the Barramundi Orogeny between 1900 and 1800 Ma. This orogeny occurred between two major sedimentary basin forming cycles and was accompanied by widespread, intracontinental, felsic volcanic-plutonic activity, dated mainly at 1880–1840 Ma (Etheridge et al. 1987; Wyborn 1988). In this chapter for convenience, the older of these basin forming cycles is referred to informally as Early Proterozoic, and the younger, post-orogenic period is termed Middle Proterozoic.

Worldwide, no major sediment hosted Zn-Pb-Ag deposits formed before this orogenic event. In contrast, the Middle Proterozoic (in particular the Statherian) was an important period for formation of large, mainly stratiform, sediment-hosted Zn-Pb-Ag deposits. The largest Australian examples of such deposits, commonly termed 'shale-hosted', are *Broken Hill* in the Willyama Inlier, *Mount Isa*, *Hilton* and the newly discovered *Century* and *Cannington* in the Mount Isa Inlier, and *McArthur River* (or *HYC*) in the McArthur Basin

Table 1. Principal Australian Zn-Pb-Ag-(Cu-Au) deposits, original geological resources

Age	Tectonic unit/host	Deposit	Million tonnes	% Zn	Metal % Pb	grades g/t Ag	% Cu	g/t Au	Discovery	Production	Reference
Archaean	Yilgarn Craton, WA	Scuddles main mine	11	12	1	89	–	1.1	1979	1991	Mill et al. (1990)
	Volcanic	Scuddles copper zone	5	–	–	7	2.6	0.2			Company release 1984, 1990
		Gossan Hill copper zone	12	–	–	12	3.4	0.1	1972	est. 1995	Mill et al. (1990)
		Gossan Hill zinc zone	2	11	–	92	0.5	1.7			Company release 1990
		Teutonic Bore (closed)	1	16	1	203	4.1	–	1976	1981	Greig (1984)
Orosirian (E. Proterozoic)	Pine Creek Inlier, NT	Woodcutters mine	6	13	6	120	–	–	1964	1987	Company release 1991
	Sediment	Browns	10	0	6	–	0.2	–	1954	est. 2000	Fraser (1975)
Statherian (M. Proterozoic)	Willyama Inlier, NSW Volcanic, sediment	Broken Hill Main Iodes	280	11	10	180	0.1	0.1	1883	1885	Various estimates
Statherian	Mt. Isa Inlier East, QLD	Pegmont	10	3	7	–	–	–	1975	est. 2000	Legge et al. (1984)
Statherian	(? Cover Sequence 2)	Dugald River	43	13	2	42	–	–	1881	est. 1998	Company release 1993
Statherian	Sediment	Cannington	47	5	11	470	–	–	1991	est. 2000	Company release 1993
Statherian	Mt. Isa Inlier West,	Mt. Isa Zn-Pb-Ag mine	124	7	6	148	–	–	1923	1931	Forrestal (1990)
Statherian	QLD and McArthur	Hilton mine	102	11	5	96	–	–	1947	1990	Forrestal (1990)
Statherian	River, NT	Hilton North	53	12	6	102	–	–	1985	est. 2000	Forrestal (1991)
	(Cover Sequence 3)	Century	120	10	2	36	–	–	1990	1995	Company release 1991
Statherian	Dolomitic siltstone and	Lady Loretta	8	18	9	125	–	–	1969	est. 1995	Hancock and Purvis (1990)
Statherian	shale	HYC	227	9	4	41	0.2	–	1955	est. 1995	Plumb et al. (1990)
Statherian											
Statherian?	Bangemall Basin, WA Dolomitic siltstone	Abra	200	–	2	6	0.2	6% Ba	1981	est. 2000	Boddington (1990)
M. Cambrian	Tasmania	Rosebery mine	23	14	4	142	0.6	2.8	1893	1905	Green (1990)
M. Cambrian	Volcanic, shale	Hellyer mine	17	14	7	167	0.4	2.5	1983	1986	McArthur and Dronseika (1990)
M. Cambrian		Que River mine (closed)	3	13	7	195	0.7	3.3	1974	1981	McArthur and Dronseika (1990)
M. Cambrian		Hercules mine (closed)	2	18	6	176	0.4	2.9	1894	1900	Lees et al. (1990)
Cambrian	North Queensland	Thalanga mine	6	12	4	99	2.2	0.6	1975	1900	Gregory et al. (1990)
Ordovician	Volcanic, phyllite	Balcooma-Dry River South	1	10	4	98	0.9	0.8	1985	est. 2000	Huston and Taylor (1990)
Silurian	Lachlan Fold Belt	Woodlawn mine	6	14	6	89	1.7	–	1968	1978	McKay and Hazeldene (1987)
	NSW, Victoria	Woodlawn mine stockwork	4	–	–	–	1.9	–			
Silurian	Shale, volcanic	Captains Flat (closed)	4	10	6	56	0.7	1.7	1874	1937	
Silurian		Benambra (Wilga + Currawong)	14	5	1	34	2.1	0.9	1976	est. 1995	Davis (1990)
Silurian											Company release 1991
E. Devonian	Cobar Basin, NSW	Elura mine	32	9	6	109	–	–	1974	1983	Schmidt (1990)
E. Devonian	Turbidite	CSA mine	18	3	1	23	1.7	–	1872	1905	Scott and Phillips (1990)
E. Devonian		Peak mine	5	2	2	21	0.7	7.0	1981	1992	Hinman and Scott (1990)
M. Devonian	Lennard Shelf, WA	Cadjebut mine	3	14	5	–	–	–	1984	1988	Murphy (1990)
M. Devonian	Carbonate and	Blendevale	20	8	3	17	142 g/tCd		1978	est. 1995	Murphy (1990)
M. Devonian	evaporite	Twelve Mile Bore	3	10	3	35	–	–	1985	est. 2000	Murphy (1990)
E. Carbonif	Bonaparte Basin WA	Sorby Hills	16	1	5	56	–	–	1971	est. 2010	Lee and Rowley (1990)

Fig. 1. Archean terranes and zinc-copper-silver deposits of Australia

(Fig. 2). *Broken Hill* is the largest single Zn-Pb-Ag-Ag deposit known anywhere, with an original resource of the order of 280 million tonnes (Mt) (Table 1).

Although large stratiform sediment-hosted Zn-Pb-Ag (and barite) deposits formed in North America, Europe and elsewhere in the Early and Middle Palaeozoic, no such deposits are known in the Australian Phanerozoic. However, there are significant deposits of discordant sediment-hosted deposits at *Elura*, *CSA* and *The Peak* (Fig. 3), within Early Devonian flysch which have been folded and sheared during inversion of the Cobar Basin.

There are also numerous Palaeozoic Zn-Pb-rich volcanogenic massive sulphide deposits throughout eastern Australia (Fig. 3), commonly associated with carbonaceous tuffaceous shales in predominantly felsic volcanic sequences. Important examples of Cambrian age include *Rosebery, Que River* and *Hellyer* in Tasmania, and *Thalanga*, *Reward* and *Liontown* in north Queensland. Late Silurian examples include *Woodlawn*, *Captains Flat* and *Benambra* in southeastern Australia.

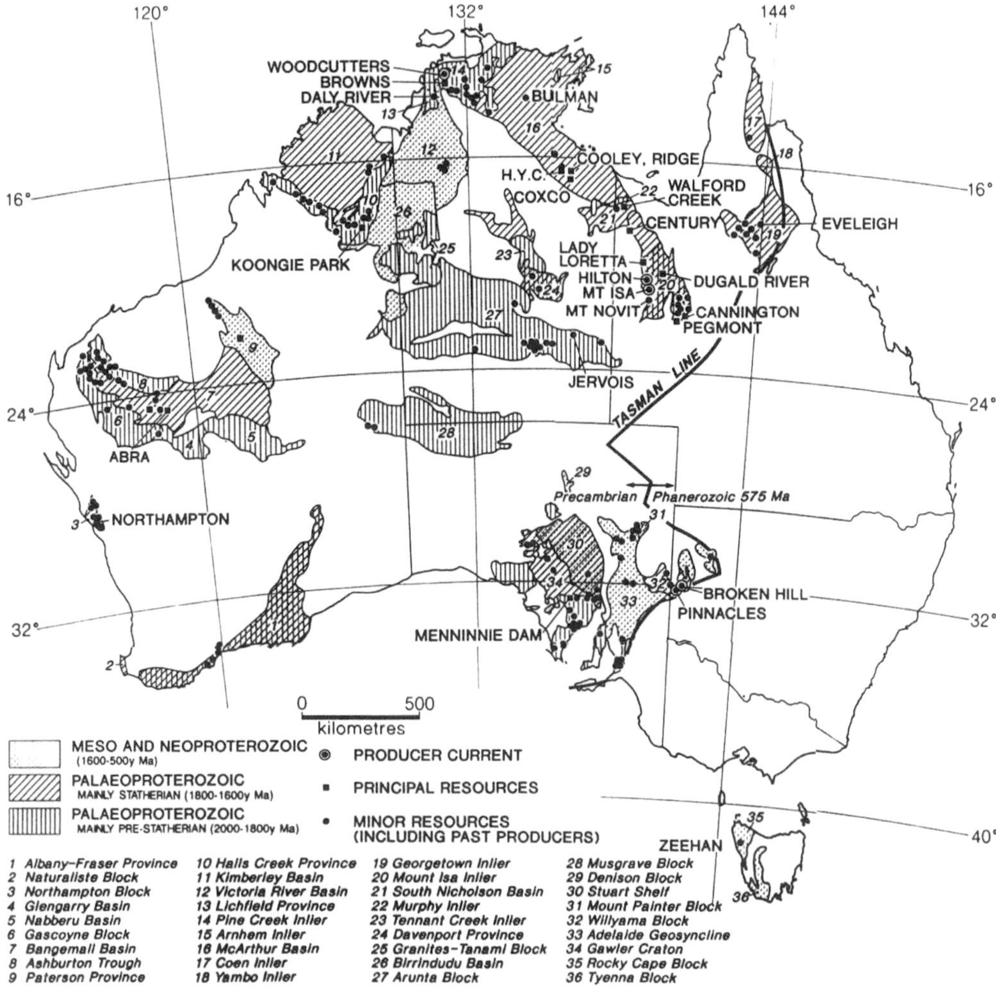

Fig. 2. Proterozoic terranes and lead-zinc-silver (copper) deposits of Australia. Note the age divisions used (after the Subcommission on Precambrian Stratigraphy): Neoproterozoic 1000–625 Ma; Mesoproterozoic 1600–1000 Ma; Palaeoproterozoic 2500–1600 Ma (including Statherian 1800–1600 Ma and Orosirian 2050–1800 Ma)

In Western Australia, several significant Mississippi Valley type carbonate hosted Zn-Pb deposits, including *Cadjebut*, *Blendevale*, *Admiral Bay* and *Sorby Hills*, have been discovered in Middle Palaeozoic strata.

2 Early Proterozoic

The Early Proterozoic (pre-Statherian) basins in Australia (Fig. 2) contain intracratonic rift accumulations of sedimentary and subordinate volcanic strata,

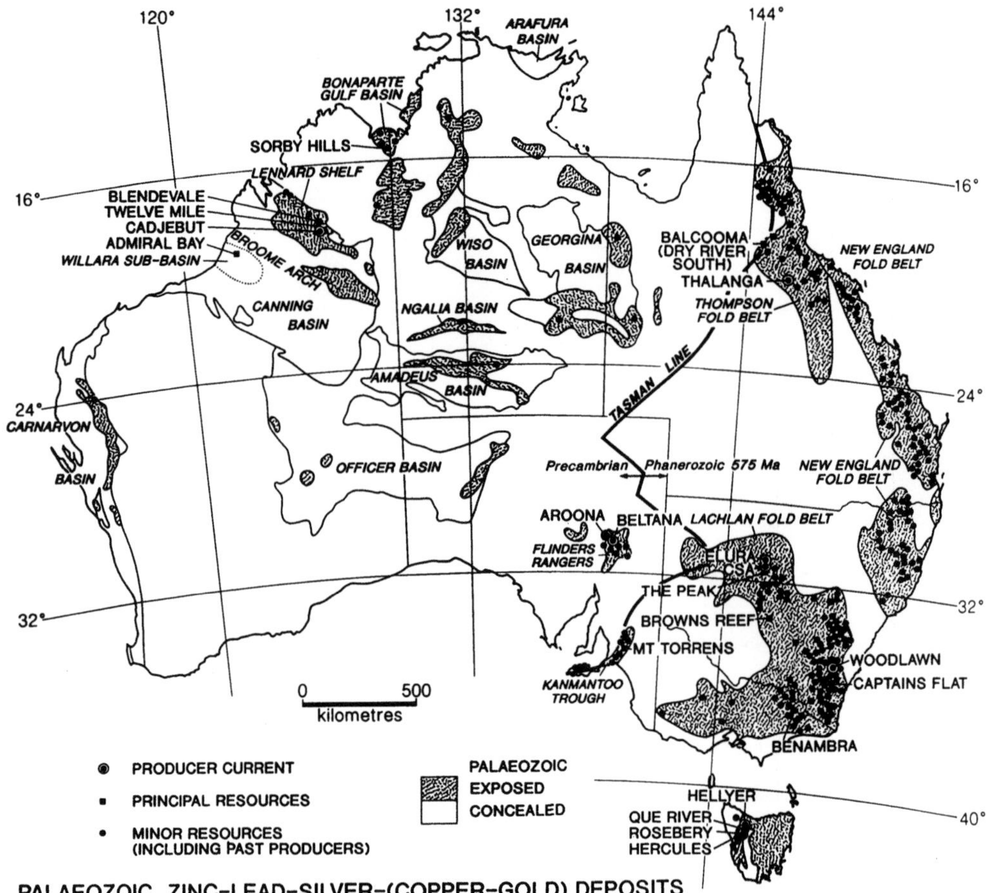

Fig. 3. Palaeozoic terranes and zinc-lead-silver-(copper)-(gold) deposits of Australia

up to 14 km thick, which formed during initial extension of the Archaean craton after approximately 2000 Ma (Etheridge et al. 1987). These accumulations were subsequently deformed, metamorphosed and intruded by voluminous granites and porphyries, from approximately 1880 Ma (Fig. 4).

The sedimentary environments were broadly similar to those of the Middle Proterozoic, and numerous small mineral deposits are present, but the Early Proterozoic basin sequences do not contain world-class sediment hosted Zn-Pb-Ag deposits. This lack of large deposits is probably a reflection of crustal fractionation processes. We consider, following Lambert (1989), that K-, Pb- and U- rich igneous rocks were largely responsible for the upward fractionation of Pb into the crust. The earliest widespread emplacement of such high heat producing igneous rocks in the upper crust was during the Late Archaean, and they became particularly abundant after the Early Proterozoic (Ferguson et al. 1980). Their weathering and erosion products probably were not incorporated as generally significant components into basin sediments until the Middle

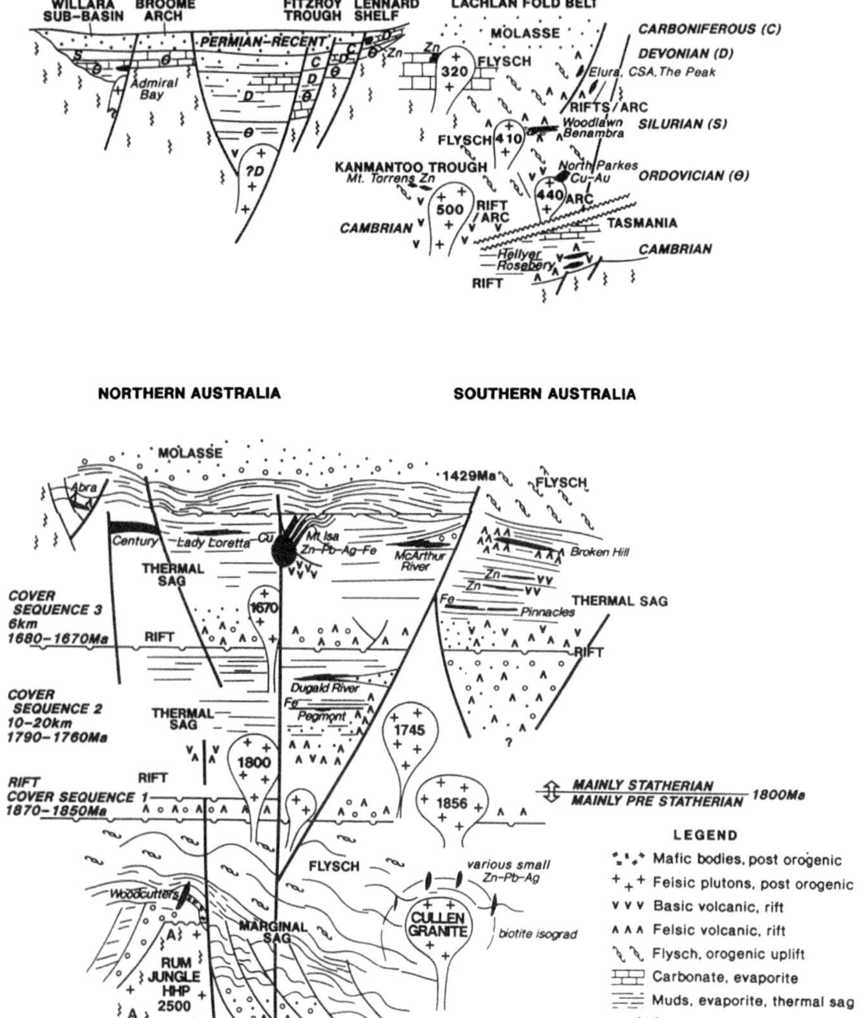

Fig. 4. Schematic section illustrating tectonic settings and ages of major Australian zinc-lead-silver-(copper)-(gold) deposits

Proterozoic, when chloride-rich basinal brines, capable of transporting Pb and Zn at significant concentrations under moderate to low temperatures also evolved (Lambert 1989). Similarly, Sawkins (1989) suggested that major sediment-hosted Zn-Pb-Ag deposits required a significant contribution of Pb from the leaching of rift sediments derived from Pb-enriched anorogenic igneous rocks, which were only abundant after about 1870 Ma in northern Australia (Wyborn et al. 1988).

The moderately sized *Woodcutters* Zn-Pb-Ag deposit (Table 1), comprising discordant massive sulphide pods warrants discussion in this regard. This, and the nearby *Browns* prospect occur in the Early Proterozoic Pine Creek Inlier (Figs. 2 and 4), adjacent to the Rum Jungle uranium field. The basement of uraniferous Late Archaean granite and gneiss in this region is overlain by arkose, sandstone, conglomerate and dolomite. At *Woodcutters*, these are in turn overlain by a 300-m-thick sequence containing finely foliated, graphitic mudstone interbedded with decreasing proportions of dololutite up section, and local sericitic tuff marker beds (Taube 1984). The formation of this and other Zn-Pb-Ag deposits in the region can be related to extraction of metals by fluids moving through the basement-derived Early Proterozoic strata, and from the basement itself. The uranogenic nature of the Pb at *Woodcutters* (G. Carr, pers. comm. 1991) is consistent with this model. The elevated thermal gradients generated by the presence of highly radioactive rocks would have assisted the fluid circulation and metal leaching process.

The *Woodcutters* massive sulphides filled dilatational zones along major vertical fractures which may be listric at depth. We consider that these periodically opened and breached the stratigraphic sequence and that the Zn-Pb-Ag mineralization precipitated where the metalliferous brines encountered reactive graphitic strata. The presence of significant levels of Ni at Woodcutters and the existence of nearby U, Ni, Co mineralization suggest a low temperature oxidized fluid association. The mineralizing event may have been related to the tectonism associated with the ca. 1870 Ma orogeny or indeed later events. Smolonogov and Marshall (1992) give details and suggest a protracted span of ore genesis from diagenesis to regional deformation.

3 Middle Proterozoic (Statherian) Deposits of Northern Australia

3.1 McArthur Basin

The Middle Proterozoic McArthur Basin (Fig. 2) sequence has been reviewed recently by Jackson et al. (1987) and Plumb et al. (1990). The little deformed sequences is essentially unmetamorphosed, as attested by the presence of autochthonous live oil in its upper section.

The intracratonic McArthur Basin volcanic and sedimentary sequence, in the central Batten Trough, is inferred to be up to approximately 12 km thick. It accumulated in a series of rift sub-basins over relatively short intervals, during a 200 million year period of block rotation and E-W crustal extension (Plumb et al. 1990). Thinner sequences were deposited on marginal shelves.

Basal blanket sandstone, formed under shoreline to shallow marine conditions, alternates with volcanic-lutite-carbonate sequences formed under peritidal to lacustrine conditions (Plumb et al. 1990). D. Leaman (pers. comm. 1992) has inferred, using gravity and magnetic interpretation, that in places mafic and/or felsic volcanics comprise a large proportion of the lower rift sequences; this has implications for possible metal sources. The early rift deposits were overlain unconformably by the stromatolitic carbonate-sandstone-siltstone associations of the lower part of the McArthur Group which formed in hypersaline peritidal to continental sabkha environments. The HYC Pyritic Shale, which hosts the *McArthur River* deposit, is towards the top of this group. This unit is highly variable in thickness, locally exceeding 1000 m, and comprises pyritic, tuffaceous carbonaceous shale and siltstone, and dolomitic beds. Organic matter has increased maturation levels where the sediments have been affected by hydrothermal activity associated with base metal mineralization (Crick 1992). Zircon dating of potash feldspar-rich tuff bands from the mineralized unit has yielded an age of 1690–1670 Ma (Page 1981).

The mineralized unit developed in near-shore evaporitic environments, comprising lakes and exposed mudflats (Muir 1983; Logan et al. 1990) within the Batten Trough. Movement on the major NNW Emu Fault zone, and subsidiary faults, appears to have controlled the subsidence of depressions in which the thickest shale-siltstone intervals of this unit accumulated. The broad structural environment has been envisaged by Muir (1983) as a pull-apart basin between two strike slip faults. The fine-grained nature of the tuff bands implies that volcanic activity was distal.

3.1.1 **McArthur River** ***Deposit***

Development of the *McArthur River* (HYC) deposit, described by Lambert (1976, 1983), and by Logan et al. (1990), and discovered over 38 years ago, has been delayed because of metallurgical problems caused by the predominance of extremely fine-grained ore minerals, commonly less than 5 μ across; development plans are being considered on an identified resource of 47.7 Mt, 15.7% Zn, 6.5% Pb, 66 g/t and a feasibility study is in progress. The fine-grain size is a reflection of the unmetamorphosed nature of the deposit. Seven stacked shale-siltstone ore zones over an interval averaging 55 m thick contain the major sulphide minerals, pyrite, galena and sphalerite (with low iron contents); these are accompanied by minor chalcopyrite, arsenopyrite and marcasite.

Much of the mineralization comprises very thin pyrite-rich laminae intercalated with slightly more diffuse sphalerite-rich bands and galena-rich schlieren. The sulphide bands exhibit widespread slumping and other soft-sediment structures. There is also coarser grained galena and sphalerite, and abundant pyrite overgrowths in the ore.

The mineralized zones are separated by interbeds which vary from pyritic, bituminous and partly nodular dolomitic shales in the lower parts, to predominantly dolomitic arenites and talus breccias, containing occasional clasts of laminated pyrite-sphalerite-galena mineralization in the upper parts of the deposit. For several hundred metres above the ore, highly pyritic, carbonaceous, tuffaceous shales with negligible to low base metal concentrations are present. Elevated contents of Fe and Mn occur in dolomites for some 20 km laterally from, and many hundreds of metres above and below, the deposit (Lambert and Scott 1973).

In other sub-basins within the Batten Trough, low-grade stratiform mineralization, or metal anomalism, occurs at the same stratigraphic level as the

McArthur River deposit. In the dolomite units to the east of the *McArthur River* deposit, there is scattered, mainly medium- to coarse-grained, fracture and void-filling Pb-Zn mineralization which shows a progressive increase in Cu content towards the Emu Fault Zone (Williams 1978).

3.1.2 Ore Genesis

The Pb in galena in the McArthur mineralization is isotopically homogeneous (Richards 1975) and plots on the average (orogene) growth curve of Doe and Zartman (1979). This is consistent with the derivation of the metals from the basin strata by homogenized basinal brines, although it does not preclude contribution of metals from the felsic magmas which sourced the tuffites in the sequence. At the very least, the distal igneous activity would have been accompanied by high geothermal gradients, conducive to large-scale fluid circulation and metal leaching in the basin strata.

Eldridge et al. (1991) studied the S-isotopic compositions of the McArthur mineralization using the ion microprobe technique. While the resolution of this technique is much better than other methods used for S-isotope analyses, it is not good enough to analyze the predominant fine-grained sulphides. Analyzing individual crystals larger than $25\,\mu$, Eldridge et al. (1991) found that the $\delta^{34}S$ values of early pyrite (from -13 to 15 per mill) are lower on average than later pyrite (-5 to 45 per mill), while galena and sphalerite have a more restricted range of S-isotope compositions (-5 to 9 per mill), and are not in isotopic equilibrium with adjacent pyrite.

Both early and late pyrite evidently incorporated biogenic sulphide. The S-isotope compositions for early and late pyrite are within the range characteristic for disseminated pyrite in Middle and Late Proterozoic sedimentary strata (Lambert and Donnelly 1991). They imply closed system bacterial reduction of sulphate, in basins with limited access to the open ocean, and/or in near surface sediments where influx of sulphate-bearing water was restricted. The early pyrite incorporated biogenic sulphide generated under conditions of greater replenishment of sulphate, but could also incorporate components of the hydrothermal sulphide which formed the associated sphalerite and galena.

The S-isotopic data and the textures of the coarser sulphides were interpreted by Eldridge et al. (1991) in terms of a "permeation" theory. Microbiological reduction of seawater sulphate below the sediment-water interface was considered to have produced an early pyrite in the porous sediment column. The system then became closed to sulphate influx during microbiological production of further pyrite. Later, Zn-Pb-Ag-Cu mineralising brines with CO_2 and H_2S migrated from the Emu Fault Zone into the deposition sites with Fe precipitating as marcasite and ankerite, and Zn and Pb as sphalerite and galena.

Williams (1990) suggested a similar model involving

"two distinct and spatially separate mineralizing fluids that pass into and spread through the unconsolidated sediments beneath the HYC lake or lagoon. The uppermost fluid, which sits beneath sulphate- and nutrient-rich surface waters is a reduced fluid rich in iron and sulphur. It deposits iron monosulphides in the sediments. The deeper denser fluid is rich in lead and zinc, but poor in reduced sulphur. When previously formed iron monosulphides encounter this fluid, they are dissolved and galena and sphalerite are precipitated. Ferrous iron released by these reactions moves upwards and is either reprecipitated as iron monosulphides in the domain of the upper hydrothermal fluid or as pyrite in the surface water domain by way of microbial sulphate reduction."

We accept important elements of such models: a source of at least some of the Fe separate from that of the Zn and Pb, permeation of some moderately low temperature hydrothermal fluids through the sediments, and the importance of bacterial sulphate reduction in the depositional basin. However, it is unlikely that the very near-surface permeation processes could have occurred without formation of a significant component of synsedimentary (exhalative) mineralization. The range of data available for the McArthur River mineralization appear consistent with a sedimentary-diagenetic model like that proposed by Lambert (1983). Such a model is in accord with the isotopic data and explains the soft-sediment structures of the sulphide laminae and the clasts of laminated sulphides in the inter-ore beds, which could not have been generated by entirely post-sedimentary mineralization processes.

In this model, sphalerite and galena laminae plus early iron sulphide laminae (as iron monosulphide and/or pyrite) formed from hydrothermal exhalations which carried sulphide derived from thermochemical reduction of evaporitic sulphate at depth in the basin. The mineralizing fluids emanated into both the water column and bottom sediments of the depositional depression, where high levels of bacterial sulphate reduction were occurring. Additional Fe was introduced from a separate source. Pyrite formed where biogenic sulphide reacted with iron monosulphide and other reactive ions, modifying the isotopic composition of the laminated iron sulphides, and resulting in the common nucleation of the overgrowth pyrite on earlier-formed grains. The coarser, late sphalerite and galena precipitated from the metalliferous fluids within the sulphidicorganic muds.

The edge effects of syndepositional growth faults are reflected by the banded ore passing laterally into, and interbedded with, talus breccia. This talus indicates intermittent vertical movement on marginal faults and rapid deepening of the sub-basin depocentre. Subject to the hydrothermal metalliferous brines emanating at relatively low temperatures, around 150 °C or less, the deepening of the basin could have prevented boiling if water depths of several tens of metres or more were achieved locally, particularly in the presence of a dense brine layer in the depositional basin. The evidence for rapid vertical movements, and the nature, isotopic compositions and metal zonation of mineralization in the carbonate fault block to the east of the McArthur deposit, support ascent of the ore fluids by dilational fault pumping resulting from movement on the Emu and related faults. A short period of tensional rifting appears to have been superimposed on the overall "thermal sag phase" at this time which enabled the breaching of the overall sedimentary sequence and provided access for the rise of metal brines.

3.2 Mount Isa Inlier

Three broad tectonic units are present in the intracratonic Mt. Isa Inlier: the Western Fold Belt, the central Kalkadoon-Leichhardt Belt and the Eastern

Fold Belt (Blake et al. 1990). The Western Fold Belt includes the Lawn Hill Platform, on which the *Century* deposit occurs, and the Leichhardt River Fault Trough, which contains the *Mount Isa*, *Hilton* and *Hilton North* deposits. The Eastern Fold Belt has been subjected to greater deformation, metamorphism and metasomatism than the Western Fold Belt and hosts significant Zn-Pb-Ag mineralization at *Dugald River* and *Pegmont*, and the recently discovered *Cannington* deposit. Segments of the Eastern Fold Belt appear to be allochthonous and thrust west onto the Kalkadoon-Leichardt block.

The Middle Proterozoic succession comprises three "cover sequences" which accumulated after, and partly as a result of, intense compression of basement rocks in the major orogeny at 1890–1870 Ma (Wyborn 1990).

Cover sequence 1 formed just prior to the Statherian in the interval 1870–1850 Ma, and is represented by metamorphosed felsic volcanics some +1000 m thick. It is intruded by and coeval with granites considered to have been derived by the melting of a mafic rock which underplated the crust at 2300–2000 Ma (Wyborn 1990). No significant Zn-Pb-Ag deposits are known within this lower sequence.

Cover sequence 2, formed during a major crustal extension event in the interval 1790–1760 Ma. The sequence is 10–20 km thick and comprises bimodal volcanics, with mantle derived tholeiite, and crustal derived felsic igneous rocks (Wilson 1987). These are overlain by metasedimentary rocks including slate and calc-silicate units (former shale, calcareous and evaporitic rocks) which host the *Dugald River* Zn-Pb-Ag deposit. Sheppard and Main (1990) have suggested the *Dugald River* deposit is at the same stratigraphic level as the *Mount Isa* deposit (cover sequence 3) on the basis of some lithological similarities; the sulphide lode is locally discordant. *Pegmont* and the recently discovered *Cannington* deposit are distinguished by being in iron-rich metamorphosed sequences of uncertain correlation (Fig. 2).

Cover sequence 3, some 6 km thick, accumulated in the Leichhardt River Fault Trough during the short time interval, 1680–1670 Ma. This was a period of subsidence accompanying renewed extension. An initial rift succession of bimodal volcanics and hematitic sandstone and conglomerates is overlain by a sequence of sandstone, siltstone and shale. These are overlain by the Mount Isa Group, dated at 1670 Ma (Page 1981), which hosts the *Mount Isa*, *Hilton* and *Hilton North* deposits. It comprises mainly siltstone and shale with some thin tuff beds. To the northwest, on the Lawn Hill Platform, the shale-siltstone succession, which hosts *Century* and the small high-grade deposit at *Lady Loretta*, is richer in dolomites and stromatolites than the Mount Isa Group.

Granitic intrusions, at around 1670 Ma, appear to be genetically related to the tuffs in the mineralized unit at Mount Isa. The felsic volcanics of the 1800 to 1620 Ma period (Wyborn et al. 1987) are all anorogenic and differ from the earlier Proterozoic felsic igneous rocks in being more enriched in incompatible elements (K and other large ion lithophile elements, including Pb).

The two major tectonic extension phases which have been recognized in this region both comprise early rift deposits of coarse clastics and bimodal volcanics, overlain by more extensive thermal sag phase deposits of finer clastics and carbonates. The stratiform Zn-Pb-Ag mineralization occurs in weakly metamorphosed pyritic and dolomitic shale-siltstone strata deposited during the final sag phase.

Basin deposition ceased with regional compression events at 1550 Ma which caused variable deformation and metamorphism up to amphibolite facies, with major strike slip faulting. Inferred movements are up to 200 km and juxtapose allochthonous terranes of the Eastern and Western Fold Belts. In general there is an absence of deep water flyschoid sequences related to the culminating orogenic phase.

Only the major deposits of the Western Fold Belt are discussed here.

3.2.1 **Mount Isa-Hilton** *Area*

The Mount Isa Group accumulated on a thick sequence of continental flood basalts and associated shallow water sediments. The *Mount Isa*, *Hilton* and *Hilton North* deposits are hosted by a thin-bedded dolomitic, variably carbonaceous and pyritic, siltstone and mudstone of the Urquhart Shale, up to 1000 m thick and extending for 24 km along strike. Thin K-rich tuff aand mudstone beds are marker horizons in this unit. These tuffs are K-rich, like the coeval tuffs in the McArthur area, probably reflecting reactions between glassy pyroclastic material and K-rich brines around the depocentre of alkaline lake environments (van den Heuvel 1969; Lambert and Scott 1973). The overall sedimentary environment was one of fault controlled, saline lakes with associated shallow water deposition, intermittent subaerial emergence and evaporation, and persalinity of groundwater (Neudert and Russell 1981). The deposits are more deformed and metamorphosed (greenschist facies) than those at McArthur River and have a slightly coarser grain size. The underlying and overlying unmineralized siltstone sequences are similar to the mineralized Urquhart Shale, except for generally lower pyrite and carbon contents, and albite at the expense of K-feldspar.

In the Zn-Pb-Ag ore, zones of sulphide-rich bands, of sphalerite, galena, pyrite and pyrrhotite, alternate with sulphide free layers, centimetres to tens of centimetres thick. Each sulphide layer has laminations up to several hundred microns thick but the sulphide-free layers are unlaminated. As noted by Neudert (1986), barren shale shows an alternation of layers deposited from turbulent suspension currents (laminated) and settling from suspension columns (unlaminated), and most of the sulphide orebodies are hosted by slope to basin facies in a saline lake. In the associated playa facies there is evidence of frequent subaerial exposure.

At *Mount Isa* the orebody, occurring on the west-dip limb of a major anticline, is 1.6 km long, extends 1.2 km down dip and is concentrated over a stratigraphic interval some 650 m thick. There is at least local fault repetition of the sequence. A prominent shear zone separates the ore sequence from highly altered basic volcanics below. There are 14 en echelon Zn-Pb-Ag ore lenses. At their southern and deepest extent the lenses interfinger with non-stratiform, highly metasomatized and fractured 'silica-dolomite', the principal host to Cu mineralization, which occurs closely adjacent to, but not overlapping, the Zn-Pb-Ag mineralization. The highest Pb and Ag grades occur close to silica dolomite. Zn is more laterally persistent with the northern ores becoming increasing Zn-rich, and there are elevated concentrations of Zn laterally for several kilometres away from ore.

At the *Hilton* deposit, which is mineralogically similar to the Zn-Pb-Ag ore at *Mount Isa*, some 20 km to the south, there are seven economic ore horizons. The thinner Urquhart Shale (110 to 250 m thick) at the *Hilton* deposit may be a function of an originally thinner sequence and/or deformation. The ore shows pinch and swell, with sulphide mobilization into breccia, boudins and piercement structures, and faults and post-ore basic dykes affect ore continuity.

3.2.2 **Century-Lawn Hill** *Area*

The *Century* deposit, on the Lawn Hill Platform, was discovered in 1990 some 100 years after the initial discovery of the nearby Silver King deposit in the Lawn Hill mineral field in 1887 (Fig. 2). Cambrian limestone and dolomite unconformably overlie, truncate and conceal most of the area, but two outcrops of the oxidized ore remained unnoticed even though small Ag-Pb veins had been worked only 20 m away.

Limited information is available from media releases and public lectures presented by CRA geologists (Main 1992; Waltho 1992). A resource of 120 Mt, grading 10.2% Zn, 1.5% Pb, 36 g/t Ag, has been calculated to a depth of 300 m.

Century is in a turbidite sequence in the upper part of a rift and sag phase. At the base of the sequence, rhyolitic volcanics and conglomerates pass up into hematitic lithic sandstone. Unconformably above, there is mature shallow water carbonaceous and carbonate bearing siltstone, with some sandstone and greywacke, of the 8.5-km-thick McNamara Group, which contains the *Century* deposit and is equivalent to the Mount Isa Group.

The orebody is in the Lawn Hill Formation, in the nearly flat northwest limb of a syncline in an area of domal folds. It lies adjacent to a regionally prominent NW-trend fracture (Termite Fault), is bounded north and south by normal faults, and is divided into two zones by a major normal fault. The mineralized zone is some 40 m thick with ore concentrated in black shale, and dolomitic siltstone. It is underlain by a footwall shale and siltstone unit, some 180 m thick, and a variably pyritic unit, over 250 m thick.

The mineralized zone has a distinctive stratigraphy with unmineralized dolomitic siltstone and carbonaceous mudstone interlayered with carbonaceous shale beds which have regular sulphide mineral assemblages and Zn : Pb : Ag ratios. An example of the ore grade is found in a lower unit of mineralized carbonaceous shale, some 13 m thick, which contains 13% Zn, 1% Pb, 10 g/t Ag.

The mineralization at *Century* comprises fine, diffuse laminae of sphalerite, pyrite and galena. Sphalerite grains show atoll-like shapes with overdrape sediment; two forms are noted, one normal and the other porous with high silver and pyrobitumen in the matrix. The sphalerite is generally fine-grained and has low iron and manganese (<1%), most likely a reflection of the low metamorphic grade. Pyrite, commonly 5–10μ across, is finer grained than sphalerite. Galena has unusually low Ag. Sphalerite and galena also occur in irregular veinlets and in later small veins and there are some clasts of mineralization within a matrix of dolomite. Sphalerite content is highest in the lowest mineralized unit while pyrite tends to increase towards the edges of the mineralization, both vertically and laterally. "The mineralization appears to transgress bedding/banding on a deposit to microscopic scale" (Main 1992).

The ore zone is overlain by a deep water turbiditic sequence of dolomitic siltstone and shale some 80 m thick in turn overlain by sandstone; the dolomite in these overlying units and the mineralized sequence appears to be secondary.

The adjacent *Lawn Hill* mineral field contains discordant siliceous Zn-Pb-Ag veins in a sequence of sandstone, quartzite, siltstone, shale, slate, tuff and felsite.

3.2.3 Ore Genesis

The Pb-isotope compositions of *Mount Isa* and *Hilton* Zn-Pb-Ag mineralization are indistinguishable; they are homogeneous, conformable to the growth curve, and similar to those of the *McArthur River* deposit (Gulson 1985). Again they are in accord with the metals being leached from large volumes of the underlying basin strata in the Middle Proterozoic, with possible contributions from the coeval felsic magmas. Discordant mineralization appears to have precipitated in feeder channels for the stratiform deposits. Solomon et al. (1991) noted that the post 1870 Ma high heat producing granites which pre-date (and are in part coeval with) the *Mount Isa* Zn-Pb-Ag ore (and can be inferred to occur in the basement of the McArthur Basin), could have been important in generating the large deposits. They suggested that basin fluids moving down a regional hydraulic gradient were "thermally driven" by such granites to provide a much larger flux of metal bearing brines than would otherwise be available from simple basin compaction.

The $\delta^{34}S$ values for *Mount Isa* Zn-Pb-Ag ore, overall, are slightly higher than for the *McArthur River* deposit, and they indicate a degree of isotopic equilibration between all sulphide minerals during deposition or metamorphism (Smith et al. 1978; Heinrich et al. 1989). They are also consistent with sulphide produced by abiological plus biological reduction of sulphate, dissolved from underlying evaporitic strata.

A variety of studies have addressed the timing of the *Mount Isa* mineralization and the general conclusions reached in a representative range of these are outlined in the remainder of this section. The unravelling of the timing of ore formation in this area is complicated by the greenschist facies regional metamorphism, and particularly by the significant levels of deformation and metasomatism.

3.2.4 Cu Ore

Mount Isa is one of the very few stratiform sediment-hosted Zn-Pb-Ag deposits known to have major Cu mineralization immediately adjacent. Finlow-Bates and Stumpfl (1979) considered that the broadly transgressive Cu and the stratiform Zn-Pb-Ag ores at *Mount Isa* were the product of one hydrothermal system, with the Cu forming in feeder zones beneath stratiform Zn-Pb-Ag ore. Subsequent studies have supported the epigenetic formation of the Cu mineralization, during a late deformation phase (for example, Perkins 1984, 1990; Valenta et al. 1990). Most recently, Waring (1991) concluded that the Zn-Pb-Ag deposit is metamorphosed, folded and overprinted by a locally intense S_3 cleavage, whereas pseudomorphs and cross-cutting relationships provide evidence for separate and distinctly later formation of the Cu mineralization, in the retrogressive phase following peak regional metamorphism. Further evidence for the separate formation of the Cu mineralization is provided by the Pb-isotope differences between the Zn-Pb-Ag sulphide and the Cu sulphide ores (Gulson et al. 1983). It is feasible that pyritic, Zn-Pb-Ag mineralized strata provided a reactive host for precipitation of Cu from oxidizing metamorphic fluids (Heinrich et al. 1989; McGoldrick and Keays 1990).

3.2.5 Zn-Pb-Ag Ore

The genesis of the stratiform Zn-Pb-Ag mineralization, which was discovered well before the adjacent Cu ore, has been keenly debated. Broadly speaking, early concepts of selective replacement of sedimentary components as Zn-Pb-Ag fluids passed through the host unit, were overtaken by synsedimentary models from the late 1950s to the early 1980s. Recently, several studies have argued that the Zn-Pb-Ag ore is entirely diagenetic or epigenetic, particularly on the basis of evidence for shallow water to emergent conditions during sedimentation and the sulphide textures and structures which indicate that at least some components of the ore formed after sedimentation. As emphazised by Lambert and Scott (1973), care is needed in extrapolating from post-sedimentary sulphide textures within the ore, as these could reflect predictable post-depositional processes within essentially syngenetic mineralization, rather than entirely epigenetic mineralization.

The available data imply that the stratiform *Mount Isa* Zn-Pb-Ag ore formed in a similar manner to the *McArthur River* deposit, which has many comparable features and for which we have favoured a sedimentary-diagenetic model, above. Distinctive features in the Mount Isa ore can be explained largely by overprinting by metamorphism, metasomatism (notably the Cu-mineralizing event) and deformation.

The alternative models published in recent years will now be discussed. Neudert (1984) argued on sedimentological grounds that the emergent to shallow water strata within the host sequence precluded synsedimentary exhalative processes because he believed that these require a deep water pelagic setting. He considered the ore fluids would have boiled under the relatively shallow water conditions he documented, and assumed that this would have prevented metal sulphide precipitation. The model he favoured involved diagenetic or later emplacement of galena and sphalerite under a sufficient thickness of sediments to prevent boiling. However, as for *McArthur River*, there is no evidence for or against boiling in the fine-grained laminated ore at

Mount Isa. Again water depths of only several tens of metres could have prevented boiling of highly saline ore fluids.

McGoldrick and Keays (1990) proposed precursor Zn-Pb-Ag-Cu-Co formation under diagenetic conditions, largely on the basis of evidence for infilling of primary porosity and for chemical fronts within the deposit, and the variability of sulphide mineralogy and concentration in some siltstone bands. The same authors argued on geochemical grounds for a seawater-derived, cool oxidized ore fluid, which was reduced by pre-existing sulphides or organic compounds in a diagenetic environment. Subsequent redistribution by a "zone refining" of the stratiform precursor, during deformation and metamorphism, was considered to have produced the current metal zoning.

Neither study produced unequivocal evidence against the synsedimentary formation of much of the laminated ore.

Entirely epigenetic models have been argued for the Mount Isa Zn-Pb-Ag-Cu system. Perkins (1991) considered that galena, sphalerite and pyrrhotite replaced metasomatic dolomite, phyllosilicate and K-feldspar. Laing (1990) believes that deformation created dilatation of dolomitic siltstone bedding planes and brecciation as the result of movement over the underlying greenstones along thrust ramp structures (Bell et al. 1988). Syntectonic hydrothermal fluids of high temperature, introduced along the truncating fault, precipitated replacive Cu-ore at the base and emanated up and formed the lower temperature Zn-Pb-Ag ore largely by replacement of the bedded, variably dolomitic, siltstone. These totally metamorphogenic models are at odds with the several lines of evidence, above, for separate formation of the Cu and Zn-Pb-Ag deposits. Furthermore, they could not be applied to the McArthur River deposit, which is unmetamorphosed.

At the *Century* deposit there are clearly post-depositional sulphides but the presence of fine sulphide laminae, and clasts of mineralization in the dolomitic matrix suggest that a component of the mineralization formed at the sediment-water interface.

We make the point that chloride rich brines being channelled from basins undergoing deformation and/or metamorphism have the potential to dissolve and reprecipitate sulphides in the pathway. Consequently stratiform sulphides can be modified slightly or even, when transported away from the primary stratiform site, modified completely into discordancy.

4 Middle Proterozoic (Statherian) Deposits of Southern Australia

4.1 Broken Hill Inlier

The huge *Broken Hill* Zn-Pb-Ag deposit occurs in the Willyama Supergroup (Stevens et al. 1988), which consists of various sedimentary schist and gneiss units, metamorphosed up to granulite facies. The Supergroup is more than 7 km thick; neither the top nor base is exposed and there are no unconformities.

In the lowest observable part of the Willyama Supergroup, there is a rift sequence with immature fluvio-deltaic and lacustrine sediments and bimodal volcanics. Fe-rich stratiform lodes of quartz-magnetite and quartz-pyrite, with or without garnet or sillimanite, are common; some contain minor Zn-Pb-Ag or Fe-Cu-Co mineralization.

Evaporitic, deltaic and lacustrine conditions prevailed during deposition of the overlying Broken Hill Group, with shallow water turbidity deposits containing terrigenous and acid volcanic debris, and occurrence of widespread acid and mafic volcanism. The Group has three principal periods of Zn-Pb-Ag mineralization each associated with increased volcanic activity as interpreted from an increase occurrence of garnet-plagioclase gneiss (probably originally rhyodacite) and in some cases from increased amphibolite (probably original mafic volcanics (Plimer and Lottermoser 1988).

At *Broken Hill* the Broken Hill Group is up to 700 m thick and the rocks are now gneisses of various types. The garnet-biotite quartzo-feldspathic gneisses and mafic gneisses (amphibolites) are variably enriched in base metals, with several hundred ppm Pb and Zn being common (Stevens et al. 1988). Numerous small (<100 000 t) stratiform Zn-Pb-Ag sulphide deposits, mainly located below the *Broken Hill* ore zone, occur as lenses associated with quartz-gahnite rock; some calc-silicate lenses contain small stratiform scheelite occurrences (Stevens et al. 1988). The mine sequence of exhalite and pelite at the *Pinnacles* deposit has an anomalous REE pattern and there is a systematic change in REE from footwall to hanging wall, reflecting changes in T, fO_2, pH of ore-forming fluid (Parr 1992).

The garnetiferous quartzo-feldspathic Hores Gneiss at the top of the Broken Hill Group is, in places, split into lenses by intercalated metasediments. One zone contains the 8-km-long *Broken Hill* orebody and minor stratabound Zn-Pb-Ag-Cu-Au-W deposits. These metasediments are interpreted by Stevens et al. (1988) as shelf muds and silts. Along strike, in areas of low metamorphic grade, the Hores Gneiss is clearly a porphyritic rhyodacitic volcaniclastic (Willis et al. 1988; Page and Laing 1990). The depositional age of the *Broken Hill* host gneiss-metasediment has been measured by zircon dating at 1690–1680 Ma (Page 1991). Haydon and McConachy (1987) and Wright et al. (1987), in contast, reject the presence of primary acid volcanics in the sequence and interpret quartzo-feldspathic gneisses as arkosic sediments deposited in fluvio-deltaic wedges interfingering with marine cycles. The *Broken Hill* orebody is, in their opinion, resident in a shallow marine sand unit.

Stratigraphically above the Hores Gneiss there are predominantly carbonaceous metasediments and a marked absence of volcanics and exhalites. Leyh (1984) suggested that this reflected a rapid deepening of the basin and a change in the chemistry of sedimentation. This may represent a sag phase of basin development (Stevens et al. 1988) but Zn-Pb-Ag is essentially absent perhaps by virtue of the lack of volcanics and associated heating of the sequence. The environment then changed to one of pre-orogenic turbidite deposition in shallow to deep water, with low energy graphitic pelite, well sorted sandstone and impure dolomite, which is essentially unmineralized.

The entire sequence lacks clastics or pyroclastics coarser than sand grain size and has only minor carbonates. The overall environment was considered by Stevens et al. (1988) to be of a rift, with an as yet unidentified eastern margin, open to marine incursions and surrounded by a landscape of low relief, with most of the sediments being derived from the reworking of ash and volcanic piles within the rift. A facies change is evident near Broken Hill, perhaps reflecting original NE to SW irregularities in the depositional trough.

The Broken Hill Inlier underwent intense deformation at around 1660 Ma. Three major episodes of folding have been recognised with the first producing large, reclined to recumbent folds; *Broken Hill* occurs in an area of downward facing folds (Stevens et al. 1988). Metamorphism at the *Broken Hill* mine, which is in an amphibolite to granulite zone (Binns 1964; Phillips 1980), is dated at 1600 Ma (Page 1991); this isochemical metamorphism produced a coarsening of sulphide and silicate grain size (Vaughan and Stanton 1986). Intrusive rocks are relatively minor but some post-tectonic alkali feldspar-rich granites and pegmatites were intruded at about 1490 Ma and basic and ultrabasic bodies were emplaced at about 560 Ma (Stevens et al. 1988).

4.2 *Broken Hill* Area

At *Broken Hill*, eight ore lenses with distinctive chemical and mineralogical characteristics occur now as an inverted stack in the overturned sequence. There is an increase in Pb/Zn ratios and Ag upwards. The geometrically lower "Zn lodes" have gradational boundaries with the host garnet-quartz-rich rocks. The upper "Pb lodes" have sharp contacts with fine-grained Mn-garnet 'sandstone', pelite and psammite. These quartz-garnet and garnet rocks are siliceous and differ from the normal psammites elsewhere in the sequence. We concur with the view that the orebody is closely associated with felsic volcanics and diverse clastic and exhalative sedimentary strata (Stevens et al. 1988).

Recent studies by Mackenzie and Davies (1990) have shown that the stratigraphically lowest (structurally highest) lode, C-lode, is of wide extent, lenticular and appears to be a stratigraphic precursor to the other main lodes rather than being a feeder zone. They demonstrated that the ores of the southern part of the lode system, separated laterally several hundred metres from the main ores, belong to a coherent sequence of lenses, implying a stratigraphic control on the localization of sulphide accumulation. Distinct pulses of mineralization are suggested by the distinctive gangue mineralogy, metal grades and trace element association for each ore lens. The "Zn lodes" give way laterally to a quartz-gahnite rock extending for 25 km along strike, with associated small stratiform Zn-Pb-Ag occurrences. The presence of gahnite and the paucity of iron sulphides imply low sulphur availability in the sequence.

4.3 Ore Genesis

Models for the genesis of the *Broken Hill* ore have been wide ranging partly because each theory must initially decipher the complex and multiple deformation and metamorphic events that have obliterated or transformed initial geology. Two general models have been aired, both involving pre-metamorphic mineralization. One of these proposes an essentially exhalative origin (Stanton 1972; Johnson and Klingner 1975; Plimer 1985 and elsewhere). The other involves epigenetic infill of pore spaces in an inhalation model based on models for sandstone lead deposits as found at *Laisvall*, Sweden (Haydon and McConachy 1987; Wright et al. 1987). The ore fluids have been considered to be volcanogenic (Stanton 1972); brines from the sedimentary pile (Johnson and Klinger 1975; Haydon and McConachy 1987; Wright et al. 1987); downwardly convecting seawater linked with the onset of rapid subsidence and driven by heat from high level magma chambers (Russell 1983; James et al. 1987); and mantle derived fluids (Plimer 1985; 1986; Lottermoser 1989).

The large size of the *Broken Hill* deposit relative to other deposits in the district has been considered to be a function of the greater size of the seafloor depression and the greater concentrations of the exhalative ore solutions at *Broken Hill* (Barnes 1980).

The mantle metasomatic ore fluid is a provocative suggestion based on the interpretation of various mineralogical and geochemical features in terms of "carbonatite" fluids, which are considered to have ascended along mantle-tapping fractures (Plimer 1985, 1986). However, the Pb isotope composition of the *Broken Hill* ore (Gulson et al. 1985) is too radiogenic for a mantle derivation; it falls on the growth curve for the average crust (Doe and Zartman 1979). Furthermore, the direct derivation of Pb from the mantle is not consistent with the empirical evidence that the formation of the major Zn-Pb-Ag deposits required prior fractionation of Pb into the upper crust (Sect. 2).

The $\delta^{34}S$ values of sulphide minerals at *Broken Hill*, concentrated around 0 per mill, are difficult to interpret because of the isotopic homogenization that is likely to have occurred during the high grade metamorphism. Spry (1987) concluded that the S was derived either from inorganic reduction of seawater sulphate mixed with magmatic S (of mantle derivation), or by low temperature biological reduction of sulphate from contemporaneous seawater.

5 Subdivision of Sediment-Hosted Zn-Pb-Ag Deposits

Plimer (1986) classified *Broken Hill*, and similar deposits such as the subeconomic *Pegmont* deposit in the eastern Mount Isa Inlier, and *Aggeneys*, in South Africa as Broken Hill type. He views this type of relatively S-poor deposit as having formed in deep water distal turbidite-volcanic sequences, which underwent high-grade metamorphism. It contrasts with the Mount Isa type which formed in shallower water environments, in rifts which were not subjected to such intense deformation and metamorphism.

Beeson (1990) reviewed Zn-Pb-Ag sediment-hosted deposits in metamorphic terrains and subdivided the Broken Hill type deposits into iron formation hosted types (e.g. *Aggeneys*, South Africa; *Pinnacles*, NSW) and iron-poor calc-silicate, psammopelite types (e.g. *Broken Hill*). He considers that the iron-rich deposits occur in sequences formed in the rift phase of basin evolution under subaerial to shallow water, oxidizing, soda-rich conditions, and were accompanied by local Cu-sulphide deposition. A change from Na to K-rich, sometimes hypersaline conditions, favoured the formation of Zn-Pb-Ag deposits of the calc-silicate type later in the sequence.

6 The Role of Magmatic Activity and Rifting

The association between magmatism and the genesis of major stratiform Zn-Pb-Ag deposits is commonly dismissed as insignificant because the host sequence is predominantly of sedimentary origin. However, in the Proterozoic Zn-Pb-Ag deposits of Australia, there is increasing evidence that magmatic activity and/or high heat producing granites acted at the very least as a heat engine driving large-scale fluid convection in the rift strata. In addition, fractionating magmas derived from crustal melting could have released metal- and sulphide-bearing fluids. These would have been incorporated into basinal brines and/or seawater circulating within the thick piles of rift-fill sedimentary and volcanic strata.

It has become clear recently that the major Zn-Pb-Ag deposits of northern and southern Australia formed pene-contemporaneously in the short interval about 1690–1670 Ma. This implies they were related to the same very widespread extensional event. Etheridge et al. (1987) argued that this occurred over zones where there had been underplating of Archaean crust, which is largely unexposed in northern and southern Australia, at approximately 2000 Ma.

There is a range of geological, stable isotopic and palaeomagnetic evidence for the existence of a supercontinent through much of the Proterozoic (Lambert and Donnelly 1992). Such a supercontinent explains the paucity of Middle and Late Proterozoic volcanogenic massive sulphide deposits and porphyry deposits, these being styles of mineralization associated with igneous activity characteristic of active continental margin environments. The Proterozoic supercontinent was subjected to major extensional events, but these did not lead to its breakup and many of the intracratonic rifts which formed were largely or entirely closed off from the open ocean for signifiant periods (Lambert and Donnelly 1992). The breakup of the supercontinent marked the end of the Proterozoic.

The evolutionary model favoured here for the Australian sediment-hosted Zn-Pb-Ag deposits has similarities to the rift models of Large (1988) and Sawkins (1989).

7 Palaeozoic of Eastern Australia

The fact that no large stratiform sediment-hosted Zn-Pb-Ag deposits have been found in the Palaeozoic of eastern Australia may be explained by the lack of known thick black shale sequences. Elsewhere such sequences have associated major stratiform sediment-hosted Zn-Pb-Ag deposits, particularly in the Cambrian to Devonian of the Yukon and British Columbia, and of the Middle Palaeozoic of Europe. In contrast, Australia's Early Palaeozoic contains important volcanogenic Zn-Pb-Ag-Cu-Au deposits, and discordant, sediment-hosted Zn-Cu-Pb-Ag deposits. The volcanogenic massive sulphide deposits are associated with rhyolitic to andesitic volcanic units, and commonly with locally developed tuffaceous (carbonaceous) shales.

7.1 Volcanogenic Massive Sulphide Deposits

In Tasmania, late Middle Cambrian (ca. 500 Ma) volcanism during post collisional relaxation rifting (Crawford and Whitford 1992) was synchronous with the formation of a world class Zn-Pb-Cu-Ag-Au province (Fig. 3), including *Rosebery, Hercules, Que River, Hellyer* (volcanogenic massive sulphide deposits) and *Mount Lyell1* (subvolcanic vein and disseminated mineralization). The varied host sequence of submarine lava and volcaniclastic sediments (Cas et al. 1992) and the ores formed in a very short interval of a few million years, but hydrothermal activity, which may or may not be related to the ore-forming fluids, continued intermittently for up to 60 Ma after host rock and ore deposition and affected the basalts overlying the ore zone (Perkins et al. 1992). The Pb-isotope compositions of the Cambrian Tasmanian deposits are more radiogenic than those of mineralization on the mainland (Gulson et al. 1991), implying that Tasmania was an allochthonous block. No significant volcanic Zn-Pb-Cu-Ag-Au deposits are known in the Cambrian sequences of the southern part of mainland Australia, but the Late Cambrian to Middle Ordovician volcanic arcs of north Queensland host Cu-Zn-Pb-Ag deposits at *Thalanga, Highway, Reward, Liontown and Balcooma* (Withnall et al. 1991; Fig. 3).

The Ordovician to Devonian Lachlan Fold Belt of New South Wales and Victoria contains significant volcanogenic massive sulphide Cu-Zn-Pb deposits, and discordant, sediment-hosted Zn-Pb-Ag-Cu-Au deposits. Reconstruction of the Ordovician of southeastern Australia suggests a cratonic margin in the west with a marginal sea, a central volcanic arc of dominantly shoshonitic character, with associated Cu-Au at *North Parkes*, and open ocean in the far east. This major tectonic cycle of a Cambrian-Ordovician rift and orogenic flysch phase was followed by a rift, flysch and molasse phase in the Silurian to Upper Devonian. In the Middle to Late Silurian, thick silicic submarine volcanics and turbidites accumulated in a number of graben-like basins. Late Silurian massive sulphide deposits occur at *Woodlawn, Captains Flat* and *Benambra* (Fig. 3). Early Carboniferous orogeny converted the Lachlan Fold Belt into a neocraton (Suppel and Schiebner 1990).

The volcanics associated with many of the Australian deposits include ignimbrites suggesting shallower depositional basins than for the Japanese Kuroko deposits. It appears that these Australian deposits formed at and near the seafloor in fault-controlled sub-basins within the generally shallow water basins (e.g. Bain et al. 1987). The moderately positive $\delta^{34}S$ values of their constituent sulphides are consistent with reduction of sulphate in the seawater-dominated ore fluids (e.g. Ayres et al. 1979). Igneous activity could have heated the ore fluids and provided metals to the circulating seawater.

The Late Silurian deposits in the Lachlan Fold Belt of southeastern Australia have remarkably constant Pb isotopic compositions, which imply crustal derivation (Carr et al. 1991), either by leaching of rocks with the same average composition and age throughout this region, or from the Late Silurian magmas. Deformation and metamorphism were followed by a major, short period (410 to 400 Ma) of regional heating and some probable redistribution of earlier formed precious and base metals during widespread granite emplacement.

In Tasmania evidence is accumulating for chemical reworking of Cambrian ores during Devonian magmatism; at *Rosebery* local dissolution of Pb-Zn ore and replacement with an iron sulphide rich assemblage are evident (Solomon et al. 1987).

7.2 Cobar Basin Deposits

The mineralized Siluro-Devonian Cobar Basin lies within the Lachlan Fold Belt (Fig. 3). It is the northern part of the larger Darling Basin, in which a 7-km-thick sequence of rift and sag phase turbiditic trough sediments accumulated; some shelf carbonates and volcanics overlie earlier Ordovician marginal sea deposits (Suppel and Schiebner 1990).

Major polymetallic sediment-hosted mineral deposits occur in distal, fine-grained turbidite facies overlying the rift strata. The main examples are *Elura* (Pb-Zn-Ag-As), *CSA* (Cu-Zn-Pb-Ag lenses) and *The Peak* (Pb-Zn-Ag lenses in a predominantly Au-Cu deposit) (Fig. 3). The Cobar Basin has the most important concentration of base metals in the Phanerozoic of eastern Australia, amounting to a combined total of 85 Mt of ore containing 2.6 Mt Zn, 1.6 Mt Pb, 1.0 Mt Cu, 4000 t Ag and 70 t Au (Schmidt 1990).

At the *Elura* mine, seven vertical pods of massive sulphide lie discordantly in a deformed, deep marine, turbiditic sequence of mudstone and siltstone, and within tight domes strung along an anticlinal trend. At the *CSA* mine, steeply dipping, discordant, foliation-parallel and plunge persistent submassive to massive iron and base metal sulphide lenses occur in a host sequence of thin bedded metasiltstone; a structureless chert bed 30 m thick stratigraphically above the mine is a probable tuff. The consistent easterly dip of the cleavage and the W dip of the beds become disrupted in the ore zones as if in a shear zone roughly parallel to cleavage (Scott and Phillips 1990). Sulphides occupy cleavage planes with sphalerite in massive banded form or in veins. *The Peak* Zn-Pb-Ag mineralization has similarities with the *CSA* and *Elura* deposits, but has a significiant Au-Cu overprint. Five discordant sulphide-bearing "lenses", within a sequence of sandy and shaly turbidite, lie adjacent to an infaulted and apparently basement derived rhyolitic core, deep in the deposit. *The Peak* deposit lies close to the base of deposition of the Late Silurian to Early Devonian sediments.

Glen (1991) and Glen et al. (1992) proposed development of the Cobar Basin as a left-lateral, transtensional rift which formed in the Late Silurian to Early Devonian (410–400 Ma) before undergoing inversion during a right-lateral transpresional regime, with thin-skinned folding and thrusting above high-level detachments, and a sag phase of passive subsidence. Drummond et al. (1992) using seismic data propose that the Cobar basin formed as the result of the movement of upper crust along a detachment at a depth of 25 km, over a ramp in the detachment. The basin is asymmetric, with the western margin steeper than the eastern margin, and is cut by offsetting major reactivated ENE transfer faults one of which passes through the Cobar area. The controlling structures focussing metal-bearing fluids were splay faults derived from a major crustal terrain boundary which is inferred to extend for hundreds of kilometres to the southeast of Cobar. Left-lateral movement in the Late Silurian also controlled en echelon granite emplacement to the south of the Cobar Basin. Although volcanics comprise a very minor part of outcrop they can be inferred to be significant at depth based on the increased gravity anomaly in the Cobar Basin (Glen 1991) and on the wider occurrence of volcanics and high level granite to the south and east. On the western margin of the Cobar Basin there is a post-rift sag phase of upwardly shallowing sediments, including carbonates (Glen 1991).

7.3 Genesis of the Cobar Basin Deposits

These discordant deposits appear to have formed during late Early Devonian metamorphism. Muscovite, intimately intergrown with quartz and sulphide in the CSA mine, has been dated at 390 Ma (Binns 1985) which accords with the Early Devonian age for the end of cleavage development in the host siltstone (Glen 1991).

The NNW-strike fault at the eastern margin of the Cobar Basin experienced movement and metamorphism at the close of the Early Devonian. The Zn-Pb-Ag deposits occur along secondary faults which are inferred to splay off major faults penetrating deep into the central basin (Glen 1991). All deposits, with the exception of *Elura*, occur in fault-related high strain zones characterized by steep regional cleavage and down-dip elongation, and Glen (1991) surmised that the anticline at *Elura* is developed over a blind thrust. The major reactivated transfer fault through the Cobar area may have focussed the mineralizing fluids.

Features reflecting passive replacement of ore into the host rock at *Elura* have been highlighted by Schmidt (1990). At the contact between sulphide and host, unrotated and variably mineralized beds extend for some metres into the ore. Porosity control on the fluids is demonstrated by the mineralization in sandstones for up to 150 m from the massive sulphide, and textures in siliceous ore which are identical to those in the adjacent country rocks. Sulphide and siderite alteration haloes extend for 5 to 70 m into the host strata and there is a wider halo in which chlorite, albite, detrital biotite and Fe-Ti oxide have been destroyed, and calcite converted to ankerite.

There is ore zoning, with siliceous ore forming an outer annulus, a massive pyrite type forming an inner annulus, and the pyrrhotite ore forming a core zone. All types contain pyrite, sphalerite, and galena. A primary metal zoning shows Zn, Pb, Cu enriched at depth and Ag, As (Au) in cap zones. The $\delta^{34}S$ values for pyrite, sphalerite, pyrrhotite and galena range from 4.7 to 12.6 per mill, which is in accord with a source from underlying sediments, including volcanogenic sulphide mineralization (Seccombe 1990). Seccombe (1990) accepts an origin by metamorphic dewatering of sediments and transport of Fe, Zn, Pb and Ag in the low salinity CO_2—CH_4 bearing fluids by chloride complexing. Deposition may have been promoted by transient pressure and temperature drop during periods of vein openings.

Deformed syngenetic models proposed for the mineralization in the Cobar district (Marshall and Sangameshwar 1982) have been rejected for the Elura deposit in favour of the replacement models (de Roo 1989; Schmidt 1990; Lawrie 1991), whereby metal-bearing brines, channelled through fractured anticlinal axes, caused metasomatism of hinge zones by banded sulphide during the dominant regional deformation. Siliceous ore at the margins represents an early arrested stage while the central pyrrhotite, with equilibrium textures, represents an advanced stage. Periodic release of overpressure by brecciation triggered metasomatism by carbonate and sulphide; the brecciation is attributed to dome collapse, following dilatational rupture due to shear zone faults (de Roo 1989). Rocks below *Elura* are inferred to be 2 km of Early Devonian turbidite and volcanics inferred from exposures further south, an unknown thickness of multiply deformed Ordovician turbidite and a Proterozoic craton. Rocks above Elura at the time of formation are surmised as some 7 km of turbidite and fluviatile sediments.

For the *CSA* mineralization, Scott and Phillips (1990) favour a genesis of ore coeval with folding, and shear zones acting as conduits for hydrothermal fluids expelled from the sedimentary pile. Pervasive silicification and chloritization were followed by repeated sulphide and quartz deposition and cleavage development.

Hinman and Scott (1990) suggest *The Peak* deposit is epigenetic, and formed during a regional cleavage forming event in a zone of inhomogeneous high strain developed adjacent to slices of competent "basement". They identify three mineralizing events. Initially, Zn-Pb-Ag and silica replacement formed in areas of ductile dilation, particularly around the rigid volcanic core. Later, after the whole zone became rigid by silicification, onset of high strain caused brittle fracture, and introduced Au-Cu and pyrrhotite. A final phase of magnesium metasomatism locally produced banded to massive black chlorite-sphalerite-galena (and widespread chlorite replacement), perhaps by redistribution of pre-existing Zn-Pb-Ag.

There is a concentration of deposits near the intersection of a major ENE transfer fault and N thrust faults along the eastern basin margin; a linked fault system is proposed as fluid pathways throughout the basin's history by Glen et al. (1992).

The common occurrence of Zn-Pb-Ag-(Cu) mineralization in the Silurian volcanics throughout the Lachlan Fold Belt, the presence of felsic volcanics in *The Peak* mine and the similarity of the Pb isotopes between the Cobar deposits and Late Silurian volcanogenic mineralization elsewhere in the Lachlan Fold Belt suggest to us that Late Silurian volcanogenic Zn-Pb-Ag-Au-Cu mineralization was an important source of the metals for the Cobar Basin deposits. Had the Early Devonian ore fluids leached the ore metals predominantly from Late Silurian and older clastic sediments, the Pb isotopic composition of the Cobar deposits should be more radiogenic than the Late Silurian volcanogenic deposits of the Lachlan Fold Belt. This is because the isotopic composition of the rock Pb in the underlying strata would have changed progressively as associated U and Th decayed. In contrast, Pb in mineralization would have retained its isotopic composition essentially unchanged over time because it has U and Th contents that are negligible in comparison with Pb contents.

8 Palaeozoic Deposits of Western Australia

Of the Palaeozoic basins of Western Australia, the Canning Basin is considered the most Zn-Pb-Ag "prone" with deposits of significance along the *Admiral Bay Fault* and on the *Lennard Shelf* (Fig. 3). The neighbouring Bonaparte Gulf Basin, not discussed here, also has mineralization at *Sorby Hills* and elsewhere. The mineralization in this region is carbonate-hosted.

8.1 Willara Sub-basin

The Willara Sub-basin contains a complete cycle of basin fill sediments reflecting a rift and sag evolution. Sedimentation commenced in the Ordovician with basal shoreline sand passing up to marine shale and limestone (up to 1500 m) then basin evaporite culminating with terrestrial red beds in the Devonian. A Silurian salt sequence, up to 1350 m thick characterized by flowage and solution brecciation, contains some Zn-Pb-Ag at its base. The Broome Arch, on the northeast flank of the Willara Sub-basin (Fig. 3), has been a positive feature since the Early Ordovician; vertical displacements along the Admiral Bay Fault, on its southern margin, are some 4000 m.

Exploratory drilling for oil on the north flank of the Willara Sub-basin, in the southern Canning Basin, and subsequent mineral exploration have established medium grade Zn-Pb-Ag mineralization along 19 km of the Admiral Bay Fault, at depths below 1280 m and over drilled intervals of up to 459 m (Connor 1990). The depths of the mineralization are too great for economic recovery at the present time.

The *Admiral Bay Fault* mineralization, comprising sphalerite, galena and minor chalcopyrite, is mainly below the evaporite bearing sequence, in Ordovician limestone where dolomitization has produced porosity and brecciation. A lower zone stockwork has multiple age veins of gangue barite, dolomite, calcite, siderite and quartz with galena, chalcopyrite, fluorite, pyrite, hematite, magnetite, sphalerite and hydrocarbon. This stockwork underlies upper zone mineralization where gangue and base metal sulphides form a cement in the porous limestone and dolomite, thus forming extensive stratabound deposits with a zinc content higher than in the stockwork zone. The gangue is dolomite, barite, pyrite and hydrocarbon; baritic replacement of anhydrite is evident. At least three episodes of barite veining, and two species of sphalerite, dolomite and hydrocarbon indicate multistage paragenesis (Connor 1990). Maximum depth of burial was in the Triassic prior to the Fitzroy Movement.

8.2 Lennard Shelf

On the Lennard Shelf, a sequence of Devonian shelf carbonates contains numerous lead-zinc occurrences including *Cadjebut*, *Blendevale*, *Twelve Mile Bore*, *Fossil Downs*, *Narlara* and *Wagon Pass* (Fig. 3). Combined resources are some 25 Mt at 12% Pb + Zn and the first significant production from the region started in 1988 at *Cadjebut* (Murphy 1990).

The shallow marine carbonates, containing classic Late Devonian carbonate reefs, are in places covered by basinal facies shale. At the *Cadjebut* mine Pb-Zn mineralization is hosted by carbonate-evaporative units at the base of the Devonian sequence near the unconformity on Precambrian basement. Fault displacements are common, particularly near the major basin margin faults separating the Lennard Shelf from the 18-km-deep Fitzroy Trough to the south where deposition commenced in the Early Ordovician (Shaw 1991). The mineralization of galena and sphalerite, with subordinate marcasite and pyrite, is hosted by various reef facies and is spatially associated with faults which appear to have been active during reef formation.

8.3 Ore Genesis

The carbonate-hosted mineralization of northwestern Australia is epigenetic, stratabound, has both replacement and open space fill sulphides, and it is clearly related to faults. It is of the so-called Mississippi Valley type.

Mory and Dunn (1990) suggest than at an early stage in ore genesis, during the Ordovician, metal-rich brines were exhaled, ponded and became incorporated into the sediments; later these same brines and metals were remobilized, and formed the Zn-Pb mineralization during a phase of rifting.

At the *Cadjebut* mine an early banded stratiform Zn-rich ore replacing banded primary evaporite, and a later Pb-rich breccia ore filling collapse structures, have been described by Tompkins (1992) who relates the mineralization to fluid flow focussed by continental scale compressional tectonics in the Middle Carboniferous. A contrary view is held by Wallace and McManus (1992) who describe one major sulphide-precipitating event, associated with primary porosity in the early diagenetic carbonates, occurring in the Devonian (350 ± 15 Ma).

Present-day fluids in the Canning Basin are hypersaline, slightly acid (pH 5), Ca-rich and Mg/SO_4 depleted and have similarities with the Central Mississippi Salt Dome Basin.

Chemical modelling by Jaireth et al. (1990) demonstrated that a single acidic fluid, originating from a clay-carbonate source at temperatures >150 °C, could react with calcium rich rocks and cause carbonate dissolution, dolomitization and chloritization associated with base metal sulphide precipitation.

Metals for the Lennard Shelf deposits were probably moved in highly saline brines from deep in the compacting sedimentary pile in the Fitzroy Trough. They migrated largely along basin margin faults into the prepared carbonate reefs on the Lennard Shelf. Fluid inclusions of Zn-Pb-Ag deposits indicate minimum sulphide precipitation temperatures of 70–100 °C, and an association with hydrocarbons (Etminan et al. 1984). At least part of the mineralization appears to have formed during the Late Devonian-Early Carboniferous (Arne et al. 1988; McManus and Wallace 1992). The anomalously radiogenic lead (Vaasjoki and Gulson 1986) indicates that the mineralizing brines incorporated some metals from the underlying Proterozoic basement.

Tompkins (1992) suggests that her recognition of evaporite at *Cadjebut* can constrain the model for the formation of MVT deposits to a single fluid mechanism with in situ reduction of oxidized sulphur acquired at the site of deposition.

9 Concluding Remarks

In this brief review, recent observations and ideas have been highlighted in an attempt to emphasize the key genetic processes involved in the origin of sediment hosted Zn-Pb-Ag deposits. The Zn-Pb-Ag mineralizing events in Australia were brief, with principal episodes as follows:

Pre-Statherian	– minor deposits, discordant, formation, ca. 1870 Ma?
Statherian	– major, stratiform, ca. 1690 Ma
Cambrian	– major, volcanogenic, ca. 500 Ma
Silurian	– minor, volcanogenic, ca. 410 Ma
Devonian	– minor/major, epigenetic emplacement, ca. 390 Ma
	– major, carbonate replacement ca. 350/330 Ma

The Archaean volcanogenic deposits of the Yilgarn craton, which have not been discussed here, are Zn-rich and typically have low Pb contents and are not associated with sedimentary strata.

It appears that two features which evolved in the Early to Middle Proterozoic were essential for the formation of sediment-hosted Zn-Pb-Ag deposits. There were a Pb-enriched upper crust and thick rift sequences containing evaporites. In contrast to volcanogenic Cu-Zn mineralization, which is uncommon worldwide in the Middle and Late Proterozoic, the major sediment hosted Zn-Pb-Ag deposits formed at lower temperatures and from much more chloride-rich fluids.

The process of intracratonic crustal evolution involved cycles of K-rich felsic volcanics and associated granites, formed during the early rifting and heating of crust, followed by subsidence, mainly related to cooling. During subsidence, thick piles of sediments accumulated. These fined upwards into very shallow water evaporitic sequences, which are the setting for the world-class stratiform Zn-Pb-Ag deposits in the mid-Proterozoic as sedimentation spread to platform areas.

In the Palaeozoic of Eastern Australia, the discovery of large shale-hosted stratiform Zn-Pb-Ag deposits is considered to be unlikely in the apparent absence of thick black shale rift sequences; but the western margin of the Cobar basin shows promise for carbonate hosted types.

The Cambrian volcanism in Tasmania produced large Zn-Pb-Ag-Cu-Au deposits, some associated with shales. Volcanism in the Silurian and Devonian of mainland Australia produced relatively small deposits, again commonly with a shale association. Later, orogeny produced turbiditic flysch sequences which host some unusual bedding-discordant massive sulphide deposits. These formed under structurally permissive conditions during wrench faulting and may be sourced from underlying mineralized Siluro-Devonian volcanic-related mineralization.

In the Palaeozoic of northwestern Western Australia, epigenetic, carbonate- and evaporite-hosted deposits are associated mainly with faults which appear to have been active during reef formation, adjacent to thick sequences of rift and basin fill sediments, which include evaporites.

Acknowledgements. We thank Drs. David Mackenzie, Leigh Schmidt and Lesley Wyborn for their valuable comments and North Broken Hill Peko Limited's exploration group, Geopeko, for permission to publish this paper.

References

Arne DC, Green PF, Duddy IR, Gleadow AJW, Lambert IB, Lovering JF (1988) Regional thermal history of the Lennard Shelf, Canning Basin, from apatite fission track analysis: implications for the formation of Pb-Zn ore deposits. Aust J Earth Sci 36:495–513

Ayres DE, Burns MS, Smith JW (1987) Sulphur-isotope study of the massive sulphide orebody at Woodlawn, NSW. Aust J Earth Sci 26:197–201

Bain JHC, Wyborn D, Henderson GAM, Stuart-Smith PG, Abell RS, Solomon M (1987) Regional geological setting of the Woodlawn and Captains Flat massive sulphide deposits NSW, Australia. In: Proc Pacific Rim Congr 87. Australasian Institute of Mining and Metallurgy, Melbourne, pp 25–28

Barnes RG (1980) Types of mineralization in the Broken Hill Block and their relationship to stratigraphy. Rec NSW Geol Surv 20(1):33–70

Beeson R (1990) Broken Hill-type lead-zinc deposits in an overview of their occurrence and geological setting. Trans Inst Min Metall 99:B163–B175

Bell TH, Perkins WG, Swager CD (1988) Structure controls on development and localization of syntectonic mineralization at Mt. Isa, Queensland. Econ Geol 83:69–85

Binns RA (1964) Zones of progressive regional metamorphism in the Willyama complex, Broken Hill district, New South Wales. J Geol Soc Aust 11:283–330

Binns RA (1985) Age of the CSA mineralization at Cobar, NSW. CSIRO Div Miner Geochem Res Rev 1985:26–27

Blake DH, Etheridge MA, Page RW, Stewart AJ, Williams PR, Wyborn LAI (1990) Mount Isa Inlier – regional geology and mineralization. In: Hughes FE (ed) Geology of the mineral deposits of Australia and Papua New Guinea. Australasian Institute of Mining and Metallurgy, Melbourne, pp 915–925

Boddington TDM (1990) Abra lead silver copper gold deposit. In: Hughes FE (ed) Geology of the mineral deposits of Australia and Papua New Guinea. Australasian Institute of Mining and Metallurgy, Melbourne, pp 659–664

Carr GR, Dean JA, Gulson BL, Suppel DW (1991) Lead isotope signatures of Ordovician to Permian mineralization in the Lachlan Fold Belt and exploration implications. Geol Soc Aust Abstr 29:6

Cas RAF, Allen RF, Hutton D (1992) Volcaniclastic megaturbidites in deep marine basins in the lateral equivalents of subaerial ignimbrite? Examples from the Palaeozoic of southeastern Australia. Geol Soc Aust Abstr 32:133–134

Connor AG (1990) Admiral Bay zinc-lead deposit. In: Hughes FE (ed) Geology of the mineral deposits of Australia and Papua New Guinea. Australasian Institute of Mining and Metallurgy, Melbourne, pp 1111–1114

Crawford AJ, Whitford DJ (1992) Petrogenesis and tectonic significance of the Hellyer Basalt, Mount Read volcanics, W. Tasmania. Geol Soc Aust Abstr 32:198–199

Crick IH (1992) Petrological and maturation characteristics of organic matter from the Middle Proterozoic McArthur Basin, Australia. Aust J Earth Sci 39:501–519

Davis LW (1990) Silver-lead-zinc-copper mineralization in the Captains Flat-Goulburn Synclinal Zone and Hill End Synclinorial Zone. In: Hughes FE (ed) Geology of the mineral deposits of Australia and Papua New Guinea. Australasian Institute of Mining and Metallurgy, Melbourne, pp 1375–1384

de Roo JA (1989) The Elura Ag-Pb-Zn mine in Australia. Ore genesis in a slate belt by syndeformational metasomatism along hydrothermal fluid conduits. Econ Geol 84:256–278

Doe BR, Zartman RE (1979) Plumbotectonics 1. The Phanerozoic. In: Barnes HL (ed) Geochemistry of hydrothermal ore deposits. 2nd edn. Wiley, New York, pp 22–70

Drummond BJ, Glen RA, Goleby BR, Wake-Dyster KK, Palmer D (1992) Deep seismic of the Cobar Basin II: ramp basin controlled by a mid-crustal detachment. Geol Soc Aust Abstr 32:45

Eldridge CS, Williams N, Compston W, Walshe JL (1991) The HYC deposit at McArthur River Northern Territory, Australia: a non-exhalative model. Bureau of Mineral Resources, Geologh and Geophysics. SGEG Ore Fluids Conf Rec 1990/95:15–16

Etheridge MA, Rutland RWR, Wyborn LA (1987) Orogenesis and tectonic process in the Early to Middle Proterozoic of northern Australia. In: Kroner A (ed) Proterozoic lithospheric evolution. Geodynamic Ser 17, Am Geophys Union, Washington DC, pp 115–130

Etminan H, Lambert IB, Buchhorn I, Chaku S, Murphy GC (1984) Research into diagenetic and mineralising processes Lennard Shelf Reef complexes, WA. In: Purcell P (ed) The Canning Basin, WA. Proc Geol Soc Aust/Pet Expl Soc Aust Symp, Perth WA, pp 447–453

Ferguson J, Chappell BW, Goleby AB (1980) Granitoids in the Pine Creek geosyncline. In: Ferguson J, Goleby AB (eds) Uranium in the Pine Creek geosyncline. International Atomic Energy Agency, Vienna, pp 73–90

Finlow-Bates T, Stumpfl EF (1979) The copper and lead-zinc-silver orebodies of Mt. Isa mine, Queensland. Products of one hydrothermal system. Soc Geol Belg Ann 102:497–517

Forrestal PJ (1990) Mount Isa and Hilton silver-lead-zinc deposits. In: Hughes FE (ed) Geology of the mineral deposits of Australia and Papua New Guinea. Australasian Institute of Mining and Metallurgy, Melbourne, pp 927–934

Fraser WJ (1975) The embayment line of mineralization, Rum Jungle. In: Knight CL (ed) Economic geology of Australia and Papua New Guinea, vol 1. Metals. Australasian Institute of Mining and Metallurgy, Melbourne, pp 271–277

Glen RA (1991) Inverted transtensional basin setting for gold and copper and base metal deposits at Cobar, NSW. BMR J Aust Geol Geophys 12:13–24

Glen RA, Drummond BJ, Goleby BR, Palmer D, Wake-Dyster KD (1992) Deep seismic of the Cobar Basin I. Structure and implications for mineral prospectivity. Geol Soc Aust Abstr 32:43

Green GR (1990) Palaeozoic geology and mineral deposits of Tasmania. In: Hughes FE (ed) Geology of the mineral deposits of Australia and Papua New Guinea. Australasian Institute of Mining and Metallurgy, Melbourne, pp 1207–1223

Gregory PW, Hartley JS, Wills KJA (1990) Thalagna zinc-lead-copper-silver deposit. In: Hughes FE (ed) Geology of the mineral deposits of Australia and Papua New Guinea. Australasian Institute of Mining and Metallurgy, Melbourne, pp 1527–1537

Greig DD (1984) Geology of the Teutonic Bore massive sulphide deposit, Western Australia. Aust Inst Min Metall Proc 289:147–156

Gulson BL (1985) Shale hosted lead-zinc deposits in northern Australia: lead isotope variations. Econ Geol 80:2001–2012

Gulson BL, Perkins WG, Mizon KJ (1983) Lead isotope studies bearing on the genesis of copper orebodies at Mount Isa Queensland. Econ Geol 78:1466–1504

Gulson BL, Porritt PM, Mizon KJ (1985) Lead isotope signatures of stratiform and stratabound mineralisation in the Broken Hill Block, NSW, Australia. Econ Geol 80:488–496

Gulson BL, Solomon M, Vaasjoki M, Both R (1991) Tasmania adrift. Aust J Earth Sci 38:249–250

Hancock MC, Purvis AH (1990) Lady Loretta silver-lead-zinc deposit. In: Hughes FE (ed) Geology of the mineral deposits of Australia and Papua New Guinea. Australasian Institute of Mining and Metallurgy, Melbourne, pp 943–948

Haydon RC, McConachy GW (1987) The stratigraphic setting of Pb-Zn-Ag mineralization at Broken Hill. Econ Geol 82:826–856

Heinrich CA, Andrew AS, Wilkins RW (1989) A fluid inclusion and stable isotope study of synmetamorphic copper ore formation at Mt. Isa, Australia. Econ Geol 84:529–550

Hinman MC, Scott AK (1990) The Peak gold deposit, Cobar. In: Hughes FE (ed) Geology of the mineral deposits of Australia and Papua New Guinea. Australasian Institute of Mining and Metallurgy, Melbourne, pp 1345–1351

Hughes FE (1990) Geology of the mineral deposits of Australia and Papua New Guinea. Australasian Institute of Mining and Metallurgy, Melbourne, 1828 pp

Huston DL, Taylor TW (1990) Dry River copper and lead-zinc-copper deposits. In: Hughes FE (ed) Geology of the mineral deposits of Australia and Papua New Guinea. Australasian Institute of Mining and Metallurgy, Melbourne, pp 1519–1526

Jackson MJ, Muir MD, Plumb KA (1987) Geology of the southern McArthur Basin, Northern Territory. Bur Miner Resour, Geol Geophys Bull 220:173 pp

Jaireth S, Etminan H, Heinrich CA (1990) Lennard Shelf: Bur Miner Res Geol Geophy, Rec 1990/38, 58 pp

James SD, Pearce JA, Oliver RA (1987) The geochemistry of the Lower Proterozoic Willyama Complex volcanics, Broken Hill Block, NSW. In: Beckinsale TC, Rickard D (eds) Chemistry and mineralization of Proterozoic volcanic suites Geol Soc Lond Spec Publ 33:395–408

Johnson IR, Klingner GD (1975) The Broken Hill ore deposit and its environment. In: Knight CL (ed) Economic geology of Australia and Papua New Guinea, vol 1. Metals. Australasian Institute of Mining and Metallurgy, Melbourne, pp 476–491

Laing WP (1990) Evidence for a replacive origin for the Mount Isa Pb-Zn orebodies in a hydrothermal Cu-Pb-Zn deposit. Mount Isa Inlier Geology Conference. Victorian Institute of Earth and Planetary Science, Melbourne, pp 64–66

Lambert IB (1976) The McArthur zinc-lead-silver deposit: features, metallogenesis and comparisons with some other stratiform ores. In: Wolf KE (ed) Handbook of stratabound and stratiform ore deposits, vol 6. Elsevier, Amsterdam, pp 535–585

Lambert IB (1983) The major stratiform lead-zinc deposits of the Proterozoic. In: Medaris LG, Byers CW, Mickelson DM, Shanks WC et al. (eds) Proterozoic geology: selected papers from an international Proterozoic symposium. Geol Soc Am Men 161:209–226

Lambert IB (1989) Precambrian sedimentary sequences and their mineral and energy resources. In: Price RA (ed) Original and evolution of sedimentary basins and their energy and mineral resources. International Union of Geodesy and Geophysics and American Geophysical Union, Washington. Geophys Monogr 48(3):169–173

Lambert IB, Donnelly TH (1990) The palaeoenvironmental significance of trends in sulfur isotope compositions in the Precambrian: a critical review. In: Herbert HK, Ho SE (eds) Stable isotopes and fluid processes in mineralization. Spec Publ 23, Geol Dept Univ Western Australia, pp 260–268

Lambert IB, Donnelly TH (1991) Global oxidation and a supercontinent in the Proterozoic: evidence from stable isotopic trends. In: Schidlowski M (ed) Early organic evolution: implications for mineral and energy resources. Springer, Berlin Heidelberg New York

Lambert IB, Scott KM (1973) Implications of geochemical investigations within and around the McArthur zinc-lead-silver deposit, Northern Territory. J Geochem Explor 2:307–330

Large D (1988) The evaluation of sedimentary basins for massive sulphide mineralization. In: Friedrich GH, Herzig PM (eds) Base metal sulphide deposits. Springer, Berlin Heidelberg New York, pp 3–11

Lawrie KC (1991) Metal zoning and fluids in the Elura orebodies NSW: implications for the formation of syntectonic massive sulphide orebodies. Bur Miner Resour Geol Geophys Rec 1990/95:52

Lees T, Khin Zaw, Large RR, Huston DL (1990) Rosebery and Hercules copper-lead-zinc deposits. In: Hughes FE (ed) Geology of the mineral deposits of Australia and Papua New Guinea. Australasian Institute of Mining and Metallurgy, Melbourne, pp 1241–1247

Legge PJ, Haslam CO, Taylor S (1984) Lead-zinc-silver exploration and development in Australia. Proc Aust Inst Min Metall 289:119–135

Leyh WR (1984) Shear zone and lineament control of mineralization, Broken Hill, New South Wales. Geol Soc Aust Abstr 12:332–334

Logan RG, Murray WJ, Williams N (1990) HYC silver-lead-zinc deposit, McArthur River. In: Hughes FE (ed) Geology of the mineral deposits of Australia and Papua New Guinea. Australasian Institute of Mining and Metallurgy, Melbourne, pp 907–911

Lottermoser BG (1989) Rare earth element study of exhalites within the Willyama Supergroup, Broken Hill Block, Australia. Mineral Depos 24:92–99

Mackenzie DH, Davies RH (1990) Broken Hill lead-silver-zinc deposit at ZC mines. In: Hughes FE (ed) Geology of the mineral deposits of Australia and Papua New Guinea. Australasian Institute of Mining and Metallurgy, Melbourne, pp 1079–1084

Main J (1992) The Century zinc deposit, North West Queensland. Geol Soc Aust Abstr 32:55
Marshall B, Sangameshwar SR (1982) Commonality and differences in ores on the Cobar Super-Group, NSW. In: Symposium on Geology and Mineralization in Lachlan Fold Belt, Sydney. Geol Soc Aust Abstr 6:15–16
McArthur GJ, Dronseika EV (1990) Que River and Hellyer zinc-lead-silver deposits. In: Hughes FE (ed) Geology of the mineral deposits of Australia and Papua New Guinea. Australasian Institute of Mining and Metallurgy, Melbourne, pp 1229–1239
McGoldrick PJ, Keays RR (1990) Mount Isa copper and lead-zinc-silver ores: coincident or cogenesis? Econ Geol 85(3):641–650
McKay WJ, Hazeldene RK (1987) Woodlawn Zn-Pb-Cu deposit, NSW, Australia: an interpretation of ore formation from field observations and metal zoning. Econ Geol 82:141–164
McManus A, Wallace MW (1992) Age of Mississippi Valley-type sulphides determined using cathodoluminescence and cement stratigrphy, Lennard Shelf, Canning Basin, Western Australia. Econ Geol (in press)
Mill JHA, Clifford BA, Dudley RJ, Ruxton PA (1990) Scuddles zinc-copper deposit at Golden Grove. In: Hughes FE (ed) Geology of the mineral deposits of Australia and Papua New Guinea Australasian Institute of Mining and Metallurgy, Melbourne, pp 583–590
Mory AJ, Dunn PR (1990) Bonaparte, Canning, Ord and Officer basins – regional geology and mineralization. In: Hughes FE (ed) Geology of the mineral deposits of Australia and Papua New Guinea. Australasian Institute of Mining and Metallurgy, Melbourne, pp 1089–1096
Muir MD (1983) Depositional environments of host rocks to northern Australian lead-zinc deposits, with special reference to McArthur River. In: Sangster DE (ed) Short course in sediment hosted stratiform lead-zinc deposits. Miner Assoc Can 8:141–174
Murphy GC (1990) Lennard Shelf lead-zinc deposits. In: Hughes FE (ed) Geology of the mineral deposits of Australia and Papua New Guinea. Australasian Institute of Mining and Metallurgy, Melbourne, pp 1103–1109
Neudert MK (1984) Are the Mount Isa lead-zinc ores really syngenetic? Geol Soc Aust Abstr 12:402–404
Neudert MK (1986) Sedimentology of the Middle Proterozoic Mt Isa Group, Queensland Australia. In: Sediments down under. 12th Int Sedimentological Congr 24–30 August 1986, p 228, Abstr Canberra: Inter Sedimentological Congr
Neudert MK, Russell RE (1981) Shallow water and hypersaline evaporites from the Middle Proterozoic Mount Isa sequence, Queensland. Nature 293:284–286
Page RW (1981) Depositional ages of the stratiform base metal deposits at Mt Isa and McArthur River, Aust. based on U-Pb zircon dating of concordant tuff horizons. Econ Geol 76:648–658
Page RW (1991) Depositional age of volcanic precursors of the 'Potosi' gneiss, Broken Hill Group, BMR Res Newslett 14:3–4
Page RW, Laing WP (1990) Depositional age of the Broken Hill Group from volcanics stratigraphically equivalent to the Ag-Pb-Zn orebody. Geol Soc Aust Abstr 25:18–19
Parr JM (1992) Exhalite associated with Broken Hill-type mineralization: genetic implications and exploration potential. Geol Soc Aust Abstr 32:69
Perkins C, McDougall I, Compston W, Walshe JL, Gemmell JB (1992) Geochronology of the Hellyer volcanogenic massive sulphide deposit and Mt. Read volcanics, Tasmania. 29th Int Geological Congr Proc, Kyoto, Japan, August 1992 Abstr, p 779
Perkins WG (1984) Mount Isa Silica dolomite and copper orebodies: the result of a syntectonic hydrotermal alteration system, Econ Geol 79:601–637
Perkins WG (1990) Mount Isa copper ore bodies. In: Hughes FE (ed) Geology of the mineral deposits of Australia and Papua New Guinea. Australasian Institute of Mining and Metallurgy, Melbourne, pp 935–941
Perkins WG (1991) The Mt Isa Ag-Pb-Zn orebodies stratiform replacement lodes in a zoned hydrothermal alteration system. SGEG Ore fluids conference, Canberra. Bur Min Res Geol Geophys Rec 1990–95:63–64
Phillips GN (1980) Water actively changes across an amphibolite-granulite facies transition, Broken Hill, Australia. Contrib Mineral Petrol 75:377–386
Plimer IR (1985) Broken Hill Pb-Zn-Ag deposit-a product of mantle metasomatism. Mineral Depos 20(3):147–153
Plimer IR (1986) Sediment-hosted exhalative Pb-Zn deposits-products of contrasting ensialic rifting. Geol Soc S Afr Trans 89:57–73

Plimer IR, Lottermoser G (1989) Comments and replies on "Sedimentary model for the giant Broken Hill Pb-Zn deposit, Australia". Geology 16:564–568

Plumb KA, Ahmad M, Wygralak AS (1990) Mid Proterozoic basins of the northern Australian craton – regional geology and mineralization. In: Hughes FE (ed) Geology of the mineral deposits of Australia and Papua New Guinea. Australasian Institute of Mining and Metallurgy, Melbourne, pp 881–902

Richards JR (1975) Lead isotope data on three north Australian galena localities. Mineral Depos 10:185–194

Russell MJ (1983) Major sediment-hosted exhalative zinc, lead deposits: formation from hydrothermal convection cells that deepen during crustal extension. Min Assoc Can Short Course 8:251–282

Sawkins FJ (1989) An orogenic felsic magmatism, rift sedimentation, and giant Proterozoic Pb-Zn deposits. Geology 17:657–660

Schmidt BL (1990) Elura zinc-lead-silver deposit, Cobar. In: Hughes FE (ed) Geology of the mineral deposits of Australia and Papua New Guinea. Australasian Institute of Mining and Metallurgy, Melbourne, pp 1329–1336

Scott AK, Phillips KG (1990) CSA Copper lead-zinc deposit, Cobar. In: Hughes FE (ed) Geology of the mineral deposits of Australia and Papua New Guinea. Australasian Institituе of Mining and Metallurgy, Melbourne, pp 1337–1343

Seccombe PK (1990) Elura mine. Mineral Depos 25:304–312

Shaw R (1991) New K-Ar constraints on the onset of subsidence in the Canning Basin. BMR Res Newslett 14:13–14

Sheppard WA, Main JV (1990) Recent developments in the evaluation of the Dugald River zinc-lead deposit, Queensland. Geol Soc Aust Abstr 25:113–114

Smith JW, Burns MS, Croxford NJW (1978) Stable isotopic studies of the origins of mineralization at Mount Isa. Mineral Depos 13:369–381

Smolonogov S, Marshall B (1992) The Woodcutters Pb-Zn orebodies: deformed epigenetic mineralisation. Geol Soc Aust Abstr 32:68–68

Solomon M, Vokes FM, Walshe JL (1987) Chemical remobilization of volcanic-hosted sulphide deposits at Rosebery and Mt Lyell, Tasmania. Ore Geol Rev 2:173–190

Solomon M, Heinrich CA, Swift M (1991) Are high heat producing granites essential for the formation of giant sedex-type lead-zinc deposits. Bur Mineral Resour Geol Geophys Rec 1990/95:78–79

Spry PG (1987) A sulphur isotope study of the Broken Hill deposit, New South Wales, Australia. Mineral Depos 22:109–115

Stanton RL (1972) A preliminary account of the chemical relationship between sulphide lode and "banded iron formation" at Broken Hill, New South Wales. Econ Geol 67:1128–1145

Stevens BPJ, Barnes RG, Brown RE, Stroud WJ, Willis IL (1988) The Willyama Supergroup in the Broken Hill and Euriowie Blocks, New South Wales. Precambrian Res 40(1):297–327

Suppel DW, Schiebner E (1990) Lachlan Fold Belt in New South Wales – regional geology and mineral deposits. In: Hughes FE (ed) Geology of the mineral deposits of Australia and Papua New Guinea. Australasian Institute of Mining and Metallurgy, Melbourne, pp 1321–1327

Taube A (1984) Woodcutters; Darwin Conference, August 1984. AusIMM Conf Ser 13:347–356

Tompkins LA (1992) MVT mineralization at the Cadjebut mine, Lennard Shelf. Abstracts of the 1992 Selwyn Memorial Symposium, Carbonate reefs and associated oil and mineral deposits. Geol Soc Aust Vic Div Abstr, 1 pp

Valenta R, Ord A, Hobbs B (1990) Structural controls on copper mineralization in the Hilton mine. In: Mount Isa Inlier Geology Conference. Victorian Institute of Earth and Planetary Sciences, Melbourne, pp 58–59

Van den Heuvel HB (1969) Sedimentation, stratigraphy and post-depositional changes in the sediments of the upper formations of the Mount Isa Group, northwest Queensland. PhD Thesis, University of Queensland, 217 pp

Vaasjoki M, Gulson BL (1986) Carbonate-hosted base metal deposits: lead isotope data bearing on their genesis and exploration. Econ Geol 81:156–172

Vaughan JP, Stanton RL (1986) Sedimentary and metamorphic factors in the development of the Pegmont stratiform Pb-Zn deposit, Queensland. Inst Min Metall Trans/Sect B95:B94–B121

Wallace MW, McManus A (1992) Timing and age of Mississippi Valley-type sulphides in the Devonian reef complexes of the Canning Basin, Western Australia. In: Abstracts of the 1992

Selwyn Memorial Symposium, Carbonate reefs and associated oil and mineral deposits. Geol Soc Aust Vic Div, 2 pp

Waltho A (1992) A resource modelling of the Century zinc-lead deposit, North Queensland. Geol Soc Aust Abstr 32:321–322

Waring CL (1991) The distiction between metamorphosed stratiform Pb-Zn-Ag mineralization and metamorphic syntectonic Cu mineralization at Mt Isa. Bur Mineral Resour Rec 1990/95:81–82

Williams N (1978) Studies of base metal sulphide deposits at McArthur River, Norther Territory, Australia II. The sulphide-S and organic-C relationship of the concordant deposits and their significance. Econ Geol 73:1036–1056

Williams N (1990) Sediment hosted stratiform lead-zinc deposits. Geol Soc Aust Abstr 25:313–314

Willis IL, Barnes RG, Stevens BPJ, Brown RE (1988) Broken Hill. Geology 16:567

Wilson IH (1987) Geochemistry of Proterozoic volcanics, Mt Isa Inlier, Australia. In: Beckinsale TC, Richard D (eds) Geochemistry and mineralization of Proterozoic volcanic suites. Geol Soc Lond Spec Publ 33:409–423

Withnall IW, Black LP, Harvey KJ (1991) Geology and geochronology of the Balcooma area: part of an Early Proterozoic magmatic belt in North Queensland. Aust J Earth Sci 38:15–29

Wright JV, Haydon RC, McConachy GW (1987) Sedimentary model for the giant Broken Hill Pb-Zn deposit, Australia. Geology 15:598–602

Wyborn LA (1988) Petrology, geochemistry and origin of a major Australian 1880–1840 Ma felsic volcano-plutonic suite: model for intracontinental felsic magma generation. Precambrian Res 40/41:37–60

Wyborn LAI (1990) Geochemical evolution of the felsic igneous rocks of the Mt. Isa inlier and their significance for mineralization. In: Mt. Isa Inlier Geology Conf. Victoria Institute of Earth and Planetary Science, Melbourne, pp 6–7 (Abstr)

Wyborn LAI, Page RW, Parker AJ (1987) Geochemical and geochronological signatures in Australian Proterozoic igneous rocks. In: Pharaoh TC, Beckinsale RD, Rickard D (eds) Geochemistry and mineralization of Proterozoic volcanic suites. Geol Soc Lond Spec Publ 33:377–394

Wyborn LAI, Page RW, McCulloch MT (1988) Petrology, geochronology and isotope geochemistry of the post 1820 Ma granites of the Mount Isa inlier: mechanisms for the generation of Proterozoic anorogenic granites. Precambrian Res 40/41:509–541

Sediment-Hosted Pb-Zn Deposits in China: Mineralogy, Geochemistry and Comparison with Some Similar Deposits in the World

X. Song[1]

Abstract

Sediment-hosted Pb-Zn deposits in China can be classified on a geographical basis into the Langshan, Quinling, and Jinding types. These types also show a distinctive tectonic setting, age and lithology of the host rock, regional geochemical environment, mineral assemblages, and element associations. The Langshan- and Qinling-type deposits have mostly Zn/Pb ratios of 2–5 and Cu average contents of 0.5–1.4%, whereas the Jinding deposit shows a wide range of Zn/Pb ratios (0.29–17.2) and does not contain significant Cu. Low Co/Ni ratios (less than 1) and high S/Se ratios (10×10^4 to 53×10^4) are indicative of a sedimentary genesis of pyrite for the Langshan- and Qinling-type deposits. Sphalerites from Chinese sediment-hosted Pb-Zn deposits usually contain 1200 to 2300 ppm Cd except for Jinding sphalerite (2700–8400 ppm Cd). Pyrite from most Langshan-type deposits has a narrow range (26–37.3‰) of $\delta^{34}S$.

Sulphide minerals from the Qinling-type deposits show a relatively wide range of $\delta^{34}S$ values (0.6–22.5‰ or more). The main sulfide minerals of the Jinding deposit are characterized by a broad range and negative values of $\delta^{34}S$ (−0.29 to − 30.4‰). The Pb-isotope data also show differences between the Jinding deposit (mantle affinity) and the other types of deposits (upper crust and orogene signatures). A comparison of Chinese sediment-hosted Pb-Zn deposits with selected stratiform Pb-Zn deposits in the world is given.

1 Introduction

Sediment-hosted Pb-Zn deposits are the most important class of Pb-Zn deposits in China. They not only possess huge reserves of Pb and Zn, but also contain Cu, Ag, Cd, and Tl to a certain extent. In addition, pyrite, celestite, and gypsum ores associated with these Pb-Zn deposits are economically important. Geographically, they are mainly located in the Langshan-Chartai Mountains (central Inner Mongolia), in the Qinling Mountains (Gansu and Shaanxi Provinces), and in the Lanping-Simao region (Western Yunnan; Fig. 1).

[1] Institute of Mineral Deposits, Chinese Academy of Geological Sciences, Baiwanzhuang Road 26, Beijing 100037, China; present address: Institute of Mineralogy and Petrology, University of Heidelberg, 69120 Heidelberg, Germany

Spec. Publ. No. 10 Soc. Geol. Applied to Mineral Deposits
Fontboté/Boni (Eds.), Sediment-Hosted Zn-Pb Ores

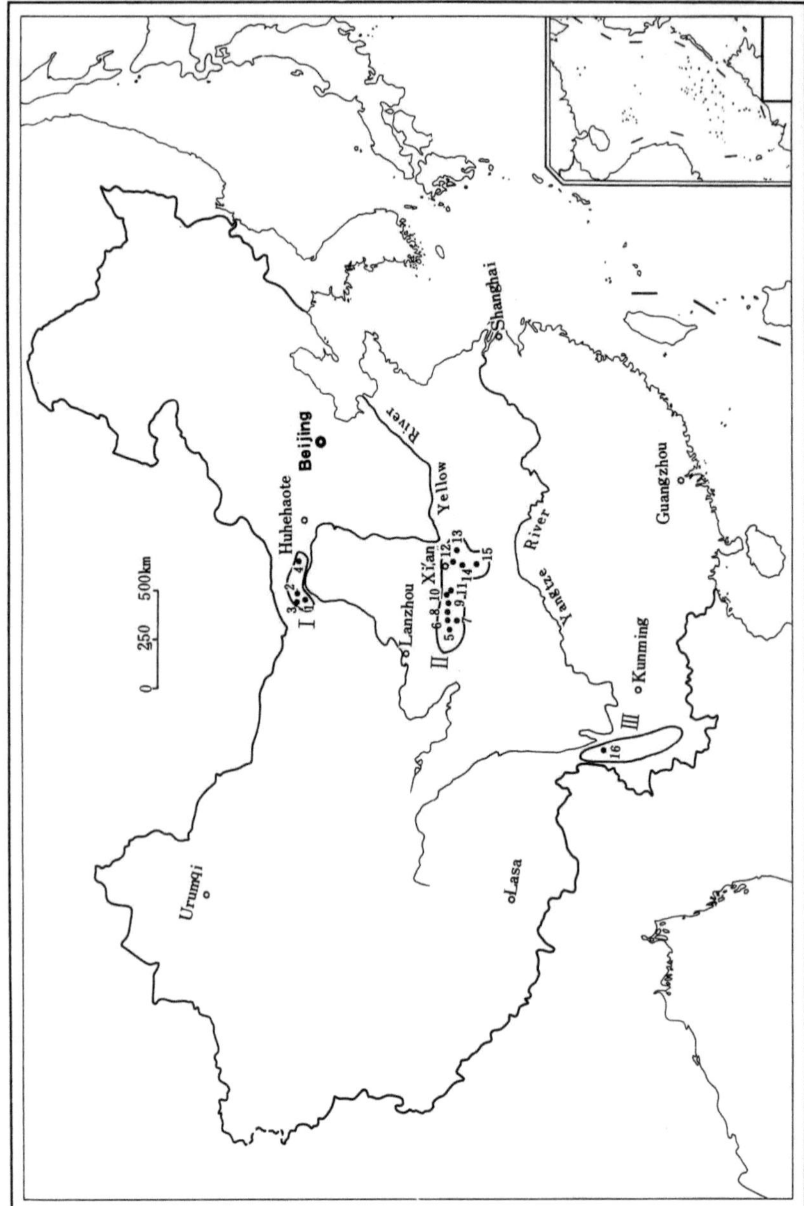

Fig. 1. Location of sediment-hosted Pb-Zn deposits in China. Metallogenic belts and deposits: **I** Langshan-Chartai Metallogenic Belt: *1* Tanyaokou; *2* Dongshengmiao; *3* Huogeqi; *4* Jiashengpan. **II** Qinling Metallogenic Belt: *5* Dengjiashan; *6* Changba-Lijiagou; *7* Bijiashan; *8* Luoba; *9* Qiandongshan; *10* Bafengshan; *11* Yinmusi; *12* Daxigou-Yindongzi; *13* Tongmugou; *14* Xidonggou; *15* Zhaojiazhuang. **III** Lanping-Simao Metallogenic Belt: *16* Jinding

Within each of these three areas, the deposits are characterized by distinctive tectonic settings, ages and lithologies of the host rock, regional geochemical environment, mineral assemblages, and element associations. Therefore, the sediment-hosted Pb-Zn deposits of China are grouped into the Langshan, Qinling, and Jinding types (Song and Cao 1990).

2 General Features of the Three Types of Chinese Sediment-Hosted Pb-Zn Deposits

2.1 Langshan Type

Four "Langshan"-type deposits of Proterozoic age have been discovered in the Langshan-Chartai Mountains, central Inner Mongolia (Fig. 2). They were formed in an aulacogen at the northern, passive continental margin of the North China Platform (Qiao et al. 1991). Host rocks include carbonaceous silty slate, micaquartz schist and slightly metamorphosed marl and argillaceous dolostone, which are embedded within a carbonaceous sandstone-argillite-carbonate sequence. The orebodies are mostly stratiform, and rarely lens-shaped. Their length along strike ranges from tens of metres up to 1300 m. The main ore structures are banded, lamellar, taxitic-banded, mottled-lamellar, disseminated, massive and wrinkled. Ore textures are commonly fine-grained, metacolloform and crystalloblastic, rarely cataclastic and interstitial. Framboidal pyrite is scarce. The ore mineral assemblage is characterized by pyrite-pyrrhotite-Fe-rich sphalerite-galena $\pm$ chalcopyrite; a typical element association is Zn $-$ Pb $\pm$ Cu $\pm$ Ag. Zoning can be seen at the district scale. This is best illustrated at the Jiashengpan deposit where compositional changes in the following order: Fe $\rightarrow$ Zn + Fe $\rightarrow$ Zn + Pb + Fe $\rightarrow$ Fe, can be seen going from west to east. The vertical zoning pattern from bottom to top is Pb + Zn $\rightarrow$ Pb + Zn + Fe $\rightarrow$ Zn + Fe $\rightarrow$ Fe (Lang and Zhang 1987). Most of the $\delta^{34}S$ values for sulphide minerals range between 20.3 and 37.3‰. This implies that the likely source of S is coeval seawater sulphate. Pb isotope data define two Pb-Pb model ages for galena from these deposits (Li 1986): 2350 Ma, which is interpreted as the age of the primary source of Pb, and 780 Ma, which represents the age of late superimposed mineralization, probably related to a certain stage of granite intrusion.

The Langshan-type Pb-Zn deposits are interpreted to be sedimentary-exhalative deposits, but some of them (e.g. the Huogeqi deposit) were affected by a late granite intrusion and by the associated hydrothermal system.

2.2 Qinling Type

The "Qinling-type" Pb-Zn deposits are mostly stratiform. They are located in the Qinling Mountains (Gansu and Shaanxi Provinces, NW China) and belong to the major Qinling Metallogenic Belt. They occur in metamorphosed, fine-grained clastic rocks (sandstone and shale) and carbonate rocks of Devonian

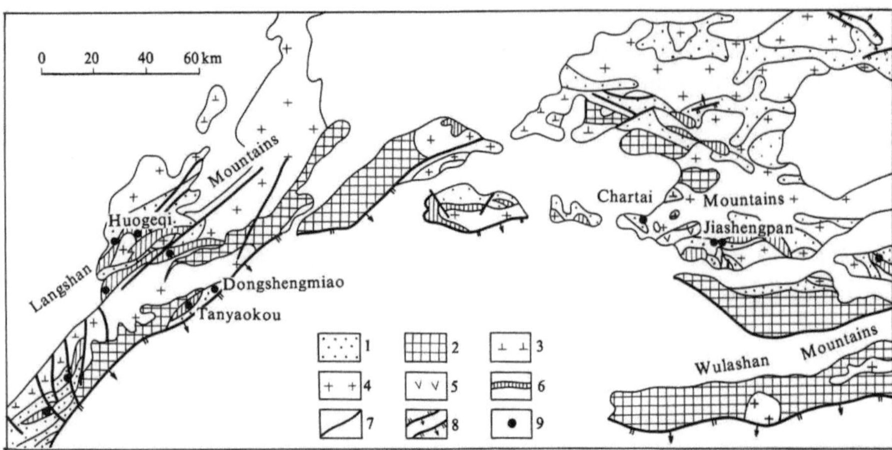

Fig. 2. Distribution of the Langshan-type Pb-Zn deposits in central Inner Mongolia (after Lang and Zhang 1987). *1* Chartai Group of Middle Proterozoic age, consisting of carbonaceous silty slate, metasandstone, micaquartz schist and slightly metamorphosed marl and argillaceous dolostone; *2* Wutai Group of Early Proterozoic age and Wulashan Group of Late Archaeozoic age; *3* intermediate to acid intrusive bodies; *4* acid intrusive bodies; *5* gabbro; *6* ore-bearing strata of the Agulugu Formation of the Chartai Group; *7* shear zone; *8* normal fault and reverse fault; *9* deposit or occurrence

age. Prominent examples include Changba-Lijiagou, Deng-jiashan, Bijiashan, Qiandongshan, Daxigou-Yindongzi and other deposits (Zhang et al. 1987).

The Qinling-type Pb-Zn deposits are concentrated in the Xihe-Chengxian, Fengxian-Taibai, Zhashui-Shanyang, Zhen'an-Xunyang and other orefields (Fig. 3). They were formed in several graben basins of the South Qinling active continental margin in the collision zone between the Yangtze and North China plates (Gu and Lin 1990). Their host rocks include micaschist, calcareous phyllite, impure crystalline limestone, ferruginous dolostone, chert and albite- and scapolite-bearing hornfels. The host rocks originally belonged to a siliciclastic-carbonate sequence deposited in a miogeosynclinal environment. The Changba-Lijiagou deposit, the largest Qinling-type deposit in China, is hosted by biotite-quartz schist, marble and crystalline limestone. The schist-hosted orebodies are predominantly stratiform and show sharp boundaries to their country rocks, whereas the carbonate-hosted orebodies have more complex forms, such as beds, lenses, veins and pockets. Both types of orebodies are concordant with the host rock strata. Within the Qinling-type deposits, slight carbonatization, silicification, sericitization and chloritization of the host rocks can be observed. There are intrusive bodies of granite and granodiorite near some deposits.

Ore structures are mainly banded, lamellar, massive and disseminated. Ore textures are commonly fine-grained, colloidal, eutectic, metasomatic and resorbed. The principal mineral assemblage is sphalerite-galena-pyrite-pyrrhotite ± siderite, and the common element association is Zn − Pb − Ag ± Cu. Among the Qinling-type deposits, there is a unique Ag-Pb deposit, the Yindongzi deposit in the Zhashui-Shanyang orefield. The Qinling-type ores

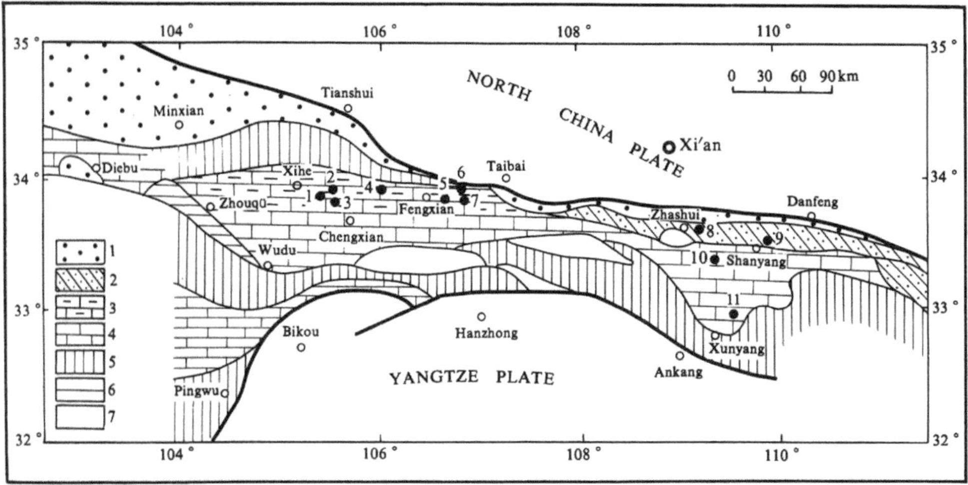

Fig. 3. Location of the Qinling-type Pb-Zn deposits and distribution of sedimentary rocks of Devonian age in the Qinling Mountains, China (after Zhang et al. 1987; Du 1986). Locations: *1* Dengjiashan; *2* Changba-Lijiagou; *3* Bijiashan; *4* Luoba; *5* Qiandongshan; *6* Bafangshan; *7* Yinmusi; *8* Daxigou-Yindongzi; *9* Tongmugou; *10* Xidonggou; *11* Zhaojiazhuang. Sedimentary rocks: *1* sandstone + siltstone; *2* siltstone + argillaceous carbonate rocks; *3* shale + siltstone + carbonate rocks + silica rock; *4* carbonate rocks; *5* siltstone + argillite; *6* argillite; *7* Pre-Devonian strata

generally contain recoverable amounts of Cd, Ga and In. In general, zoning is not well developed, but in the Daxigou-Yindongzi deposit there is an apparent horizontal zoning eastward as follows: barite + magnetite → siderite + Cu → Pb + Ag. The S-isotopic composition of the Qinling-type ores is characterized by relatively heavy sulphur, with a wide range of values (from 0.6 to 27.81‰ $\delta^{34}S$). Pb-Pb model ages of ore lead range between 330 and 550 Ma.

Most of the Qinling-type deposits are generally interpreted as sedimentary-exhalative (Zhang et al. 1987).

2.3 Jinding Type

This type is named after the largest Pb-Zn deposit in China, the Jinding Pb-Zn deposit, West Yunnan, SW China. The Jinding deposit is tectonically situated in the Lanping-Simao back-arc rift valley, west of the suture line between the India Plate and the Eurasian Plate, marked by the Lancang River deep fault zone (Tan 1991). Host rocks of the Jinding deposit are mainly calcareous quartz sandstones of Early Cretaceous age, and calcirudites of Eocene age. These calcirudites are included in a continental gypsiferous-saline red-bed formation. Among the orebodies of the deposit, both stratabound and fault-controlled (thrusts) types occur. Orebodies hosted by calcareous sandstone are bedded to lens-shaped and consist mainly of disseminated ores with an average Zn/Pb ratio of 3:1. Some of them, e.g. orebodies Nos. 1 and 2, extend about 1000 m along strike, and are continuous down-dip for hundreds of metres. In contrast, the orebodies hosted by calcirudite are characterized by stockwork, breccia, and banded ores with an average Zn/Pb ratio of 4:1, and occur in the form of lenses, columns, and irregular veins. The largest one is 700 m long, 200 m wide and tens of metres thick.

Cemented, oolitic, and framboidal textures are common in the sandstone-hosted ores, whereas resorbed and metasomatic textures can be usually observed in the calcirudite-hosted ores. The principal ore mineral assemblage consists of sphalerite-galena-pyrite-marcasite. The development of an oxidized ore consisting of limonite, smithsonite, gunningite and cerussite is a typical feature of this deposit. Moreover, a special kind of high-grade Pb-Zn ore, formed by secondary enrichment and composed of galena and sphalerite with up to 80 vol% sulphides, has been discovered in the district. Forty percent of the total reserves is covered by oxidized ore. In addition, there are celestite and gypsum beds of economic importance in the Jinding deposit. The common element association is Zn-Pb-Sr-Ba-Cd-Ag. Pb-Pb model ages range between 40 and 70 Ma, approximately corresponding to the ages of the ore-bearing strata.

Genetically, the Jinding Pb-Zn deposit is considered to be sedimentary-diagenetic, but it probably underwent late modification by action of higher temperature fluids (Bai et al. 1985).

3 Mineralogy of Sediment-Hosted Pb-Zn Deposits in China

Ore minerals, gangue minerals and secondary minerals in the oxidized zone of sediment-hosted Pb-Zn deposits in China are listed in Table 1.

4 Geochemistry of Sediment-Hosted Pb-Zn Deposits in China

4.1 Minor Elements

The Cu, Pb, Zn, Ag, Au, Ba, and Sr average contents in ore-bearing units for the three metallogenic belts of sediment-hosted Pb-Zn deposits are listed in Table 2.

4.2 Zn/Pb Ratio and Cu and Ag Contents

Zn/Pb ratios and Cu and Ag contents for sediment-hosted Pb-Zn deposits in China are listed in Table 3.

The Zn/Pb ratios for Langshan-type deposits mostly range between 2.93 and 2.99, but the Huogeqi deposit, which has a higher Cu content, has very low Zn/Pb ratios (0.34 to 1.01).

The Zn/Pb ratios for Qinling-type deposits commonly range from 2.06 to 5.19, with two exceptions. One corresponds to the Tongmugou deposit, where the highest Zn value (21.7%) and an equally high Zn/Pb ratio (10.96) have

Table 1. Minerals of sediment-hosted Pb-Zn deposits in China

Deposit type and deposit	Ore minerals		Gangue minerals		Minerals of the oxidized zone
	Major	Minor	Major	Minor	
Langshan type					
Tanyaokou	Pyrite, chalcopyrite, sphalerite, galena, pyrrhotite	Arsenopyrite, marcasite, magnetite	Calcite, dolomite, quartz, mica, chlorite	Barite, plagioclase, tremolite, siderite	Limonite, jarosite, malachite, lepidocrocite, azurite
Dongshengmiao	Pyrite, pyrrhotite, Fe-rich sphalerite, galena, chalcopyrite	Marcasite, magnetite, arsenopyrite, gold, silver	Quartz, sericite, dolomite, calcite	Chlorite, barite	
Huogeqi	Chalcopyrite, galena, Fe-rich sphalerite, pyrrhotite, pyrite, magnetite	Cubanite, bornite, arsenopyrite	Quartz, sericite, diopside, tremolite, calcite	Chlorite, biotite, clinozoisite	Limonite, malachite, jarosite, azurite, cerussite, anglesite
Jiashengpan	Pyrite, galena, pyrrhotite, Fe-rich sphalerite	Arsenopyrite, chalcopyrite, marcasite	Dolomite, sericite, quartz, diopside, tremolite	Calcite phlogopite	Limonite, jarosite, smithsonite
Qinling type					
Changba-Lijiagou	Pyrite, galena, sphalerite	Pyrrhotite, chalcopyrite, arsenopyrite, tennantite, boulangerite	Quartz, calcite	Barite, muscovite, biotite, tremolite, chlorite, epidote, fluorite, plagioclase	Limonite, jarosite, smithsonite, hemimorphite, cerussite
Dengjiashan	Sphalerite, galena	Pyrite, chalcopyrite, bournonite, tetrahedrite, boulangerite	Quartz, calcite	Barite, ankerite, gypsum, celestite	
Qiandongshan	Galena, sphalerite	Pyrite, arsenopyrite, chalcopyrite, tetrahedrite	Ankerite, quartz	Ferrug. calcite, illite, sericite, chlorite, montmorillonite	Limonite, smithsonite, cerussite

Table 1. *Continued*

Deposit type and deposit	Ore minerals		Gangue minerals		Minerals of the oxidized zone
	Major	Minor	Major	Minor	
Yinmusi	Sphalerite, galena	Pyrite, chalcopyrite, pyrrhotite, arsenopyrite, tetrahedrite, marcasite	Quartz, calcite, ankerite, dolomite	Chlorite, sericite	Cerussite, limonite
Bafangshan	Sphalerite, galena, chalcopyrite	Pyrite, arsenopyrite, pyrrhotite	Quartz, calcite, ankerite	Sericite, chlorite	
Daxigou-Yindongzi	Siderite, magnetite, galena, Ag-bearing galena, sphalerite	Pyrite, chalcopyrite, pyrrhotite, arsenopyrite, tennantite, freibergite, acanthite	Sericite, quartz	Barite, thuringite, albite	Limonite, malachite, martite
Tongmugou	Sphalerite, pyrrhotite	Galena, chalcopyrite, pyrite, tetrahedrite, magnetite, siderite, arsenopyrite	Sericite, quartz	Biotite, muscovite, scapolite, zoisite	
Jinding type					
Jinding	Sphalerite, galena, pyrite marcasite	Hematite, chalcopyrite, freibergite, silver, argentite	Calcite, quartz	Ba-bearing celestite, Sr-bearing barite, dolomite, anhydrite	Limonite, smithsonite, gunningite, cerussite

Table 2. Average contents of minor elements in ore-bearing strata of the three Pb-Zn metallogenic belts in China (ppm)

Metallogenic belt[a]	Age[b]	Cu	Pb	Zn	Ag	Au, ppb	Ba	Sr
Langshan-Chartai Belt								
East part	Pt_2	7.58	17.64	17.6	0.14	0.31	258.6	70.2
Central part	Pt_2	15.4	13.92	32.25	0.08	0.69	404.6	84.7
West part	Pt_2	20.5	23.4	69.3	0.06	0.70	nd	nd
Qinling Belt								
Xihe-Chengxian orefield	D_{2+3}	14	23	57	nd[c]	nd	273	189
Fengxian-Taibai orefield	D_{2+3}	19.5	43.0	71.3	nd	1.54	410.4	255.4
Zhashui-Shan yang orefield	D_{2+3}	20	45	68	0.064–0.31	nd	435	109
Lanping-Simao Belt								
Jinding district	K_1	48	11	76	0.4	0.2–1	296	48
	E	34	62	135	0.4	0.2–1	288	91
Abundance in the upper crust		25	20	71	0.05	1.8	550	350

[a] Data sources: Langshan-Chartai Belt (Inner Mongolia Institute of Geology 1990, unpubl. data); Xihe-Chengxian and Zhashui-Shanyang orefields (Gu and Lin 1990); Fengxian-Taibai orefield (Wei and Lu 1990); Jinding district (Gao 1989); abundances in the upper crust (Taylor and McLennan 1985).
[b] Pt_2 = Middle Proterozoic; D_{2+3} = Middle and Late Devonian; K_1 = Early Cretaceous; E = Eogene.
[c] nd = No data.

been recorded. The other is the Yindongzi Ag-Pb deposit with the lowest Zn value (0.81%) and Zn/Pb ratio (0.35%).

The Jinding-type deposit shows a wide range of Zn/Pb ratios varying with orebodies and ore types.

Regarding the contents of Cu and Ag, Table 3 shows that (1) the Jiashengpan deposit of the Langshan type, the Changba, Qiandongshan, Yinmusi and Tongmugou deposits of the Qinling type as well as the Jinding deposit contain no or very little Cu (0.08% for the Yinmjsi deposit); (2) the Huogeqi, Dongshengmiao, and Bafangshan deposit display average Cu contents >1%; (3) the Yindongzi, Heigou, and Tongmugou deposits, which belong to the Zhashui-Shanyang orefield, have Ag grades of 107.03, 55.02 and 42.35 ppm, respectively; (4) the Jinding deposit, the largest Pb-Zn deposit in China, also has high Ag grades ranging from 2.6 to 100 ppm; (5) the Changba and Qiandongshan deposits, i.e. the two Cu-poor Pb-Zn deposits of the Qinling type, have quite high Ag contents (14.61 and 23.55 ppm, respectively); (6) Ag contents for Langshan-type Pb-Zn-(Cu) deposits are generally very low, but the cuprous pyrite ores of Dongshengmiao have high Ag values, averaging 18.77–39.63 ppm.

Figure 4 shows Cu − Pb + Zn − Ag × 10^3 ratios of sediment-hosted Pb-Zn ores in China and in other countries for reference. Most points are close to the Pb + Zn − Ag × 10^3 line. This means that sediment-hosted Pb-Zn deposits in China are also poor in Cu except for some Langshan- and Qinling-type deposits (e.g. Huogeqi, Dongshengmiao and Bafangshan). Some Qingling-type deposits (e.g. Yindongzi, Yindonggou and Heigou) have high values of Ag ×

Table 3. Zn/Pb ratio and contents of Cu and Ag for sediment-hosted Pb-Zn ores in China

Deposit	Zn/Pb	Cu, wt%	Ag, ppm
I. Langshan type			
Tanyaokou	2.95	0.98	–
Dongshengmiao	3.99	1.18	18.77–39.63
Huogeqi	0.34–1.01	0.56–1.35	3.7–9.0
Jiashengpan	2.93	–	–
II. Qinling type			
Changba	4.96	–	14.61
Bijiashan	2.07	0.65	5.0
Qiandongshan	4.41	–	23.52
Yinmusi	5.19	0.08	–
Bafangshan	2.67	1.17	–
Yindongzi	0.35	0.56	107.03
Heigou	2.06–2.49	0.87	55.02
Tongmugou	10.96		42.35
III. Jinding type			
Jinding	0.29–17.2	–	2.6–100

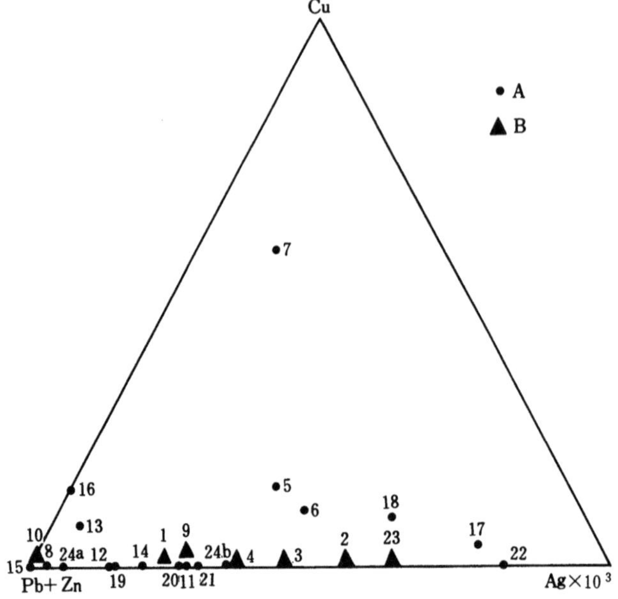

Fig. 4. Weight ratios of Cu, Pb + Zn and Ag × 10^3 of sediment-hosted Pb-Zn deposits in the world. *1* McArthur River; *2* Mt. Isa; *3* Broken Hill; *4* Sullivan; *5* Dongshengmiao; *6* Tanyaokou; *7* Huogeqi; *8* Jiashengpan; *9* Rammelsberg; *10* Meggen; *11* Yeshuihe; *12* Changba; *13* Bijiashan; *14* Qiandongshan; *15* Yinmusi; *16* Bafangshan; *17* Yindongzi; *18* Heigou; *19* Tongmugou; *20* Xidonggou; *21* Yuexigou; *22* Yindonggou; *23* Largentiere; *24* Jinding. *a* Sandstone-hosted ore; *b* calcirudite-hosted ore. *A* Chinese deposits; *B* Foreign deposits

Table 4. Average contents and ratios of minor elements for pyrite from several Chinese sediment-hosted Pb-Zn deposits (in ppm)

Deposit No.[a]	n[b]	Co	Ni	Co/Ni	S%	Se	S/Se	Ag
1	21	330	320	1.03	51.47	5	10.3×10^4	nd
2	5	32	128	0.25	54.06	1.9	28.5×10^4	7.4
3	5	114	207.6	0.55	51.27	4.3	11.9×10^4	nd
4	13	14.9	23.1	0.64	57.35	2.96	19.9×10^4	2.11
5	30	42.3	193.3	0.22	52.34	2.5	21.0×10^4	nd
6	2	60	380	0.15	52.66	nd	10.0×10^4	nd
7	5	173.8	181.2	0.96	47.83	2.8	17.0×10^4	nd
8	3	nd	nd	0.07	53.00	1	53.0×10^4	nd
9		91.1	127.6	0.7	nd	nd	53.0×10^4	12.64

[a] 1, Tanyaokou; 2, Dongshengmiao; 3, Huogeqi; 4, Jiashengpan; 5, Bijiashan; 6, Dengjiashan; 7, Qiandongshan; 8, Bafangshan; 9, Yindongzi. Data sources: 1 + 3, Li et al. (1981, unpubl.); 2, Xia (1990); 4, Lang and Zhang (1987); 5 + 7, Zhu (1989, unpubl.); 6, Li (1984, unpubl.); 8, Lü and Wei (1990); 9, Hu (1989).
[b] n = Number of samples averaged; nd = not determined or no data.

10^3. The calcirudite-hosted ore of the Jinding deposit has a higher Ag $\times$ 10^3 value than the sandstone-hosted ore.

4.3 Contents and Ratios of Minor Elements in Pyrite

Table 4 reports Co, Ni, S, Se, and Ag contents as well as Co/Ni, and S/Se ratios for pyrite from several Chinese sediment-hosted Pb-Zn deposits.

Although pyrite shows a broad range of Co (14.9–330 ppm) and Ni (23.1–380 ppm) contents, its Co/Ni ratio varies slightly, and is generally lower than 1. The low Co/Ni ratio indicates a sedimentary origin for pyrite from sediment-hosted Pb-Zn deposits (Song and Zhang 1986). The low Se content, ranging from 1 to 5 ppm, is additional evidence for a sedimentary origin of pyrite. Because of the low Se content, the S/Se ratios in pyrite are very high, ranging from 10×10^4 to 53×10^4. The few available data indicate that the Ag content of pyrite from these deposits is relatively low (2.11–12.64 ppm).

4.4 Contents and Ratios of Minor Elements in Sphalerite

Table 5 shows the Fe, Cd, Ga, Ge, In, and Ag average contents, as well as Ga/In and Ge/In ratios in sphalerite from several sediment-hosted Pb-Zn deposits in China. The following conclusions can be drawn: (1) sphalerite (marmatite) from the Jiashengpan Pb-Zn deposit has the highest Fe average content (10.36%). The average content of Fe in sphalerite from Qinling-type Pb-Zn deposits ranges between 2 and 5%. (2) Sphalerite from most Chinese sediment-hosted Pb-Zn deposits shows a range of 1200–2300 ppm Cd. Only Jinding sphalerites show higher contents, ranging from 2700 to 8400 ppm. (3) Sphalerites from Qinling-type deposits have the highest Ga content (17–

Table 5. Average contents and ratios of minor elements for sphalerite from several Chinese sediment-hosted Pb-Zn deposits (in ppm)

Deposit No.[a]	n[b]	Fe%	Cd	Ga	Ge	In	Ag	Ga/In	Ge/In
1	6	10.36	1198–1360	nd	nd	nd	1.58		
2	16	3.54	1420	3.4	1.4	3.2	35	1.06	0.44
3	4	5.81	1225	nd	nd	0.3	29.2		
4	4	1.7	1940	50.8	55.5	2.6	nd	19.5	21.35
5	14	nd	1875	35.9	9.8	10	nd		
6	20	3.09	2333	26.4	35.3	3	44.4	8.8	11.7
7	8	5.04	nd	25.38	7.92	4.25	nd	5.97	1.86
8	3	3.50	1443	22	5.7	1	44.1	22.0	5.7
9		4.19	nd	17	7	1	nd	17	7
10		nd	2700–8400	nd	nd	nd	59.5		

[a] 1, Jiashengpan; 2, Changba; 3, Lijiagou; 4, Dengjiashan; 5, Bijiashan; 6, Qiandongshan; 7, Bafangshan; 8, Yinmusi; 9, Yindongzi; 10, Jinding. Data sources: 1, Lang and Zhang (1987); 2 + 3 + 6 + 8, Li H (1986); 4 + 9, Zhu (1989, unpubl.); 5, Li S (1984, unpubl.); 7, Lü and Wei (1990); 10, Zhao et al. (1980, unpubl.); 11, Hannak (1981).
[b] n = Number of samples averaged; nd = not determined or no data.

Table 6. Contents of organic carbon and boron for host rocks within and outside sediment-hosted Pb-Zn deposits in China

Deposit and orefield[a]	Organic carbon, wt%[b]		Boron, ppm
	Range	Mean	Mean
1. Tanyaokou	0.1–0.39	0.22 (5)	74.39 (28)
2. Dongshengmiao	0.13–0.68	0.19 (12)	197.16 (25)
3. Huogeqi		0.10 (12)	119.33 (18)
4. Xihe-Chengxian orefield			
a) Lijiangou (L)	0.12–0.89	0.49 (12)	nd[c]
b) Bijiashan (B)	0.10–0.77	0.47 (93)	nd
c) Outside the L + B deposit	0.06–0.09	0.08 (8)	nd
5. Fenxian-Taibai orefield			
a) Qiandongshan (Q)	0.09–0.19	0.15 (28)	146.52 (29)
b) Yinmusi (Y)	0.13–0.26	0.19 (17)	nd
c) Outside the Q + Y deposit	0.13–0.17	0.15 (20)	75 (4)
6. Zhashui-Shanyang orefield			
a) Daxigou (D)	0.019–1.050	0.47 (4)	134.08 (12)
b) Yindongzi (YD)		0.46 (15)	nd
c) Tongmugou (T)		0.97 (9)	nd
d) Outside the D + YD + T deposit	0.16–0.37	0.25 (8)	91.18 (17)
7. Jinding (J)	0.28–2.95	1.63 (23)	nd
Outside the J deposit	1.12–3.52	2.91 (6)	nd
8. Shales		0.80	100
9. Sandstones			35
10. Carbonate rocks			20

[a] Data sources: 1 to 3, Li et al. (1981, unpubl.); 4 to 6, Gu and Lin (1990) and Chen and Lin (1990); 7, Hu (1988) and (1989); 8 to 10, Pettijohn (1975) and Turekian and Wedepohl (1961). Langshan-Chartai Metallogenic Belt = 1 + 2 + 3; Qinling Metallogenic Belt = 4 + 5 + 6; Lanping-Simao Metallogenic Belt = 7.
[b] In parentheses: number of samples averaged.
[c] nd = No data.

50 ppm) and quite low In content (1–4.25 ppm). As a result, they have a high Ga/In ratio, generally ranging between 6 and 22. (4) Qinling sphalerites have a wide range of Ge contents (1.4–55.5 ppm). Their Ge/In ratio ranges, in general, from 1.86 to 21.35. (5) As for Ag contents, Jinding sphalerites display the highest Ag average content (59.5 ppm), higher than Qinling sphalerites (30–45 ppm) and Jiashengpan sphalerites (1.58 ppm).

4.5 Organic Carbon

Several researchers have emphasized the role of organic matter in the formation of mineral deposits (e.g. Hu 1989, Kribek 1991; Sawlowicz 1991). Specifically, Hu (1989) reported analyses of organic carbon in the host rocks of sediment-hosted Pb-Zn deposits in China (Table 6). From his work the following conclusions can be drawn:

1. In comparison with Jinding- and Qinling-type deposits, the Langshan-type deposits have rather low contents of organic C, averaging 0.1–0.22%.
2. For the Qinling-type deposits, such as the Lijiagou, Bijiashan, Daxigou, Yidongzi, and Tongmugou deposits, the average contents of organic C of host rocks within the deposits are one to five times higher than those of corresponding rocks around and outside the deposits. This may imply that some relation exists between the enrichment of metals and organic C.
3. By contrast, for the Jinding deposit, the average content of organic C (1.63%) of host rocks within the deposit is much lower than that (2.91%) of corresponding rocks around the deposit. Hu (1989) ascribed this to depletion of organic C in host rocks during the formation of the ore deposit.

4.6 Boron

Table 6 shows that host rocks of sediment-hosted Pb-Zn deposits in China yield relatively high average contents of B, ranging from 74–197 and 134–147 ppm for the Langshan-Chartai and the Quinling metallogenic Belts, respectively.

As in the case of organic C, the B contents of host rocks from these deposits are higher than those from around them. This may suggest that B is an ore indicator.

5 S, and Pb-Isotope Compositions of Sediment-Hosted Pb-Zn Deposits in China

5.1 S-Isotope Composition

The S-isotope composition of sulphides from the considered deposits is reported in Table 7. Pyrite from Langshan-type Pb-Zn deposits has the highest $\delta^{34}S$, suggesting that S was derived from seawater sulphate. However, $\delta^{34}S$

Table 7. $\delta^{34}S$ values for some sulphide minerals from sediment-hosted Pb-Zn deposits in China (in ‰)

Deposit[a]	Pyrite		Sphalerite		Galena	
	Range	Mean	Range	Mean	Range	Mean
1. Tanyaokou	23.5 to 37.3	31.54				
2. Dongshengmiao	26.0 to 35.1	32.33				
3. Huogeqi		10.88		16.50		14.16
4. Jiashengpan		26.02		23.40		20.34
5. Changba		22.50		19.30		17.70
6. Qiandongshan		8.40	5.3 to 10.0	7.90	0.6 to 4.90	2.80
7. Yinmusi			9.7 to 12.2	11.40	5.9 to 9.60	7.90
8. Bafangshan		6.50	6.0 to 13.9	9.60	0.6 to 4.90	2.80
9. Daxigou-Yindongzi	10.3 to 17.9	13.50	11.7 to 14.7	13.20	5.5 to 18.5	12.80
10. Tongmugou		17.50	16.6 to 19.9		15.50 to 18.6	
11. Jinding	−3.84 to −21.4		−1.7 to 14.6		−0.29 to −30.4	

[a] Data sources: 1 + 2, Li Z (1986); 3, Jiang (1983); 4, Lang and Zhang (1987); 5, Yang and Miao (1986); 6 + 7, Wei and Lu (1990); 8, Lü and Wei (1990); 9, Du (1986); 10, Wang (1987); 11, Shi (1983, unpubl.).

values vary within various deposits, i.e. pyrite from Tanyaokou, and Dongshengmiao deposits show the highest $\delta^{34}S$ values (approximately 32‰) and that from the Huogeqi deposit, the lowest (approximately 11‰). According to Sangster's (1976) discussion on the influence of open and restricted circulation on the isotopic composition, the two former deposits could have been formed in a closed to semi-closed sedimentary basin, where circulation of seawater was restricted, and sulphate left in solution was progressively enriched in the heavy S-isotope due to the lack of supply of new ocean water. As a result, pyrite produced by the reduction in sulphates is rich in $\delta^{34}S$. By contrast, the Huogeqi deposit was formed in a relatively open basin, where sulphates might be supplemented by new ocean water. Consequently, pyrite produced by the reduction in sulphates in the latter deposit has a lower $\delta^{34}S$ value than that from Tanyaokou and Dongshengmiao (Li Z 1986).

Pyrite from Qinling-type deposits shows rather low $\delta^{34}S$ values (6.5–22.5‰), suggesting that it was formed in relatively open basins. This has been confirmed by research on sedimentary facies and environments of the Devonian in the Qinling region (Du 1986).

As far as the Jinding deposit is concerned, the negative values and the wide range of $\delta^{34}S$ (−0.29 to −30.4‰) are an indication of biogenetic processes, including bacterial reduction of sulphates and action of other biogenic sulphurs in host rocks during diagenesis and epigenesis.

In addition, Table 7 shows that in several Chinese sediment-hosted Pb-Zn deposits the expected trend $\delta^{34}S_{pyrite} > \delta^{34}S_{sphalerite} > \delta^{34}S_{galena}$ is found. The trend indicates that S-isotope fractionation approaches equilibrium, in agreement with observations of mineral intergrowths of these deposits.

Table 8. Pb isotope ratios for galena from Chinese sediment-hosted Pb-Zn deposits

Deposit[a]	$^{206}Pb/^{204}Pb$[b]	$^{207}Pb/^{204}Pb$	$^{208}Pb/^{204}Pb$
Langshan type			
1. Tanyaokou	15.434 (1)	15.105 (1)	35.338 (1)
2. Dongshengmiao	15.166 – 15.656	14.91 – 15.422	35.30 – 36.682
	15.275 (8)	15.150 (8)	35.726 (8)
3. Huogeqi	16.016 – 17.193	14.774 – 15.491	33.849 – 36.945
	16.866 (9)	15.331 (9)	36.343 (9)
4. Jiashengpan	15.889 – 16.081	15.158 – 15.455	35.112 – 36.271
	15.986 (5)	15.363 (5)	35.831 (5)
Qinling type			
5. Changba-Lijiagou	17.884 – 18.459	15.420 – 15.874	38.007 – 38.373
	18.049 (9)	15.606 (9)	38.184 (9)
6. Qiandongshan	18.0349 – 18.1899	15.5829 – 15.7750	38.0739 – 38.7216
	18.1072 (7)	15.6809 (7)	38.4318 (7)
7. Bafangshan	18.0179 – 18.1494	15.5567 – 15.8406	37.9943 – 38.6565
	18.0813	15.6481 (4)	38.2455 (4)
8. Yindongzi	18.001 – 18.315	15.550 – 15.851	37.965 – 38.935
	18.118 (8)	15.645 (8)	38.300 (8)
9. Tongmugou	18.014 – 18.134	15.553 – 15.694	37.946 – 38.420
	18.063 (5)	15.601 (5)	38.166 (5)
Jinding type			
10. Jinding	18.1729 – 18.4018	15.3454 – 15.5487	37.8031 – 38.3572
	18.2493 (21)	15.4470 (21)	38.0739 (21)

[a] Data sources: 1 + 2 + 3, Li Z (1986); 4, Lang and Zhang (1987); 5, Yang and Miao (1986); 6, Dai (1987, unpubl.); 7, Lü and Wei (1990); 8, Chen and Lin (1990); 9, Wang (1987); 10, Wang (1984).

[b] In parentheses: number of samples averaged; $\frac{15.166 - 15.656}{15.275\ (8)} = \frac{\text{range}}{\text{mean}}$.

5.2 Pb-Isotope Composition

The Pb isotope ratios for galena from several Chinese sediment-hosted Pb-Zn deposits have been investigated by many researchers, (Table 8). Table 8 also shows:

1. All the $^{206}Pb/^{204}Pb$, $^{207}Pb/^{204}Pb$, and $^{208}Pb/^{204}$ ratios in galenas from Tongmugou, Changba-Lijiagou, and Jinding deposits display a maximum range of less than 0.6. This indicates a derivation of lead largely from a single source.
2. $^{206}Pb/^{204}$, and $^{207}Pb/^{204}Pb$ ratios for galena from Yindongzhi, Bafangshan, Qiandongshan, Dongshingmiao, and Jiashengpan deposits exhibit a narrow range, with maximum variations of less than 0.6, but the $^{208}Pb/^{204}Pb$ ratios

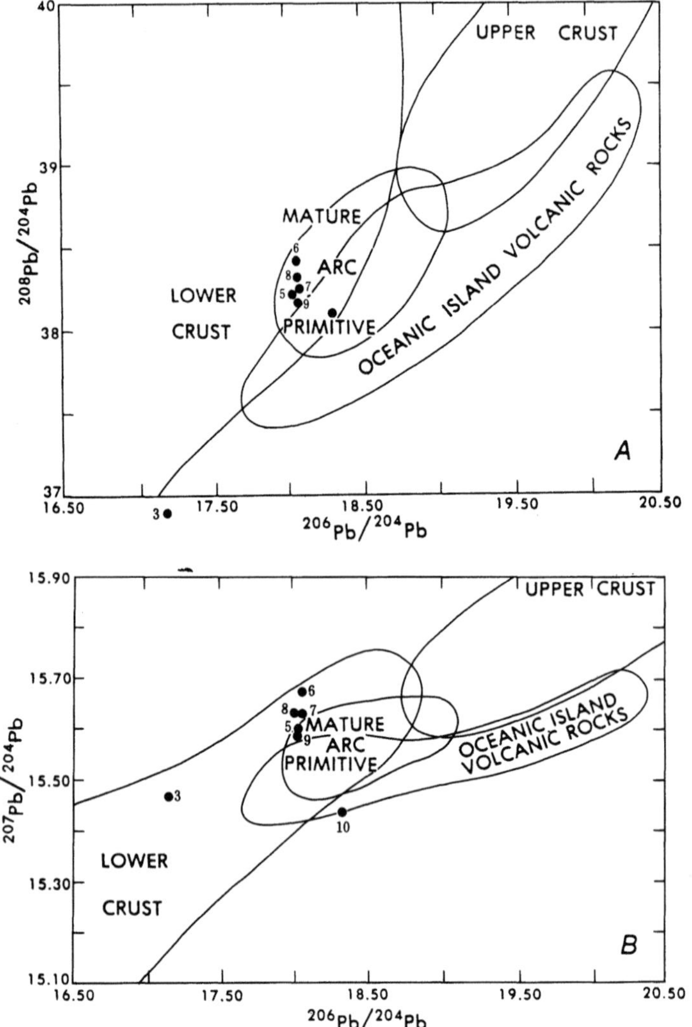

Fig. 5. Summary plots of lead isotopic ratios for galena from investigated Chinese sediment-hosted Pb-Zn deposits. The *points* represent average values for each deposit. The *numbers* correspond to deposit numbers in Table 8. *Solid lines* delimit the geologic environments. (After Zartman and Doe 1981)

span a broad range, up to 0.6–1.4. This implies that the source of ^{207}Pb and ^{206}Pb is relatively homogeneous, but that of ^{208}Pb is multiple, i.e. one source had very high Th/U.

3. The broad range of Pb isotopic ratios for Huogeqi galenas is indicative of complex sources of Pb, including Pb from country rocks as well as magmatic Pb.

The average ratios of Pb isotopes for galena from Chinese sediment-hosted Pb-Zn deposits are plotted in Fig.5. The diagram shows that the value for the Huogeqi deposit is within the lower crust field; the values for the Qinling-type deposits are enclosed in the orogene field, indicating some mixing and the Jinding deposit is plotted in the mantle field.

The mantle affinity of Jinding is consistent with its tectonic setting in the Lancang River deep fault zone.

6 Comparisons with Some Similar Sediment-Hosted Pb-Zn Deposits in the World

Characteristics such as age, tectonic setting, host rocks, organic association with organic compounds, ore structure, metal association, Zn/Pb ratio, Cu and Ag contents and S and Pb isotopes for selected sediment-hosted Pb-Zn deposits in China and other countries are listed in Table 9. The following conclusions can be drawn:

1. Most Chinese sediment-hosted Pb-Zn deposits occur in pericratonic basins, whereas most of those outside China typically occur in intracratonic basins. The Jinding deposit, the largest Pb-Zn deposit in China, is peculiar, as it is situated in a back-arc rift environment near the suture between two plates. In addition, both Chinese and foreign sediment-hosted Pb-Zn deposits are related to syndepositional faults.
2. In terms of ages, the Chinese deposits range from Proterozoic to Devonian, Early Cretaceous and Eocene. The most important foreign deposits are of Proterozoic age.
3. The association with organic carbon is a common feature for both the Chinese and foreign deposits.
4. Ore structures of the Chinese deposits are generally similar to those of the corresponding deposits in other countries, but the Jinding ores have complex structures, such as stockwork and brecciated ores.
5. There is an apparent difference in the S-isotope composition between the Chinese and foreign Pb-Zn deposits of Proterozoic age, as the Chinese sulphide ores are richer in heavy sulphur with respect to deposits such as McArthur River and Sullivan. In particular, all sulphide minerals (galena, sphalerite, pyrite and pyrrhotite) from Sullivan have slightly negative $\delta^{34}S$ averages (−3.6, −0.6, −2.6 and −1.9‰, respectively, Campbell et al. 1978). The S-isotope composition of the Chinese deposits of Devonian age is similar to that of Rammelsberg ores. However, the former deposits have a narrower range and lower averages than the latter. In Jinding deposit, the youngest sediment-hosted Pb-Zn deposit in the world, is unparalleled with its high negative values and wide range of $\delta^{34}S$ values (−0.3 to −30.4‰).
6. The Pb isotope composition in the selected foreign sediment-hosted Pb-Zn deposits is characterized by homogeneity, whereas most Chinese deposits show inhomogeneity and multiple sources for Pb. The Pb isotopic homogeneity of the Jinding deposit may be considered to have a deep and single source with mantle affinity.

Table 9. Characteristics of selected sediment-hosted Pb-Zn deposits

Deposit or deposit type	Langshan type	McArthur River	Sullivan	Qinling type	Rammelsberg	Jinding	Largentière
Age	Middle Proterozoic (1512–1612 Ma)	Middle Proterozoic (1650 Ma)	Middle Proterozoic (−1430 Ma)	Middle-Upper Devonian	Middle Devonian	Early Cretaceous to Eocene	Early Triassic
Tectonic setting	Aulacogen at passive continental margin of the North China Platform; syndepositional fault	Intracratonic basin; deposit located in rifted trough next to marginal syndepositional fault	Early stages of intracratonic basin; near major faults; breccia in footwall	Graben basins at active continental margin in collision zone between plates; synsedimentary faults	Intracratonic basin; synsedimentary faulting	Back-arc rift valley next to suture line between plates; deposit located near a thrust	Near structural and depositonal margin of intracratonic basin; over local margin of Permian red bed graben; syndepositional faulting
Host rocks	Carbonaceous silty slate, marl, argillaceous dolostone, micaquartz schist	Carbonaceous, dolomitic, pyritic siltstone, dolostone, slump breccias and turbidites	Argillite, siltstone, conglomerate	Calcareous phyllite, impure crystalline limestone, ferruginous dolostone, silica rock micaschist, albite- and scapolite-bearing hornfels	Black shale, interbedded sandstone and tuff	Calcareous quartz sandstone and calcirudite; footwall and hanging wall red beds	Arkosic sandstone to conglomerate, partly fluvial; footwall red conglomerate to shale and hanging wall shallow water marl to siltstone
Organic association	0.1–0.7% (averaging 0.2%) $C_{org.}$ in host rocks	0.4% $C_{org.}$ in ore; 1% in unmineralized equivalents	Sparse graphite in host rocks	0.1–1.0% $C_{org.}$ within deposits; 0.05–0.4% outside deposits	Carbonaceous shale host	0.3–3.0% $C_{org.}$ within deposit; 1.1–3.5% outside deposit	Dark argillaceous hanging wall sediments; low $C_{org.}$ in ore sandstone
Ore structure	Banded, lamellar, disseminated, massive	Banded	Massive, banded	Banded, lamellar, massive, disseminated	Massive, banded, disseminated	Disseminated, massive, stockwork, brecciated, banded	Disseminated, veins

Metal association	Zn-Pb ± Cu ± Ag	Zn-Pb-Ag-Cu	Pb-Zn-Ag-Sn-Cu	Zn-Pb-Ag ± Cu	Zn-Pb-Cu-Ag	Zn-Pb-Sr-Ba-Ag	Pb-Zn-Ag
Zn/Pb ratio	3–4, except for Huogeqi	2.5	0.9	2–5, except for Yindongzi and Tongmugou	2.1	0.3–17	0.2
% Cu	0.6–1.4	0.2	0.1	0.1–1.2	1	–	–
g/t Ag	4–40	45	68	5–55, rarely 107	103	2.6–100	80
S-isotope for sulphides, $\delta^{34}S$ %	11–37; limited range, peaked shape, strongly positive, average values in general	−3.9 to +14.3; limited range, peaked shape, slightly positive, average values	−10.4 to +4.7; limited range, peaked shape, averaging −2.2	0.6 to 27.8; quite wide range, peaked shape, averaging 15–25	−13 to +21, wide range, peaked shape, averaging 15–20	−0.3 to −30.4 wide range, scattered	
Pb isotope ratios:							
206/204	15.17–17.19	16.07–16.16	16.526	17.88–18.46	18.23–18.26	18.17–18.40	
207/204	14.77–15.49	15.37–15.48	15.504	15.42–15.87	15.61–15.63	15.35–15.55	
208/204	33.85–36.95	35.57–35.90	36.195	37.95–38.94	38.17–38.22	37.80–38.36	
	Nonhomogeneous, multiple sources	Homogeneity, suggesting either homogeneous source or thorough mixing of the Pb	Homogeneous	Relatively homogeneous, not excluding multiple sources	Homogeneous and single source	Homogeneity, suggesting either homogeneous source or thorough mixing of the Pb	
Major references	Song and Cao (1990); Li Z (1986); Qiao et al. (1991); Lang and Zhang (1987)	Gustafson and Williams (1981); Lambert (1976)	Gustafson and Williams (1981); Campbell et al. (1978, 1980); Cumming and Richards (1975)	Zhang et al. (1987); Yang and Miao (1986); Gu and Lin (1990); Song and Cao (1990); Lü and Wei (1990); Chen and Lin (1990)	Gustafson and Williams (1981); Hannak (1981); Wedepohl et al. (1978)	Bai et al. (1985); Hu (1988); Wang (1984); Tan (1991)	Gustafson and Williams (1981); Samama (1976)

Acknowledgements. Prof. P. Lattanzi, Prof. L. Fontboté and Dr. A. Jiang are gratefully acknowledged for critically reviewing the manuscript and suggesting many worthwhile improvements. The author is also thankful to Prof. G.C. Amstutz and Dr. R.A. Zimmermann for helpful discussion and language improvement of this paper. The Alexander von Humboldt Foundation, FRG, is thanked for generously providing a research fellowship for the author's stay in Heidelberg where he continued to work on this paper. Sincere gratitude is due to Mrs. I. Beck of the Institute of Mineralogy and Petrography, University of Heidelberg for help in typing the revised manuscript.

References

Bai J, Wang C, Na R (1985) Geological characteristics of the Jinding lead-zinc deposit in Yunnan with a special discussion on its genesis. Mineral Depos 4(1):1–9 (in Chinese)

Cambpell FA, Ethier VG, Krouse HR, Both RA (1978) Isotopic composition of sulphur in the Sullivan orebody, British Columbia. Econ Geol 73:248–268

Campbell FA, Ethier VG, Krouse HR (1980) The massive sulfide zone: Sullivan orebody. Econ Geol 75:916–926

Chen D, Lin B (1990) A geochemical study of metallogenesis of the Devonian stratiform-stratabound Pb-Zn ore belt in South Qinling. In: Zang B (ed) Contributions to regional geochemistry of Qinling and Dabashan Mountains. China Geoscience University Press, Wuhan, pp 168–191 (in Chinese)

Cumming GL, Richards JR (1975) Ore lead isotope ratios in a continuously changing Earth. Earth Planet Sci Lett 28:155–171

Du D (1986) Research of the Devonian system of the Qinling-Bashan region within the territory of Shaanxi. Xi'an Chiao-Tung University Press, Xian (in Chinese)

Gao J (1989) Preliminary study of the genetic relationship between the evaporite formation and lead-zinc deposit of Jinding in West Yunnan Province, China. Earth Sci 14:513–521 (in Chinese)

Gu X, Lin B (1990) Geochemistry of the Devonian ore-bearing strata from the South Qinling. In: Zhang B (ed) Contributions to regional geochemistry of Qinling and Dabashan Mountains. China Geoscience University Press, Wuhan, pp 153–167 (in Chinese)

Gustafson LB, Williams N (1981) Sediment-hosted stratiform deposits of copper, lead, and zinc. Econ Geol 75th Anniversary Vol: 139–178

Hannak WW (1981) Genesis of the Rammelsberg ore deposit near Goslar/Upper Harz, FRG. In: Wolf KH (ed) Handbook of strata-bound and stratiform ore deposits, vol 9. Elsevier, Amsterdam, pp 551–642

Hu M (1988) A preliminary study of the organic geochemistry of the stratabound deposits. Earth Sci 13:195–202 (in Chinese)

Hu M (1989) Hydrothermal maturation of indigenous organic matters and their significance in the metallogenic processes of the Jinding Pb-Zn deposit, Yunnan Province. Earth Sci 14:503–512 (in Chinese)

Jiang X (1983) A discussion on the genesis and ore-forming mechanism of the Huogeqi Cu-Pb-Zn deposit. Mineral Depos 2(4):1–9 (in Chinese)

Kribek B (1991) Organic matter of syngenetic and epigenetic uranium deposits in the Bohemian Massif. In: Pagel M, Leroy JL (eds) Source, transport and deposition of metals. Balkema, Rotterdam, pp 539–544

Lambert IB (1976) The McArthur zinc-lead-silver deposit: features, metallogenesis and comparison with some other stratiform ores. In: Wolf KH (ed) Handbook of strata-bound and stratiform ore deposits, vol 6. Elsevier, Amsterdam, pp 535–585

Lang D, Zhang X (1987) Geological setting and genesis of the Jiashengpan Pb-Zn-S ore belt, Inner Mongolia. Mineral Depos 6(2):39–54 (in Chinese)

Li H (1986) Minor elements in sphalerite and their significance in geology. Geol Prospect 22(10): 42–46 (in Chinese)

Li Z (1986) S, Pb, C, and O isotopic compositions and ore genesis of the stratabound polymetallic sulfide deposits in Middle Inner Mongolia, China. Geochimica 1:13–22 (in Chinese)

Lü R, Wei H (1990) The geological characteristics and genetic investigation of Bafangshan stratabound polymetallic ores in Shaanxi Province. J Xian Coll Geol 12(4):10–17 (in Chinese)

Pettijohn FJ (1975) Sedimentary rocks. Harper & Row, New York; Evanstone, San Francisco

Qiao X, Yiao P, Wang Ch, Tan L, Zhu S, Zhou S, Zang Y (1991) Sequence stratigraphy and tectonic environment of the Chartai group, Inner Mongolia. Acta Geol Sin 65(1):1–15 (in Chinese)

Samama JC (1976) Comparative review of the genesis of the copper-lead sandstone-type deposits. In: Wolf KH (ed) Handbook of strata-bound and stratiform ore deposits, vol 6. Elsevier, Amsterdam, pp 1–18

Sangster DF (1976) Sulfur and lead isotope in strata-bound deposits. In: Wolf KH (ed) Handbook of strata-bound and stratiform ore deposits, vol 2. Elsevier, Amsterdam, pp 219–226

Sawlowicz Z (1991) The relationship between copper mineralization and organic matter in the Kupferschiefer. In: Pagel M, Leroy JL (eds) Source, transport and deposition of metals. Balkema, Rotterdam, pp 539–554

Song X, Cao Y (1990) Mineralogy and geochemistry of sediment-hosted Pb-Zn deposits in China. In: Abstr 13th Int Sedimentological Congr, 26–31 Aug 1990, Nottingham, p 514

Song X, Zhang J (1986) Minor elements in pyrites of various genetic types from China. Bull Inst Mineral Depos CAGS 18:166–175 (in Chinese)

Tan X (1991) Metallogenic geological setting of major nonferrous metal deposits in western Yunnan. Yunnan Geol 10(1):11–43 (in Chinese)

Taylor SR, McLennan SM (1985) The continental crust: its composition and evolution. Blackwell, London, pp 45–49

Turekian KK, Wedepohl KH (1961) Distribution of the elements in some major units of the earth's crust. Sci Am Bull 72(2):175–192

Wang Q (1987) Geological characteristics and genesis of Tong-Mugou Zn deposits, Shanyang, Shaanxi. Contrib Geol Mineral Resour Res 2(2):54–64 (in Chinese)

Wang W (1984) Pb isotope data. Geol Yunnan 3(3):294–309 (in Chinese)

Wedepohl KH, Delevaux MH, Doe BR (1978) The potential source of lead in the Permian Kupferschiefer beds of Europe and some selected Paleozoic mineral deposits in the Federal Republic of Germany. Contrib Mineral Petrol 65:273–281

Wei H, Lu R (1990) An exploratory study of the source of mineralized materials in the stratabound Pb-Zn-(Cu) ore deposits of the Fengtai ore field in the Qinling Montains. J Xian Coll Geol 12(4):10–17 (in Chinese)

Xia X (1990) Genetic mineralogy research for pyrite of polymetallic sulfide deposit in Dongshengmiao, Inner Mongolia. Mineral Resour Geol 4(4):47–53 (in Chinese)

Yang S, Miao Y (1986) Geological characteristics of the Changba-Lijiagou Pb-Zn deposit in Gansu province. Mineral Depos 5(2):15–23 (in Chinese)

Zartman RE, Doe BR (1981) Plumbotectonics – the model. In: Zartman RE, Taylor SR (eds) Evolution of the upper mantle. Tectonophysics 75. Elsevier, Amsterdam, pp 135–162

Zhang H, Li F, Sun N (1987) The basic geological features and genesis of Qinling type of stratabound lead-zinc sulfide deposit. Bull Xian Inst Geol Mineral Resour CAGS 19:83–96 (in Chinese)

Peridiapiric Metal Concentration: Example of the Bou Grine Deposit (Tunisian Atlas)

J.J. Orgeval

Abstract

The Bou Grine deposit, at the edge of the Jebel Lorbeus diapir in the Tunisian Atlas, provides a good example of the close association between mineralization and diapirism. With a geological potential of about 1 Mt (million tonnes) Zn metal, it shows three main economic ore types: (1) lenticular orebodies in a Triassic-Cretaceous transition zone; (2) stratiform mineralization in organic-rich limestone laminites of the Cenomanian-Turonian Bahloul Formation; and (3) high-grade orebodies (>25% Zn + Pb) cutting the Cenomanian to middle Turonian series. The deposit shows the close association between a semi-massive orebody, resulting from a complex structural evolution of the area, and a host layer showing early metallic mineralization. The location of the deposit on a high within a depression is paradoxical but very characteristic of its association with diagenetic processes.

Lead and strontium isotope studies reveal fundamental characteristics related to the role of circulating fluids in such particular areas. A narrow range of lead isotope compositions at the deposit indicates a single metal source, whereas an Sr isotope study shows anomalous $^{87}Sr/^{86}Sr$ values in carbonates from the mineralized zone and the ore deposit itself.

The Bou Grine deposit thus provides a very good model for understanding the mechanisms and processes of Zn-Pb metal concentration in a diapiric setting. The economic importance of this kind of deposit would appear to be underestimated and a closer examination of areas affected by diapirism near lead-zinc provinces is obviously warranted.

1 Introduction

The economic interest of diapiric structures is well known and has often been described in relation to salt, sulphur and petroleum products (oil and gas). The relationship between Pb-Zn mineralization and massive salt domes is not so well known, although European authors pointed out this possibility more than a century ago, especially for North Africa (Levat 1894), and in the 1930s, diapir-related Pb-Zn-Ba occurrences were revealed during exploration drilling for oil and/or sulphur (Hanna and Wolf 1934).

BRGM, Avenue de Concyr, BP 6009, 45060 Orléans cedex 2, France

Spec. Publ. No. 10 Soc. Geol. Applied to Mineral Deposits
Fontboté/Boni (Eds.), Sediment-Hosted Zn-Pb Ores

Metallogenic investigations in northern Tunisia, carried out since 1978 as part of a joint ONM (Office National de Mines de Tunisie) and BRGM mineral exploration program, have revealed an apparently new economic base-metal potential, spatially close to diapirs but hosted by the nearby cover rocks. The host rocks of this "new" mineralization show features that reflect the "diapir effect" during their sedimentation and diagenesis, i.e. a strong structural, sedimentological and geochemical control. Exploration of these new targets resulted, in 1981, in the discovery of the Bou Grine deposit (Orgeval et al. 1981) whose economic potential, using a cutoff of 4% Zn, was estimated in 1985 at 7.3 Mt assaying 2.4% Pb and 9.7% Zn, or 880 000 t of metal. Such figures put Bou Grine among the major Tunisian Pb-Zn deposits and among the second ranking Pb-Zn deposits of the Maghreb (after the Touissit-Bou Becker district in Morocco/Algeria). In 1992, the future exploitation company (Société Minière de Bou Grine) announced estimated mineable reserves of 5 Mt at 2.6% Pb and 11.7% Zn (Mining Journal 1992).

Several publications have proposed syntheses and genetic models for the diapir/base-metal deposit association (Price et al. 1983; Orgeval et al. 1985a; Rouvier et al. 1985; Kyle and Price 1986; Machel 1989). Our contribution is based on Bou Grine and its immediate environment; it presents the major features of this particular type of peridiapiric deposit, and proposes a metallogenic model based on multi-disciplinary studies carried out since 1981.

2 Regional Geology and Mineralization

The Bou Grine deposit lies at the edge of the Lorbeus diapir in the Domes area (southern Tunisian Atlas) which, oriented NE-SW and covering almost 8000 km^2 (Figs. 1 and 2), is characterized by a large number of Triassic diapirs aligned parallel to the regional folding and piercing Cretaceous cover rocks. The Domes area is separated from the Nappes zone to the northwest by a belt that still shows signs of tangential tectonism (the Medjerda thrusts), and is bordered by the Central Tunisian Trough area (Tertiary grabens) and the central Tunisian carbonate platform to the southeast.

Most of the Domes area, which formed a trough during Early Cretaceous times and remained one of the permanent features of a depositional structure resulting from deep basement movements, is covered by Cretaceous pelagic sedimentary rocks, in places overlain by Tertiary phosphate-bearing formations that were deposited in elongate basins. These rocks are pierced by Triassic diapirs derived from thick salt formations that include psammitic clays, psammites, mottled clays, carbonate lenses and evaporites. The Jurassic has only rarely been observed and is nowhere seen overlying the diapirs.

The Cretaceous shows a major discontinuity in the upper part of the Albian, enabling it to be divided into: (1) a megasequence extending from the Early Cretaceous to the late Albian, and (2) a megasequence extending from the terminal Albian (Vraconian) to the Palaeocene. Figure 3 shows (after Marie et al. 1984) the north-south variations of the major lithological units of the second megasequence, with strong facies and thickness variations close to

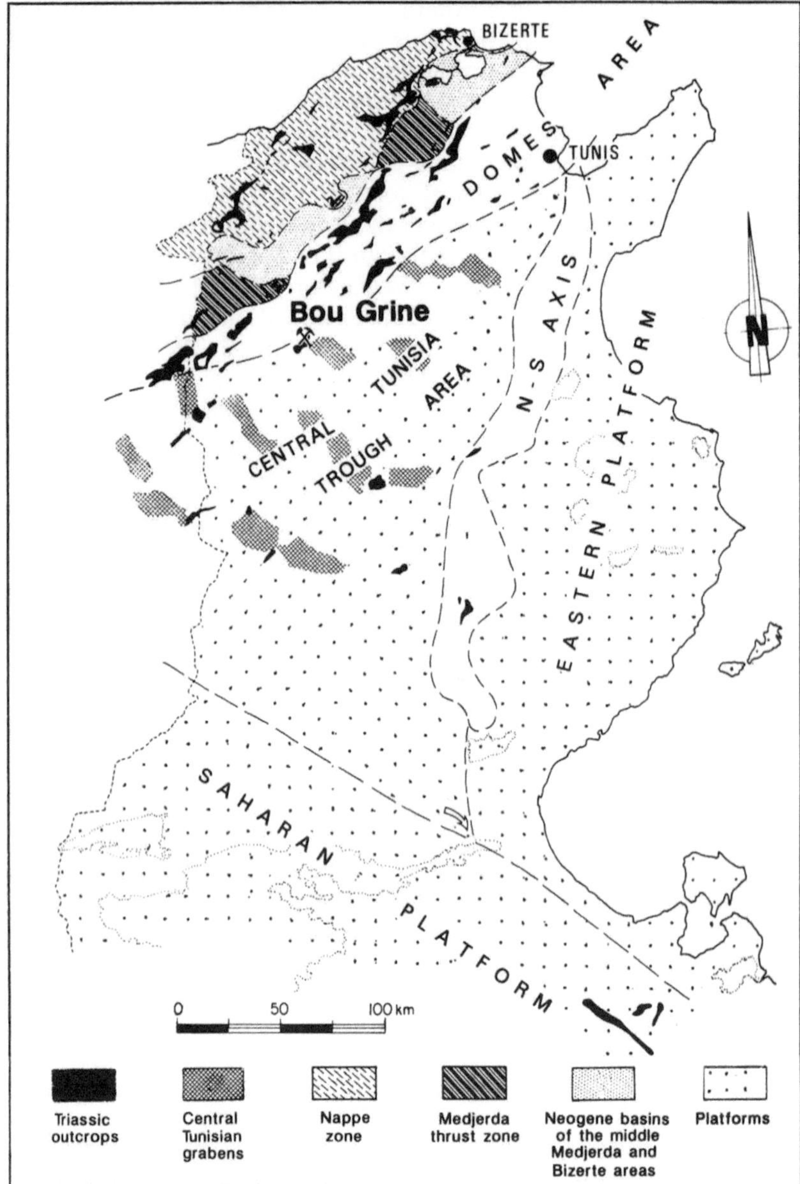

Fig. 1. Structural sketch map of Tunisia showing the location of the Domes (or diapirs) area. (After Perthuisot 1978)

the diapirs reflecting the halokinetic movements that began in the Aptian (Perthuisot 1978).

The Pb-Zn (Ba, Sr, F) mineralization of the Domes area (Fig. 4) occurs under various morphologies, mainly in the Cretaceous cover rocks at the edges

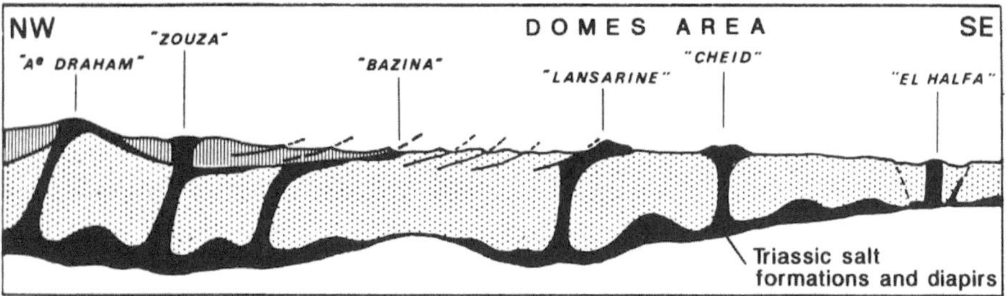

Fig. 2. Simplified cross section, from the Nappe zone to central Tunisia, showing the distribution of Triassic structures and the progressive dying out of the tangential tectonics from northwest to southeast. The Domes area corresponds to the Lansarine and Cheid diapirs. (After Perthuisot 1978)

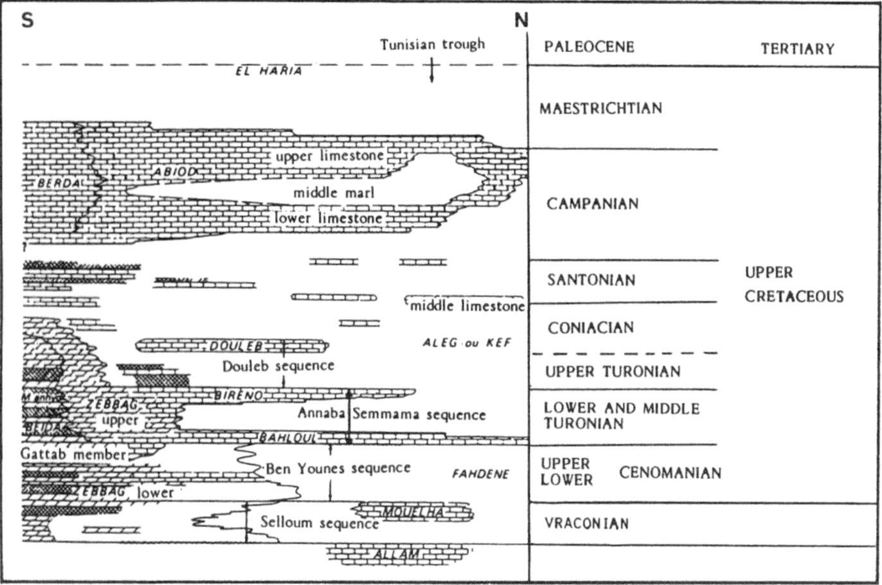

Fig. 3. Variations in the major lithological units of the Middle and Upper Cretaceous from the south to the north of Tunisia. (After Marie et al. 1984)

of the Triassic diapirs and less commonly in the Tertiary cover rocks (Sainfeld 1952; Monthel et al. 1986). The major economic deposits can be classified into four major groups (Table 1): (1) Pb-Zn (Ba-Sr) Fe deposits at the Triassic/cover rock contact; (2) Pb-Zn-Ba-F and Fe-Pb deposits with features similar to the Mississippi Valley-type deposits in the Aptian reef formations; (3) stratiform Pb-Zn concentrations and bodies localized mainly in the Cretaceous cover rocks; and (4) vein-type deposits. Bou Grine belongs to the first and third groups and is found in the second megasequence, associated with a black shale formation.

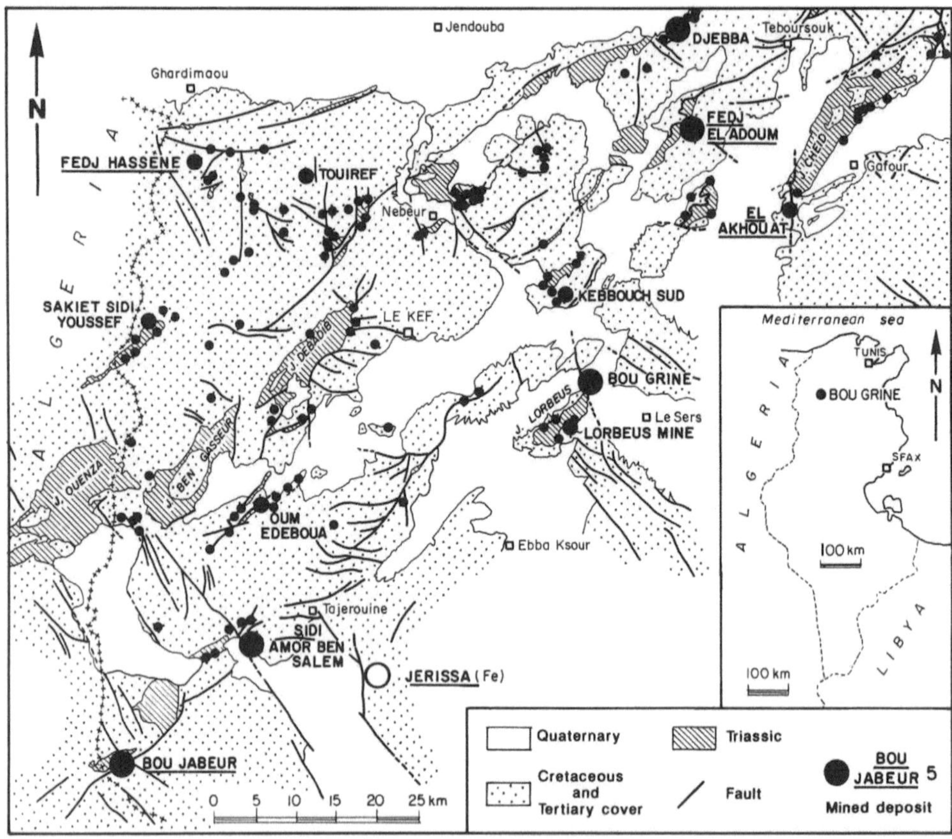

Fig. 4. Location map of Pb-Zn and Ba-Sr (F) deposits and occurrences. (After Monthel et al. 1986)

Two important regional features may help to understand the metallogenic process of the Cretaceous period.

1. Cenomanian-Turonian sedimentation followed two stages of rifting related to the Africa-Europe drift movement in the Gibraltar-Messine transform zone: the one from Neocomian to Early Aptian, and the other from late Aptian to Albian (Guirand and Maurin 1991). These resulted first in a N-S extension that persisted from the Jurassic until Early Aptian, giving rise to E-W half-grabens as well as to volcanism in the Pelagian Sea. The instability then increased until the Albian with the appearance of a NW-SE graben and normal faults along the same direction. It was in this setting that the first diapirs were formed (Perthuisot 1977, 1978).
2. The "E2 event" or one of the "Cretaceous oceanic anoxic events" (Schlanger and Jenkyns 1976; Jenkyns 1980; Graciansky et al. 1982; Herbin et al. 1986).

Table 1. Main deposit types in the Domes area, Tunisia

Deposit groups	Ores	Examples
Lenticular bodies in Neogene detrital deposits	Pb-Zn-As	Sidi Bou Aouame Jalta Jebba EL Haouria
Vein-type deposits in the Cenomanian-Turonian and the Abiod formation (Campanian-Maestrichtian)	Pb-Zn	EL Akhouat Lorbeus J. Hallouf Fedj Assene
Bodies in Turonian limestones	Zn-Pb	Bou Grine
Stratiform concentrations in the Bahloul Formation (Cenomanian-Turonian)	Zn-Pb	Bou Grine
Bodies in the Aptian reef formations	Pb-Zn-Ba-F	Bou Jabeur Sidi Amor
Deposits at the Triassic-Cretaceous contact	Pb-Zn-Ba-Sr	Fedj EL Adoum Kebbouch South Bou Grine

Although formation of black shales is fairly common in the Tethys-Cretaceous series, two horizons are particularly rich in organic matter: one at the Aptian-Albian transition (event E1) and the other at the Cenomanian-Turonian transition (event E2). Factors evoked to explain anoxic conditions during sedimentation of these horizons are (1) the large Cenomanian transgression, and (2) a salinity stratification of the waters, which may have been temporary and repetitive. It can be suggested, following Busson (1984a), that the effect of a large marine transgression was to reduce the influx of continental materials (both solid and liquid) to give a condensed series in which organic matter was deposited and preserved due to isolation of the bottom by hypersaline brines resulting from salinity stratification. Busson (1984b) also emphasized the effects of brine stagnation and the possibility, during structural evolution, that they may have overspilled from one basin to another. These factors are considered to have been active in the Tunisian trough which was, in addition, affected by halokinetic movements.

3 The Bou Grine Deposit

The Bou Grine deposit lies at the northern end of the Jebel Lorbeus diapir, which consists of a core of Triassic salt formations surrounded by Cretaceous and Tertiary rocks. It contains three types of economic base-metal concentrations that will be described in more detail in Section 3.4: (1) lenticular bodies at the transition between Triassic and Cretaceous; (2) stratiform mineralization in the organic-rich limestone of the Cenomanian-Turonian Bahloul Formation; and (3) a semi-massive orebody (hanging wall orebody) cutting the Cretaceous series.

3.1 Exploration

The Bou Grine ore occurrences were known for some time before their reassessment by OMN-BRGM in 1978. The earliest known workings date back to Roman times, whilst in the present century, between 1901 and 1930, mining was focused on calamine mineralization (some 3300 t extracted) along a north-striking fracture in the Cretaceous (Berthon 1922; Sainfeld 1952).

More recently, in 1968, ONM carried out a study of Bou Grine and decided to begin exploration with a detailed geological survey. Then, in 1970–71, ONM drilled four holes to test the mineralized Triassic/Cretaceous contact (Massin 1972), one of these intersecting highly oxidized Pb-Zn mineralization in the Cretaceous (later classified as karstic bodies).

Bou Grine was taken up again in 1978 by the ONM-BRGM team of the Central North Tunisian Project following a re-interpretation of the geology and types of mineralization (Gharbi et al. 1979; Caia et al. 1980). The choice was guided by the results of previous work on the borders of Triassic areas (Bolze 1952; Sainfeld 1952; Nicolini 1968, 1970) and on (1) the apparently favourable position of the Lorbeus massif in relation to the Medjerda thrust zone; (2) the numerous mineral occurrences distributed through a large section of the stratigraphic column; (3) the existence of mineralization at the Triassic/Cretaceous boundary, defined thereafter as the "transition zone"; (4) the presence of disseminated sphalerite associated with organic matter in the Bahloul Formation limestone; and (5) the large potential of the mineralized structures demonstrated by surface observation (Orgeval 1981). The aim was to show quickly, through fieldwork, the mining interest of the disseminated sphalerite which constitutes a new type of concentration, and this was achieved in 1979 through: (1) geology and surface exploration, including lead and zinc reagent prospecting (see Fig. 5); (2) soil geochemistry over an area of 2.0 × 0.5 using a 50 × 25 m grid (Leduc and Orgeval 1979); and (3) underground surveys at 1:250 scale with systematic horizontal channelling sampling which revealed several important mineralized lenses.

Marker beds were defined for an accurate drill-hole survey of the deposit, which began in 1980 and continued without interruption until 1984, with 72 surface holes (totalling 20700 m) and 16 underground holes (totalling 1800 m); at the end of this stage, the reserves, using a cutoff of 4% Zn, were estimated at 7.3 Mt of ore assaying 2.4% Pb and 9.7% Zn (Orgeval et al. 1985a). Then, in 1985–1986, an additional six holes (totalling around 2000 m) were drilled in the central part of the deposit to confirm the hanging wall orebody. Concurrently with the drilling program, ore processing tests were carried out on core samples.

3.2 Host Rock Sedimentology and Palaeogeography

The lithostratigraphic succession at Bou Grine, from Triassic to Palaeocene, was studied in detail for the exploration work (Monciardini and Orgeval 1987).

The *Triassic* succession forms a steeply dipping homocline of interbedded gypsum, dolomite, variegated marl and finely cross-bedded clayey sandstone

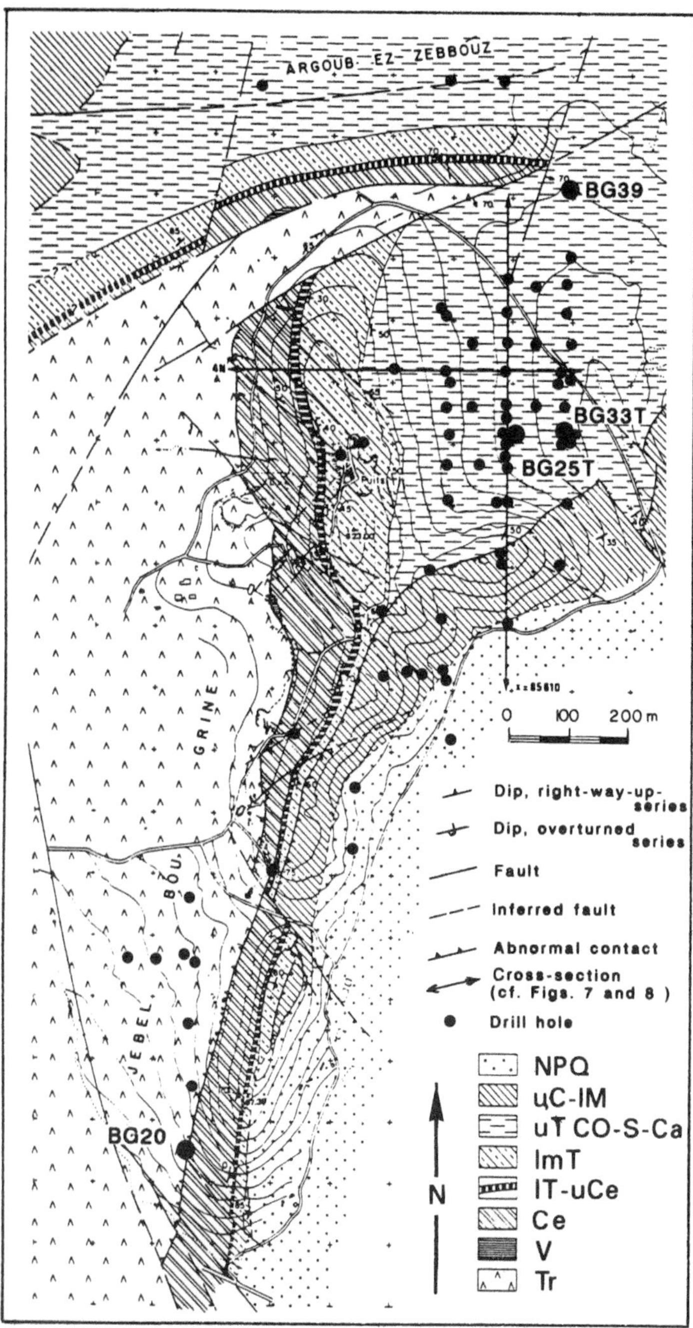

Fig. 5. Detailed geological map of Jebel Bou Grine showing drill-hole locations. *NPQ* Neogene, Pliocene, Quaternary: clayey sand, molasse, scree; *uC-lM* Upper Precambrian – Lower Maastrichtian (Abiod): massive chalky limestone; *uT-Co-S-Ca* Upper Turonian, Coniacian, Sanonian, Lower Campanian (Aleg or Kef): marl; *lmT* Lower and Middle Turonian: clayey limestone; *lT-uCe* lowermost Turonian, uppermost Cenomanian: platy bituminous limestone = Bahloul facies; *Ce* Cenomanian: clayey limestone and marl; *V* Vraconian: marl; *Tr* Triassic and transition zone with the Cretaceous: argillite, gypsum, anhydrite, calcareous sandstone and carbonate facies

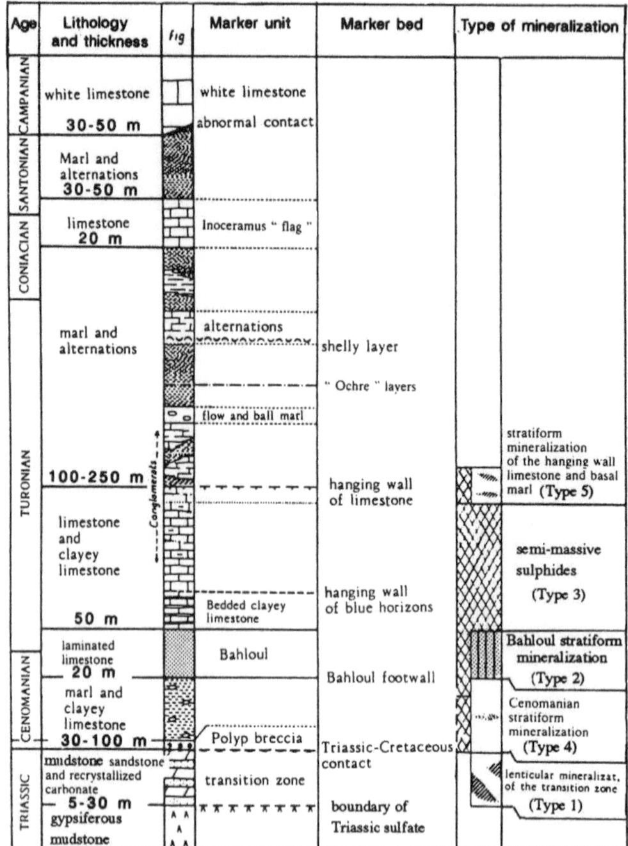

Fig. 6. Stratigraphic column at Jebel Bou Grine, showing lithology, marker beds, and mineralization

(Pervinquière 1903; Perthuisot 1977). In certain areas, a so-called transition zone, defined by Sahli et al. (1981) and consisting of partly carbonatized Triassic sulphate layers, is found on top of the Triassic evaporite sediments (Figs. 6, 7, 8). This carbonatization is related to Cretaceous fluid circulation (Lenindre et al. 1979; see below). The most striking feature of the transition zone is the presence of lenticular horizons of recrystallized carbonates ("X limestone"), showing pseudomorphism of the precursor sulphate layers (Fig. 9B) and sulphate-bearing argillite (Fig. 9C).

Where visible, the base of the *Upper Cretaceous* is generally highly tectonized and characterized by a several metres thick monogenic breccia (Triassic fragments). At Bou Grine, however, it forms a sedimentary contact marked by a breccia containing Triassic clasts and glauconite (Fig. 9A); the nature of this late Albian basal bed is an important argument for early diapir formation, as envisaged by Perthuisot (1977). Overlying the basal bed is an early Cenomanian succession of marl and nodular clayey limestone of variable thickness in which early sedimentary breccias indicate instability of the submarine basement and the probable existence of shoals with reduced sedimentation.

The Cretaceous continues with the Cenomanian-Turonian *Bahloul Formation*, which consists of 20 m of laminated, dark-brown to black limestone

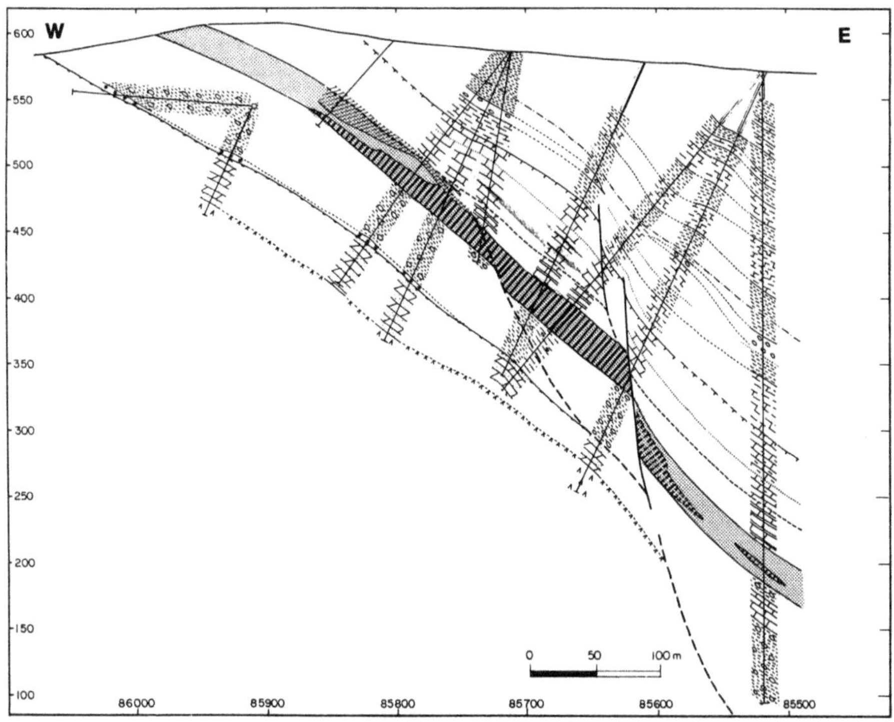

Fig. 7. Bou Grine, vertical E-W geological section 4N showing drill holes (see Fig. 5 for location, and Fig. 6 for explanation)

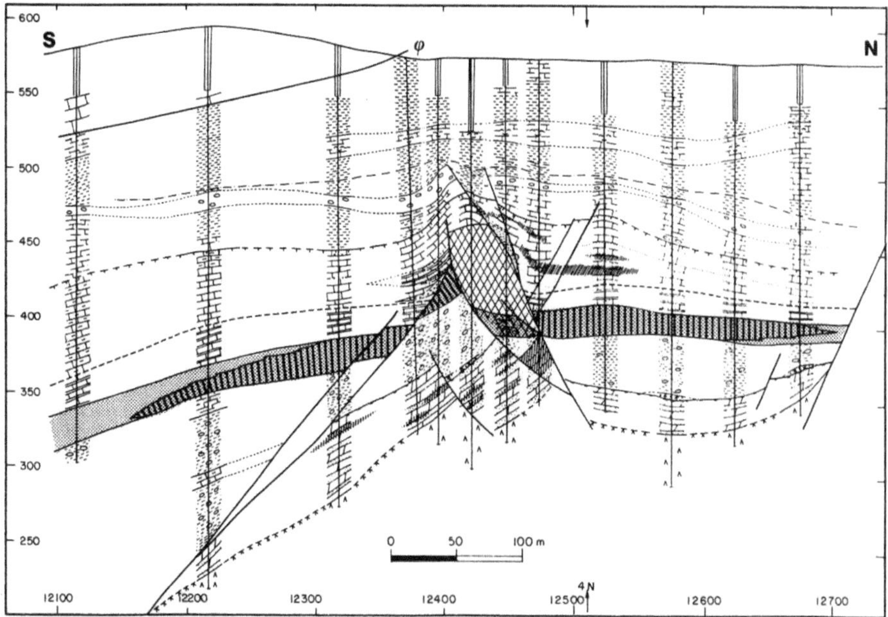

Fig. 8. Bou Grine, N-S geological section inclined 65 °W in the drilling plane (abscissa: 85 610) (see Fig. 5 for location, and Fig. 6 for explanation)

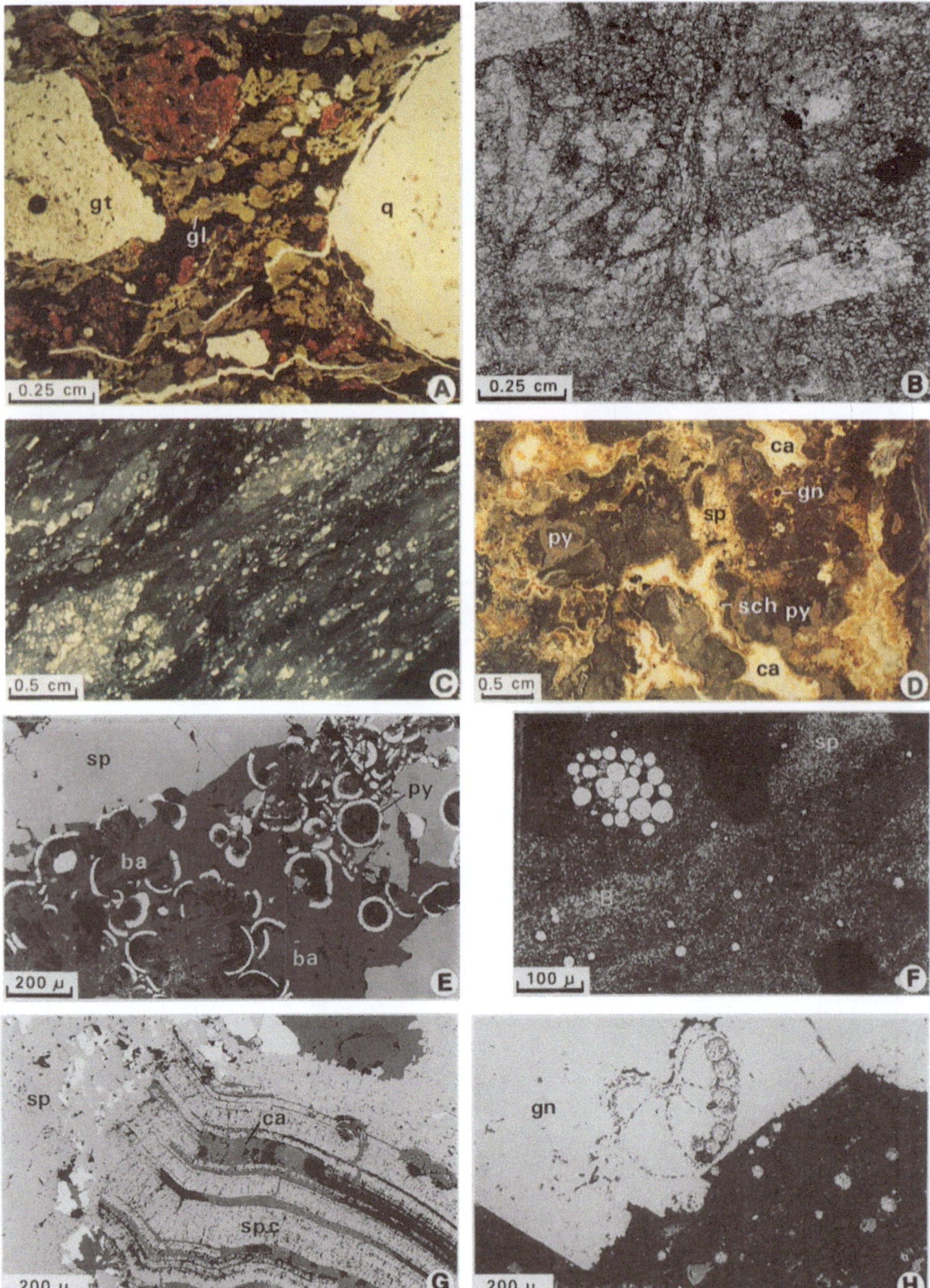

Fig. 9. Mineralogical aspects of the transition zone and ore types. **A** Sedimentary breccia of the basal Vraconian-Cenomanian containing reworked Triassic sandstone (*gt*) and broken quartz (*q*) in a micritic cement with glauconite (*gl*). **B** Sulphate pseudomorph in carbonate beds of the transition zone. **C** Sulphate-bearing argillite with carbonate lenses and layers of fine-grained micaceous sandstone in the transition zone. **D** Type 1 ore showing pyrite lenses (*py*), sphalerite (*sp*), schalenblende (*sch*), galena (*gn*) and terminal calcite (*ca*) in the Triassic-Cretaceous transition. **E** Type 1 ore (natural light) showing broken concentric pyrite grains with included baryte lamellae, surrounded by sphalerite (*sp*) and baryte

rich in organic matter (Burollet 1956) and which hosts the main zinc mineralization at Bou Grine. The laminated facies is a biomicrite with planktonic foraminifera where the lamination is due to rhythmic alternation of two microfacies: laminae with small monospecific foraminifera, reflecting severely confined conditions, and laminae with normal pelagic microfauna (Orgeval et al. 1981). The Bahloul facies, in particular the hanging wall, was affected by synsedimentary movements at Bou Grine as expressed notably by calcite-filled tension fissures deformed by tight microfolds at the time of sediment compaction; detailed study has enabled the formation of these veinlets to be correlated with Turonian diapiric movements (Bles 1982). In the orebody, these fissures have a sulphide filling and constitute valuable indicators for disseminated mineralization; for this reason, they were very early termed "informers".

The transition from limestone laminites of the Bahloul Formation to more massive limestone is gradual via bedded clayey limestones. The 50-m-thick lower and middle Turonian carbonate constitutes the first ridge surrounding the Triassic diapir. This is followed by the upper Turonian-lower Campanian *Aleg Formation* which mostly consists of blue marl, locally gypsiferous. Carbonate sedimentation returned with deposition of the white chalky upper Campanian-middle Maastrichtian *Abiod Limestone*. Overlying the Abiod Formation, the upper Maastrichtian-Palaeocene *El Haria Marl* only crops out on the east side of the massif, in the Lorbeus Mine area.

3.3 Structural Development

Horizontal and vertical serial sections, respectively at 25-m and 50-m intervals from available (mainly drilling) data, show that the Bou Grine deposit was affected by synsedimentary faulting, striking northeast and northwest. This would have occurred on the slope of an early diapiric structure (Bles 1982; Giot et al. 1982; Giot and Monciardini 1984), which acted as an unstable shoal:

In the Vraconian-basal Cenomanian, this structural uplift of the structure is expressed by erosion of the Triassic substratum and the location of coral colonies.

In the Cenomanian, abrupt variations in thickness and facies, unconformities and conglomerate horizons indicate varying rates of sedimentation on a slope undergoing frequent readjustment.

In the Cenomanian-Turonian, the sedimentation of the Bahloul laminite facies expresses an episode of relative quiescence; the formation shows disturbances (slide marks, slump structures) mainly at the hanging wall and in the bedded clayey limestone capping. Structural analysis has enabled one to define a northwest-southeast extension phase and indicates a sedimentary slope to the southeast. This phase of extension is indicated mainly by fractures and

←

(*ba*). **F** Type 2 ore (natural light) showing framboid pyrite clusters in a clayey-carbonate matrix with light-coloured microsphalerite. **G** Type 3 ore (natural light) showing sphalerite crystals (*sp*) cutting banded colloform sphalerite (*sp.c*) partly replaced by calcite (*ca*). **H** Type 3 ore (natural light) showing fossil replaced by galena (*gn*) within a clayey-carbonate matrix. Note the presence of framboid pyrite in the fossil chambers

tension gashes deformed during compaction, giving rise to the sphalerite-galena "informers" (see above).

In the lower and middle Turonian, intraformational conglomerate horizons and graded calcarenites, which indicate deposition similar to that of turbidites, express a high instability at this time. Structural analysis here shows a phase of N-S extension.

At the base of the upper Turonian, stratigraphic hiatuses in the passage zone from limestone to marl sedimentation, and intraformational conglomerate at the base of the Aleg marl, indicate continuing instability.

The diapiric instability, which undoubtedly played an important role in deforming the deposit, constitutes new data that have not yet been found elsewhere in Tunisia. Nevertheless, the influence of north-south tangential slip of Oligocene-Miocene age, which could have locally intensified older features, cannot be excluded.

Most of the economic mineralization shows tectonic control (Figs. 7 and 8). The type 1 mineralization (see Sect. 3.4) is hosted by the Triassic/Cretaceous boundary, marked by a major north-striking fault. The type 2 mineralization is located in Cenomanian-Turonian rocks, dipping 30–40°E north of Jebel Bou Grine, then steepening and overturning towards the south, so that the Triassic structurally overlies the Cretaceous. The type 3 mineralization is controlled by subvertical faults with four main strike directions: N30°E, N70°E5, N140°E, and N160°E.

Another important structural feature is the detachment of the Abiod Formation from the Aleg Fm.: in the southern part of the Jebel the subvertical Abiod is in contact with Turonian limestone.

3.4 Ore Types

Orgeval et al. (1986, 1989), on the basis of geological, morphogical and mineralogical criteria (Laforet 1979, 1981), defined five types of ore, of variable importance, at Bou Grine (see Figs. 6, 7, 8).

Type 1, occurs in the Triassic to Cretaceous transition zone. It is characterized by abundant iron sulphides associated with sphalerite and galena, and by celestite-baryte. The average Pb + Zn content is about 8.5% and can locally reach 35%. The mineralization is situated in the central and southern parts of the deposit, either as impregnations and fissure-vein disseminations, or as massive sulphide lenses (Fig. 9D,E). This type is best exemplified by the Fedj el Adoum deposit (Laatar 1980).

Type 2, corresponds to stratiform mineralization in the Bahloul Formation. It displays a single paragenesis with predominant fine sphalerite (20–200 μm in size) and accessory galena, pyrite and marcasite. The Pb + Zn content averages about 10%, but can locally exceed 20%, and the ore is rich in organic matter (3.0–7.5% organic carbon). This facies probably formed during a stage of early diagenesis; later evolution led to the disappearance of primary structures, with enrichment of the mineralization and a moderate increase in the size of the sulphide grains. Enrichment is indicated by mineralized microfractures cutting laminae, massive sphalerite layers, amoeboid pyrite masses, and sulphide-cemented fractures and breccia. The last evolutionary stage, which is similar to the hanging wall orebody described below, is formed by mineralized fractures and cavities filled with sphalerite concretions showing dendritic galena, the final filling being of calcite (Fig. 10A,B,I).

Type 3, or hanging wall orebody, cuts Cretaceous rocks from the Cenomanian to middle Turonian and is characterized by a Pb + Zn content >20%. It shows increasingly evolved structures from bottom to top: mainly breccia with recognizable components in the lower part; then a facies where sulphides permeate the pre-existing matrix ("breccioid facies"), which represents the main part of the mineralization; and, at the top, a distinct facies of sphalerite-bearing marl (Fig. 9G,H; Fig. 11A–F).

Type 4, corresponding to stratabound impregnations in Cenomanian clayey limestone and marl, is only of minor economic interest. It consists of pyrite and sphalerite disseminations in clayey joints and in the surrounding Cenomanian limestone beds.

Type 5. consisting of stratabound impregnations in the lower and middle Turonian limestone up to the base of the Aleg Marl, forms lenticular sphalerite and pyrite impregnations. The average Pb + Zn content is about 7.5%.

The distribution of the economic mineralization and the geometric relationships between the main hanging wall orebody (type 3) and the other ore types is indicated in Fig. 8.

1. In the lower, southern part of the deposit, the lenticular type 1 mineralization occurs in a fault-uplifted Triassic block.
2. The stratiform type 2 mineralization is symmetrically distributed to either side of the main orebody; in the central part, over a distance of 250 m in a north-south direction. It occurs throughout the entire thickness of the formation (i.e. over 20 m); it then thins out both north and south (see Fig. 8) and downdip (see Fig. 7).
3. The lower and middle Turonian stratabound mineralization (type 5) forms preferentially on the periphery of the main orebody in the north (see Fig. 8).

The relationships between tonnage and grade distribution of the three economic types of mineralization are shown in Fig. 12.

3.5 Petrography of the Main Ore Type

A study of the different ore facies shows an evolution from an early, pre-compaction phase of mineralization through successive phases of ore enrichment. Chemical analyses support the concept of an almost continuous evolution

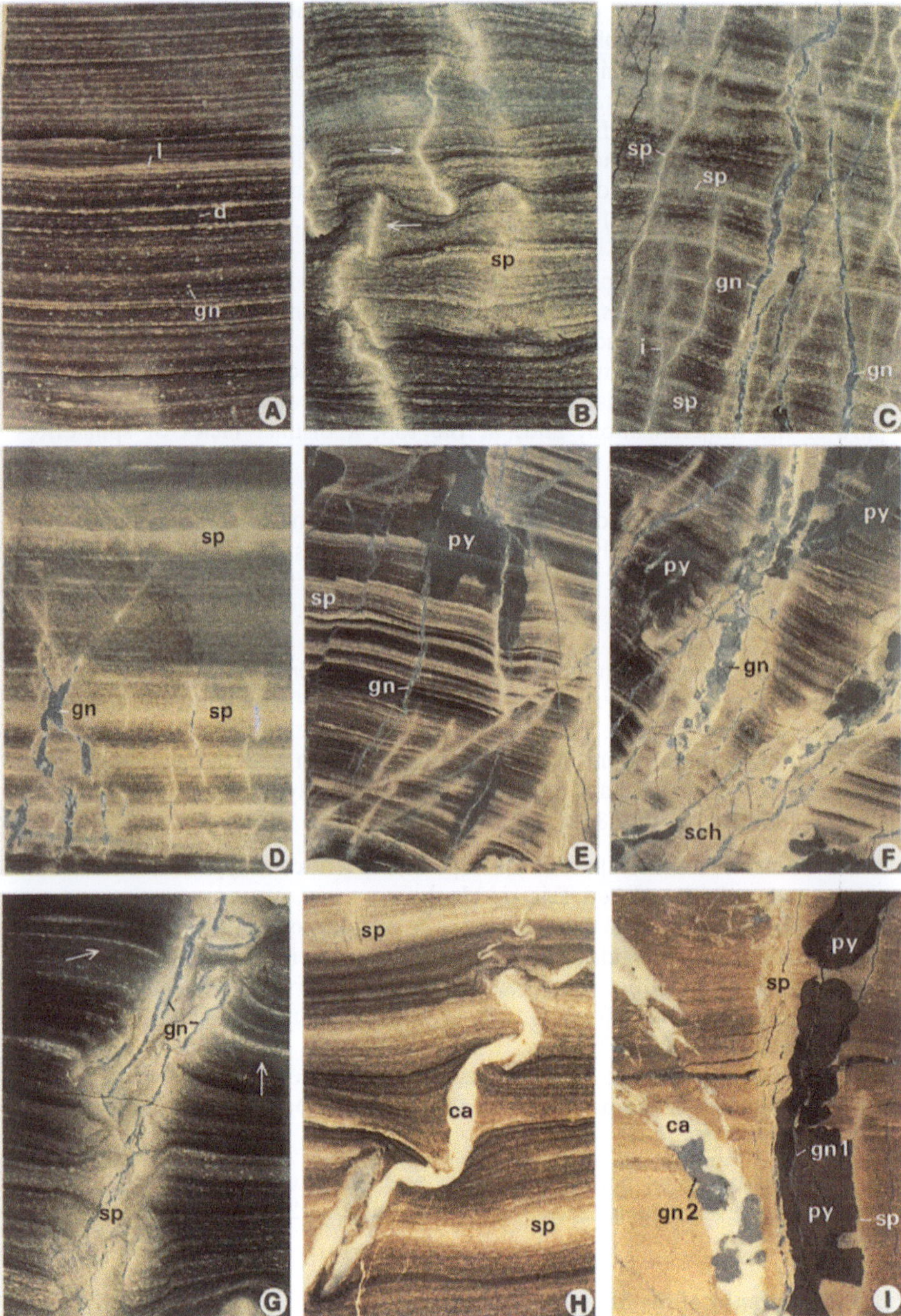

Fig. 10. Examples of type 2 ore at different stages of development (×0.5). **A–F** Bahloul ore, its evolution and enrichment. **A** Characteristic appearance of the Bahloul ore with alternating millimetre- to centimetre-thick dark (*d*) and light (*l*) laminae. The disseminated mineralization is

from a zinc-bearing layer to an economic concentration of zinc, reflecting intense diagenesis and fluid circulation.

The *zinciferous laminites* forming the stratiform type 2 ore consist of a mineralized bioclastic clayey-carbonaceous calcimicrite whose major features are (Lenindre 1979): (1) a regular, bedded structure, in places laminar; (2) numerous foraminifera and other bioclasts as components; and (3) mineralization that is, in part, affected by early diagenetic structures (Fig. 13A,B). *Sphalerite* is present as microcrystals disseminated along the bedding planes or coalescent as flattened nodules; it appears to have been remobilized and redeposited as concentrations in fissures and late fractures, where it occurs as crystalline and/or concretionary (schalenblende) sphalerite. *Pyrite* occurs as framboids and as disseminated microcrystals (Fig. 9F), and also forms lenticular bodies of aggregated crystals; it is also remobilized in fractures and develops large patches of late amoeboid pyrite (Fig. 10E). *Galena* occurs in euhedral crystals along the bedding and in scattered crystals in the fractures, associated with calcite and schalenblende. *Calcite* is observed in fissures deformed during compaction (Fig. 10H) and in all the late fractures; it also shows pseudomorphism of rare sulphate crystals (baryte and anhydrite).

The early development of the mineralization before and/or during compaction of the mineralized layers is reflected in the deformation of the mineralized structures: (1) convoluted laminar beds deformed by intraformational slip, with flow structures (Fig. 13C); (2) compaction structures around rigid cores, i.e. pyrite nodules, galena crystals (Fig. 13D); (3) deformed infilled fissures (Marteau 1987).

The phases of sphalerite development are thus: (1) early, regularly disseminated deposition of microsphalerite crystals in a slightly compacted sedi-

---

still not very evident, but forms thin continuous layers. Cubes of euhedral galena (*gn*) are seen in the dark laminae. Grade = 0.13% Pb; 7.20% Zn. **B** As in **A**, with the presence of "informers" (compacted synsedimentary tension gashes). Note the sphalerite enrichment (*sp*) forming a lighter coloured fringe on either side of the "informers" (*arrows*) – fracture drainage effect, also well seen **G**. The enrichment and the fracture mineralization occurred before compaction; deformation moulding the indurated parts. **C** Enrichment (grade = 0.30% Pb; 13.52% Zn) through individualization of layers of massive sphalerite (*sp*), "informers" (*i*) and numerous microfractures with sphalerite (*sp*) and galena (*gn*). **D** Enrichment through regular sphalerite impregnation along the bedding (*sp*) with late fractures containing sphalerite (*sp*) and galena (*gn*). **E** Development of amoeboid pyrite (*py*) in facies already heavily impregnated with sphalerite, notably in the bedding (*sp*). The partial replacement of the zinc ore by fine pyrite crystals preserves the earlier textures (laminations, "informers"). The pyrite invasion is along the cross-cutting fractures with galena (*gn*). Grade = 6.19% Pb; 19.60% Zn. **F** Fracturing, brecciation and recrystallization of a (fine) sulphide-rich laminated ore. Sphalerite enrichment begins to obliterate the original bedding. The fractures contain concretions of schalenblende (*sch*), galena (*gn*) and pyrite (*py*). Grade = 7.60% Pb; 30.24% Zn). This evolution foreshadows the massive ore of type 3 (hanging wall ore). **G** Compacted mineralized drainage fractures. Some laminae to the side of the fractures show greater sphalerite enrichment (*arrows*) close to the factures. In the centre, sphalerite (*sp*) coalesces with galena (*gn*). **H** Calcite-filled fractures deformed during compaction in highly mineralized laminated facies with fine disseminations in the dark carbonaceous clayey matrix. **I** Detail showing three types of fractures. In the *middle*: sphalerite concentration (*sp*) along an old "informer". On the *right*: development of amoeboid pyrite (*py*) along a galena-filled fracture (*gn1*) between two "informers" (*sp*). On the *left*: late fracturing with calcite (*ca*) and galena (*gn2*)

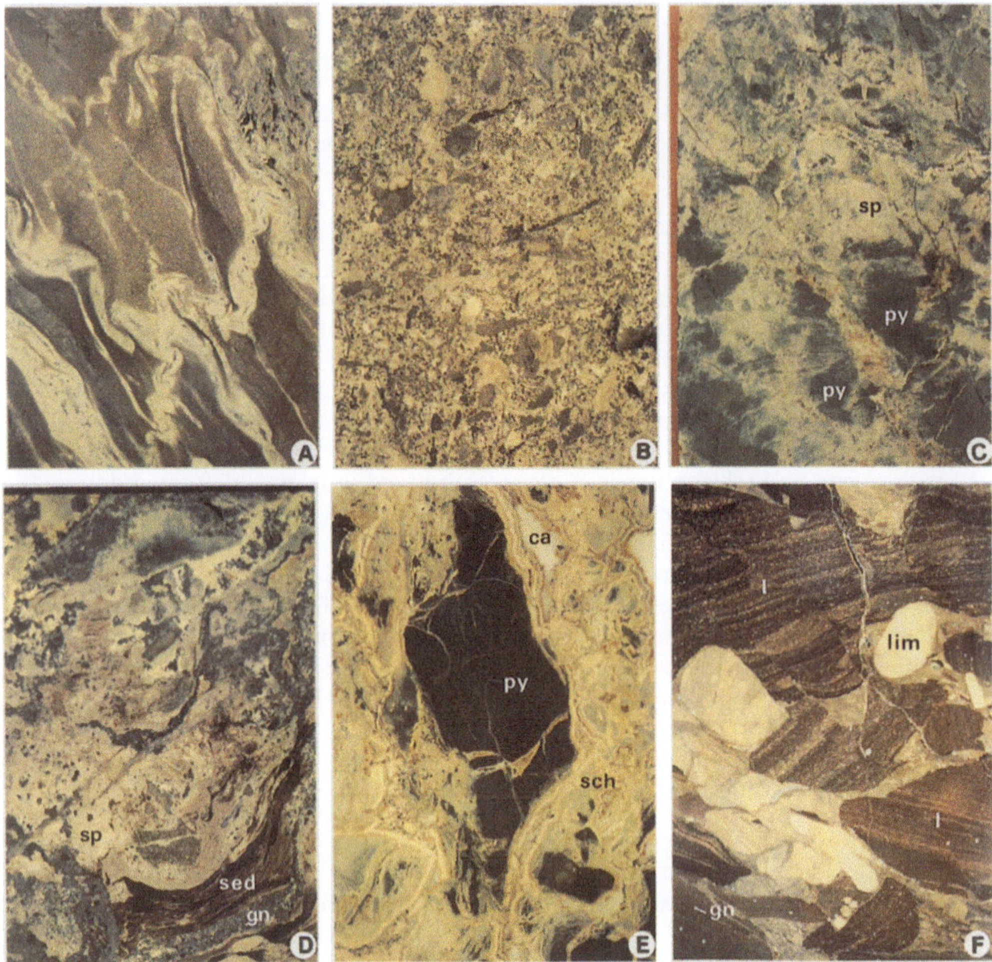

Fig. 11. Type 3, or hanging wall ore: facies from the top to the bottom of the body (×0.5). **A** Zinciferous marl showing intense deformation – flow structure. **B** Microbreccia with reworked mineralized clasts. This facies alternates in places with the zinciferous marl. **C** Breccioid facies representative of the richest parts of the orebody (5.68% Pb; 43.20% Zn). The photo shows a pseudobreccia with pyrite-marcasite (*py*) "clasts", developed at the expense of black marl, in a matrix of fine sphalerite (*sp*) and galena. **D** Facies similar to that of **C**, showing fine black sediments (*sed*) (inner sediments?) altered by massive sphalerite (*sp*) and galena (*gn*). **E** Concretionary ore cementing varied fragments and filling voids: cavity facies with schalenblende (*sch*) and middle crystalline sphalerite around a pyritized fragment (*py*). Calcite (*ca*) fills the last spaces. **F** Breccia facies at the base of the orebody with centimetre-sized clasts of barren beige limestone (*lim*) and mineralized laminites (*l*) of type 2 ore showing fine impregnation of euhedral sphalerite and galena (*gn*). The cement is composed of sphalerite and galena.

ment; (2) almost simultaneous concentration on structures, textures and discontinuities; (3) later concentration and deformation of the mineralization due to compaction and subsequent fracturing; and (4) diagenetic remobilization, either passive or provoked by circulation of mineralizing fluids.

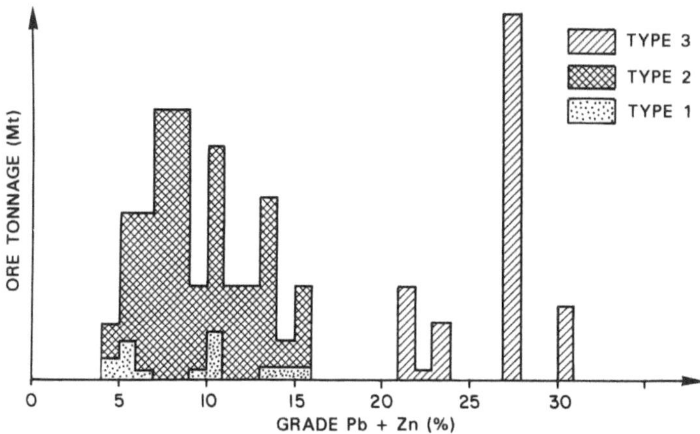

Fig. 12. Tonnage-grade distribution of the three economic types of mineralization

Petrographic evidence also provides insight into the small-scale effects of mineralizing fluids: Figs. 10G and 13E reveal an increased sphalerite concentration near drainage fissures; Fig. 13F illustrates selective high-grade mineralization controlled by a small fissure cutting the bedding. Subsequent dislocation, brecciation and fracturing of the concentrations in the original laminae is also observed (Fig. 10I).

3.6 Association of Organic Matter and Ore

The association of organic matter with Pb-Zn sulphides in sedimentary environments is well known and has been demonstrated at both local and regional scales. As deposit scale, evidence has been provided that the organic matter commonly associated with the ore may have participated directly in the sulphide precipitation through biogenic or abiogenic reduction of sulphate sulphur (Germanov 1965; Connan and Orgeval 1973, 1977; Nooner et al. 1973; Pering 1973; Powell and MacQueen 1984; Disnar and Sureau 1990). It was this association that was used as an exploration guideline in the Dome area, where the Triassic diapiric extrusions are a possible source of sulphate and the adjacent Cenomanian-Turonian organic-rich Bahloul Formation (4–5% TOC) is a favourable site for sulphate reduction and mineral deposition, and led to the discovery of the Bou Grine deposit in 1978–1980 (Orgeval et al. 1986).

Studies of the organic matter in the laminites from boreholes of the Bou Grine deposit by Montacer et al. (1986, 1988) and Montacer (1989) show a high total organic carbon content for the Bahloul Formation (4–5% TOC), a low maturity (Rock-Eval pyrolysis [Tmx] = 423 °C) and a marine planktonic origin (Fig. 14). The high value of the oil production index (around 30%) indicates the presence of migrated oil in the deposit with some features of biodegradation; the gas chromatography of the saturate hydrocarbons (Fig. 15) shows an unimodal distribution of the high-level n-alkanes ^{17}C–^{19}C, whilst the

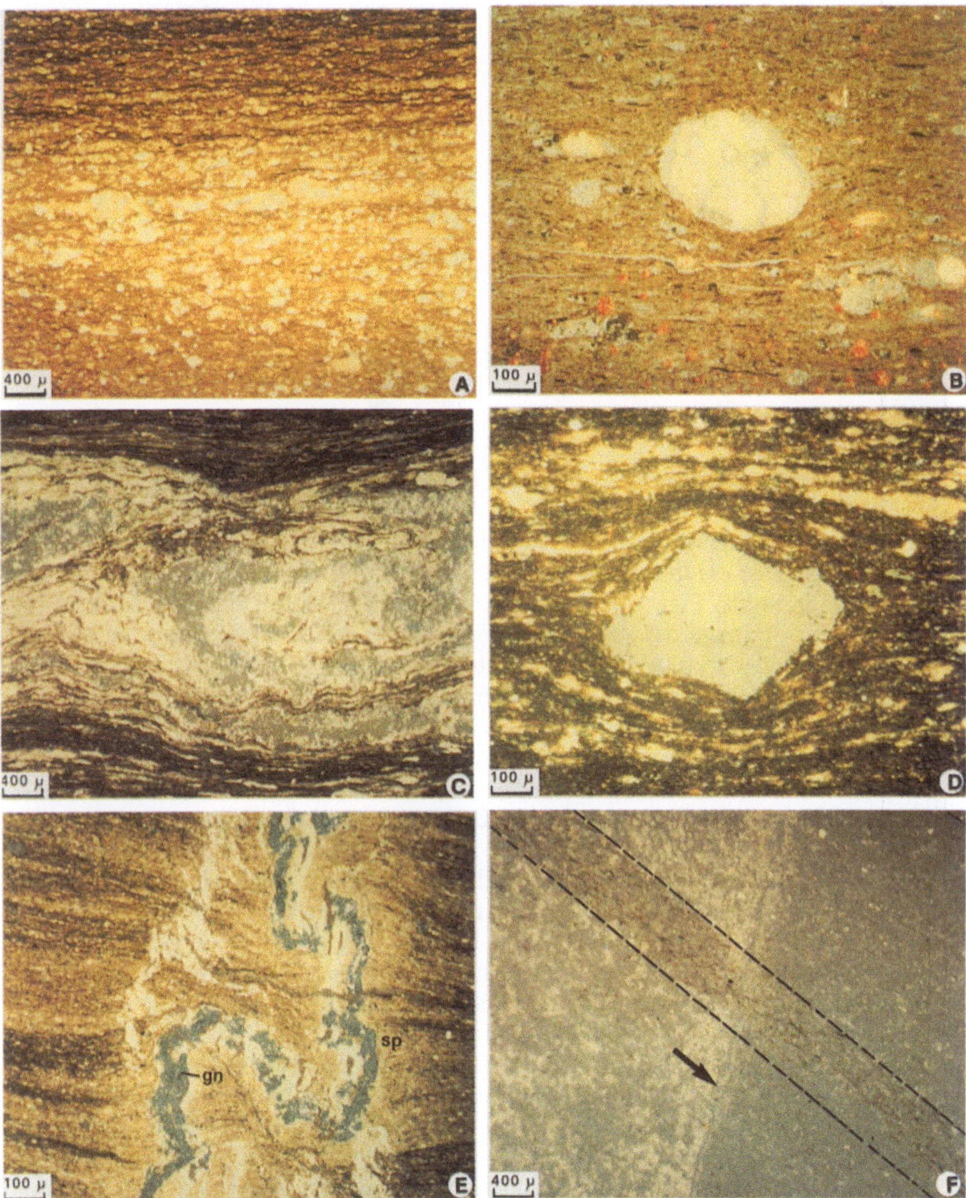

Fig. 13. Examples of diagenetic evolution of the ore types with indications of drainage and fluid circulation. **A** Laminae with microsphalerite mineralization replacing the foraminifera. **B** Detail of foraminera replaced by sphalerite. **C** Convolute laminae with microsphalerite mineralization deformed by intraformational slip. **D** Dark laminae with layers and lenticular bodies of microsphalerite compacted around an euhedral galena crystal. **E** Set of light and dark laminae impregnated with microsphalerite. Drainage along a previously compacted fracture. The fracture infill comprises a core of crystalline sphalerite (*sp*) and galena (*gn*). **F** Sphalerite enrichment (*light grey* in the *left half of the photo*) controlled by a subvertical fracture indicating lateral circulation of the mineralized fluid

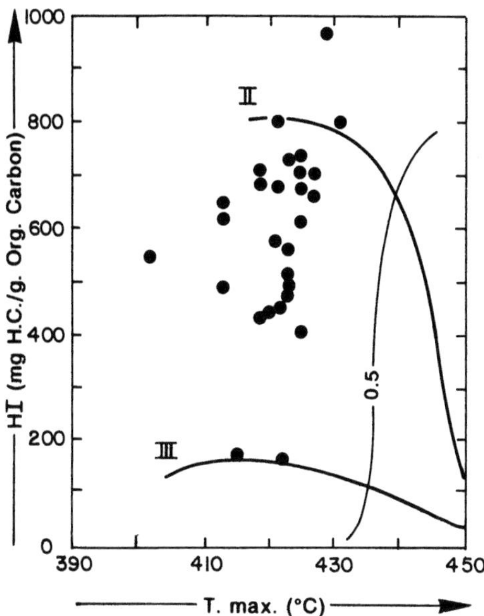

Fig. 14. Rock-Eval pyrolysis data from representative boreholes: *HI* vs. *Tmax* diagram. (After Montacer et al. 1988)

isoprenes (pristane-phytane) show peaks close or equal to those of the n-alkanes n-^{17}C and n-^{19}C. Such figures are associated with immature marine material (Tissot et al. 1977), whose biodegradation is indicated by a preferential elimination of the n-paraffins in particular (Williams and Winters 1969; Connan et al. 1975; Disnar et al. 1986).

The presence of considerable quantities of steranes and triterpenes and the similarities of their distribution in mineralized and unmineralized material indicates a uniform composition of the organic matter in the study area and is evidence of intense biological activity in this part of the Bahloul Formation (Montacer et al. 1988). Further evidence of such activity is given by the presence of homohopanes up to C35 (Ourisson et al. 1979, 1982). The intense biological activity would have developed at the expense of the organic matter in the Bahloul Formation, and the sulphates from the adjacent Triassic diapir (or from brines) may have produced the hydrogen sulphide responsible for both the first ore deposition and its later remobilizations.

3.7 Fluid Inclusions

Studies have been made of the fluid inclusions in the sphalerites of ore types 1, 2 and 3, and also of those of the calcites (Orgeval and Legendre 1989). The main results (Fig. 16) show that: (1) the primary inclusions in the sphalerites are mainly aqueous biphase xenomorphs with a high coefficient of infill (low gas phase) and a high salinity (10–17 wt% eq. NaCl) reflecting homogenization

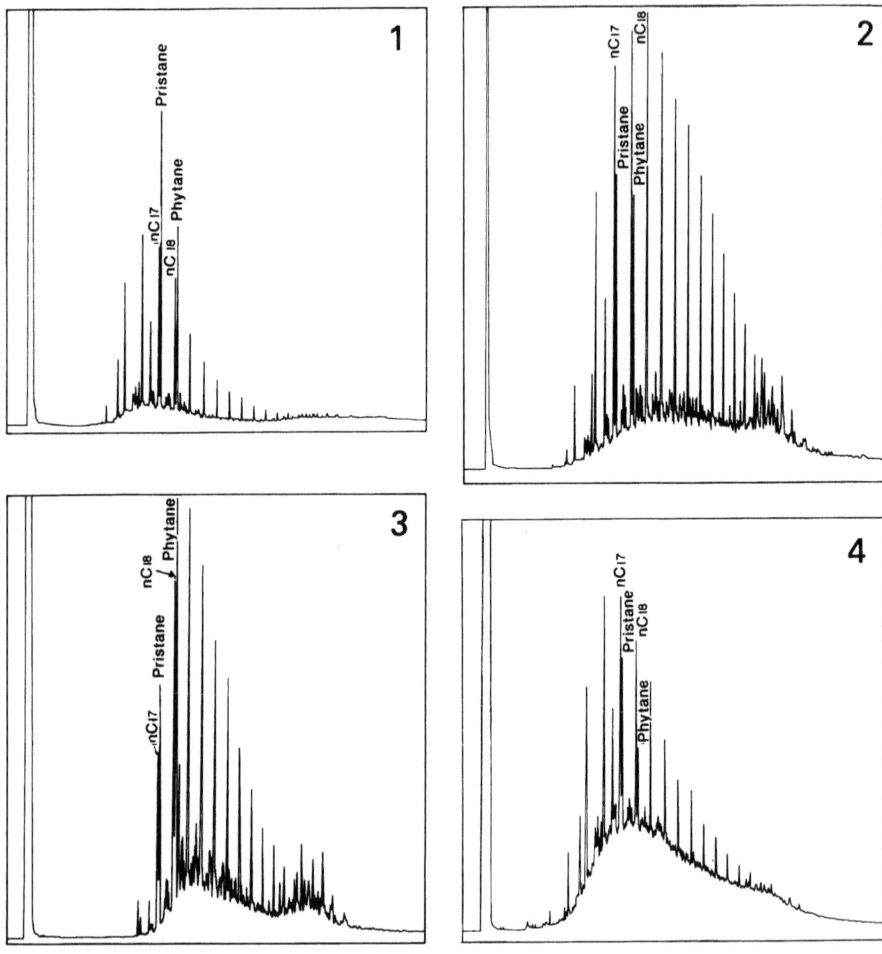

Fig. 15. Gas chromatographic traces of saturated hydrocarbons (after Montacer et al. 1988). Borehole sample: **1** BG20; **2** BG25T; **3** BG33T; **4** BG39

temperatures of around 75–100 °C; (2) the sphalerites of the type 3 ore also contain "necking down" structures, indicating evolution within the mineralization; (3) the calcites contain monophase tablet inclusions of various shapes, and ragged edged triphase xenomorphic inclusions that show a gaseous phase, an aqueous phase and hydrocarbons; these inclusions have a low salinity (0.2–0.4 wt% eq. NaCl) with homogenization temperatures of around 50–60 °C. The studies have thus revealed two distinct families of low-temperature inclusions, which agrees well with the parameters derived from other studies of the deposit, especially of organic matter. The values obained on the type 1 ore in the Triassic-Cretaceous transition zone are of the same order as those obtained by Charef and Sheppard (1987) on the nearby Fedj el Adoum deposit.

The thermal effect of the Triassic salt diapirism is seen in the numerous diagenetic minerals: different types of quartz, dolomite, magnesite, albite,

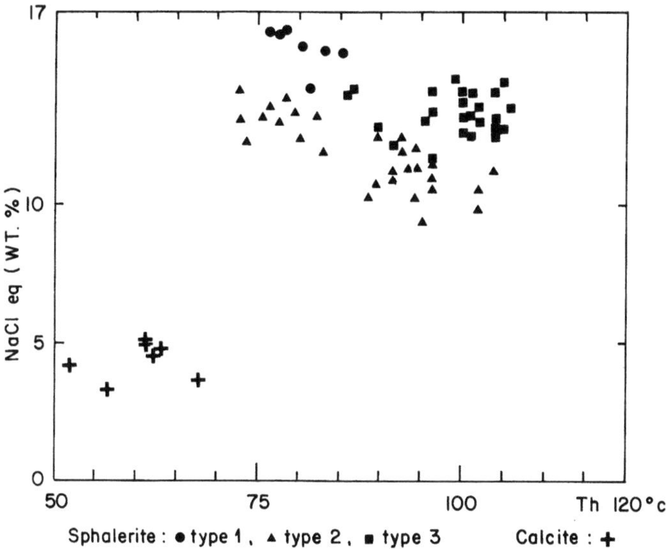

Fig. 16. Fluid inclusions in sphalerite and calcite plotted on a homogenization (Th°C) vs. salinity (wt% eq. NaCl) diagram

potassium feldspar, phengite, phlogopite and chlorite (Perthuisot 1978; Perthuisot et al. 1978). Quartz and dolomite fluid inclusion studies by Perthuisot et al. (1978) show the presence of NaCl-saturated (or near-saturated) brines with variable amounts of KCl as well as CO_2, nitrogen and liquid hydrocarbons. Microthermometric measurements of the hypersaline inclusions (with no CO_2, N_2 or hydrocarbons) have enabled the formation temperatures and pressures of these minerals to be determined, and thus the thermal gradient index for the final phase of diapiric rise (Touray 1989), i.e. the early quartz and dolomite reveal the conditions of maximum burial of the Triassic series with temperatures between 200 and 300 °C and pressures of 0.6 to 1.7 kbar, and the late quartz reveals moderate pressures (<0.6 kbar) with temperatures between 150 and 240 °C. One must, however, differentiate between the fluid inclusions of the diagenetic minerals in the Triassic and the inclusions of the carbonates and sulphides (sphalerite) of the ore deposit. Thus, the Vraconian conglomerate at the base of the Bou Grine Cretaceous series contains reworked broken early quartz (Fig. 9A); here, Perthuisot (1978) provides information concerning the absolute dating of several diagenetic minerals (K/Ar meaurements on potassium feldspars and micas) that give an age of 110 Ma (late Aptian). Thus, part of the thermal effect is older than the deposit.

3.8 Isotope Geochemistry

Investigations on stable (S, C, O) and radiogenic (Sr, Pb) isotopes have provided the final elements for understanding the peridiapiric Bou Grine

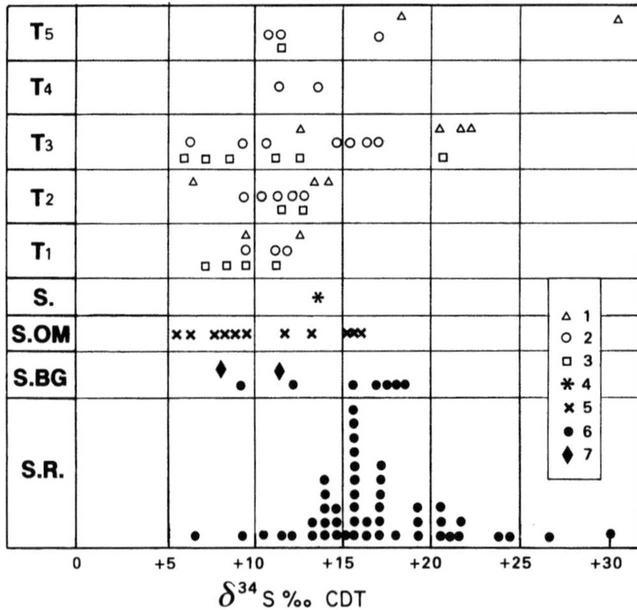

Fig. 17. Sulphur isotope compositions of the Tunisian Dome area: sulphate material and Bou Grine orebodies. *S.R.* Regional sulphates; *S.BG* Bou Grine sulphates; *S.OM* sulphur from Bou Grine deposit organic matter; *S.* sulphur from the Triassic/Cretaceous contact; *T1* type 1: lenticular orebodies in the Triassic-Cretaceous transition zone; *T2* type 2: stratiform mineralization in the limestone laminites of the Bahloul Formation; *T3* type 3: high-grade orebodies cutting the Cretaceous series; *T4* type 4: stratabound impregnations in the Cenomanian clayey limestone and marl; *T5* type 5: stratabound impregnations in the Turonian marl-limestone alternations. *1* Pyrite; *2* sphalerite; *3* galena; *4* sulphur; *5* sulphur (OM); *6* sulphate; *7* baryte

deposit, which appears to reflect a remarkable metallogenic continuity from syndiagenesis to epigenesis.

Sulphur isotope analyses of the sulphates (gypsum, anhydrite and celestite-baryte), sulphides (pyrite, sphalerite, galena) and organic matter from the deposit area can be summarized as follows (Fig. 17).

1. Regionally, the Triassic and Triassic/Cretaceous transition zone sulphates have a broad isotopic distribution from 6 to 30‰, with an average close to 15‰, a value similar to that of Triassic marine sulphates (Claypool et al. 1980). The values to either side of this mean reflect the carbonatization of sulphates at the "transition zone", as well as enrichment through bacterial metabolism.
2. At the deposit, where more than 40 measurements were made on sulphides, the values lie between 6 and 30.4‰. These values show great variability, a large part being close to or greater than the values of the Triassic sulphates and even the Cretaceous sulphates (18‰; Claypool et al. 1980). It can thus be concluded that part of the sulphur originated from bacterial reduction of sulphates possibly contained in the sediments, as indicated by pseudomorph structures.

3. The sulphur isotopes measured in the organic matter extracted from the different ore types and the barren country rocks show values of between 6 and 15‰. The carbon isotopes measured on these samples give "oceanic anoxic event" values (−23.8 to −27.6; Schlanger and Jenkyns 1976).

The S-isotope studies for the Bou Grine deposit give a wide range of positive values (6 to 30‰), some being higher than the Triassic marine sulphate values (12 to 22‰) and the Cretaceous sulphate values (18‰) (Claypool et al. 1980; Ghazban et al. 1990). The question is thus raised as to the origin of the reduced sulphate; was it only from the diapir (thus Triassic) or was there incorporation of younger sulphur from the Cretaceous cover rocks overlying the diapir? The large range of sulphur isotope values in the sulphides could also be explained by a closed system evolution, i.e. a system already isolated in relation to the diapir. This is a distinct possibility because during the transition from Cenomanian to Turonian the Lorbeus diapir was well beneath the Cretaceous cover. As in many studies of diapiric systems, several possible origins for the sulphur need to be considered in relation to the geochemical evolution of the system (sulphur from biogenic reduction of diapiric sulphate,

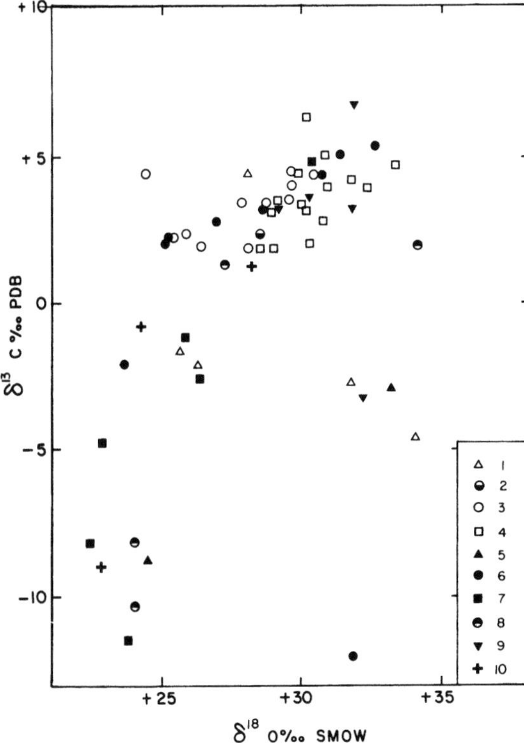

Fig. 18. Oxygen and carbon isotope compositions from the Bou Grine ore deposit. Whole rock: *1* Triassic and transition zone rocks; *2* Cenomanian rocks; *3* Bahloul limestone laminae; *4* post-Bahloul Cretaceous series. Carbonates (calcite from fractures associated with the five main ore types); *5* type 1; *6* type 2; *7* type 3; *8* type 4; *9* type 5; *10* other

sulphur from Cretaceous sediments, sulphur reduced from the formation waters) in a succession of environments with varied geochemical characteristics.

Carbon and oxygen isotope analyses of the carbonates (mineralized country rocks and calcite in veinlets and fractures, but not necessarily associated with the mineralization) fall into several large fields (Fig. 18):

1. For the mineralized country rocks, it is possible to distinguish between (a) the Triassic and transition zone rocks, with $\delta^{13}C$ (PDB) = −5 to 0 and $\delta^{18}O$ (SMOW) = 25.5 to 34, and (b) the Cretaceous rocks with $\delta^{13}C$ (PDB) = 1.5 to 6.3 and $\delta^{18}O$ (SMOW) = 24.3 to 33.3, where the Bahloul Formation (mineralized, immediate environment of the deposit, and regional) gives $\delta^{13}C$ (PDB) = 1.7 to 4.5 and $\delta^{18}O$ (SMOW) = 24.3 to 30.5
2. For the veinlet and fracture calcite, two main groups are distinguished: (a) calcite in fissures associated with the Bahloul type 2 stratiform mineralization and with the type 5 disseminated mineralization in marl and limestone, where $\delta^{13}C$ (PDB) = 1.8 to 5.3 and $\delta^{18}O$ (SMOW) = 25.1 to 32.6 and (b) calcite in fissures within the type 3 massive mineralization and in the type 1 pyrite lenses, where $\delta^{13}C$ (PDB) = −11.5 to −1.0 and $\delta^{18}O$ (SMOW) = 22.5 to 26.5.

The $\delta^{8}O$ values show that the oxygen of the host rock carbonates derive from sea-water. The $\delta^{13}C$ values of the Cretaceous country rocks also indicate a marine source. On the other hand, the carbonate facies and the transition zone (representing carbonates which show pseudomorphism of the evaporites) have $\delta^{13}C$ values lower than those of the Cretaceous country rocks. According to Holland and Malinin (1979), two sources of carbon for hydrothermal solutions are CH_4 and CO_2. These authors show that the $\delta^{13}C$ is between 0 and −9‰ if derived from CH_4, whereas a CO_2 origin gives between −20 and 40‰ (Thode et al. 1954). Another origin for the carbon is seawater, giving $\delta^{13}C$ values of −5 to 5‰.

As regards the $\delta^{13}C$ values in the calcite associated with the mineralization, two groups stand out. The one shows $\delta^{13}C$ in equilibrium with the country rock, which is the case for the calcite from the type 2 and type 5 ores, particularly from the compacted fissures or "informers" in the early ore. The other is the calcite from the type 1 and type 3 ores, and part of the high-grade type 2 ore, where the $\delta^{13}C$ is between 0 and −12‰. Thus, could there have been two possible sources of carbon? This is interesting and supports the idea that the mineralization evolved in situ from syndiagenesis to epigenesis.

Lead isotope, analyses made on sulphides (mainly galena) in 32 deposits and occurrences of northern Tunisia (Calvez et al. 1986) have revealed an overall compositional field ($^{206}Pb/^{204}Pb$ = 18.57–18.87; $^{207}Pb/^{204}Pb$ = 15.65–15.75; $^{208}Pb/^{204}Pb$ = 38.65–38.97) within which the five mineralization types from Bou Grine define a distinct narrow field ($^{206}Pb/^{204}Pb$ = 18.71–18.77; $^{207}Pb/^{204}Pb$ = 15.66–15.70; $^{208}Pb/^{204}Pb$ = 38.81–38.97). This would indicate a single source for the metals (Fig. 19) and could support remobilization of lead in a closed system.

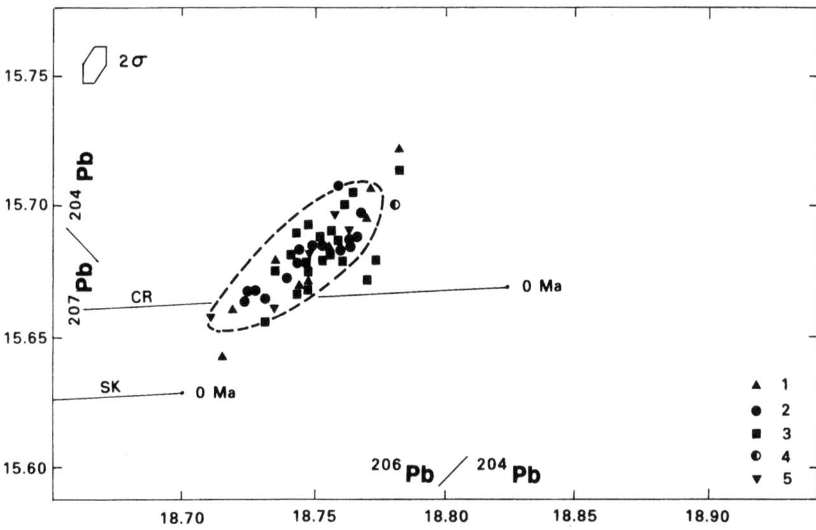

Fig. 19. Lead isotope data for the Bou Grine ore samples plotted on a $^{206}Pb/^{204}Pb$ vs. $^{207}Pb/^{204}Pb$ diagram. Types 1 to 5, see legend Fig. 18

Strontium isotope analyses have been performed on carbonates and host rocks along a vertical profile from the transition zone facies at the base to the overlying Upper Turonian (shelly layer) and along a horizontal profile of the Bahloul Formation from the distal zone (50 km) to the centre of the deposit (Fig. 20).

1. The $^{87}Sr/^{86}Sr$ signature for the non-mineralized Bahloul Formation is <0.708, close to that of coeval seawater (0.7074; Koepnick et al. 1985). The same signature is found in the shelly limestone overlying the deposit, and is thus considered as the "normal" signature of the Cretaceous carbonates in Tunisia.
2. The mineralized carbonates and the Bahloul Formation rocks in the deposit and its immediate environment contain "anomalous", more radiogenic $^{87}Sr/^{86}Sr$ signatures (0.7083–0.7087).
3. The $^{87}Sr/^{86}Sr$ signatures of the carbonate pseudomorphs (after Triassic sulphates) of the "X limestone" in the "transition zone" at the base of the deposit are similar to those observed in the mineralized Bahloul Formation.
4. Slight differences in $^{87}Sr/^{86}Sr$ ratios exist between the carbonates associated with the different types of ore, those of type 1 ore (Triassic/Cretaceous contact) being the more radiogenic (identical to those of the "transition zone" carbonates) and those of the fissures associated with the late mineralization showing the lowest values (Calvez and Orgeval 1987).

The data indicate that: (1) "anomalous" isotope compositions are encountered in the mineralized zones, and (2) the fluids acquired their "anomalous" isotope compositions by leaching rocks with a radiogenic $^{87}Sr/^{86}Sr$ signature, possibly the Triassic diapiric formations. These results are similar to those of Posey (1986) and Posey et al. (1987) from an isotopic study of Hockley dome

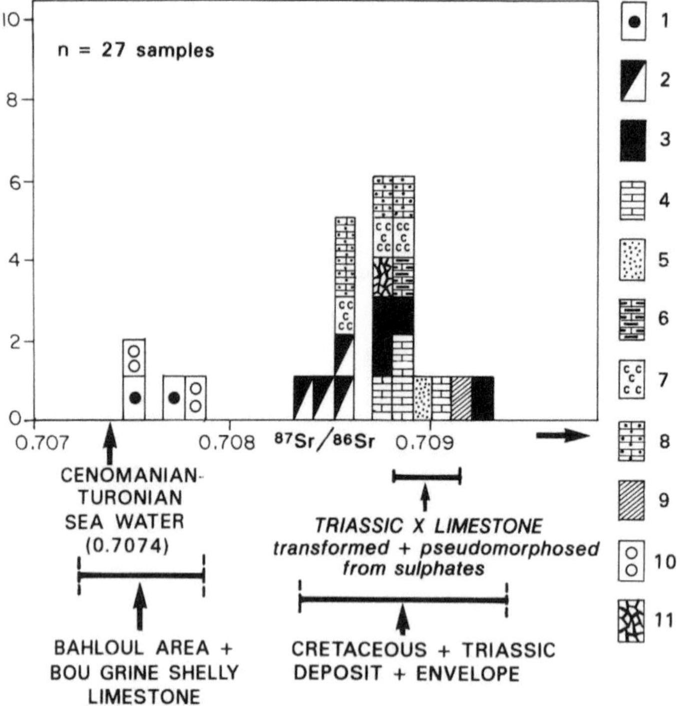

Fig. 20. Histogram of $^{87}Sr/^{86}Sr$ ratio variations in the host and country rocks of the Bou Grine deposit. *1* Bahloul Formation; *2* Bahloul envelope; *3* Bahloul mineralization; *4* X limestone (transition limestone); *5* graded horizons; *6* bedded limestone; *7* Cenomanian limestone; *8* hanging wall limestone; *9* blue horizons; *10* shelly limestone; *11* polyp breccia

cap rock, and show that for deposits like Bou Grine the strontium isotope signature can reveal the influence of fluid circulation and can thus be used as an indicator of potential metal concentration (Reesmann 1968; Hedge 1974; Kessen et al. 1981; Lange et al. 1983; Medford et al. 1983; Barbieri et al. 1984).

In spite of various ore types, the Pb isotopes indicate that the lead was derived from a single source, unless one of the sources was far more preponderant than the others. This, in addition to the fact that the deposit shows a multi-stage development within a closed environment away from outside influence, agrees well with the concept of a large initial metal stock. The Sr isotopes clearly show the deposit "effect" and its associated palaeocirculations, both laterally in the Bahloul Formation and vertically up to the sedimentary level which seals the synsedimentary faults, i.e. the shelly beds (Figs. 6, 7, 8), and marked the end of the system (Calvez and Orgeval 1988).

4 Elements of a Metallogenic Model

The Bou Grine deposit differs from models proposed for base-metal deposits associated with diapiric structures, in particular their cap rocks (Price et al.

1983; Price and Kyle 1983, 1986; Orgeval et al. 1985a,b, 1986, 1989; Rouvier et al. 1985; Charef and Sheppard 1987; Posey et al. 1987; Montacer et al. 1988; Perthuisot and Rouvier 1990). An important difference is the early type 2 mineralization hosted by limestone laminites of the Cenomanian-Turonian transition with a metal weight representing 50% of the total deposit tonnage (Orgeval et al. 1981). On the other hand, the main features of the Bou Grine deposit, from the geological situation to geochemical associations, include characteristics of many of the geological models proposed for MVT deposits as summarized by Noble (1963), Beales and Jackson (1966), Pelissonnier (1967), Connan and Orgeval (1973, 1977), Ohle (1980), Anderson and MacQueen (1982), Sangster (1983), Powell and MacQueen (1984), Sverjensky (1984), as well as in the recent compilation of Sangster (1990).

4.1 Discussion

Field observations and petrographic and geochemical studies have provided some indications and a few answers concerning the origin of economic Pb-Zn concentrations in a diapiric setting. Palaeogeographic and palaeostructural events provided favourable conditions for early fixing of a metal stock whose origin is to be looked for in connection with (1) the existence of deep faults, whose role was to enable diapirism by deforming the substratum, and (2) the development of a subsident sedimentary basin with local formation of black shale facies.

An active role of organic matter in the early fixing and precipitation of metal sulphides has often been evoked; the development of our understanding of this process is provided by Winters and Williams (1969), Connan and Orgeval (1973, 1977), Connan et al. (1975), and more recently Disnar and Sureau (1990) and Disnar (1990). One of the major features of the Bou Grine deposit is the high organic content in the laminites of the Bahloul Formation (4–5% TOC), which is of purely marine origin and shows weak maturity at the end of diagenesis, whereas the massive type 5 mineralization only contains 0.6% TOC. Hydrocarbon enrichment at certain levels suggests early migration towards beds initially poor in organic content, thus forming a reservoir; these levels were affected by fluid circulations that elsewhere gave rise to early microsphalerite mineralization.

The numerous traces of biological activity (e.g. initial biodegradation affecting certain carbon compounds; Montacer 1989), almost certainly of bacterial origin, indicate that sulphate reduction in the laminite facies was biologically controlled (Montacer et al. 1988), with the evolution of the organic matter probably occurring by in situ maturation of the source rock, since the Triassic provided only a very weak thermal source (indicated elsewhere by the formation of various diagenetic minerals) possibly due to an early loss of heat stored by the salt formations being caused by the thinness, if not absence, of Lower Cretaceous deposits (Perthuisot 1978).

The Bou Grine model must take into account the early association of microsphalerite ($<10\,\mu m$) and pyrite framboids which were affected by contemporaneous diagenetic deformation (compaction). This is an important chronological factor: Hill and Wedow (1972) used similar factors to establish a chronology in the East Tennessee Zinc District. Another point to be considered concerns simple parageneses that differ both spatially and regionally: e.g. the type 1 ore at Bou Grine is characterized by the presence of baryte and baryte-celestite, but at regional scale shows variability in element distribution: Zn = Pb at Fedj el Adoum; Fe > Zn > Pb at Bou Grine; Fe at Kebbouch Sud; Ba, Sr (and F) at Doggra; to the northwest, some parageneses described by Hatira et al. (1990) for the Sakiet diapir deposit reflect a Tertiary volcanic influence. The multi-phase model must therefore contain several mechanisms, or more precisely several channels, for supplying these elements from a single initial stock (as shown by the lead isotope ratios).

A close relationship between the initial metal stock and the sedimentological and petrographic environment is supported by isotope studies. In particular, these show that the sulphur isotope developed in a closed system. They also show that the deposit and its aureole are characterized by radiogenic values of the strontium isotopes, indicating palaeocirculations whose effects are also revealed by the migration of organic compounds and by the diagenetic facies of the mineralization. Although these are distinctive characteristics of the Bou Grine deposit, the geological and chronological elements (Fig. 21) were essential for the hydrodynamism of the system.

The source of the metal has been debated in relation to several hypotheses, some of which are classic for MVT deposits. Thus:

1. Deep fractures, proposed by Perthuisot (1978) to have initiated the diapiric movements, could have enabled the rise of deep fluids, circulating preferentially along the diapirs.
2. The Triassic has often been advocated as possibly playing a role in supplying the metals, i.e. in leaching and transporting a metal stock. Certainly, the

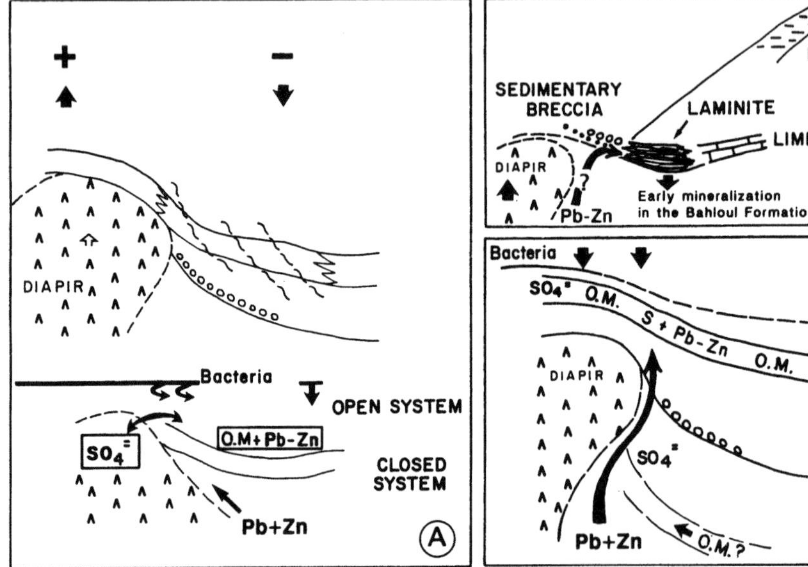

Fig. 21A,B. Elements of a metallogenic model

baryte and celestite paragenesis (with some rare sulphides) may have been associated with the Triassic formations. On the other hand, this paragenesis if found at different levels in the stratigraphic column (Boulhel 1985).

3. The Cretaceous sediments themselves are carriers of metallic elements expulsed during compaction of the series. This relates to a hypothesis, often put forward (Beales and Jackson 1966; Jackson and Beales 1967; Connan and Orgeval 1976, 1977), for which the principal "components" are observed at Bou Grine: i.e. the metal source is the sedimentary series, with both metals and hydrocarbons being transported, during compaction, by chloride (calcic and sodic) brines (Carpenter et al. 1974; Price and Kyle 1986); the reservoir would have been a site of confluence between ascending brines containing metals and hydrocarbons and surface waters rich in H_2S (from biogenic reduction of sulphates in the presence of organic matter). This model would explain the formation of the early diagenetic microsphalerite in the Bhaloul Formation. The limiting factors of this model are that it requires (a) a certain maturation of the organic matter and (b) a shallow-water environment for the addition of the sulphate-reducing bacteria. The Cretaceous evolution of the Domes area, with the presence of black shale horizons, particularly in the Cenomanian culminating in the Bahloul series, was obviously favourable for the fixation and redistribution of metals according to such a model.
4. The development of stratified brines in restricted basins, having conditioned the black shale-type deposits, and the preservation of organic matter, could also have contributed to the addition of metals. Busson (1984a,b) describes in detail the system of brine development during the Cenomanian over extensive coastal areas formed by the Cenomanian transgression. Noting the clayey-gypsiferous character of the resultant deposits and the fact that the brines did not deposit sodium chloride, he concluded that the heavy brines flowed from the upstream zones towards the semi-closed basins of the Atlantic. In this way the brines could have infiltrated the peridiapiric basins of the Triassic in northern Tunisia.

4.2 Metallogenic Model

The *Bou Grine model*, which has required the incorporation of all these hypotheses to describe the formation of the deposit and to understand its environment, begins with the initiation of a diapiric system through the movement of a deep Middle Cretaceous structure, and consequent sedimentary differentiation that structured the future site of the deposit. This was followed by a confrontation of environmental conditions: (1) rising of the salt mass: shoals and emersion (reefs, erosion, solution and redeposition, etc.; Fig. 21A); (2) transgression over an irregular surface, especially in the peridiapiric trough (rim syncline); (3) repetitious pulsation of the diapir, with migration of the sedimentary basin causing (a) chemical stratification in the water with, in particular, isolation of the bottom (Fig. 21A) and (b) abrupt thickness and facies variations in the deposits (Fig. 21B). Then came synsedimentary movement giving rise to (1) breccias, conglomerates, sorting, etc.; (2) tilting of panels by listric faults; and (3) overlapping of sedimentary bodies (sheets or wedges) due to progradation or to basal truncation from "upstream" to "downstream". The deposit site seems, from the beginning, to have been a submarine slope at the edge of a structural high which, other than for a few local diapirs, remained immersed at such a depth that a true shallow-water carbonate platform could never develop; the mineralization is related, therefore, to the formation of the few salt diapirs with production and, above all, preservation of organic matter in a peridiapiric depression (evolving into a rim syncline?) and very early concentration of metal (notably zinc). The system developed with reconcentration always at the same point – the diapir – which (1) provided easy passage for the fluids (Fig. 21C; Pelissonnier 1962, 1967; Ulrich et al. 1984; Perthuisot et al. 1987; Perthuisot and Rouvier 1990); (2) played an intensified role in providing a favourable (but not necessarily concentrating) environment; and (3) formed traps. Without the diapiric traps to enable a convergence of the mineralizing metallogenic elements from diagenesis to epigenesis, the Cretaceous Zn would have remained a simple anomaly in a sedimentary series (as in the black shales of the French Toarcian).

Thus, we have a combination of events that could lead, from the sedimentary and diagenetic standpoints, to conditions suitable for fixing metals. However, as Shanks et al. (1987) have pointed out for the Selwyn Basin and the "Early Cambrian anoxic event" "an additional source of zinc and lead is probably required to turn an anoxic basin into a Pb-Zn province".

5 Conclusions

The Cretaceous province of northern Tunisia is apparently only just beginning to reveal its potential. Although the rifting and the "E2 event" provided an overall favourable setting, it was the local conditions (synsedimentary movement, variation of facies, etc.) resulting from the halokinetics that created the metallogenic trap. The Bou Grine Pb-Zn mineralization reveals the importance of diapiric control in the environment and in trapping sulphides. Three types of economic deposits occur at this site: type 1 is a lenticular pyritic body in the Triassic-Cretaceous transition zone, where the X limestone reflects sulphate-carbonate pseudomorphs (limited cap rock type); type 2 is a basin-type accumulation of zinciferous limestone laminites formed in a specific peridiapiric site (rim syncline?) with early, extremely fine sphalerite crystals associated with framboidal pyrite; type 3 is a high-grade massive orebody in a carbonate reservoir – this accumulation was controlled by a conjugate fracture network and reworked the type 2 ore.

Apart from its diapiric associations, which nevertheless differ from those of evolved cap rock deposits, could the Bou Grine deposit, in fact, be classed within the vast and imprecise MVT group? It is contained in a basin site; it is associated, indirectly, with deep fractures of the substratum which also controlled the diapirism. Dolomitization is absent (other than in the Triassic and the transition zone). The mineralization was early with its initial phase characterized by extreme fineness. It reflects an association of hydrocarbon source rock and zinc-bearing beds. Structures and details of the type 2 ore in the laminites suggest continuous syndiagenetic to epigenetic circulation. Many of these characteristics are those of the sedimentary exhalative stratiform Zn-Pb sulphide deposits (Sangster 1983). In addition, the presence of volcanic rocks is reported, from oil drilling, in Cretaceous series of the same age in the Pelagian Sea, some 150 km to the east of Bou Grine (although no relationship has yet been established).

The overall economic importance of such deposits would appear to be underestimated and a closer examination of areas with diapirism and near lead-zinc provinces is obviously warranted.

Acknowledgements. This work, since its inception, has had the constant support of M. Zidi, who was then Director General of the Office National des Mines de Tunisie (ONM), and later of A. Attia and M. Zerelli, who succeeded him in this post. I should like to offer them my sincere appreciation and thanks. I should also like to thank Mr. R. Sahli, Project Head and Director of the Mineral Deposit section of the ONM, who collaborated in all the field operations.

Mention needs to be made of all my BRGM colleagues (too numerous to list) who, in both the field and laboratory, have formed part of the ONM-BRGM multi-disciplinary team that made this study possible, and of M. Tixeront, BRGM Director for Arab countries, who provided permanent support for the project.

Finally, I should like to especially thank two people who have made significant contributions to the preparation of this document: Prof. Lluis Fontboté of the University of Geneva, and H. Zeegers, Head of the BRGM Department of Exploration.

The preparation for this article (BRGM contribution No. 92038) was made possible by the financial support of BRGM research project RM 19.

References

Anderson GM, MacQueen RW (1982) Ore deposits modes-6 Mississippi Valley-type lead-zinc deposits. Geosci Can 9:108–117

Barbieri M, Masi U, Tolomeo L (1984) Strontium geochemical evidence for the origin of the barite deposits from Sardinia, Italy. Econ Geol 79:1360–1365

Beales FRW, Jackson SA (1966) Precipitation of lead-zinc ores in carbonate reservoir as illustrated by Pine Point ore field, Canada. Trans Inst Min Metall 75B:278–285

Berthon L (1922) L'industrie minérale en Tunisie. Direction générale des transports publics, Service des mines, Tunis, 272 pp

Bles JL (1982) Etude structurale de la mine de Bou Grine, du sondage BG25 et du flanc nord de l'anticlinal de Lorbeus (Région de Kef, Tunisie). BRGM Rep 82 SGN 752 GEO, 24 pp

Bolze J (1952) Diapirisme et métallogénie en Tunisie. XIXe Int Geol Congr, Alger XII, XII:91–104

Boulhel S (1985) Composition chimique, fréquence et distribution des minéraux de la série barytine-celestine dans les gisements de fluorine de Hamman Jedidi et Hamman Zriba, Jebel Guelbi. Bull Mineral 108:403–420

Burollet PF (1956) Contribution à l'étude stratigraphique de la Tunisie centrale. Ann Min Géol (Tunis) 18:350

Busson G (1984a) La sédimentation épicontinentale des bassins salins accomplis et des bassins évaporitiques avortés: effet sur la sédimentation océanique contemporaine et contigüe dans le cas du Crétacé saharien et nord-atlantique. CR Acad Sci Paris 209 II(5):213–216

Busson G (1984b) Relation entre les gypses des plate-formes du Nord-Ouest africain et les black shales de l'Atlantique au Cenomanien inferieur moyen. CR Acad Sci Paris 298 II(28):801–804

Caia J, Orgeval JJ, Sahli R (1980) Bilan des résultats du projet Pb-Zn Tunisie Centre NOrd et éléments d'orientation pour une prospection générale de la zone des Dômes. BRGM Rep RDM/AAO 906, 12 pp

Calvez JY, Orgeval JJ (1987) Signature isotopique $^{87}Sr/^{86}Sr$ "anormales" des carbonates et roches du gisement de Bou Grine (Tunisie): traceur de circulation de fluides et indicateur de proximité de minéralisation. BRGM Principal Sci Tech Results 1987:127–129

Calvez JY, Orgeval JJ (1988) Pb and Sr isotopic studies of the carbonate hosted Zn-Pb deposit of Bou Grine (Tunisia) and its environment. Genetic implications and possible use in exploration. Géochimie et Cosmochimie, Paris 29/8–2/9, 1988. In: Chemi Geol Spec Issue 70(1/2):133 (Abstr)

Calvez JY, Orgeval JJ, Marcoux E (1986) Etude isotopique du plomb du gisement de Bou Grine (zone des Dômes, Tunisie) et comparaison avec quelques données de la province tunisienne. BRGM Principal Sci Tech Results 1986:121–123

Carpenter AB, Trout ML, Pickett EE (1974) Preliminary report on the origin and chemical evolution of lead and zinc-oil field brines in central Mississippi. Econ Geol 69:1191–1206

Charef A, Sheppard S (1987) Pb-Zn mineralization associated with diapirism: fluid inclusion and stable isotope (H, C, O) evidence for an origin and evolution of the fluids at Fedj el Adoum, Tunisia. Chem Geol 61:113–134

Claypool GE, Molsen WT, Kaplan IR, Sakai M, Zak I (1980) The age curves of sulfur and oxygen isotopes in marine sulfate and their mutual interpretation. Chem Geol 28:199–260

Connan J, Orgeval JJ (1973) Les bitumes des minéralisations barytiques et sulfurées de Saint-Privat (bassin de Lodève, France). Bull Cent Rech Pau-SNPA 7(2):557–585

Connan J, Orgeval JJ (1976) Un exemple de relation hydrocarbures-minéralisations: cas du filon de barytine de St-Privat (bassin de Lodève, France). Bull Centre Rech Pau SNPA 10(1): 359–374

Connan J, Orgeval JJ (1977) Un exemple d'application de la géochimie organique en métallogénie: la mine des Malines (Gard, France). Bull Cent Rech Explor Prod Elf-Aquitaine 1:59–105

Connan J, Le Tran K, Van Der Weide B (1975) Alteration of petroleum in reservoirs. Proc 9th World Petrol Congr, vol 2. Applied Science Publishers, London, pp 171–178

Disnar JR (1990) Apport de la géochimie organique à la compréhension de la genèse des gîtes Zn-Pb en environnement sédimentaire. In: Mobilité et concentration des métaux de base dans les couvertures sédimentaires. Manifestations, mécanismes, prospection (March 1988). BRGM Doc 183:249–268

Disnar JR, Sureau JF (1990) Organic matter in ore genesis: progress and perspectives. Org Geochem 16(1–3):577–599

Disnar JR, Gauthier B, Chabin a, Trichet J (1986) Early biodegradation of ligneous organic materials and its relation to ore deposition in the Trèves Zn-Pb orebody (Gard, France). In: Leythaeuser D, Rullkotter J (eds) Advances in organic geochemistry 1985. Org Geochem 10:1005–1013

Germanov A (1965) Geochemical significance of organic matter in hydrothermal process. Geochim Int 2:643–652

Gharbi M, Orgeval JJ, Sahli R (1979) Note sur les principaux résultats de l'opération de Coopération TCN pour l'année 1978–1979: quelques perspectives techniques. ONM/BRGM Rep, 6 pp

Ghazban F, Schwarcz H, Ford D (1990) Carbon and sulfur isotope evidence for in situ reduction of sulfate, Nanisivik lead-zinc deposit, Northwest Territories, Baffin Island, Canada. Econ Geol 85:360–375

Giot D, Monciardini C (1984) Etude sédimentologique et biostratigraphique du profil de sondages 3N; Bou Grine (Tunisie). BRGM Rep 84 SGN 123 GEO, 58 pp

Giot D, Monciardini C, Le Strat P (1982) Etude lithostratigraphique du sondage de Bou Grine 25 – Projet TCN. BRGM Rep 82 SGN GEO ES 098, 16 pp

Graciansky PC de, Brosse E, Deroo G, Herbin JP, Montadert L, Müller C, Sigal J, Schaaf A (1982) Les formations d'âge crétacé de l'Atlantique Nord et leur matière organique: paléographie et milieu de dépôts. Rev Inst Fr Petrôl 37(3):275–337

Guirand R, Maurin JC (1991) Le rifting en Afrique au Crétacé inférieur: synthèse structurale, mise en évidence de deux étapes dans la génèse des bassins, relation avec les ouvertures océaniques peri-africaines. Bull Soc Geol Fr 162(5):811–823

Hanna MA, Wolf AG (1934) Texas and Louisiana salt dome caprock minerals. Am Assoc Petrol Geol Bull 8:212–225

Hatira N, Perthuisot V, Rouvier H (1990) Les minéraux à Cu, Sb, Ag, Hg des minerais à Pb-Zn de Sakiet Koucha (diapir de Sakiet Sidi Youssef, Tunisie Septentrionale). Mineral Depos 25:112–117

Hedge CE (1974) Strontium isotopes in economic geology. Econ Geol 69:823–825

Herbin JP, Montadert L, Müller C, Gomez R, Thurow J, Weidmann J (1986) Organic-rich sedimentation at the Cenomanian-Turonian boundary in oceanic and coastal basins in the

North Atlantic and Tethys. In: Summerhayes, Shackleton (eds) North Atlantic paleo-oceanography. Geol Soc Spec Publ 21:389–422

Hill WT, Wedow H Jr (1972) An early Middle Ordovician age for collapse breccias in the East Tennessee Zinc District as indicated by compaction and porosity features. Econ Geol 66:725–734

Holland HD, Malinin SD (1979) The solubility and occurrence of non-ore minerals. In: Barnes H (ed) Geochemistry of hydrothermal ore deposits. Springer, Berlin Heidelberg New York, p 142

Jackson SA, Beales FW (1967) An aspect of sedimentary basin evolution: the concentration of Mississippi Valley type ores during late stage of diagenesis. Can Petrol Geol Bull 50:383–433

Jenkyns HC (1980) Cretaceous anoxic events: from continents to oceans. Geol Soc Lond J 137:171–188

Kessen KM, Woodruff MS, Grant NK (1981) Gangue mineral $^{87}Sr/^{86}Sr$ ratios and the origin of Mississippi Valley-type mineralisations. Econ Geol 76:913–920

Koepnick RB, Burke WH, Denison RE, Hetherington EA, Nelson HF, Otto JB, Waite LE (1985) Construction of the seawater $^{87}Sr/^{86}Sr$ curve for the Cenozoic and Cretaceous: supporting data. Chem Geol 58:55–81

Kyle JR, Price PE (1986) Metallic sulphide mineralization in salt-dome cap rocks, Gulf Coast, USA. Trans Inst Min Metal B95:B6–B16

Laatar E (1980) Gisements de plomb-zinc et diapirisme du Trias salifère en Tunisie septentrionale. Les concentrations péridiapiriques du district minier de Nefate-Fedhj el Adoum. Thesis, Univ P et M Curie, Lab Géologie Appl, 280 pp

Laforet C (1979) Etudes minéralogiques. BRGM Rep SGN/MGA 79/190

Laforet C (1981) Etudes minéralogiques. BRGM Rep SGN/MGA 81/1261

Lange S, Chaudhuri S, Claver N (1983) Strontium isotopic evidence for the origin of barites and sulfides from the Mississippi Valley-type ore deposits in southeast Missouri. Econ Geol 78:1255–1261

Leduc C, Orgeval JJ (1979) Projet TCN: prospection géochimique tactique du secteur de Bou Grine: résultats et interpétation. ONM-BRGM Rep, 15 pp (5 maps)

Lenindre YM (1979) Mission TCN: étude petrographique de 29 échantillons. BRGM Rep ES(01) 01-79

Lenindre YM, Laforet C, Sureau JF (1979) Projet TCN: essai d'interprétation des processus diagénetiques. BRGM Rep 79 SGN 633 GEO, 14 pp

Levat D (1894) Gisements de phosphate de chaux et gisements de calamine de la Tunisie. 23rd Congr Assoc Française Avancement Sci, Caen, CR, pp 420–431

Machel HG (1989) Relationships between sulfate reduction and oxidation of organic compounds to carbonate diagenesis, hydrocarbon accumulations, salt domes, and metal sulfide deposits. Carbonates Evaporites 4(2):137–151

Marie J, Trouve P, Desforges G, Dufaure P (1984) Nouveaux éléments de paléographie du Crétacé de Tunisie. Comp Fr Pétrol Notes Mém 19, 37 pp

Marteau P (1987) Etude petrographique d'échantilllons minéralisés du Crétacé supérieur de Tunisie (Mine de Bou Grine et région avoisinants). BRGM Rep 87 GEO SED 033, 44 pp

Massin JM (1972) Mine de Bou Grine: étude géologique et minière. ONM Rep, 40 pp

Medford GA, Maxwell RJ, Armstrong RL (1983) $^{87}Sr/^{86}Sr$ ratio measurements on sulfides, carbonates, area fluid inclusions from Pine Point, Northwest Territories, Canada: an $^{87}Sr/^{86}Sr$ ratio increase accompanying the mineralization. Econ Geol 78:1375–1378

Mining Journal (1992) ICF funding for Tunisian mine. March 6:162

Monciardini C, Orgeval JJ (1987) Apport de la biostratigraphie à la connaissance du gisement (Zn-Pb) de Bou Grine (Tunisie). BRGM Principal Sci Tech Results 1987:125–126

Montacer M (1989) Etude de la matière organique de la formation Bahloul (Cenomano-Turonien) dans l'environnement sédimentaire du gisement Zn-Pb de Bou Grine (Nord-Ouest de la Tunisie). Thesis, Univ Orléans, 211 pp

Montacer M, Disnar JR, Orgeval JJ (1986) Etude préliminaire de la matière organique de la formation Bahloul dans l'environnement du gisement Zn-Pb de Bou Grine (Tunisie). BRGM Principal Sci Tech Results 1987:121

Montacer M, Disnar JR, Orgeval JJ, Trichet J (1988) Relationship between Zn-Pb ore and oil accumulation processes: example of the Bou Grine deposit (Tunisia). Org Geochem 13(1–3):423–431

Monthel J, Lagny Ph, Orgeval JJ (1986) Inventaire des gisements et indices de la zone des Dômes de l'Atlas Tunisien et de ma bordure septentrionale. BRGM Rep 86 TUN 1127 GMX, 37 pp (1 map)
Nicolini P (1968) Gisements plombo-zincifères de Tunisie. Symposium sur les gisements de plomb-zinc en Afrique. Ann Min Géol (Tunis) 23:270–240
Nicolini P (1970) Gîtologie des concentrations minérales stratiformes. Gauthiers-Villars, Paris, 792 pp
Noble EA (1963) Formation of ore deposits by water of compaction. Econ Geol 58:1145–1156
Nooner DW, Updegrove WS, Flory DA, Oro J, Mueller G (1973) Isotopic and chemical data of bitumens associated with hydrothermal veins from Windy Knoll, Derbyshire, England. Chem Geol 11:198–202
Ohle EL (1980) Some considerations in determining the origin of ore deposits of the Mississippi Valley type. Part II. Econ Geol 75:161–172
Orgeval JJ (1981) Bilan des travaux 1977–1981. ONM-BRGM Rep, 18 pp
Orgeval JJ, Legendre O (1989) Etude des inclusions fluides des minéralisations Zn-Pb de Bou Grine (Atlas Tunisien). BRGM Principal Sci Tech Results 1987:117–118
Orgeval JJ, Giot D, Sahli R, Gharbi M, Le Nindre YM, Monciardini C (1981) Caractérisation de concentrations sulfurées associées au faciès de calcaire Bahloul (Cénomanien supérieur – Turonien inférieur) du secteur de Jebel Bou Grine (Gouvernorat du Kef, Tunisie). Ist Nat Earth Sci Congr, Tunis, Abstr Univ Tunis, p 148
Orgeval JJ, Sahli R, Lagny Ph, Monthel J, Abdelaoui F, Gharbi M, Giot D, Karoui J, Leduc C (1985a) Le gisement Pb-Zn de Bou grine (Atlas Tunisien): synthèse géologique. BRGM Rep 85 DAM 043 GMX, 157 pp
Orgeval JJ, Giot D, Karoui J, Monthel J, Sahli R (1985b) Le gisement Pb-Zn de Bou Grine (Atlas tunisien), un nouveau type de concentration Pb-Zn liée à l'activité d'un haut fond diapirique dans le sillon tunisien au Crétacé supérieur. In: BRGM Principal Sci Tech Results 1985:120–121
Orgeval JJ, Giot D, Karoui J, Monthel J, Sahli R (1986) Le gisement de Bou Grine (Atlas tunisien). Description et historique de la découverte. Chron Rech Min 482:5–32
Orgeval JJ, Giot D, Karoui J, Monthel J, Sahli R (1989) The discovery and investigation of the Bou Grine Pb-Zn deposit (Tunisian Atlas). Chron Rech Min Spec Iss: 53–68
Ourisson G, Albrecht P, Rohmer M (1979) The hopanoids paleochemistry and biochemistry of a group of natural products. Pure Appl Chem 51:709–729
Ourisson G, Albrecht P, Rhomer M (1982) Predictive microbial biochemistry from molecular fossil to procaryotic membranes. Trends Biol Soi: 236–239
Pelissonnier H (1962) Un facteur de concentration métallique: l'étranglement. CR Acad Sci Paris 225:2792–2794
Pelissonnier H (1967) Analyses paléohydrogéologiques des gisements stratiformes de plomb-zinc-baryte-fluorite du type "Mississippi Valley". Econ Geol Monogr 3:234–252
Pering KL (1973) Bitumens associated with lead, zinc and fluorite ore minerals in North Derbyshire, England. Geochim Cosmochim Acta 37:401–417
Perthuisot V (1977) Un exemple de diapirisme polyphasé en Tunisie: le Djebel Lorbeus. Géol Médit VI (4):465–476
Perthuisot V (1978) Dynamique et pétrogenèse des extrusions triasiques en Tunisie septentrionale. Thesis, Ecole Normale Sup Paris, 312 pp
Perthuisot V, Rouvier H (1990) Les relations métal-soufre-eau-hydrocarbures-microorganismes et la genèse des concentrations de sulfures et de soufre des diapirs évaporitiques. In: Proc Meet Mobilité et concentration des métaux de base dans les couvertures sédimentaires, Paris-Orléans (1988). Doc BRGM 183, pp 269–278
Perthuisot V, Guillaumou N, Touray JC (1978) Les inclusions fluides hypersalines et gazeuses des quartz et dolomies du Trias évaporitique Nord-tunisien. Essai d'interprétation géodynamique. Bull Soc Geol Fr 7, XX(2):145–155
Perthuisot V, Hatina N, Rouvier H, Steinberg M (1987) Concentration métallique (Pb-Zn) sous un surplomb diapirique: Exemple de Jebel Bou Khil (Tunisie Septentrionale), Bull Soc Geol Fr 8, III(6):1153–1160
Pervinquière L (1903) Etude géologique de la Tunisie centrale. de Rudeval FR (ed), Paris, p 359
Posey HH (1986) Isotope geochemistry of Hockley Dome cap rock. In: Seni SJ, Kyle R (eds) Comparison of cap rocks, mineral resources and surface features of salt domes in the Houston diapir province. Geol Soc Am Meet, San Antonio, Texas, pp 185–208

Posey HH, Kyle JR, Jackson TJ, Hurst SD, Price PE (1987) Multiple fluid components of salt diapirs and salt dome cap rocks, Gulf coast, USA Appl Geochem 2:523–534

Powell TG, MacQueen RW (1984) Precipitation of sulfide ores and organic matter: sulfate reactions at Pine Point, Canada. Science 224:63–66

Price PE, Kyle JR (1983) Metallic sulphide deposits in Gulf Coast salt-dome caprocks. Gulf Coast Assoc Geol Soc Trans 33:189–193

Price PE, Kyle JR (1986) Genesis of salt dome hosted metallic sulfide deposits: the role of hydrocarbons and related fluids. In: Dean WE (ed.) Proc Symp on Organics and ore deposits. Denver Region. Exploration Geologists Society, pp 171–184

Price PE, Kyle JR, Wessel GR (1983) Salt dome related zinc-lead deposits. In: Kisvarsanyi G, Grant SK, Pratt WP, Koenig JW (eds) Proc Vol Int Conf on Mississippi Valley type lead-zinc deposits. University of Missouri, Rolla, MO, pp 558–573

Reesmann RH (1968) The Rb-Sr analyses of some sulfide mineralization. Earth Planet Sci Lett 5:23–26

Rouvier H, Perthuisot V, Mansouri A (1985) Pb-Zn deposits and salt-bearing diapirs in southern Europe and North Africa. Gisements de Pb-Zn et diapirs salifères: proposition d'un modèle pour la mise en place des concentrations diapiriques. Econ Geol 80(3):666–687

Sahli R, Giot D, Orgeval JJ, Le Nindre YM, Sureau F, Snoep J (1981) Schéma des principaux événements diagénétiques affectant les séries sulfato-carbonatées associées aux minéralisations plomb-zinc du Jebel Kebbouch, Jebel Lorbeus et Jebel Dogra (Gouvernorat du Kef, Tunis). Essai d'interprétation diagénétique. 1st Natl Earth Sci Congr, Tunis, Univ Tunis, Abstr, p 89

Sainfeld P (1952) Les gîtes plombo-zincifères de Tunisie. Ann Min Géol (Tunis) 9:285

Sangster DF (1983) (ed) Sediment-hosted stratiform lead-zinc deposits. Mineralogical Association of Canada, Short course handbook, vol 8, 309 pp

Sangster DF (1990) Mississippi Valley type and Sedex lead-zinc deposits: a comparative examination. Tran Inst Min Metal B:21–42

Schlanger SO, Jenkyns HC (1976) Cretaceous oceanic anoxic events: causes and consequences. Geol Mijnbouw 55(3–4):179–184

Shanks WC, Goochuff LG, Jilson GA, Jennings DS, Modene JC, Ryan BD (1987) Sulfur and lead isotope studies of stratiform Zn-Pb-Ag deposits, Anvil Range, Yukon: basinal brine exhalation and anoxic bottom water mixing. Econ Geol 82:600–634

Sverjensky DA (1984) Oil field brines as ore-forming solutions. Econ Geol 79:23–37

Thode HG, Wanless RK, Wallouch R (1954) The origin of native sulphur deposits from isotope fractionation studies. Geochim Cosmochim Acta 5:286–298

Tissot B, Pelet R, Roucaché J, Combag A (1977) Utilisation des alcanes comme fossiles géochimiques indicateurs des environments géologiques. In: Campas R, Goni J (eds) Advances in organic geochemistry 1975. Enadimsa, Madrid, pp 117–154

Touray JC (1989) Etude d'inclusions fluides et modelisation de la genese des gites de type "Mississippi Valley". Chron Rech Min 495:21–30

Ulrich MR, Kyle JR, Price PE (1984) Metallic sulfide deposits in the Winnfield salt dome, Louisiana: evidence for episodic introduction of metalliferous brines during cap rock formation. Gulf Coast Assoc Geol Soc Trans 34:435–442

Williams JA, Winters JC (1969) Microbial alteration of crude oil in the reservoir. In: Symp Petideum Transformation in Geologic Environments. Am Chem Soc 158th Natl Meet New York, pp 7–12

Exploration and Economics

Lithogeochemical Investigations Applied to Exploration for Sediment-Hosted Lead-Zinc Deposits

N.G. Lavery[1], D.L. Leach[2] and J.A. Saunders[3]

Abstract

Sediment-hosted stratiform massive sulphide (Sedex) deposits and Mississippi Valley-type (MVT) deposits differ in morphology, but there is considerable overlap in their host rock lithology, regional setting, size, grade, metal ratios, age of host rocks, lead isotopes, sulphur isotopes, and genesis. Recent advances in the understanding of basin-derived formation fluids provide a framework for interpreting the lithogeochemical signatures of these deposits. The composition of the ore fluids for both types of deposits is controlled in part by fluid-rock reactions within sedimentary basins, but the nature of the sediments in the basins, the source of the fluids, the thermal gradient in the basin, the migratory history of the fluids, and the physical and chemical conditions in the depositional environment cause variations in the lithogeochemical patterns present. Studies of modern sedimentary formation waters indicate that diagenetic fluid-rock reactions in basins can liberate and dissolve significant amounts of metals. The most metal-enriched formation waters are brines with more than 200 000 ppm total dissolved solids. Some brines contain more than several hundred ppm combined Fe, Zn, Pb, and Ba, and minor amounts of Cu, Ag, Mn, Cd, As, Sb, Ni, Co, Mo, and Tl.

Copper, Ag, As, Sb and Ba are reported in more than 60% of the Sedex deposits studied; Mn, Cd, Ni, Co, Au, Hg, Bi, B, Sn, Mo, F, Tl, Ge and Sr are reported in more than 20%; and V, Ga, In, Se, Cr and NH_4^+ are reported in more than 10% of the deposits. Geochemical variables which have received *some* exploration attention and which show promise as guides to Sedex deposits include: Tl, F, NH_4^+, Hg, Pt, Pd, Zr, Cl-bearing minerals, rare earth elements, stable isotopes, and carbonaceous material.

MVT districts were formed by ore-forming fluids which affected large volumes of the crust. Widespread and isolated geochemical anomalies are, therefore, to be expected. Although there is a great diversity in deposit characteristics, it is possible to establish exploration criteria for additional deposits within a district or region. Geochemical analysis of insoluble residues has been used to define geochemical patterns that coincide with known ore trends and with ore-related geological features in the Southeast Missouri districts. Studies of rock alteration or identification of minerals precipitated from the brines can

[1] Minerals Exploration Consultant 5545 Skyway Drive, Missoula, MT 59801, USA
[2] United States Geological Survey Box 25046 Federal Center, Denver, CO 80225, USA
[3] Department of Geology, Auburn University, Auburn, AL 36849, USA

Spec. Publ. No. 10 Soc. Geol. Applied to Mineral Deposits
Fontboté/Boni (Eds.), Sediment-Hosted Zn-Pb Ores

provide evidence of pathways used by the ore fluids. Hydrothermal dolomitization is the most important type of alteration; silicification is associated with ore in some districts, and alteration of feldspars near ore has been identified in the Southeast Missouri lead district. Organic geochemistry and fluid inclusion studies are in progress and may yield practical exploration guides to MVT deposits.

The presence of Cu, Mo, Pb, Zn, Ni, Co, Ag, Sb, As, Cd, and Ba in oil field brines/well scales, in MVT deposits, and in Sedex deposits lends credence to the use of these elements in lithogeochemical surveys. Many of the element associations described are based on empirical observations only. Continuing research on the relationship between element concentrations and the geological features in which they are measured will improve our understanding of these associations and will thereby improve the effectiveness of the exploration process.

1 Introduction

Sediment-hosted stratiform lead-zinc massive sulphide deposits and Mississippi Valley-type lead-zinc deposits comprise more than half of the world's known lead and zinc resources. Individual sediment-hosted stratiform massive sulphide deposits are large, and some contain as much lead and zinc as an entire Mississippi Valley-type district. Both types of deposits continue to be the subject of intense exploration efforts because they are economically attractive targets. Large tonnage sediment-hosted stratiform massive sulphide deposits occupy relatively small areas. The 190 million t McArthur River deposit has an areal extent of only 1.5 km^2, and the 56 million t Mount Isa deposit has an areal extent of only 0.3–1 km^2 (Muir 1983). Mississippi Valley-type deposits are even smaller. Assuming that most deposits with surface exposures have been discovered, exploration for new deposits is challenging and requires proficient use of all available exploration tools, including lithogeochemistry.

Govett and Nichol (1979) defined lithogeochemistry as the determination of the chemical composition of bedrock material, with the objective of detecting distribution patterns of elements that are spatially related to mineralization. With recent improvements in analytical techniques and advances in the understanding of ore systems, lithogeochemistry must now be considered not as a separate discipline in the exploration process, but rather one which is integral with isotope geochemistry, fluid inclusion geochemistry, rare earth geochemistry, mineralogy, stratigraphy, and tectonics. Geological information dictates where lithogeochemical techniques should be used and which elements should be evaluated. For example, recognition of a pyrrhotite-bearing organic-rich sedimentary horizon, a barite-bearing horizon, or a vent breccia in the search for sediment-hosted stratiform massive sulphide deposits, or carbonate reef environments in the search for Mississippi Valley-type deposits is important, and can be useful in the design of lithogeochemical surveys. Effective geochemical sample collection is no longer a rote exercise, but rather a highly skilled pursuit which requires a thorough understanding of *all* technical aspects of the minerals exploration business.

Lithogeochemistry can be used during reconnaissance exploration. The availability of only broadly spaced samples, however, commonly precludes the recognition of lithogeochemical *patterns*, and requires the explorationist to rely on the interpretation of element concentration levels alone. Recognition of favourable rock units within the stratigraphic package, knowledge of background geochemical levels in those rocks, and interpretation of the data as a function of rock type are essential. A lithogeochemical study of the Broken Hill deposit in Australia (Plimer 1979) documented the importance of treating each lithostratigraphic unit as a discrete sample population. Geochemical patterns in rock units which are laterally equivalent to ore are the signatures most commonly sought in reconnaissance exploration.

In district-scale or deposit-scale exploration programs, data are commonly available for a great number of accurately located samples, and both lateral and vertical lithogeochemical patterns can be recognized. In *all* geochemical investigations, anomalously high or anomalously low concentrations are only valuable when they form a geologically meaningful pattern, or are a part of an hypothesized geologically meaningful pattern.

Primary lithogeochemical dispersion patterns – systematic variations of element concentrations – are the result of one or more geochemical processes. Modelling the processes can lead to an improved understanding of ore systems, which in turn can lead to development of more effective lithogeochemical techniques. Close interaction between minerals explorationists and research geochemists can improve our ability to discover new ore deposits. For example, the geochemistry of metal-rich formation waters in oil fields in Mississippi, USA, provides a valuable link between ore-forming processes and the products of ore formation.

Sediment-hosted stratiform massive sulphide deposits and Mississippi Valley-type deposits differ in morphology, but there is considerable overlap in their host rock lithology, regional setting, size, grade, metal ratios, age of host rocks, lead isotopes, sulphur isotopes, and genesis (Sangster 1990). Composition of the ore fluids for both types of deposits is controlled in part by fluid-rock reactions within sedimentary basins, but the nature of the sediments in the basin, the source of the fluids, the thermal gradient in the basin, the migration history of the fluids, and the physical and chemical conditions in the depositional environment all cause variations in the concentrations of elements present and in the lithogeochemical patterns which the explorationist must be prepared to interpret. In addition to the above genetic variables, hydrothermal alteration, metamorphism, tectonic deformation, and erosion challenge the minerals explorationist to reconstruct broken, folded, scattered, and erosion-remnant pieces of lithogeochemical patterns to achieve ore discovery.

The objectives of this contribution are (1) to document the value of lithogeochemistry as an integral tool in the exploration process; (2) to prompt the use of different and perhaps new lithogeochemical techniques; and (3) to pose questions, the anwers to which might lead to breakthroughs in the ability to find ore. The chapter consists of three sections: sediment-hosted stratiform lead-zinc massive sulphide deposits (N.G. Lavery), Mississippi Valley-type carbonate-hosted lead-zinc deposits (D.L. Leach), and geochemistry of metal-rich formation waters (J.A. Saunders).

2 Sediment-Hosted Stratiform Lead-Zinc Massive Sulpide Deposits

2.1 Introduction

"Sediment-hosted stratiform massive sulphide" is an accurate, descriptive, non-genetic term for the deposits being discussed. Most current investigators agree that the deposits were formed by sedimentary-exhalative processes. "Sedex", a contraction of the genetic term, was proposed by Carne and Cathro (1982) and will be used below. The features of Sedex deposits have been described by Large (1980, 1981a, 1983), Gustafson and Williams (1981), and Sangster (1990), and will be summarized here. Sedex deposits are tabular, stratiform lenses of fine-grained pyrite (or pyrrhotite), sphalerite and galena, with economically important concentrations of copper, silver, tin, and barite in some deposits. They occur worldwide, and in both Proterozoic and Palaeozoic rocks. Sedex deposits commonly contain more than 50 000 000 t of ore, and have an average lead + zinc grade of 12 wt%. The deposits occur in both autochthonous and allochthonous chemical and clastic sediments (Large 1983). Many Sedex deposits formed in deep water, euxinic, third-order structural basins in tensional tectonic regimes (Large 1983). Other deposits, however, formed in environments of high clastic input, or in shallow water, evaporitic environments. The author agrees with the widely held concept that the metal-bearing fluids originated in clastic sedimentary basins, and that the fluids migrated upward along tectonic disruptive zones and were deposited in depressions which allowed the accumulation rather than the dispersal of the metals. Although some Sedex deposits have feeder vents directly beneath ore, the majority of them do not, indicating that the solutions in many systems migrated laterally before the metals were deposited. Lydon (1983) discussed in detail the chemical parameters which control the origin, transport, and deposition of metals in Sedex deposits.

2.2 Lithogeochemical Investigations

2.2.1 Lithogeochemical Data

Table 1 lists Sedex deposits for which geochemical data have been reported in the English language literature. Table 1 is a "working list" to which individual readers can add, based on their own experience and data availability. It shows only those elements *reported to occur* in a particular ore system, and not necessarily all of the elements which *do* occur. In addition, the table references data determined with a variety of analytical techniques with differing accuracies and sensitivities. With these caveats in mind, the tabulation has value for those charged with the job of finding ore.

The locations of the deposits listed in Table 1 are shown in Fig. 1. Because sulphide mineralization occurs in all of the deposits (with the exception of Franklin-Sterling), sulphur is not listed as a variable. Major oxide elements (with the exception of Mn) are listed only if an author specifically mentioned patterns within or around ore formed by one or more of them. Data for those

Table 1. Elements which form lithogeochemical patterns within or around Sedex deposits; elements which occur in Sedex systems

Deposit	Reference	Element patterns within or around ore	Elements which occur in the ore system
Red Dog Alaska, USA	Lange et al., 1985	Cu, Pb, Zn, Ag, Sb, As, Cd, Ba, Si	
	Moore et al., 1986	Pb, Zn, Ba, Si, Al, Ca, Mg, K, Fe, Ti	Cu, Ag, Sb, Cd, NH_4^+
Drenchwater Alaska, USA	Lange et al., 1985		Cu, Pb, Zn, Ag, Sb, As, Cd, Ba, F
Lik Alaska, USA	Sterne et al., 1984		Cu, Pb, Zn, Ag, Sb, Ba, Mn, NH_4^+
Howards Pass Yukon Territory Canada	Benton, 1984	NH_4^+	Pb, Zn
	Goodfellow, 1984	Cu, Mo, Pb, Zn, Ni, Sb, As, Ba, F, V, Si, Ca, K, P	Co, Ag, Hg, Cd
	Goodfellow and Jonasson, 1986	Cu, Mo, Pb, Zn, Ni, Co, Ag, Sb, As, Hg, Cd, Se, Cr, Ba, F, V, Ca, Mg, K, Fe, P, NH_4^+, C (org.), $\partial^{13}C$	B, Mn
	Goodfellow et al., 1983	Cu, Mo, Pb, Zn, Ni, Co, Ag, Sb, As, Hg, Cd	
	Williams et al., 1987	NH_4^+	Pb, Zn
Tom Macmillan Pass District Yukon Territory, Canada	Goodfellow and Rhodes, 1990	Cu, Pb, Zn, Ag, Sb, As, Hg, Se, Ba, Sr, Si, Ca, Mg, Fe, Mn, $\partial^{34}S$	Cd
	Large, 1980	Pb, Zn, Ag, Ba, C (org.)	Cu
	Large, 1981b	Pb, Zn, Ba, V, Zr, Si, Fe, C (org.)	Cu, Ag, Mn
Jason Macmillan Pass District Yukon Territory, Canada	Gardner and Hutcheon, 1985	Cu, Pb, Zn, Ag, Ba, Fe	Sb, Cd, F, Sr, Mn
	Longstaffe et al., 1982	^{18}O	Pb, Zn, Ba
	Smee and Bailes, 1986	Cu, Pb, Zn, As, Ba, Si, Ca, Fe, Mn	Ag
	Turner, 1990	Pb, Zn, Ba, Si, Ca, Fe	Cu, Ag, As, Hg, F
Faro Anvil District Yukon Territory, Canada	Aho, 1969		Au, Cu, Pb, Zn, Ag, Sb, As
	Tempelman-Kluit, 1972	Pb, Zn	Cu, Ag, Sb, As

Table 1. *Continued*

Deposit	Reference	Element patterns within or around ore	Elements which occur in the ore system
Vangorda	Aho, 1969		Au, Cu, Pb, Zn, Ag, Sb, As, Ba
Anvil District	Tempelman-Kluit, 1972		Cu, Pb, Zn, Ag, As, Ba
Yukon Territory, Canada			
Swim	Aho, 1969		Au, Cu, Pb, Zn, Ag, Sb, As, Ba
Anvil District	Tempelman-Kluit, 1972		Au, Cu, Pb, Zn, Ag, Sb, As, Ba
Yukon Territory, Canada			
Cirque	Carne and Cathro, 1982		Pb, Zn, Ag, Ba
Gataga District	Macintyre, 1982		Pb, Zn, Ag, Ba
British Columbia, Canada			
Driftpile	Macintyre, 1982		Pb, Zn, Ba
Gataga District			
British Columbia, Canada			
Sullivan	Ethier et al., 1976	Pb, Zn, B, Na	Au, Cu, Ni, Co, Ag, Sn, Sb, As, Cd
British Columbia, Canada	Freeze, 1966	Pb, Zn, Ag, Sn, Sb, As, B, Na, Fe	W
	Hamilton et al., 1981	Pb, Zn, Ag, Sn, B, Na	Cu, As, W
	Hamilton et al., 1983	Pb, Zn, Ag, Sn, B, Na, Fe, ^{18}O	Cu, Sb, As, Cd, In, Bi, Tl, W, Mn
	Large, 1980		Cu, Pb, Zn, Ag, Sn, As, B, Na
	Nesbitt and Longstaffe, 1982	^{18}O	Pb, Zn
	Nesbitt et al., 1984	^{18}O	Pb, Zn
	Ransom, 1977	Pb, Zn, Ag, Sn, B, Fe, Na	W
Franklin-Sterling	Beeson, 1990	Mn	Zn
New Jersey, USA	Metsger et al., 1958		Zn, Mn, Cl
	Ridge, 1952		Cu, Pb, Zn, Mn
Edwards-Balmat	Beeson, 1990	Mn	Pb, Zn, B
New York, USA	Doe, 1962	Ni, Co, Cr, Mn, Ti	Cu, Pb, Zn, Ag, Cd, In, Ga, Ge, Ba, V, Sr
	Lea and Dill, 1968		Cu, Pb, Zn, As, Ba
Black Angel	Pedersen, 1980		Au, Cu, Pb, Zn, Ag, Sn, Sb, As
Greenland			

Bleikvassli Nordland, Norway	Vokes, 1963		Cu, Mo, Pb, Zn, Ni, Ag, Sn, Sb, As, B
Tynagh Ireland	Derry et al., 1965		Cu, Pb, Zn, Ag, Sb
	Large, 1980	Zn, Mn	Cu, Pb, Ni, As, Ba
	Russell, 1974	Mn	Pb, Zn
	Russell, 1975	Pb, Zn, Ba, Fe, Mn	Cu, Ni, Ag, As
	Schultz, 1966		Cu, Pb, Zn, Ag, Sb, As, Ba
Silvermines Ireland	Large, 1980	Pb, Zn	Cu, As, Ba
	Taylor, 1984	Cu, Pb, Zn, Ni, Ag, Sb, As, Bi, Ge, Ba, Fe	
	Taylor and Andrew, 1978	Pb, Zn, Ag, Ba, Fe	Cu, Sb, As
Meggen Germany	Amstutz et al., 1971	Pb, Zn, Ba, Fe	
	Gwosdz and Krebs, 1977	Cu, Pb, Zn, Ni, Co, Sb, As, Ba, Mn	
	Krebs, 1981	Cu, Pb, Zn, Ni, Co, Sb, As, Tl, Ba, Si, Fe, Mn	Au, Ag, Sn, Cd
	Large, 1980	Zn, As, Tl, Ba, K, Mn, C (org.)	Cu, Pb, Ag
Rammelsberg Germany	Hannak, 1981	Cu, Pb, Zn, Sr, Si, Mn	Au, Pt, Pd, Mo, Ni, Co, Ag, Sn, Sb, As, Hg, Cd, In, Bi, Ga, Tl, Ge, Ba, V, W
	Krebs, 1981	Cu, Pb, Zn, Ag, As, Ba, Sí, Fe	Au, Ni, Co, Sn, Sb, Hg, Cd, Bi, Tl, Mn
	Large, 1980	Cu, Pb, Zn, Ba, Si, K, Fe	Mo, As, Bi
Rauhala Finland	Iisalo et al., 1989	Au, Cu, Pb, Zn, Ag, Sb, As, Ba, Si, Ca, Mg, Na, K, Mn, Ti, P	Ni, Co, Sn, Te, Hg, Cd, Bi
Zhairem Kazakhstan, USSR	Smirnov and Gorzhevsky, 1977		Pb, Zn, Ni, Co, Ag, Sb, As, Hg, Cd, In, Bi, Ga, Tl, Ge, Ba
Filizchai Azerbaidzhan, UUSR	Smirnov and Gorzhevsky, 1977		Cu, Pb, Zn, Co, Ag, Te, Sb, As, Bi
Mirgalimsai Kazakh, UUSR	Smirnov and Gorzhevsky, 1977		Cu, Mo, Pb, Zn, Co, Ag, Sb, As, Cd, Bi, Ga, Tl, Ge, Ba, F, Mn
Kholodnina Northern Baikal, USSR	Simrnov and Gorzhevsky, 1977		Cu, Pb, Zn

Table 1. *Continued*

Deposit	Reference	Element patterns within or around ore	Elements which occur in the ore system
Saladipura Rajasthan, India	Sarkar et al., 1980		Cu, Pb, Zn, As, Cd, Mn, Cl
McArthur River Queensland, Australia	Benton, 1984	NH_4^+	Pb, Zn
	Carr et al., 1986		Pb, Zn, Ag, Hg
	Croxford and Jephcott, 1972		Au, Cu, Pb, Zn, Ni, Co, Ag, Sb, As, Cd, Ga, Tl, Ba, Cl, V, Sr
	Corbett et al., 1975		Cu, Mo, Pb, Zn, Ni, Co, Ag, As, Hg, Ga, Tl, Cr, Ba, Y, V, Zr, Sc, Rb, Sr, B, Th, Li
	Lambert,1976	Pb, Zn, As, Hg, Tl, Fe, Mn, C (org.)	Cu, Ni, Co, Ag, Sb, Cd, Se, Ba, V, Sr, K, Cl
	Lambert and Scott, 1973	Pb, Zn, Hg, Mn	Cu, Ni, Co, Ag, Sb, As, Se, Tl, Ge
	Large, 1980	Pb, Zn, Tl, Fe, Mn, C (org.)	Cu, As, Ba
	Ryall, 1981		Pb, Zn, Ag, Hg
	Williams et al., 1987	NH_4^+	Pb, Zn
Mount Isa Queensland, Australia	Carr et al., 1986		Pb, Zn, Ag, Hg
	Croxford, 1974		Cu, Pb, Zn, Ni, Co, As
	Croxford and Jephcott, 1972		Tl
	Large, 1980	Pb, Zn	Cu, Co, Ag, Sb, Ba, B
	Mathias and Clark, 1975	Pb, Zn, Ag	Cu, Co, Sb, As, B
	O'Meara, 1961		Cu, Pb, Zn, Ni, Ag, Sb, As, Bi
	Ryall, 1979		Pb, Zn, Hg
	Ryall, 1981		Pb, Zn, Ag, Hg
	Smith and Walker, 1971		Cu, Mo, Pb, Zn, Ni, Co, Ag, Sb, As, Cd, Ge, Ba, Y, V, Zr, Sc, Sr, Mn
	Stanton, 1962		Cu, Pb, Zn, Ag
	Wilson and Derrick, 1976		Au, Cu, Pb, Zn, Ag, Cd
Hilton Queensland, Australia	Mathias and Clark, 1975		Pb, Zn, Ag
	Mathias et al., 1973		Cu, Pb, Zn, Ag, Sb, As

Dugald River Queensland, Australia	Carr et al., 1986		Pb, Zn, Ag, Hg
	Connor et al., 1982	Pb, Zn	Cu, Ni, Co, Ag, Sb, As, Hg, Cd, Cr, Ba, Mn
	Knight, 1965		Pb, Zn, Ag
	Whitcher, 1975		Cu, Pb, Zn, Ag, Sb, As
Lady Loretta Queensland, Australia	Carr et al., 1986		Pb, Zn, Ag, Hg
	Cox and Curtis, 1977		Cu, Pb, Zn, Ag, Sb, Ba
	Louden et al., 1975		Cu, Pb, Zn, Ag, Sb, Ba
Broken Hill N.S.W., Australia	Beeson, 1990	Mn	Pb, Zn, B
	Both, 1973		Cu, Pb, Zn, Co, Ag, Sn, Sb, Cd, Se, Bi, Mn
	Carr et al., 1986		Pb, Zn, Ag, Hg
	Gustafson et al., 1950		Au, Cu, Mo, Pb, Zn, Ni, Co, Ag, Sb, As, Hg, F, W, B, Mn
	Johnson and Klinger, 1976		Cu, Mo, Pb, Zn, Ni, Co, Ag, Sb, As, Hg, Cd, Bi, F, W, Mn
	Lottermoser, 1989	REE	Pb, Zn, Ag, Ba, F, B
	Lottermoser, 1991	Au, Cu, Pb, Zn, Ag, Sb, As, F, W, Si, Ca, Fe, Mn, P, REE	Ba, B, U, Th
	Plimer, 1979	Cu, Pb, Ni, Sb, As, Bi, F, Rb, Sr, Si, Ca, Mg, Na, K, ΣFe, Mn, Ti, P	Zn, Co, Ag, Hg, Cd
	Plimer, 1982		Pb, Zn, F
	Plimer, 1983	B	Cu, Pb, Zn, Ag, As, F, Mn
	Plimer, 1986	Cu, Pb, Zn, Ni, Ag, Sb, As, Bi, F, Si, Ca, Fe, Mn, P	Ba, B
	Plimer, 1988	B	Pb, Zn
	Ryall, 1979		Cu, Pb, Zn, Ag, Sb, As, Hg, Cd, Bi, F, Mn
	Ryall, 1981		Pn, Zn, Ag, Hg
	Stanton, 1976	Cu, Pb, Zn, Fe, Mn	F, Sr, P
Aggeneys District Southern Africa	Beeson, 1990	Mn	Pb, Zn, B
	Ryan et al., 1986	Cu, Pb, Zn, Mn	Ni, Co, Ag, Sb, As, Bi, Ba, F, B
El Aguilar Argentina	Sureda and Martin, 1990	Ba, Si, Fe, Mn	Au, Cu, Mo, Pb, Zn, Ni, Co, Ag, Sn, Sb, As, Bi, F, B

Fig. 1. Distribution of sediment-hosted stratiform massive sulphide deposits: *1* Red Dog; *2* Drenchwater; *3* Lik; *4* Howard's Pass; *5* Tom; *6* Jason; *7* Faro; *8* Vangorda; *9* Swim; *10* Cirque; *11* Driftpile; *12* Sullivan; *13* Franklin-Sterling; *14* Edwards-Balmat; *15* Black Angel; *16* Bleikvassli; *17* Tynagh; *18* Silvermines; *19* Meggen; *20* Rammelsberg; *21* Rauhala; *22* Zhairem; *23* Filizchai; *24* Mirgalimsai; *25* Kholodnina; *26* Saladipura; *27* McArthur River; *28* Mount Isa; *29* Hilton; *30* Dugald River; *31* Lady Loretta; *32* Broken Hill, N.S.W; *33* Aggeneys district; *34* El Aguilar

elements that "occur" in Sedex systems are important because they *might* form geologically meaningful patterns around ore and *might* become valuable lithogeochemical tools with additional documentation. Elements that occur as components of mineral phases as well as those that have been determined by chemical analysis are included. For example, if xanthoconite was referenced by an author, then both silver and arsenic are listed, and if gudmundite was referenced, antimony is listed. Boron is listed if tourmaline is present.

Data on Rauhala, Finland, are included in Table 1 because the ore consists of thin stratiform lenses of pyrrhotite, sphalerite, chalcopyrite, and galena in metaturbidites. The high Cu content of the ore, the lateral metal zonation in the deposit [(As, Au) $\gg$ Cu $\gg$ Zn $\gg$ (Pb, Ag, Sb)], and the depletion of Ti, Mn, Mg, Ca, Na, and P in the footwall and hanging wall of the deposit are, however, atypical of Sedex deposits (Iisalo et al. 1989).

Data on element ratios which form geologically meaningful patterns within or around ore are listed in Table 2.

Table 2. Element ratios which form patterns within or around Sedex deposits

Deposit	Reference	Ratios
Red Dog Alaska, USA	Moore et al., 1986	Zn/(Zn + Pb)
		(Zn + Pb + Fe)/Fe
		Fe/(Zn + Pb + Fe)
		Ag/(Ag + Zn)
		Ag/(Ag + Pb)
Howards Pass Yukon Territory Canada	Goodfellow, 1984	Pb/(Pb + Zn)
	Goodfellow and Jonasson, 1986	Pb/(Pb + Zn)
		Zn/(Pb + Zn)
		Hg/Zn
		Ni/Fe
		S/C
	Goodfellow et al., 1983	Pb/(Pb + Zn)
		S/C
Tom Macmillan Pass District Yukon Territory Canada	Goodfellow and Rhodes, 1990	Pb/(Pb + Zn)
		Hg/Zn
		Ag/Pb
		Ba/(Ba + Al)
		Si/(Si + Al)
		$^{87}Sr/^{86}Sr$
	Large, 1980	Pb/Zn
	Large, 1981b	V/Cr
Jason Macmillan Pass District Yukon Territory Canada	Gardner and Hutcheon, 1985	Pb/(Pb + Zn)
	Smee and Bailes, 1986	Pb/(Pb + Zn)
	Turner, 1990	Pb/(Pb + Zn)
Sullivan British Columbia Canada	Ethier et al., 1976	Pb/Zn
	Freeze, 1966	Pb/Zn
		Ag/Pb
	Hamilton et al., 1981	Pb/Zn
		Ag/Pb
	Large, 1980	Pb/Zn
	Ransom, 1977	Pb/Zn
		Ag/Pb
Silvermines Ireland	Large, 1980	Pb/Zn
	Taylor, 1984	Pb/Zn
	Taylor and Andrew, 1978	Zn/Pb
Rammelsberg Germany	Hannak, 1981	Sr/Ba
Mount Isa Queensland Australia	Large, 1980	Zn/Pb
Dugald River Queensland Australia	Connor et al., 1982	(K + Na + Ba)/Ba
		(Cu + Mg + Mn)/Mn
Broken Hill New South Wales Australia	Lottermoser, 1991	U/Th
	Plimer, 1979	Mg/(Mg + ΣFe)
		K/(K + Na)
		Ca/(Ca + Na)
		K/(Rb × 10^2)
		Ca/(Sr × 10^2)
		Rb/Sr
		K/Ba
Aggeneys District Southern Africa	Ryan et al., 1986	Cu/(Pb + Zn)

2.2.2 Data Interpretation

Some of the data in Tables 1 and 2 are quite familiar to minerals explorationists. Lateral zoning of Cu ≫ Pb ≫ Zn ≫ Ba is well documented within and beyond the limits of economic mineralization in many Sedex deposits (Large 1980, 1983); Cu, As, and Sb have been documented in and close to vent features (Smee and Bailes 1986); and the ratio Pb/(Pb + Zn) has commonly been used to define vectors toward the central part of ore systems. Another well-known lithogeochemical signature is Mn dispersion such as that described at Tynagh by Russell (1974) and at Meggen by Gwosdz and Krebs (1977). It is important to note that Mn can also be depleted directly above the postulated original brine pool (Smee and Bailes 1986), and that a systematic pattern of Mn enrichment might not be contiguous with ore grade Pb-Zn mineralization (e.g. at Tynagh, Russell 1975). Smee and Bailes (1986) successfully used the distribution of mineral facies, Cu and As concentrations, Pb/(Pb + Zn), and Ba and Mn concentrations to define vectors toward the heat source (vent) in the Jason deposit.

Other aspects of the data in Tables 1 and 2 which might not be as familiar to some explorationists are discussed below. Table 3 shows the number of deposits listed in Tables 1 and 2 in which each of the elements (other than Pb, Zn, major oxides) occurs.

Copper, Ag, As, Sb and Ba are reported in more than 60% of the deposits; Mn, Cd, Ni, Co, Au, Hg, Bi, B, Sn, Mo, F, Tl, Ge, and Sr are reported in more than 20% of the deposits; and V, Ga, In, Se, Cr, and NH_4^+ are reported in more than 10% of the deposits. The above elements are all associated with Pb-Zn mineralization, and should be considered for use in

Table 3. Percent occurrence of elements in 34 Sedex deposits

Element	No. deposits (%)	Element	No. deposits (%)
Copper	31 (91)	Vanadium	6 (18)
Silver	30 (88)	Gallium	5 (15)
Arsenic	28 (82)	Indium	4 (12)
Antimony	28 (82)	Selenium	4 (12)
Barium	25 (74)	Chromium	4 (12)
Manganese	19 (56)	Ammonium	4 (12)
Cadmium	17 (50)	Chlorine	3 (9)
Nickel	16 (47)	Zirconium	3 (9)
Cobalt	15 (44)	Tungsten	3 (9)
Gold	12 (35)	Tellurium	2 (6)
Mercury	11 (32)	Yttrium	2 (6)
Bismuth	11 (32)	Rubidium	2 (6)
Boron	9 (26)	Thorium	2 (6)
Tin	8 (24)	Scandium	2 (6)
Molybdenum	8 (24)	Lithium	1 (3)
Fluorine	7 (21)	Platinum	1 (3)
Thallium	7 (21)	Palladium	1 (3)
Germanium	7 (21)	Uranium	1 (3)
Strontium	7 (21)	REE	1 (3)

lithogeochemical orientation surveys. Some elements which have been reported in less than 20% of the deposits perhaps have not been adequately evaluated as lithogeochemical guides to Sedex deposits.

The decision as to which elements to use in an exploration program should be based on an understanding of the Sedex model, as expressed by the geology of the search area. For example, if a barite-rich horizon is present, the explorationist should consider the possibility that the barite formed in an oxidizing environment laterally equivalent to sulphide mineralization. Evaluation of the whole rock copper content of the barite-rich horizon might not be of much value, but measurement of the Sr content of the barite and evaluation of the distribution of Sr/Ba in the barite might be useful. Large (1980) determined that submarine exhalative barite is characterized by a low Sr content (0.3–1.5%), whereas epigenetic vein-type barite usually contains greater than 1% and often more than 1.5% Sr. Hannak (1981) documented a decrease in the Sr/Ba ratio toward the margin of ore at Rammelsberg. In barite-bearing systems, the trace element content of barite might provide environmental clues and yield valuable vectors toward the central part of sulphide-mineralized systems. It is to be noted that Ba rather than barite is the key exploration signature, and that Ba can occur in barian carbonate minerals, in barian feldspar, as well as in barite (Goodfellow and Rhodes 1989).

Several geochemical variables which have not received much attention, but which show promise as exploration guides for Sedex deposits are highlighted below.

Thallium has been documented at Sullivan (Hamilton et al. 1983), McArthur River (Lambert 1976), Mount Isa (Croxford and Jephcott 1972), Meggen (Krebs 1981), Rammelsberg (Hannak 1981), Zhairem (Smirnov and Gorzhevsky 1977), and Mirgalimsai (Smirnov and Gorzhevsky 1977). At McArthur River, anomalously high values of Tl (>10 ppm) occur in pyritic shales near the base of the H.Y.C. Member for at least 7 km from the ore deposit (Lambert 1976). Thallium might be a valuable pathfinder element in other ore systems as well.

Lavery (1985) documented the value of fluorine as a pathfinder element in the search for volcanic-hosted massive sulphide deposits. Although fluorite is commonly considered not to occur in Sedex deposits (Maynard 1991), it does indeed occur. Fluorite occurs at Drenchwater (Lange et al. 1985), Jason (Gardner and Hutcheon 1985; Turner 1990), Mirgalimsai (Smirnov and Gorzhevsky 1977), Broken Hill, N.S.W. (Johnson and Klinger 1976; Gustafson and Williams 1981; Plimer 1982), Aggeneys (Ryan et al. 1986) and El Aguilar (Sureda and Martin 1990). Fluorine is a mobile element in hydrothermal systems and may be an overlooked pathfinder element for Sedex deposits.

Ammonium has been recognized in illite at Lik (Sterne et al. 1984), and in feldspar in the hanging wall rocks at Howards Pass (Williams et al. 1987). Ammonium has also been reported at Red Dog (Moore et al. 1986) and at McArthur River (Williams et al. 1987). Ammonium may be useful in sedimentary terranes in which ore was deposited in still-water depressions over thick, organic-rich sedimentary piles, and where detrital influx was minimal (Williams et al. 1987). With additional study, NH_4^+ may become a valuable lithogeochemical exploration tool.

Very high concentrations of Hg occur in sphalerite and in tetrahedrite inclusions in galena, and high concentrations also occur in chalcopyrite, pyrrhotite, arsenopyrite, and loellingite at Broken Hill, N.S.W. (Ryall 1979). Anomalously high concentrations of Hg occur in disseminated sulphides outside the limit of economic mineralization at Broken Hill, but sulphide-free wall rocks contain low Hg concentrations. Although Hg lithogeochemical patterns might not be detectable in whole rock analyses, analysis for Hg in disseminated sphalerite and galena encountered during reconnaissance exploration might be valuable.

Anomalously high concentrations of Hg correlate positively with stratabound Zn mineralization in prospects in amphibolite facies rocks in the Edwards-Balmat district (Foose 1981). The presence of geologically meaningful lithogeochemical patterns at Edwards-Balmat and at Broken Hill, N.S.W. suggests that Hg can be a valuable pathfinder element for Sedex deposits in high-grade metamorphic terranes as well as in low-grade or unmetamorphosed terranes.

Copper, As, and Sb occur in or proximal to vent facies rocks at Jason (Smee and Bailes 1986). In addition to these commonly evaluated elements, Taylor (1984) reported Ag, Bi, Ni and Ge in vent facies rocks at Silvermines, and noted that Cu sulphosalts are common mineral species. Additional evaluation of the distribution of Bi, Ni, and Ge in Sedex deposits might be of exploration value.

Tourmaline has been identified at Sullivan (Ethier et al. 1976; Ransom 1977; Large 1980; Hamilton et al. 1981), Edwards-Balmat (Beeson, 1990), Bleikvassli (Vokes 1963), McArthur River (Mathias and Clark 1975), Mount Isa (Mathias and Clark 1975), Broken Hill, N.S.W. (Lottermoser 1989; Beeson 1990), Aggeneys district (Ryan et al. 1986; Beeson 1990), and El Aguilar (Sureda and Martin 1990). Foose (1981) determined that B in rock *and* soil samples correlates positively with stratabound Zn mineralization in a prospect in the Edwards-Balmat district, and that anomalously high B values in rock samples accurately mapped the folded extension of the Zn-bearing unit where no Zn was detectable. Analysis for boron appears to be warranted in the search for Sedex deposits.

Scapolite, a Cl-bearing aluminosilicate mineral, has been identified as a signature for shallow water Sedex depositional environments (Whitcher 1975; Muir 1983). Apatite, with highly variable amounts of F and Cl, occurs at Broken Hill, N.S.W. (Stanton 1976). Friedelite, a Cl-bearing manganese silicate occurs in the metamorphosed ore at Franklin-Sterling (Metsger et al. 1958). Recognition of these Cl-bearing minerals might be a guide to Sedex ore deposits.

Ore at Rammelsberg contains 0.04 ppm Pt and 0.02 ppm Pd (Hannak 1981). This is the only reference noted of platinum group elements occurring in Sedex deposits. Additional data from other deposits would be of interest and perhaps of economic and exploration value.

Samples containing between 40 and 100 ppm Zr occur proximal to mineralization and along strike from the Tom deposit (Large 1981b). The zirconium values may be the signature of volcanic activity, similar to the Zr-bearing tuffites at Meggen and Rammelsberg.

Lottermoser (1989) investigated the rare earth element geochemistry of exhalites at Broken Hill, N.S.W. and determined that exhalites proximal to sulphide mineralization are enriched in light rare earth elements and Eu, and depleted in heavy rare earth elements, while distal exhalites in the northern extension of the Broken Hill orebodies have distinct negative Eu anomalies. Additional investigation of rare earth element distribution as a tool in minerals exploration is warranted.

Stable isotope geochemistry may provide exploration guides as well as information on the genesis of deposits. Longstaffe et al. (1982) determined that a small but perceptible ^{18}O depletion halo exists in the clay fraction of the sedimentary rocks around the Jason deposit, and Nesbitt and Longstaffe (1982) and Nesbitt et al. (1984) determined that an ^{18}O depletion halo extends for up to 100 m above ore at Sullivan. Goodfellow and Rhodes (1990) defined systematic variations of $\delta^{34}S$ in sedimentary pyrite, galena, and barite in relation to the vent zone, and a systematic decrease in $^{87}Sr/^{86}Sr$ from the vent to the distal margins at the Tom deposit. Goodfellow and Jonasson (1986) determined that $\delta^{13}C$ values for organic carbon in the Active Member at Howards Pass are about 2‰ less than values for coeval carbonaceous cherts located 6 km away. Application of isotope geochemical techniques to the search for Sedex deposits deserves additional attention.

Finlow-Bates et al. (1977) presented evidence from Mount Isa that fine-grained pyrrhotite is the dominant iron sulphide in high-grade lead-zinc ores, and that pyrite dominates where lead and zinc are either absent or present in non-economic amounts. These authors concluded that during sedimentation the deposition of galena and sphalerite depleted a limited sulphur supply so that FeS was a stable species, and that the primary iron monosulphide which formed was, or has since become pyrrhotite. Although *some* pyrrhotite at Mount Isa is of metamorphic origin, the fact that fine-grained pyrrhotite was deposited syngenetically in a reducing environment is a mineralogical/lithogeochemical exploration clue to the chemical environment in which some Sedex deposits formed. Iisalo et al. (1989) stated that sphalerite, chalcopyrite and galena occur in association with pyrrhotite at Rauhala, and that pyrite and marcasite occur only below and above the ore horizon.

High amounts of carbonaceous material are characteristic of euxinic basins in which many Sedex deposits occur. At the Tom deposit, samples with >2% organic carbon define a locally reducing environment at the time of deposition of the ore (Large 1981b).

Within the Howards Pass basin, concentrations of elements which correlate positively with organic carbon decrease toward the ore deposit, reflecting a decrease in the amount of organic carbon present due to dilution by introduced silica and sulphides (Goodfellow et al. 1983).

The ratio V/Cr in organic-rich shales can be a clue to the chemistry of sedimentary environments. At the Tom deposit, V/Cr values <4 occur in footwall siltstones, and unusually high V/Cr values (>10) are confined to hanging wall shales (Large 1981b).

The sulphur/carbon ratio is considerably higher in the Howards Pass sub-basin than in similar rocks remote from mineralization. The limiting factor controlling the S/C ratio is apparently the supply of iron (Goodfellow et al.

1983). Similar S/C patterns might occur in other Sedex districts, and might be useful in delineating favourable depositional basins.

Little information has been published on the geochemistry of Sedex gossans, and most of this is descriptive rather than quantitative. Information is available on the following deposits: Cirque (MacIntyre 1982), Lady Loretta (Cox and Curtis 1977), Dugald River (Connor et al. 1982), and Broken Hill, N.S.W. (Johnson and Klinger 1976).

Few authors have discussed those elements for which concentrations were measured in a lithogeochemical program but which did not form geologically meaningful patterns around ore. Such data are important and should be reported. Data are available on the following deposits: Tom and Jason (Williams et al. 1987; Benton 1984), Tom (Large 1981b), Tynagh (Russell 1974), McArthur River (Lambert and Scott 1973), and Broken Hill, N.S.W. (Ryall 1979).

Only a few authors have mentioned those elements for which concentrations were measured in a lithogeochemical program but which were not detected analytically or which were present at very low, non-significant levels. Again, such data are important and should be reported. Data are available on the following deposits: McArthur River (Lambert 1976), Dugald River (Connor et al. 1982), and Meggen (Krebs 1981).

2.3 Summary

The geochemical data in Tables 1, 2, and 3 will be of value to explorationists searching for Sedex deposits. Research on the relationship between element concentrations and the geological features in which they are measured will yield additional information on the genesis of ore systems and will also make the exploration process more effective. Some features on which more geological and geochemical data are needed include pyrrhotite-bearing horizons (Campbell et al. 1980), organic-rich sedimentary horizons (Large 1981b), barite-bearing horizons (Krebs 1981), areally restricted ferroan carbonate occurrences (Goodfellow and Rhodes 1990; Turner 1990), anomalously thick stratigraphic sections of phosphatic chert (Goodfellow et al. 1983; Goodfellow and Jonasson 1986), diamictite (Goodfellow and Rhodes 1990; Turner 1990), hanging wall zones of silica enrichment (Hannak 1981), scapolite (Whitcher 1975; Muir 1983), tourmalinite (Hamilton et al. 1983), halite, anhydrite, or gypsum casts (Walker et al. 1977), vent alteration features (Hannak 1981), vent biota (Moore et al. 1986), and pyritic chimneys (Larter et al. 1981).

Lithogeochemistry is only one of the tools available to the minerals explorationist searching for Sedex deposits, but it is a powerful tool, and it holds promise for becoming an even more effective one.

3 Mississippi Valley-Type, Carbonate-Hosted Lead-Zinc Deposits

3.1 Introduction

The most important carbonate-hosted lead-zinc deposits are of the Mississippi Valley-type (MVT). MVT deposits are a similar but diverse family of epigenetic

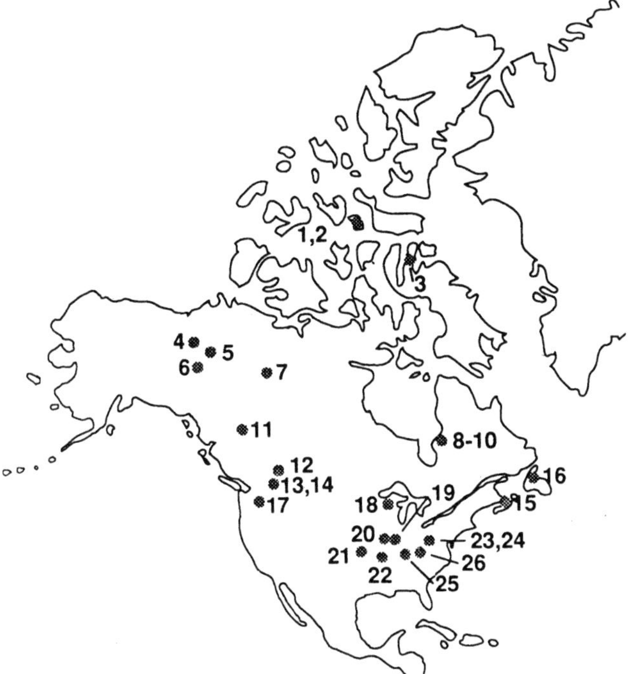

Fig. 2. Distribution of Mississippi Valley-type deposits and districts in North America: *1* Polaris; *2* Eclipse; *3* Nanisivik; *4* Gayna; *5* Bear-Twit; *6* Godlin; *7* Pine Point district; *8* Lake Monte; *9* Nancy Island; *10* Ruby Lake; *11* Robb Lake; *12* Monarch-Kicking Horse; *13* Giant; *14* Silver Basin; *15* Gays River; *16* Newfoundland Zinc; *17* Metalline district; *18* Upper Mississippi Valley district; *19* Southeast Missouri Lead District (Old Lead Belt and Viburnum Trend); *20* Central Missouri district; *21* Tri-State district; *22* Northern Arkansas district; *23* Austinville; *24* Friedensville; *25* Central Tennessee district; *26* East Tennessee district. (After Sangster 1990)

ores precipitated at low temperatures (75 to 200 °C) in platform carbonate sequences from dense basinal brines. MVT deposits characteristically occur in dolostones or large bodies of sparry dolomite that have replaced limestone; deposits wholly in limestone or sandstone are rare.

MVT deposits are found throughout the world but the largest, and most intensely studied ones occur in North America (Fig. 2) where the deposit type was first recognized more than 50 years ago (Bastin 1939).

MVT deposits owe their name to the fact that several classic districts are located in the drainage basin of the Mississippi River in central USA. Outside of North America, important MVT districts include Upper Silesia in Poland, and the Alpine district of Austria-Yugoslavia-Italy. The Lennard Shelf area of Western Australia has recently emerged as a potentially significant MVT district.

Fluorite-rich, lead-zinc deposits in carbonate rocks form an important MVT subtype, with features that clearly set them apart from other MVT lead-zinc deposits (Sangster 1990). Fluorite and barite are the dominant ore minerals, they are localized by major systems of vertical fractures, and they are commonly spatially associated with intracratonic rift systems. Important districts include Southern Illinois; Central Kentucky; Sweetwater, Tennessee; Hansonburg, New Mexico; and the English Pennines. Other types of carbonate-hosted lead-zinc deposits include the precious metal-rich replacement ores such as those in the central Colorado mineral belt, some of which might have had a magmatic component to the ore fluids (Thompson and Beaty 1990).

This discussion will focus on lithogeochemical investigations related to exploration for MVT lead-zinc deposits. Some of the lithogeochemical techniques discussed can, however, also be applied to fluorite-rich MVT deposits as well as to carbonate replacement deposits of the type found in the central Colorado mineral belt. The discussion is limited by a paucity of available studies on the lithogeochemistry of MVT deposits; the few case studies presented are from North American deposits.

Stream and lake sediment geochemistry has been useful in identifying MVT occurrences in Newfoundland (Davenport et al. 1975) and in the Northwest Territories in Canada, but only recently has lithogeochemistry been effective in the search for stratigraphically covered MVT deposits. Most exploration efforts for MVT resources are now focused on deposits beneath stratigraphic cover, with geochemical exploration being largely limited to analysis of samples obtained from drill holes.

We will first review some fundamental characteristics of MVT deposits that must be considered in designing an effective geochemical exploration program, and then we will present examples of lithogeochemical studies applied to exploration for them.

3.1.1 Importance of Scale in Lithogeochemical Investigations

MVT deposits are unique among hydrothermal ore systems in that in most cases, the ore-forming fluids affected thousands of cubic kilometres of the crust (Leach, this Vol.). Widespread and isolated geochemical anomalies are, therefore, to be expected. Geochemical studies can be directed at locating undiscovered districts where the target size may be on the order of hundreds of square kilometres, or locating new deposits within a district with targets on the order of hundreds of metres, or developing new ore within a known deposit.

Within a mineralized region, a single ore-forming event can produce districts with distinctly different mineralogy, trace and minor elements, host rock alteration, and isotope composition. The great diversity in deposit characteristics and local and regional controls on MVT mineralization preclude identification of a universally applicable list of geological and geochemical exploration criteria. However, within a district or region, it is possible to establish criteria to explore for additional deposits.

The suite of elements chosen for a lithogeochemical study depends on the target, whether exploring for deposits within a known district or in regional reconnaissance for new districts. For example, if the program is directed at locating new deposits within a district, the element suite characteristic for that district should be used. If however, the program is directed at exploring for districts in frontier areas, a larger and more general suite of elements should be chosen.

3.1.2 Mineralogy and Geochemical Signature

MVT lead-zinc deposits typically are simple mineralogically, consisting primarily of sphalerite, galena, and iron sulphides (Table 4); however, the relative proportions vary widely between districts. Combined lead-zinc grades seldom exceed 10%. A majority of districts are zinc-rich relative to lead and have Zn/(Zn + Pb) typically around 0.8 (Sangster 1990). The Southeast Missouri district is unusually rich in lead with a ratio of about 0.05. Minor associated minerals vary greatly between districts, with the consequence that potentially useful elements for geochemical exploration tend to be specific to a region or district. Copper is uncommon in MVT ores with the exception of the Southeast Missouri district. Cobalt and nickel are minor constituents in the Southeast Missouri and Upper Mississippi Valley districts. In some districts, silver, cadmium, germanium, fluorite, and barite are abundant enough to be recovered. Gangue hydrothermal dolomite is universally present and quartz may be common in some districts. Consistent, well-developed metal or element zoning is not a typical feature of MVT deposits. However, some large districts such as Southeast Missouri, Upper Mississippi Valley, and Pine Point exhibit broad trends in metal ratios (Heyl et al. 1959; Grundmann 1977; Rogers and Davis 1977; Kyle 1981). Table 4 summarizes elements that may be useful in geochemical exploration studies.

Table 4. Geochemical characteristics of selected North American MVT deposits[a]

District	Major	Minor	Trace
Viburnum Trend	Pb Zn Fe	Cu Co Ni	**Cd Ag** In Ge Ga Sb Bi **As Mo** Sn Au (1)
Old Lead Belt	Pb Zn Fe	Cu Ni Co	**Cd Ag** In Ge Ga Sb Bi **As Mo** Sn Au (1)
Central Missouri	barite Zn Pb	Fe	**Cd** Cu
Northern Arkansas	Zn	Pb Fe	**Cd** Cu Ag Ge Ga In Bi As (1)
Tri-State	Zn Pb	Fe Cu	**Cd** barite Cu Ag Ge Ga Co Ni In Sb Bi (1)
Upper Mississippi Valley	Zn Pb Fe	Cu barite	**Cd** Ag Ge Ga Co Ni In (2)
Central Tennessee	Zn	Pb Fe fluorite barite	**Cd** Cu Ag Ge Ga Co Ni Hg As (1)
East Tennessee	Zn	Pb Fe	**Cd** Cu Ag Ge Ga (1)
Austinville	Zn Pb Fe	Cu	**Cd** Ge Ga fluorite barite (1)
Friedensville	Zn	Pb	**Cd**
Timberville	Zn Fe	Pb Cu	
Gays River	Zn Pb Fe	Cu barite	**fluorite**
Newfoundland Zinc	Zn	Pb Fe	**Cd**
Polaris	Zn Pb Fe		
Nanisivik	Zn Pb Fe		**barite** Ag (3)
Pine Point	Zn Pb Fe		**Cd** Ag As Ni (3)
Gayna River	Zn Pb Fe	barite fluorite	

[a] The distinction between major and minor chemical characteristics is subjective because seldom are grades for minor metals published and there is an inherent high variability in minor metals even between deposits within a single district; typically, the minor components are less than 1% but greater than 0.1%. Trace elements are a tabulation of elements detected in various mineral phases by a variety of techniques; therefore, they may be used as general guides for the deposits. Trace elements in bold type may be useful geochemical pathfinders. Trace element data from (1) Hagni (1983); (2) Heyl et al. (1959); (3) Sangster (1968); others are unpublished data.

3.2 Lithogeochemical Investigations

3.2.1 Primary Dispersion Halos from Rock Geochemical Analysis

Geochemical analysis of rock samples obtained from surface exploration and from drilling programs can be a valuable guide to ore. The value of analyzing drill core is shown by recent exploration efforts in the Upper Silesian district in Poland. Gorecka (1991) integrated geochemical analyses of drill core with other geological data and identified new MVT resources and provided new information on ore controls in the district.

The few published studies of whole rock geochemistry applied to exploration for MVT deposits show that primary dispersion halos about MVT deposits tend to be small – on the order of metres. The small primary halos "appear" to contradict the observation that most MVT deposits form from large hydrological systems, but in reality they reflect highly localized precipitation processes. For example, in the central Tennessee zinc district, Jones (1988) determined the distribution of As, Ca, Co, F, Ga, Mg, Hg, Si, Sr, and Zn in both mineralized and unmineralized carbonate host rocks. In this study, zinc concentrations decreased to background levels at only 125 cm from sphalerite ore, whereas strontium concentrations were antithetic to zinc concentrations over the same distance. Other elements showed no apparent relationship to ore. Because of this rather small geochemical halo, Jones (1988) concluded that application of primary geochemical dispersion to exploration for MVT deposits in central Tennessee is not useful.

The dispersion of zinc about MVT deposits in the Upper Mississippi Valley district was studied by Lavery and Barnes (1971). They found the extent of the dispersion directly correlates with the size of the vein and extends to a maximum distance of 53 m. Barnes and Lavery (1977) expanded on their original work and established a series of zinc dispersion curves for the Upper Mississippi Valley district that allow for prediction of distance to, and size of, nearby ore zones (Fig. 3). This technique requires that the concentration of hydrothermally introduced zinc in a sample be computed by subtracting the background abundance from total zinc concentration. Background zinc concentration was found to be directly proportional to the clay content of the rock. Therefore, by

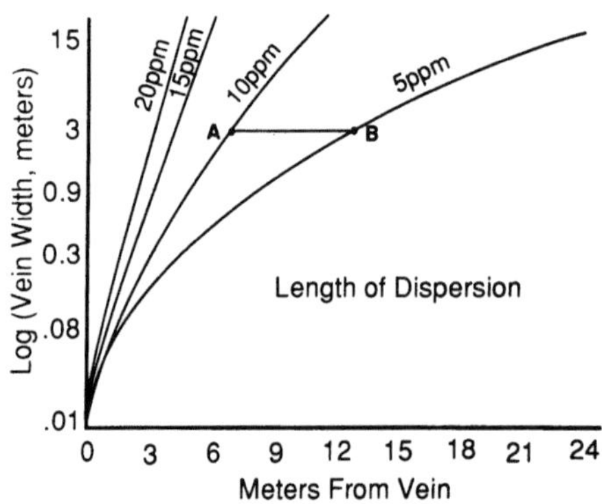

Fig. 3. Correlation of log (vein width) with length of dispersion out to four adjusted (hydrothermally introduced) zinc contents from the Upper Mississippi Valley district. *Points A* and *B* represent locations of samples with zinc contents of 10 and 5 ppm, respectively; distance to a 3-m vein is about 7.2 m from the 10 ppm sample. (Slightly modified from Barnes and Lavery 1977)

subtracting an amount of zinc proportional to the clay content in a mineralized sample from the total zinc concentration, the amount of hydrothermally introduced zinc can be computed. If the distance between two samples having 5 and 10 ppm, respectively, is 6.4 m, then the adjacent mineralization is about 3 m thick and its contact is 7.2 m from the 10 ppm sample. This technique can best be used to assist development of blind drifts and exploration drilling within known ore trends.

In the Viburnum Trend in Southeast Missouri, primary enrichment of bromine and chlorine in the host dolostones extends up to about 75 m from ore; Br/Cl ratios of dolostone samples also display an asymptotic decrease from ore (Pano et al. 1983). These halide dispersion patterns are interpreted to reflect the highly saline nature of the MVT ore fluids. Samples of host dolostone from a fence of drill core transecting the Viburnum Trend were analyzed for more than 40 elements with a sequential digestion method to determine the partition of these elements between carbonate and sulphide mineral hosts (Viets et al. 1983). Their study found that the concentration of Fe, Mn, Pb, Zn, and Cd in carbonate minerals in the Bonneterre dolostone systematically decreased away from the contact with underlying basal sandstone and toward the stratigraphic level of mineralization in the Bonneterre dolostone, over a vertical distance of about 50 m. However, Pb, Zn, Fe, Ag, Cd, Cu, and As in sulphide minerals exhibited the opposite trend, that is, concentrations increased vertically toward ore over the same distance. The metal partitioning pattern and concentration profiles are believed to reflect the ascent of the ore fluids from the basal sandstone, considered to be the primary aquifer for the ore fluid, into reduced sulphur-rich areas of the middle Bonneterre Formation that hosts ore in the Viburnum Trend. The partitioning of the metals between carbonate and sulphide minerals may be controlled by the availability of reduced sulphur to precipitate sulphide minerals. Viets et al. (1983) concluded that metal dispersion patterns in sulphide minerals provide enhanced geochemical anomalies relative to total metal concentrations in the host dolostones in the Viburnum Trend.

3.2.2 Geochemical Analysis of Insoluble Residues from Carbonate Rocks

Studies by Erickson et al. (1978, 1979, 1981, 1983) in the Ozark region of the United States demonstrated the value of geochemical analysis of insoluble residues from carbonate rocks for the assessment of potentially undiscovered MVT deposits. Geochemical analyses of insoluble residues permits the detection of trace amounts of elements whose presence in barren whole rock is unsuspected and commonly not detected by conventional whole-rock analytical methods. Using insoluble residues from visibly ore-barren drill core, Erickson et al. (1978, 1979, 1981) found a map pattern of geochemical anomalies that coincided with known ore trends in the Southeast Missouri MVT districts. In addition, outside known ore trends, geochemical anomalies from insoluble residues coincided with geological features (e.g. limestone-dolostone transitions, carbonate reef facies, and basement topographic highs) that are known ore controls in the Southeast Missouri MVT districts. On this basis, they concluded that insoluble residues are useful in identifying favourable ground for MVT deposits elsewhere in the region. As Erickson et al. (1983) pointed out (p. 575, 1983), "the resulting map patterns of distributions and abundances of trace elements permit distinction between intrinsic and epigenetic suites of elements, recognition of rock units through which metal-bearing fluids passed, and delineation of regional metal trends". The procedure described by Erickson et al. (1981) utilizes the insoluble fraction of carbonate rocks, generally from subsurface drill core, through digestion by cold HCl. Erickson et al. (1981, 1983) reported the analytical data from insoluble residues from drill core in units of anomalous metal feet (AMF), obtained by dividing the anomalous concentration of an element by the minimum concentration, multiplied by the interval of core that contains the anomalous metal content.

Figure 4 gives examples of the application of insoluble residues for the assessment of MVT mineralization in part of southern Missouri. Here, Erickson et al. (1983) determined that the epigenetic suite of elements in insoluble residues that indicate high favourability for MVT mineralization are As, Ag, Co, Cu, Mo, Ni, Pb, and Zn. Shown in Fig. 4A are the distribution of drill holes sampled and abundances of metals in insoluble residues from the Bonneterre Formation. These data, collected only from visibly sulphide-barren rocks, identify known ore trends in the Bonneterre Formation in Southeast Missouri. These data indicate the Bonneterre Formation is favourable for MVT deposits in the Rolla quadrangle but not favourable elsewhere in southern Missouri. Figure 4B shows the distribution of the epigenetic MVT suite of elements in insoluble residues from Cambrian carbonate formations that overlie the Bonneterre Formation. Assuming that insoluble residues also indicate favourable mineral potential in other carbonate rocks in the region, these data (Fig. 4B) define a new and previously unknown belt of possible MVT mineralization.

Analyses of insoluble residues provided valuable guides for undiscovered MVT deposits in the midcontinent of the United States and this approach has potential as an important exploration tool in other areas. Studies of insoluble residues from carbonate rocks in other regions are needed to further establish the effectiveness of this approach and to identify the appropriate suite of elements for those particular areas.

3.2.3 Alteration Studies

Studies of rock alteration or identification of minerals precipitated from the brines can provide evidence of pathways used by the ore fluids. The most important type of alteration accompanying the passage of MVT ore fluids is dolomitization of limestones and recrystallization of pre-ore dolostones. Hydrothermal dolomitization, in contrast to early diagenetic dolomite, it nearly ubiquitous in MVT deposits and commonly forms distinctive halos that may extend for tens or hundreds of metres from the ore (e.g. East Tennessee and Tri-State districts). In the Upper Mississippi Valley district, Agnew (1955) noted that distinct geochemical halos of silica and magnesium (reflecting dolomitization of limestones) were present about ore deposits and suggested applying this observation to exploration in the district.

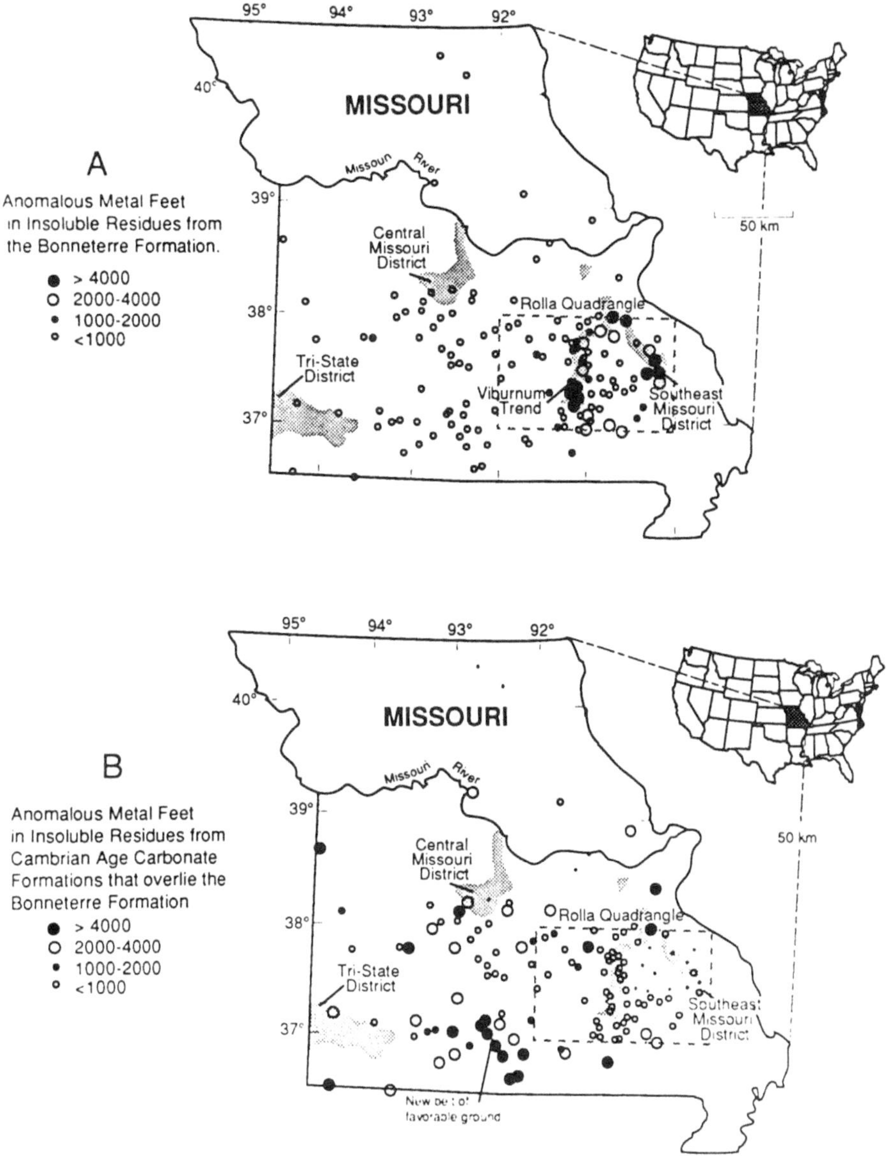

Fig. 4. A Geochemical map showing locations of drill holes sampled and the distribution and abundance of metals (As + Ag + Co + Cu + Mo + Ni + Pb + Zn) in insoluble residues of drill hole samples of the Bonneterre Formation of Cambrian age. Values are in anomalous metal feet (see text for discussion). Location of the Rolla 1° × 2° quadrangle map is shown by the *dotted line* (after Erickson et al. 1983). **B** Geochemical map showing locations of drill holes sampled and the distribution and abundance of metals (As + Ag + Co + Cu + Mo + Ni + Pb + Zn) in insoluble residues of drill hole samples of the Derby-Doe Run, Potosi, and Eminence Formations of Cambrian age. Values are in anomalous metal feet (see text for discussion). Location of the Rolla 1° × 2° quadrangle map is shown by the *dotted line*. (After Erickson et al. 1983)

Hydrothermal dolomite also occurs as a late cement filling a variety of open spaces and vuggy porosity in rocks several hundreds of kilometres away from MVT deposits (Leach et al. 1991). Commonly, the presence of hydrothermal dolomite cement in otherwise barren rocks may be the only evidence that MVT-type fluids may have traversed the rocks. This regional hydrothermal dolomite commonly contains a cathodo-luminescent microstratigraphy (Ebers and Kopp 1979; Voss and Hagni 1985) or banding that can be correlated with a stratigraphy in gangue dolomite in nearby MVT districts (Rowan 1986), providing direct evidence for widespread passage of MVT ore fluids. Notable examples of regional hydrothermal dolomite fringing MVT districts are the Ozark region (Rowan 1986; Leach et al. 1991), East Tennessee district (Ebers and Kopp 1979), and the Upper Mississippi Valley district (Calder et al. 1984).

The banding in hydrothermal dolomite cement in the Ozark region shows distinct changes near ore. Rowan (1986) found that cathodo-luminescent bands in dolomite commonly were noticeably corroded and etched near ore deposits. Hayes et al. (1989) found that the earliest stage of hydrothermal dolomite of the Ozark region has a bright yellow ultraviolet fluorescence in areas remote from known MVT deposits, but only a weak fluorescence in and near ore. The observed variations in fluorescence appear to be due to an interplay of activators, Pb and Mn, and a quencher, Fe, in the dolomite lattice. With few exceptions, this pattern of fluorescent behaviour has been a reliable indicator of ore in hundreds of drill cores throughout the Ozark region (James Palmer, pers. comm., 1990) and appears to be an important guide to ore from widely spaced exploration drill cores.

The second most common type of alteration associated with MVT ore deposition is the introduction of silica into the host rocks. In some MVT districts, extensive silicification of the host carbonates produced distinctive zones of jasperoid (Tri-State and Northern Arkansas), whereas in other districts, the surrounding host rocks contain only a modest increase in silica content. Agnew (1955) proposed that the silica content of the host rocks in the Upper Mississippi Valley district could be used as a guide to ore.

Formation of authigenic clay or feldspar minerals or the destruction of detrital K-silicates has been recognized in some districts. In the Upper Mississippi Valley, Heyl et al. (1964) recognized changes in illite crystallinity near the ore deposits. In the Southeast Missouri lead district, alteration of feldspars near ore was observed by Stormo and Sverjensky (1983). Diehl et al. (1989) recognized a large area in Southeast Missouri where feldspars in the basal sandstone were leached by the regional ore fluids and in which anomalous metals in insoluble residues occur in the overlying carbonates (Erickson et al. 1983).

3.2.4 Organic Geochemistry

Liquid or solid petroleum-type organic matter is variable in both amount and type in MVT deposits. Studies of organic material from Pine Point (Powell and Macqueen 1984) and in the Viburnum Trend (Henry 1988; Leventhal 1990) show that where organic material is closely associated with ore, it has been altered. Leventhal (1990) showed that organic material closely associated with ore in the Viburnum Trend is solid, insoluble, sulphide-rich, and hydrogen-poor. More studies are needed to determine the extent of alteration of organic material about MVT deposits and the possible application of this signature to exploration for MVT deposits.

3.2.5 Fluid Inclusion Studies

Studies of fluid inclusions in minerals from MVT deposits provide valuable information on the conditions and processes of ore deposition. Fluid inclusion studies can provide evidence of fluid flow pathways and source basins for the brines, and can identify large regions traversed by the ore fluids (Leach and Rowan 1986), but few attempts have been made to use fluid inclusion techniques directly in the exploration for MVT deposits. Fluid inclusion studies can potentially detect thermal gradients away from most types of ore deposits; however, thermal gradients about sites of MVT ore deposition are seldom present because the ore fluids were typically at or near thermal equilibrium with the host rocks (Leach and Rowan 1986; Sangster 1989). Fluid inclusion studies can provide information on the composition of the ore fluids and can identify potential changes in composition of the fluids accompanying ore deposition. Haynes and Kesler (1987) tested this concept in the East Tennessee district and found distinct differences in Na/Ca ratios of fluid inclusions in dolomite in the ore zone compared to fluid inclusions in dolomite in barren rocks. Higher Ca contents were found in fluid inclusions in ore zone dolomite as contrasted with those in dolomite not associated with ore; unfortunately, this change in composition was limited to less than 12 m from ore. Viets and Leach (1990) determined the composition of fluid inclusions in MVT minerals, both in ore deposits and distant from ore in the Ozark region of the United States and found no detectable differences.

The analysis of gases in fluid inclusions, currently used in oil-field evaluation (Roedder 1984; Barker and Sullivan 1987) is now being applied to studies of MVT deposits. Work on fluid inclusion gases in MVT minerals from the Ozark region has determined that boiling of CO_2 and other volatiles occurred during regional ore fluid migration and that mixing of fluids produced ore deposition in the mineral districts (Leach et al. 1991). Current research using fluid inclusion gas analysis might offer new ways to identify variation in gas chemistry near undiscovered MVT deposits (Landis and Hofstra 1991).

3.2.6 Field Geochemical Test

A field geochemical test for secondary zinc in carbonate rocks which has been applied successfully in Canada, Australia, and elsewhere uses a spray of "zinc zap". The zinc-zap technique uses a solution of 3% potassium ferricyanide in water and a separate solution of 3% oxalic acid and 0.5% diethyanaline in water, sequentially sprayed or dropped in equal proportions on an outcrop. The presence of zinc in carbonate rocks is indicated by a red colour.

3.3 Summary

Exploration for undiscovered MVT deposits is an enormously difficult task due in part to the great diversity in deposit characteristics, controls on ore deposition, and geochemical composition, as well as the lack of universally applicable exploration guides. Geochemical techniques and indicator elements that prove useful in one district or region may not be applicable in others. Even if a set of exploration guides can be reasonably established for a region, potential mineralized strata often extend over large regions, are usually flat-lying, and the exploration geologist must rely on widely spaced drill holes. As discussed in this study, lithogeochemical investigations in the exploration for MVT deposits are no panacea and are most useful when integrated with other studies.

Most lithogeochemical studies of MVT deposits show markedly restricted geochemical halos about ore deposits which can, however, be useful for extending known ore districts or as guides for development drilling within a deposit. In frontier areas, analysis of insoluble residues appears to be a reliable lithogeochemical technique for broad-scale geochemical reconnaissance. However, its utility in regions other than the United States midcontinent needs to be determined. Other techniques such as studies of mineral alteration, dolomite stratigraphy and fluid inclusion analysis are tools that can enhance exploration programs if carefully used in conjunction with sound geological studies.

4 Geochemistry of Metal-Rich Formation Waters

It is now widely accepted that metal-rich basinal brines were the principal ore-forming solutions in Sedex and MVT deposits. Therefore, the geochemical signature of these solutions would have been a major influence on the ore mineralogy and geochemistry, contributing to the present-day lithogeochemical expression of the deposits. The capacity of metal rich brines to become ore-forming fluids is well illustrated by the Na-Ca-Cl oil field brines in the Cheleken region (former USSR). These brines contain up to 10 mg/l Zn + Pb and deposit native lead, sphalerite, galena, and pyrite in well pipe and holding tanks at a rate analogous to ore deposit formation (Lebedev 1972). The best-documented metal-rich brines from the standpoint of their trace-metal content are the Na-Ca-Cl oil field brines hosted by Jurassic and Cretaceous formations in central Mississippi, USA. These waters have total dissolved solids contents up to 350 000 mg/l, density locally exceeding 1.2 g/ml, and contain up to several hundred mg/l Pb + Zn (Carpenter et al. 1974; Kharaka et al. 1987). Reservoir temperatures for these brines range from 90–130 °C, and salinity and metal content generally correspond to increasing temperature and depth. However, some of the deeper host units, such as the Jurassic Smackover Formation, contain significant amounts of H_2S, which results in low base metal contents. Thermochemical modelling by Kharaka et al. (1987) indicates that both the sulphide-rich (low metals) brines and the Zn-Pb-rich (sulphide-depleted) brines are approximately saturated with respect to sphalerite and galena.

The origin of the hypersaline formation waters in the Gulf Coast region of North America, in general, and in Mississippi, specifically, is controversial. For example, Land and Prezbindowski (1981) proposed that they originated by dissolution of evaporites, whereas Carpenter et al. (1974) and Kharaka et al. (1987) invoked the preservation of bittern brines remaining after evaporite deposition during the Mesozoic. Regardless of their exact origin, the involvement of evaporites apparently is a prerequisite for the formation of metal-rich sedimentary formation brines (Hanor 1979).

Zinc and lead-rich oil field brines from Mississippi contain minor amounts of several other metals, which provide information on the expected elemental association derived exclusively from sedimentary processes. Table 5 presents chemical data for a brine and associated barite-rich well scale from the Raleigh Field, Mississippi. Sverjensky (1984) chose a brine from this field to model how a basin brine could evolve to an MVT-ore-forming solution in response to water-rock reactions as it moves toward the basin margin. The brine and scale data show that Ag, Cu, Mo, Ni, Cd, and Co are present as trace constituents in the brines. Saunders and Swann (1990) reviewed available solubility data for several of these trace elements and concluded that they are generally present at levels below saturation values for the respective sulphides, and suggested that the host formations were depleted with respect to these elements. However, the brine and scale data demonstrate that sedimentary formation waters are probably capable of mobilizing enough Cu, Ag, Mo, Co, and Ni to explain their occurrence as trace constituents in MVT deposits of the Southeast Missouri and Upper Mississippi Valley districts (Heyl 1968; Snyder and Gerdemann 1968).

Table 5. Geochemistry of a typical metal-rich oil field brine, Raleigh Field, Mississippi, USA

	Brine (mg/l)	Scale (ppm)		Brine (mg/l)	Scale (ppm)
Na	65 400		Fe	346	2 900
Li	42		Pb	53	20 000
K	691		Zn	222	4 800
Mg	1 850		Cu	<0.02	450
Ca	30 100	4 100	Ag	<0.004	110
Sr	1 860	59 000	Mn	64	32
Ba	145	>570 000	Cd	0.83	2
B	87		As	<0.3	6
Cl	166 000		Sb	<0.08	12
Br	1 040		Ni		10
I	48		Co		1
SO_4	15		Mo	0.03	7
SiO_2	43		Tl	0.34	<2
Alk.	186				
TDS	268 000				
pH	5.65				

Note: Brine and well-scale data from the Central 5–6 well, Raleigh Field, Mississippi. Compiled from Kharaka et al. (1987), Saunders and Swann (1990) and Saunders and Rowan (1990)

Although the Mississippi oil field brines are some of the best studied, they may or may not be representative of metal-rich sedimentary brines in other basins and of different ages. For example, Palaeozoic basins in the midcontinent and Appalachian region of North America contain black shales locally enriched in Mo, V, Ni, Cu, and Se as well as Zn and Pb (Leventhal et al. 1983; Coveney 1989). If hypersaline formation waters encountered these metal-rich formations, or if formation waters were released from these formations during orogenesis, then the resulting brines could be enriched in some of the elements that are present as only trace constituents in Mississippi oil field brines. This hypothesis is at least partially supported by the discovery of a Mo-Zn-Pb MVT deposit hosted by Ordovician carbonates in the southern Appalachians (Foss et al. 1983). Because metal-rich black shales may themselves have been enriched by basin brines (Coveney 1989), it would appear that metals can be recycled during the history of basin development, and the resulting brine chemistry at a particular time would be a function of both the source rock composition and the tectonic history of the basin.

In addition to source rock composition, the metal content of saline, low sulphide formation waters apparently is also a function of temperature. Lydon (1983) proposed that the general enrichment of Sedex deposits in Cu and Ag relative to MVT deposits is primarily a function of temperature. For example, the common Ag and Cu sulphides are only slightly soluble at low temperature, and their solubility increases relative to sphalerite and galena with increasing temperature. This general concept is supported by the observation that MVT mineralization that formed within the centre of basins is enriched in silver relative to MVT mineralization at the basin edges (Kyle and Price 1986; Kyle et al. 1988; and Posey et al., this Vol.). This enrichment in silver apparently reflects the tapping of brines at higher temperatures.

Magmatic activity, crustal thinning, or rifting could raise the temperature of formation waters in a basin, potentially resulting in higher metal solubilities and new source rocks capable of supplying different metals. For example, the relatively high temperature brines from the Salton Sea geothermal area have relatively high concentrations of Au, Pt, and Ag, in addition to Pb and Zn (McKibben and Williams 1989).

5 Discussion and Summary

Exploration for Sedex and for MVT deposits requires a solid understanding of the geology of the search area, a familiarity with the geological features known to be associated with ore, and a thorough knowledge of the geochemical variables which have been used successfully (in a district) or which *might* be of value (in a geological environment). Many lithogeochemical signatures of Sedex and MVT deposits have been discussed above. A comparison of those elements which occur in oil field brines and/or well scales, MVT deposits, and in Sedex deposits is shown in Table 6.

The presence of Cu, Mo, Pb, Zn, Ni, Co, Ag, Sb, As, Cd, and Ba in oil field brines/well scales, in MVT deposits, and in Sedex deposits lends credence to the use of these elements in lithogeochemical surveys. Many of the element

Table 6. Comparison of reported elements; oil field brines and/or well scales – MVT deposits – Sedex deposits

Brines and/or scales	MVT	Sedex
Cu	Cu	Cu
Mo	Mo	Mo
Pb	Pb	Pb
Zn	Zn	Zn
Ni	Ni	Ni
Co	Co	Co
Ag	Ag	Ag
Sb	Sb	Sb
As	As	As
Cd	Cd	Cd
Ba	Ba	Ba
Cl		Cl
Sr		Sr
B		B
Li		Li
Tl		Tl
Br		
I		
	Au	Au
	Sn	Sn
	Hg	Hg
	In	In
	Bi	Bi
	Ga	Ga
	Ge	Ge
	F	F

associations described are based on empirical observations only. Continuing research on the genesis of Sedex and MVT deposits, however, will improve our understanding of these associations and will thereby improve the effectiveness of the exploration process. For example, geochemical data from present-day sedimentary formation waters, when coupled with new data on metal solubilities in saline solutions, and with empirical data on those elements which occur in ore systems, will be a valuable guide to those elements which should be used in a particular exploration program.

Although there are similarities between Sedex deposits and MVT deposits, each deposit (and hence each search environment) can be expected to be different. There is no single model for Sedex or for MVT deposits, and there is no one set of geological or geochemical criteria which must be satisfied for the explorationist to proceed to the next step in the exploration process. What is important is the recognition of geological and geochemical signatures in the search area which are known to be associated with the formation of ore. It is not necessary for *all* of the signatures to be present. Chayes (1980) addressed the problem generated by imposing undue constraints on the characterization of populations of "like samples" – in the present case, groups of ore deposits with similar characteristics. Chayes analyzed the effect of imposing restricted ranges of values ($x \pm s$) of multiple variables (i) on the naming of rock types.

Using a population of 1584 abyssal basalt glasses, 70% of the samples contained SiO_2 values within one standard deviation of the mean, 53% of the samples contained SiO_2 *and* Al_2O_3 values within one standard deviation of the mean of each population, and only 22% of the samples contained values of all seven of the oxide variables considered within one standard deviation of their means. Chayes' study indicates that the use of an increasingly large number of variables in the definition of a "typical" rock decreases the number of such rocks which are "typical". Chayes suggests that when all items in a group are indeed members of a class, i.e. abyssal basalts, that the use of an "any i" interpretive procedure rather than an "all i" procedure is acceptable. Using the "any i" approach, a far greater percentage of the 1584 basalt samples was correctly classified. Applied to the discipline of lithogeochemistry, the search for sediment-hosted Pb-Zn deposits can be impeded by the imposition of a large number of *necessary* variables on the definition of the "best" place to search for ore. Although Na occurs in anomalously high concentrations above the Sullivan deposit (Hamilton et al. 1983), and Mn occurs in anomalously high concentrations lateral to the McArthur River deposit (Lambert and Scott 1973), and Ba occurs in anomalously high concentrations lateral to the Gataga deposits (Muir 1983), it is not *necessary* for all of these variables to be present in a search area for exploration to proceed. Rather, it is *sufficient* for at least one geochemical signature, which is known to be related to the process or processes which have formed ore, to be present.

Cost effectiveness is always a significant factor in exploration programs. Although the determination of fluorine (or any other pathfinder element) will increase the per-sample analytical cost, if the data provide *the* key which leads to the discovery of ore, the cost is small indeed! Properly informed exploration managers should be willing to support well-designed geochemical exploration programs.

Acknowledgements. The authors are grateful for the invitation to contribute to this volume, We acknowledge the assistance of W.D. Goodfellow, D.F. Sangster, and J. Slack in providing source publications, and J.T. Nash and J.G. Viets in providing valuable reviews which improved the manuscript.

References

Agnew AF (1955) Application of geology to the discovery of zinc-lead ore in the Wisconsin-Illinois-Iowa district. Trans AIME 202:781–795

Aho AE (1969) Base metal province of Yukon. Can Inst Min Metall Trans 72:71–83

Amstutz GC, Zimmerman RA, Schot EH (1971) The Devonian mineral belt of Western Germany (the mines of Meggen, Ramsbeck and Rammelsberg). In: Guidebook 8. Int Sedimentol Congr, Sedimentology of parts of Central Europe. Kramer, Frankfurt, pp 253–272

Barker C, Sullivan GE (1987) Analysis of gases in fluid inclusions in calcite cement from the deep Smackover Formation. Am Curr Res on Fluid Inclusions (ACROFI), Jan 1–8 Socorro. University of New Mexico, Albuquerque, NM, pp 7–8

Barnes HL, Lavery NG (1977) Use of primary dispersion for exploration of Mississippi Valley-type deposits. J Geochem Explor 8:105–115

Bastin ES (ed) (1939) Contributions to a knowledge of the lead and zinc deposits of the Mississippi Valley region. Geol Soc Am Spec Pap 24, 156 pp

Beeson R (1990) Broken Hill-type lead-zinc deposits – an overview of their occurrence and geological setting. Trans Inst Min Metall 99:B163–B175
Benton LM (1984) Ammonium geochemistry of sedimentary exhalative Pb-Zn-Ag deposits: a possible exploration tool. MSc Thesis, Dartmouth College, NH, USA, 114 pp
Both RA (1973) Minor element geochemistry of sulphide minerals in the Broken Hill Lode (N.S.W.) in relation to the origin of the ore. Mineral Depos 8:349–369
Calder JS, Kaufman J, Hanson GM, Meyers WJ (1984) Two-stage cementation of the Burlington-Keokuk limestone in Illinois, Missouri, and lowa. Geol Soc Am Abstr Prog 16:462
Campbell FA, Ethier VG, Krouse HR (1980) The massive sulfide zone: Sullivan orebody. Econ Geol 75:916–926
Carne RC, Cathro RJ (1982) Sedimentary exhalative (sedex) zinc-lead-silver deposits, northern Canadian Cordillera. Can Inst Min Metall Bull 75:66–78
Carpenter AB, Trout ML, Pickett EE (1974) Preliminary report on the origin and chemical evoluation of lead and zinc-rich oil field brines in central Mississippi. Econ Geol 69:1191–1206
Carr GR, Wilmshurst JR, Ryall WR (1986) Evaluation of mercury pathfinder techniques: base metal and uranium deposits. J Geochem Explor 26:1–117
Chayes F (1980) Defining rock types by compositional ranges. Chem Geol 30:309–315
Connor AG, Johnson IR, Muir MD (1982) The Dugald River zinc-lead deposit, Northwest Queensland, Australia. Aust Inst Min Metall Proc 283:1–19
Corbett JA, Lambert IB, Scott KM (1975) Results of analyses of rocks from the McArthur River area, Northern Territory. CSIRO Minerals Research Laboratories, North Ryde, NSW Australia, Tech Commun 57
Coveney RM Jr (1989) A review of the origins of metal-rich Pennsylvanian black shales, cental USA, with an inferred role for basinal brines. Appl Geochem 4:347–367
Cox R, Curtis R (1977) The discovery of the Lady Loretta zinc-lead-silver deposit, Northwest Queensland, Australia – a geochemical exploration case history. J Geochem Explor 8:189–202
Croxford NJW (1974) Cobalt mineralization at Mount Isa, Queensland, Australia, with references to Mount Cobalt. Mineral Depos 9:105–115
Croxford NJW, Jephcott S (1972) The McArthur lead-zinc-silver deposit, N.T. Aust Inst Min Metall Proc 243:1–26
Davenport PH, Hornbrook EHW, Butler AJ (1975) Regional lake sediment geochemical survey for zinc mineralization in western Newfoundland. IGES, Vancouver, pp 555–578
Derry DR, Clark GR, Gillatt N (1965) The Northgate base-metal deposit at Tynagh, County Galway, Ireland. Econ Geol 60:1218–1237
Diehl SF, Goldhaber MG, Mosier EL (1989) Regions of feldspar precipitation and dissolution in the Lamotte Sandstone, Missouri – implications for MVT ore genesis. US Geol Surv Open File Rep 89–169:5–9
Doe BR (1962) Distribution and composition of sulfide minerals at Balmat, New York. Bull Geol Soc Am 73:833–854
Ebers ML, Kopp OC (1979) Cathodoluminescent microstratigraphy in gangue dolomite, the Mascot-Jefferson City district, Tennessee. Econ Geol 74:908–918
Erickson RL, Mosier EL, Viets JB (1978) Generalized geologic and summary geochemical maps of the Rolla 1 × 2 quadrangle, Missouri. US Geol Surv Misc Field Studies Map MF-1004-A
Erickson RL, Mosier EL, Viets JB, King SC (1979) Generalized geologic and geochemical map of the Cambrian Bonneterre Formation, Rolla 1 × 2 degree quadrangle, Missouri. US Geol Surv Misc Field Studies Map MF-1004-B
Erickson RL, Mosier EL, Odland SK, Erickson MS (1981) A favorable belt for possible mineral discovery in subsurface Cambrian rocks in southern Missouri. Econ Geol 76:921–933
Erickson RL, Mosier EL, Viets JB, Odland SK, Erickson MS (1983) Subsurface geochemical exploration in carbonate terrane – midcontinent, USA. In: Kisvarsanyi G, Grant SK, Pratt WD, Koenig JW (eds) International conference on Mississippi Valley-type deposits. Univ of Missouri-Rolla Press, Rolla, MO, pp 575–583
Ethier VG, Campbell FA, Both RA, Krouse HR (1976) Geological setting of the Sullivan orebody and estimates of temperatures and pressure of metamorphism. Econ Geol 71:1570–1588
Finlow-Bates T, Croxford NJW, Allan JM (1977) Evidence for, and implications of, a primary FeS phase in the lead-zinc bearing sediments at Mount Isa. Mineral Depos 12:143–149

Foose MP (1981) Geology, geochemistry, and regional resource implications of a stratabound sphalerite occurrence in the northwest Adirondacks, New York. US Geol Surv Bull 1519, 18 pp

Foss DW, Gatten OJ, Young RS (1983) Shiloh Church molybdenum deposit, Polk County, Georgia. Soc Min Engin of AIME preprint 83–86, 10 pp

Freeze AC (1966) On the origin of the Sullivan orebody, Kimberley, B.C. In: Tectonic history and mineral deposits of the western Cordillera. Can Inst Min Metall Spec Vol 8:263–294

Gardner HD, Hutcheon I (1985) Geochemistry, mineralogy, and geology of the Jason Pb-Zn deposits, Macmillan Pass, Yukon, Canada. Econ Geol 80:1257–1276

Goodfellow WD (1984) Geochemistry of rocks hosting the Howards Pass (XY) strata-bound Zn-Pb deposit, Selwyn Basin, Yukon Territory, Canada. In: Janelidze TV, Tralchrelidze AG (eds) Proc 6th Quadrennial IAGOD Symp. Schweizerbart'sche Verlagsbuchhandlung, Stuttgart, pp 91–112

Goodfellow WD, Jonasson IR (1986) Environment of formation of the Howards Pass (XY) Zn-Pb deposit, Selwyn Basin, Yukon. In: Morin JA (ed) Mineral deposits of northern Cordillera. Can Inst Mining Metall Spec Vol 37:19–50

Goodfellow WD, Rhodes D (1990) Geological setting, geochemistry and origin of the Tom stratiform Zn-Pb-Ag-barite deposits. In: Abbott JG, Turner RJW (eds) Mineral deposits of the northern Canadian Cordillera, Yukon-northeastern British Columbia. Geol Surv Canada Open File 2169, pp 177–241

Goodfellow WD, Jonasson IR, Morganti JM (1983) Zonation of chalcophile elements about the Howard's Pass (XY) Zn-Pb deposit, Selwyn Basin, Yukon. J Geochem Explor 19:503–542

Gorecka E (1991) Zplyw zjawisk tectonicznych na ksztaltowanie sie zloz Zn-Pb. Przegl Geol Warsaw 3:137–146

Govett GJS, Nichol I (1979) Lithogeochemistry in mineral exploration. In: Hood PJ (ed) Geophysics and geochemistry in the search for metallic ores. Geol Surv Can, Econ Geol Rep 31:339–362

Grundmann WH (1977) Geology of the Viburnum No 27 mine, Viburnum Trend, southeast Missouri. Econ Geol 72:349–364

Gustafson LB, Williams N (1981) Sediment-hosted stratiform deposits of copper, lead, and zinc. In: Skinner BJ (ed) Economic Geology, Seventy-Fifth Anniversary Volume. Economic Geology Publ Co, Lancaster Press, Lancaster, PA, pp 139–178

Gustafson JK, Burrell HC, Garretty MD (1950) Geology of the Broken Hill ore deposit, Broken Hill, N.S.W., Australia. Bull Geol Soc Am 61:1369–1438

Gwosdz W, Krebs W (1977) Manganese halo surrounding Meggen ore deposit, Germany. Trans Inst Min Metall 86:B73–B77

Hagni RD (1983) Minor elements in Mississippi Valley-type ore deposits. In: Shanks WC (ed) Cameron volume on unconventional mineral deposits. Soc Econ Geol-Soc Min Eng, AIME, New York, pp 71–88

Hamilton JM, Hauser RL, Ransom PW (1981) The Sullivan orebody. In: Thompson RI, Cook DG (eds) Field guides to geology and mineral deposits. Geol Assoc Can Annu Meet, Calgary, pp 44–49

Hamilton JM, Delaney GD, Hauser RL, Ransom PW (1983) Geology of the Sullivan deposit, Kimberley, B.C., Canada. In: Sangster DF (ed) Sediment-hosted stratiform lead-zinc deposits. Mineral Assoc Can Short Course Handbook vol 8, pp 31–83

Hannak WW (1981) Genesis of the Rammelsberg ore deposit near Goslar/Upper Harz, Federal Republic of Germany. In: Wolf KH (ed) Handbook of strata-bound and stratiform ore deposits, vol 9. Elsevier, Amsterdam, pp 551–642

Hanor JS (1979) The sedimentary genesis of hydrothermal fluids. In: Barnes HL (ed) Geochemistry of hydrothermal ore deposits, 2nd edn. Wiley, New York, pp 137–172

Hayes TS, Burruss RC, Palmer R, Rowan EL (1989) UV fluorescence of hydrothermal dolomite of the Ozark regional Mississippi Valley-type ore system. Geol Soc Am Abstr Prog 21:A3

Haynes FM, Kesler SE (1987) Fluid inclusion chemistry in the exploration for Mississippi Valley-type deposits: an example from East Tennessee, USA. Appl Geochem 2:321–327

Henry AL (1988) Alteration of organic matter in the Viburnum Trend lead-zinc district of southeast Missouri. MSc Thesis, Univ of Toronto, Toronto, Canada

Heyl AV (1968) The Upper Mississippi Valley base-metal district. In: Ridge JD (ed) Ore deposits of the United States, 1933–1967 (Graton Sales Volume), AIME, New York, pp 431–459

Heyl AV, Agnew AF, Lyons EJ, Behre CH (1959) The geology of the Upper Mississippi Valley zinc-lead district. US Geol Surv Prof Pap 424-D, 310 pp
Heyl AV, Hosterman JW, Brock MR (1964) Clay mineral alteration in the Upper Mississippi Valley district. In: Engerson E (ed) Clays and clay minerals, 12th Natl Conf, Atlanta, 1963. Macmillan, New York, pp 445–453
Iisalo E, Johanson B, Jokinen T, Kojonen K, Rasilainen K, Ruotsalainen A, Soininen H, Tornroos R, Vaasjoki M, Vasti K (1989) The Early Proterozoic Zn-Cu-Pb sulphide deposit of Rauhala in Ylivieska, western Finland. In: Kojonen K (ed) Geol Surv Finland, Spec Pap 11. Geologian Tutkimuskeskus, Espoo, pp 5–18
Johnson IR, Klinger GD (1976) Broken Hill ore deposit and its environment. In: Knight CL (ed) Economic geology of Australia and Papua New Guinea, 1. Metals. Aust Inst Min Metall Monog 5: 476–491
Jones DK (1988) A geochemical study of a breccia body in the central Tenessee zinc district. J Geochem Explor 30:197–207
Kharaka YK, Maest AS, Carothers WW, Law LM, Lamothe PJ, Fries TL (1987) Geochemistry of metal-rich brines from the central Mississippi Salt Dome basin, USA. Appl Geochem 2:534–561
Knight CL (1965) Lead-zinc lode at Dugald River. In: McAndrew J (ed) Geology of Australian ore deposits, vol 1. Aust Inst Min Metall, Melbourne, pp 247–250
Krebs W (1981) The geology of the Meggen ore deposit. In: Wolf KH (ed) Handbook of strata-bound and stratiform ore deposits, vol 9. Elsevier, Amsterdam, pp 509–549
Kyle JR (1981) Geology of the Pine Point district. In: Wolf KH (ed) Handbook of strata-bound and stratiform ore deposits, vol 9. Elsevier, Amsterdam, pp 643–741
Kyle JR, Price PE (1986) Metallic sulphide mineralization in salt dome caprocks, Gulf Coast, U.S.A. Trans Inst Min Metall 95:B6–B16
Kyle JR, Boardman S, Price PE (1988) Mississippi Valley-type Zn-Pb-Ag mineralization in carbonate rocks, Smackover Formation, southwest Arkansas. Geol Soc Am Abstr Prog 20:A39
Lambert IB (1976) The McArthur zinc-lead-silver deposit: features, metallogenesis and comparisons with some other stratiform ores. In: Wolf KH (ed) Handbook of strata-bound and stratiform ore deposits, vol 6. Elsevier, Amsterdam, pp 535–585
Lambert IB, Scott KM (1973) Implications of geochemical investigations of sedimentary rocks within and around the McArthur zinc-lead-silver deposit, Northern Territory. J Geochem Explor 2:307–330
Land LS, Prezbindowski DR (1981) The origin and evolution of saline formation water, Lower Cretaceous carbonates, south-central Texas, USA. J Hydrol 54:51–74
Landis GP, Hofstra AH (1991) Fluid inclusion gas chemistry as a potential minerals exploration tool: case studies from Creede CO, Jerritt Canyon, NV, Coeur d'Alene, ID, and MT, southern Alaska mesothermal veins, and mid-continent MVT's. J Geochem Explor 42: 25–59
Lange IM, Nokleberg WJ, Plahuta JT, Krouse HR, Doe BR (1985) Geologic setting, petrology, and geochemistry of stratiform sphalerite-galena-barite deposits, Red Dog Creek and Drenchwater Creek areas, northwestern Brooks Range, Alaska. Econ Geol 80:1896–1926
Large D (1980) Geological parameters associated with sediment-hosted, submarine exhalative Pb-Zn deposits: an empirical model for mineral exploration. Geol Jahrb 40:59–129
Large D (1981a) Sediment-hosted submarine exhalative lead-zinc deposits – a review of their geological characteristics and genesis. In: Wolf KH (ed) Handbook of strata-bound and stratiform ore deposits, vol 9. Elsevier, Amsterdam, pp 469–507
Large D (1981b) The geochemistry of the sedimentary rocks in the vicinity of the Tom Pb-Zn-Ba deposit, Yukon Territory, Canada. In: Rose AW, Gundlach H (eds) Geochemical exploration 1980. J Geochem Explor 15:203–217
Large D (1983) Sediment-hosted massive sulphide lead-zinc deposits: an empirical model. In: Sangster DF (ed) Sediment-hosted stratiform lead-zinc deposits. Mineral Assoc Can Short Course Handbook vol 8, pp 1–29
Larter RCL, Boyce AJ, Russell MJ (1981) Hydrothermal pyrite chimneys from the Ballynoe baryte deposit, Silvermines, County Tipperary, Ireland. Mineral Depos 16:309–318
Lavery NG (1985) The use of fluorine as a pathfinder for volcanic-hosted massive sulfide ore deposits. J Geochem Explor 23:35–60

Lavery NG, Barnes HL (1971) Zinc dispersion in the Wisconsin zinc-lead district. Econ Geol 66:226–242

Lea ER, Dill DB Jr (1968) Zinc deposits of the Balmat-Edwards district, New York. In: Ridge JD (ed) Ore deposits of the United States, 1933–1967 (Graton Sales Volume). AIME, New York, pp 20–48

Leach DL, Rowan EL (1986) Genetic link between Ouachita foldbelt tectonism and the Mississippi Valley-type lead-zinc deposits in the Ozarks. Geology 14:932–935

Leach DL, Plumlee GS, Hofstra AL, Landis GP, Rowan EL, Viets JG (1991) Origin of late dolomite cement by CO_2-saturated deep basin brines: evidence from the Ozark region, central United States. Geology 19:348–351

Lebedev LM (1972) Minerals of contemporary hydrotherms of Cheleken. Geochem Int 9:485–504

Leventhal JS (1990) Organic matter and the thermochemical sulfate reduction in the Viburnum Trend, Southeast Missouri. Econ Geol 85:622–632

Leventhal JS, Briggs PH, Baker JW (1983) Geochemistry of the Chattanooga Shale, DeKalb County, Tennessee. Southeast Geol 24:101–116

Longstaffe FJ, Nesbitt BE, Muehlenbachs K (1982) Oxygen isotope geochemistry of the shales hosting Pb-Zn-Ba mineralization at the Jason prospect, Selwyn Basin, Yukon. Geol Surv Can Pap 82-1C:45–49

Lottermoser BG (1989) Rare earth element study of exhalites within the Willyama Supergroup, Broken Hill Block, Australia. Mineral Depos 24:92–99

Lottermoser BG (1991) Trace element composition of exhalites associated with the Broken Hill sulfide deposit, Australia. Econ Geol 86:870–877

Louden AG, Lee MK, Dowling JF, Bourn R (1975) Lady Loretta silver-lead-zinc deposit. In: Knight CL (ed) Economic geology of Australia and Papua New Guinea, 1. Metals. Aust Inst Min Metall Monogr 5:377–382

Lydon JW (1983) Chemical parameters controlling the origin and deposition of sediment-hosted stratiform lead-zinc deposits. In: Sangster DF (ed) Sediment-hosted stratiform lead-zinc deposits. Mineral Assoc Can Short Course Handbook vol 8, pp 175–250

MacIntyre DG (1982) Geologic setting of recently discovered stratiform barite-sulphide deposits in northeast British Columbia. Can Inst Min Metall Bull 75:99–113

Mathias BV, Clark GJ (1975) Mount Isa copper and silver-lead-zinc orebodies. In: Knight CL (ed) Economic geology of Australia and Papua New Guinea, 1. Metals. Aust Inst Min Metall Monogr 5:351–372

Mathias BV, Clark GJ, Morris D, Russell RE (1973) The Hilton deposit – stratiform silver-lead-zinc mineralization of the Mount Isa type. In: Fisher NH (ed) Metallogenic provinces and mineral deposits in the southwestern Pacific. Bur Mineral Resour Geol Geophys Bull 141:33–58

Maynard JB (1991) Shale-hosted deposits of Pb, Zn, and Ba: syngenetic deposition from exhaled brines in deep marine basins. In: Force ER, Eidel JJ, Maynard JB (eds) Sedimentary and diagenetic mineral deposits: a basin analysis approach to exploration. Rev Econ Geol 5:177–185

McKibben MA, Williams AE (1989) Metal speciation and solubility in saline hydrothermal fluids: an empirical approach based on geothermal brine data. Econ Geol 84:1996–2007

Metsger RW, Tennant CB, Rodda JL (1958) Geochemistry of the Sterling Hill zinc deposit, Sussex County, New Jersey. Bull Geol Soc Am 69:775–788

Moore DW, Young LE, Modene JS, Plahuta JT (1986) Geologic setting and genesis of the Red Dog zinc-lead-silver deposit, western Brooks Range, Alaska. Econ Geol 81:1696–1727

Muir MD (1983) Depositional environments of host rocks to northern Australian lead-zinc deposits, with special reference to McArthur River. In: Sangster DF (ed) Sediment-hosted stratiform lead-zinc deposits. Mineral Assoc Can Short Course Handbook vol 8, pp 141–174

Nesbitt BE, Longstaffe FJ (1982) Whole rock, oxygen isotope results for country rocks and alteration zones of the Sullivan massive sulphide deposit, British Columbia. Geol Surv Can Pap 82-1C:51–54

Nesbitt BE, Longstaffe FJ, Shaw DR, Muehlenbachs K (1984) Oxygen isotope geochemistry of the Sullivan massive sulfide deposit, Kimberley, British Columbia. Econ Geol 79:933–946

O'Meara AE (1961) Contribution to the study of the Mount Isa copper orebodies. Aust Inst Min Metall Proc 197:163–192

Pano SV, Harbottle G, Sayre EV, Hood WC (1983) Genetic implications of halide enrichment near a Mississippi Valley-type ore deposit. Econ Geol 78:150–156

Pedersen FD (1980) Remobilization of the massive sulfide ore of the Black Angel mine, central West Greenland. Econ Geol 75:1022–1041

Plimer IR (1979) Sulphide rock zonation and hydrothermal alteration at Broken Hill, Australia. Trans Inst Min Metall 88:B161–B176

Plimer IR (1982) Fluorine-bearing minerals from Broken Hill, N.S.W. Aust Mineral 39:214–215. In: Gem & Treasure Hunter Yearbook, vol 72

Plimer IR (1983) The association of tourmaline-bearing rocks with mineralization at Broken Hill, NSW. Aust Inst Min Metall, Broken Hill Conf NSW, Australia, pp 157–175

Plimer IR (1986) Sediment-hosted exhalative Pb-Zn deposits – products of contrasting ensialic rifting. Trans Geol Soc S Afr 89:57–73

Plimer IR (1988) Tourmalinites associated with Australian Proterozoic submarine exhalative ores. In: Friedrich GH, Herzig PM (eds) Base metal sulfide deposits in sedimentary and volcanic environments. Springer, Berlin Heidelberg New York, pp 255–283

Powell TG, Macqueen RW (1984) Precipitation of sulphide ores and organic matter: sulfate reduction at Pine Point. Science 224:63–66

Ransom PW (1977) Geology of the Sullivan orebody. In: Hoy T (ed) Lead-zinc deposits of southeastern British Columbia. Geol Assoc Canada – Soc Econ Geol Annu Meeting, Vancouver, pp 7–21

Ridge JD (1952) The geochemistry of the ores of Franklin, New Jersey. Econ Geol 47:180–192

Roedder E (1984) Fluid inclusions. Rev Min 12:643 pp

Rogers RK, Davis JH (1977) Geology of the Buick Mine, Viburnum Trend, southeast Missouri. Econ Geol 72:372–380

Rowan EL (1986) Cathodoluminescent zonation in hydrothermal dolomite cements: relationship to Mississippi Valley-type Pb-Zn mineralization in southern Missouri and northern Arkansas. In: Hagni RD (ed) Process mineralogy VI. Metallurgical Society, Warrenville, PA, pp 69–87

Russell MJ (1974) Manganese halo surrounding the Tynagh ore deposit, Ireland: a preliminary note. Trans Inst Min Metall 83:B65–B66

Russell MJ (1975) Lithogeochemical environment of the Tynagh base-metal deposit, Ireland, and its bearing on ore deposition. Trans Inst Min Metall 84:B128–B133

Ryall WR (1979) Mercury in the Broken Hill (N.S.W., Australia) lead-zinc-silver lodes. J Geochem Explor 11:175–194

Ryall WR (1981) The forms of mercury in some Australian stratiform Pb-Zn-Ag deposits of different regional metamorphic grades. Mineral Depos 16:425–435

Ryan PJ, Lawrence AL, Lipson RD, Moore JM, Paterson A, Stedman DP, Van Zyl D (1986) The Aggeneys base metal sulphide deposits, Namaqualand district. In: Anhaeusser CR, Maske S (eds) Mineral deposits of southern Africa. Geol Soc S Afr, Johannesburg, pp 1447–1473

Sangster DF (1968) Some chemical features of lead-zinc deposits in carbonate rocks. Geol Surv Can Pap 68-39, 17 pp

Sangster DF (1989) Thermal comparison of MVT deposits and their host rocks. Geol Soc Am Abstr Prog 21:A7

Sangster DF (1990) Mississippi Valley-type and Sedex deposits: a comparative examination. Trans Inst Min Metall 99:B21–B42

Sarkar SC, Bhattacharyya PK, Mukherjee AD (1980) Evolution of the sulfide ores of Saladipura, Rajasthan, India. Econ Geol 75:1152–1167

Saunders JA, Rowan EL (1990) Mineralogy and geochemistry of metallic well scale, Raleigh and Boykin Church oil fields, Mississippi, USA. Trans Inst Min Metall 99:B54–B58

Saunders JA, Swann CT (1990) Trace-metal content of Mississippi oil field brines. J Geochem Explor 37:171–183

Schultz RW (1966) The Northgate base-metal deposit at Tynagh, County Galway, Ireland. Econ Geol 61:1443–1459

Smee BW, Bailes RJ (1986) The use of lithogeochemical patterns in wall rock as a guide to exploration drilling at the Jason lead-zinc-silver-barium deposit, Yukon Territory. In: Nichols CE (ed) Exploration for ore deposits of the North American Cordillera. J Geochem Explor 25:217–230

Smirnov VI, Gorzhevsky DI (1977) Deposits of lead and zinc. In: Smirnov VI (ed) Ore deposits of the USSR vol II. Pitman, London, pp 182–256

Smith SE, Walker KR (1971) Primary element dispersions associated with mineralization at Mount Isa, Queensland. Bur Mineral Resour Geol Geophys Bull 131, 80 pp

Snyder FG, Gerdemann PE (1968) Geology of the southeast Missouri lead district. In: Ridge JD (ed) Ore deposits of the United States, 1933–1967 (Graton Sales Volume), AIME, New York, pp 326–358

Stanton RL (1962) Elemental constitution of the Black Star orebodies, Mount Isa, Queensland, and its interpretation. Trans Inst Min Metall 72:69–124

Stanton RL (1976) Petrochemical studies of the ore environment at Broken Hill, New South Wales: 1 – constitution of "banded iron formation". Trans Inst Min Metall 85:B33–B46

Sterne EJ, Zantop H, Reynolds RC (1984) Clay mineralogy and carbon-nitrogen geochemistry of the Lik and Competition Creek zinc-lead-silver prospects, DeLong Mountains, Alaska. Econ Geol 79:1406–1411

Stormo S, Sverjensky, DA (1983) Silicate hydrothermal alteration in a Mississippi Valley-type deposit-Viburnum Trend, southeast Missouri. Geol Soc Am Abstr Prog 15:699

Sureda RJ, Martin JL (1990) El Aguilar mine: an Ordovician sediment-hosted stratiform lead-zinc deposit in the Central Andes. In: Fontbote L, Amstutz GC, Cardozo M, Cedillo E, Frutos J (eds) Stratabound ore deposits in the Andes. Springer, Berlin Heidelberg New York, pp 161–174

Sverjensky DA (1984) Oil field brines as ore-forming solutions. Econ Geol 79:23–37

Taylor S (1984) Structural and paleotopographic controls of lead-zinc mineralization in the Silvermines orebodies, Republic of Ireland. Econ Geol 79:529–548

Taylor S, Andrew CJ (1978) Silvermines orebodies, County Tipperary, Ireland. Trans Inst Min Metall 87:B111–B124

Tempelman-Kluit DJ (1972) Geology and origin of the Faro, Vangorda, and Swim concordant zinc-lead deposits, Central Yukon Territory. Geol Surv Can Bull 208, 73 pp

Thompson T, Beaty DW (1990) Geology and origin of ore deposits in the Leadville district, Colorado: Part 2. Oxygen, hydrogen, carbon, sulfur and lead isotope data and development of a genetic model. In: Beaty DW, Landis GP, Thompson T (eds) Carbonate-hosted sulfide deposits of the Colorado mineral belt. Econ Geol Monogr 7:156–179

Turner RJW (1990) Jason stratiform Zn-Pb-barite deposit, Selwyn Basin, Canada (NTS 105-0-1): geological setting, hydrothermal facies and genesis. In: Abbott JG, Turner RJW (eds) Mineral deposits of the northern Canadian Cordillera, Yukon-northeastern British Columbia. Geol Surv Can Open File 2169:137–175

Viets JG, Leach DL (1990) Genetic implications of regional and temporal trends in ore fluid geochemistry of Mississippi Valley-type deposits in the Ozark region. Econ Geol 85:842–861

Viets JG, Mosier EL, Erickson MS (1983) Geochemical variation of major, minor, and trace elements in samples of the Bonneterre Formation from drill holes transecting the Viburnum Trend Pb-Zn district of southeast Missouri. In: Kisvarsanyi G, Grant SK, Pratt WD, Koenig JW (eds) Int Conf on Mississippi Valley-type deposits. Univ of Missouri-Rolla Press, Rolla, MO, pp 174–187

Vokes FM (1963) Geological studies on the Caledonian pyritic zinc-lead orebody at Bleikvassli, Nordland, Norway. Nor Geol Unders 222:1–126

Voss RL, Hagni RD (1985) The application of cathodoluminescent microstratigraphy for sparry dolomite from the Viburnum Trend, southeastern Missouri. In: Hausen DM, Kopp OC (eds) Mineralogy – applications to the minerals industry. Proc Paul F Kerr Memorial Symp. AIME, New York, pp 51–68

Walker RN, Muir MD, Diver WL, Williams N, Wilkins N (1977) Evidence of major sulphate evaporite deposits in the Proterozoic McArthur Group, Northern Territory, Australia. Nature 265:526–529

Whitcher IG (1975) Dugald River zinc-lead lode. In: Knight CL (ed) Economic geology of Australia and Papua New Guinea, 1. Metals. Aust Inst Min Metall Monogr 5:372–376

Williams LB, Zantop H, Reynolds RC (1987) Ammonium silicates associated with sedimentary exhalative ore deposits: a geochemical exploration tool. J Geochem Explor 27:125–141

Wilson IH, Derrick GM (1976) Precambrian geology of the Mount Isa region, Northwest Queensland. 25th Int Geol Congr, Sydney, Australia, Guide to Excursions 5A and 5C, 44 pp

The Economics of Sediment-Hosted Zinc-Lead Deposits

F.-W. Wellmer, T. Atmaca, M. Günther, H. Kästner and A. Thormann

Abstract

Two types of sediment-hosted zinc-lead deposits are economically analyzed and compared:

1. Mississippi Valley-type (MVT) deposits in carbonate host rocks;
2. Stratiform exhalative (Sedex type) deposits predominantly in clastic sediments and less commonly in carbonates.

Other deposits taken into account for comparison are vein-type Pb-Zn deposits and volcanic-hosted deposits, which, apart from containing Zn, Pb and Ag, usually also carry Cu and Au, elements which are normally absent in MVT and not very common in Sedex-type deposits.

Sedex-type deposits are clearly defined orebodies often with natural cutoff boundaries and a relatively uniform grade distribution, whereas MVT deposits are quite irregular in shape and grade distribution, normally with economic cutoff boundaries. MVT deposits, therefore, require far denser drilling grids than Sedex-type deposits in exploration and during production. Geostatistical parameters quantitatively describing the irregularity of grade distribution are similar for MVT and vein-type deposits.

MVT deposits are rather uncomplicated deposits metallurgically and with respect to mining, thus allowing the production of clean, high-grade concentrates and the application of effective mining methods (mainly room-and-pillar). Sedex-type ores frequently have finely intergrown phases which present metallurgical problems during separation. Therefore, only lower grade selective concentrates and sometimes only mixed concentrates can be produced at acceptable recovery rates. The total average operating costs (for mining and milling) are lower for MVT than for Sedex-type deposits or even vein-type deposits. Consequently, in a grade-tonnage diagram the borderline of viability (roughly separating viable from marginal and non-viable deposits in a worldwide context) encloses lower grade and lower tonnage deposits for MVT deposits than for Sedex-type deposits. A price-risk analysis shows that the zone of the lowest price risk coincides with the largest maximum in the multimodal frequency distribution of Zn:(Zn + Pb) ratios for MVT as well as Sedex-type deposits.

Federal Institute for Geosciences and Natural Resources, Stilleweg 2, 30655 Hannover, Germany

Spec. Publ. No. 10 Soc. Geol. Applied to Mineral Deposits
Fontboté/Boni (Eds.), Sediment-Hosted Zn-Pb Ores

An economic analysis and comparison, using the net present value, the internal rate of return parameters and average grade and tonnage data for MVT, Sedex-type and volcanic-hosted deposits shows that Sedex-type deposits are the most attractive due to, on average, higher grades, despite lower costs for MVT deposits.

1 Introduction

The two main types of sediment-hosted zinc-lead deposits which are economically evaluated here and compared with other types of zinc-lead deposits are:

1. Carbonate-hosted deposits of the Mississippi Valley type (MVT), and
2. Stratiform sedimentary-exhalative deposits (Sedex type) in clastic and carbonate sediments.

The classification used in this chapter and the categorization of individual deposits closely follow the definitions of Sangster (1990). Examples of MVT and Sedex-type Zn-Pb deposits from different regions of the world and showing a wide range of ages are given in Table 1A,B. Table 1A, which lists MVT deposits, is biased towards smaller deposits due to the lack of quantitative data on large mining districts in the USA (see Sect. 6.3). For the location of the deposits, the reader is referred to Fig. 1 in Sangster (1990).

The MVT and Sedex-type Zn-Pb deposits will be compared with vein-type and volcanic-associated deposits. In the latter type, Zn or Zn and Pb are typically associated with Cu and Au, which only rarely occur in the two main types of sediment-hosted deposits. No reference will be made to metasomatic deposits (e.g. Meat Cove, Canada, or Groundhog Central Mining District, New Mexico, USA) and sandstone-hosted (impregnation-type) lead deposits (e.g. Yava, Canada, Mechernich and Maubach, Germany, or Laisvall, Sweden) (Bjørlykke and Sangster 1981; Einaudi et al. 1981).

2 Economic Importance of MVT and Sedex-Type Zn-Pb Deposits

The MVT and Sedex-type Zn-Pb deposits represent important resources and are significant suppliers of Pb and Zn. According to Tikkanen (1986), 65% of world Zn and 77% of world Pb reserves are contained in these two deposit types. The production proportions are 56 and 51%, respectively (Table 2). Tikkanen (1986) classified the Zn and Pb deposits strictly according to the type of host rock: clastic and carbonate sediments. Since Sangster's (1990) classification includes some carbonate-hosted deposits (e.g. Balmat-Edwards, USA, or Silvermines, Ireland) in the Sedex-type group, the line of separation between the MVT and Sedex-type groups as defined here, is not identical to that of Tikkanen (1986) (see first column in Table 2).

Table 1. Examples of MVT (A) and Sedex-type (B) deposits

A. MVT deposits

Deposit	Country	Tonnage (million t)	Total size (reserves + production) and ore grade Zn (%)	Pb (%)	Ag (g/t)	Cu (%)	Age of host rock
Nanisivik	Canada	7.0	14.0	2.0			Proterozoic (Helikian)
Gayna River	Canada	>50	5	(Pb + Zn)			Proterozoic (Helikian)
Austinville (16 orebodies)	USA (Virginia)	30	4.7	(Pb + Zn)			Lower to Middle Cambrian
Daniels Harbour (Newfoundland Zinc) (12 orebodies)	Canada	6	7.9				Lower Ordovician
Central Tennessee district (10–20 orebodies)	USA	>250	3–4				Lower Ordovician
Song Toh district (5 deposits)	Thailand	5.5	9–10	(Pb + Zn)	80–100		Middle Ordovician
Polaris	Canada	22	18	(Pb + Zn)			Late Ordovician
Pine Point district (36 orebodies, 87 including uneconomic deposits)	Canada	84	6.6	2.9			Middle Devonian
Robb Lake	Canada	5.5	7.3	(Pb + Zn)			Devonian
Lennard Shelf (Blendevale, Cadjebut, and Twelve Mile Bore)	Australia	25	11.5	(Pb + Zn)			Devonian
Sorby (13 orebodies)	Australia	16.2	0.6	5.25	56		Lower Carboniferous
Gays River	Canada	12	7	(Pb + Zn)			Carboniferous (Mississippian)
Salafossa	Italy	10	4.6	1			Trias (Carnian)
Bleiberg	Austria	40	5.5	(Pb + Zn)			Trias (Carnian)
San Vicente	Peru	>15	12	1			Lower Jurassic

Sources: Cerny (1989), Miller (1991), Rhodes et al. (1984), additional literature (see references for Table 1A,B), company reports, pers. comm.

Table 1. *Continued*

B. Sedex-type deposits

Deposit	Country	Tonnage (million t)	Total size (reserves + production) and ore grade Zn (%)	Pb (%)	Ag (g/t)	Cu (%)	Age of host rock
Black Angel/Marmorilik (10 orebodies)	Greenland	13.6	12.3	4.0	90		Lower Proterozoic
Gamsberg	South Africa	150	7.1	0.5			Proterozoic (~2000 m.y.)
Broken Hill	Australia	180	9.8	11.3	175	0.2	Lower to Middle Proterozoic (>1700 m.y.)
Mt. Isa (Pb-Zn)	Australia	88.6	6.1	7.1	160	0.06	Middle Proterozoic (~ 1650 m.y.)
Sullivan	Canada	155	5.7	6.6	68	–	Middle Proterozoic (~1430 m.y.)
Rosh Pinah	Namibia	150	7.1	0.5			Upper Proterozoic (>780–<920 m.y.)
Faro/Anvil	Canada	63.5	5.72	3.4	37	–	Cambrian
Howards Pass	Canada	≥100	6	1.5	?	–	Lower Silurian
El Aguilar	Argentina	30	18	(Pb + Zn)			Silurian
Cirque	Canada	40	7.8	2.2	47	–	Devonian
Rammelsberg	Germany	30	19	9	103	1	Middle Devonian
Meggen	Germany	60	10	1.3	–	–	Middle Devonian
Tynagh	Ireland	12.3	4.5	4.9	58	0.4	Lower Carboniferous
Silvermines	Ireland	18.4	7.4	2.8	21		Lower Carboniferous
Red Dog	USA/ Alaska	85	17.1	5.0	82	–	Carboniferous (Mississippian/ Pennsylvanian)
Filizchai	USSR	95	4.1	1.6	50	–	Jurassic

Sources: Gustafson and Williams (1981), Large (1983), additional literature (see references for Table 1A,B), company reports, pers. comm.

Table 2. Lead and zinc tonnage distribution of various deposit types (Tikkanen 1986)

Deposit type	Share of world reserves in %				Share of world production in %			
	Zn		Pb		Zn		Pb	
Clastic sediment-hosted (Sedex)	54	} 65	61	} 77	31	} 56	25	} 51
Carbonate-hosted (mostly MVT, some Sedex)	11		16		25		26	
Volcanic rock-hosted	23		18		30		11	
Vein-hosted	3		2		5		5	
Other/unidentified	9		3		9		33	

3 Tonnage, Grade and Age Distributions of MVT and Sedex-Type Zn-Pb Deposits

According to Eckstrand (1984), the annual production figures of 38 world-wide examples of Sedex-type deposits range from 4 to 550 million t (average 60 million t) at ore grades of 0.6 to 18% Zn (average 7.3%), 0.3 to 13% Pb (average 4%), nil to 1% Cu (average 0.1%) and trace to 180 g/t Ag (average 48 g/t). For MVT deposits Eckstrand (1984) gave only best estimates between 1 and 10 million t and grades of 5 to 10% combined Pb + Zn for most deposits, because data for individual deposits are difficult to obtain due to the fact that production figures are not released and that in many districts deposits tend to be interconnected. Typically, MVT deposits occur in clusters (Table 1A) or the "deposits" of different mines are actually interconnected sectors of one large deposit and are separated by artificial claim boundaries (e.g. Viburnum Trend, USA). For this reason Sangster (1990) combined several, single "deposits" to form a so-called MVT district in his grade-tonnage plot, which in this chapter is used as a basis to derive the borderline of economic viabiliby (Fig. 8a). This will be discussed in Section 6.3.

Sangster's grade-tonnage diagram shows that grades and tonnages of MVT and Sedex-type deposits are very similar, but combined Pb + Zn grades are slightly higher for Sedex-type deposits. In the low-tonnage field of the diagram, single MVT deposits are dominant, whereas the large-tonnage field chiefly contains large Sedex-type deposits which are only matched by a few MVT districts (e.g. Viburnum Trend, Old Lead Belt Missouri, and East Tennessee District, USA).

Grade-tonnage curves are also given in the US Geological Survey Mineral Deposit Models volume (Cox and Singer 1986) for various types of deposits. For Sedex-type deposits the mean values are 15 million t at 5.6% Zn, 2.8% Pb and 30 g/t Ag. The only MVT deposits considered are those in the Pb-Zn district of southeast Missouri and the Appalachian Zn deposits. The mean values are 35 million t at 4% Zn, 0.87% Pb and 0.48 g/t Ag.

As far as metal ratios are concerned, Sangster (1990) showed that the MVT deposits display a clear bimodal distribution, the majority of the deposits (85%) being Zn-rich [modal value of 0.8 Zn/(Zn + Pb)] and 15% having a

Table 3. Grade and tonnage data of Sedex-type Zn-Pb deposits (After Lydon 1983)

Time period	Host rock	Number of deposits	Tonnage (in million t)	Zn (%)	Pb (%)	Ag (g/t)	Cu (%)
Phanerozoic	Clastics	14	67.0	7.24	2.86	26	0.05
Phanerozoic	Carbonates	9	14.5	8.12	2.46	8	0.02
Average of all Phanerozoic	–	23	46.5	7.35	2.81	24	0.04
Proterozoic	Clastics	13	90.0	7.14	5.09	71	0.13
Proterozoic	Carbonates	2	12.2	11.51	1.78	9	–
Average of all Proterozoic	–	15	79.6	7.23	5.02	70	0.13
Average of all Sedex-type deposits	–	38	59.6	7.29	3.98	48	0.09

modal value of 0.05. The distribution of metal ratios for Sedex-type deposits is similar. For this group Lydon (1983) showed that deposits with lower Zn/(Zn + Pb) ratios tend to be hosted by predominantly clastic sediments, whereas deposits with higher ratios mainly occur in carbonate-dominated sediments. The diagrams of Sangster (1990) for metal ratios of the two deposit types are used as a basis for delineating intervals of maximum and minimum price risk (Fig. 6), as discussed below.

The epigenetic MVT deposits are found in carbonate host rocks that range in age from Early Proterozoic to Mesozoic; most occur in rocks of Palaeozoic age. The stratiform Sedex-type deposits, which are coeval with their host rocks, are abundant in both Proterozoic and Palaeozoic sediments and range in age from Early Proterozoic to Jurassic (Sangster 1990). There is insufficient data to carry out a meaningful statistical analysis of tonnages and grades as a function of age. Lydon (1983) considered two groups of Sedex-type deposits, Proterozoic and Phanerozoic ones. The grade and tonnage figures are given in Table 3.

4 Description of MVT and Sedex-Type Deposits and Economic Aspects

4.1 Geometry and Ore-Grade Distribution, Implications for Exploration

Sedex-type deposits normally consist of well-defined, laterally continuous orebodies with mostly natural grade cutoffs. The orebodies comprise planar, concordant, interbedded layers of sulphides and host rock whose lateral extent is several tens to hundreds of times their thickness. The ore grade tends to show a quite uniform distribution or a relatively steady change with only minor local variations (see Fig. 1a, example from Meggen, Germany). This fact often allows exploration holes to be drilled at a spacing of 100 m or more. It is not uncommon to find several ore lenses one above the other in a single deposit (e.g. Mt. Isa, Broken Hill, Australia, and possibly Rammelsberg, Germany). Considering a single orebody or several such layered lenses derived from one local ore-deposition event, Sedex deposits are quite often formed by local, isolated events. Again, the exceptions of clusters of deposits, sometimes even of different age, are concentrated in certain regions of the world, e.g. in the Yukon Territory, Canada, with the Cambrian Faro district (Anvil proper, Grum, Dy, Vangorda, Swim), the Early Silurian Howards Pass district (XY and Anniv and Op deposits) and the Devonian MacMillan Pass deposits (Tom and Jason deposits) (E & MJ staff 1981; Goodfellow 1984) or in northern Australia the Proterozoic McArthur River (HYC) deposit with the small Cooley and Ridge deposits (Williams 1978) and, on a larger scale, those of the Mt. Isa district in Australia of the same Proterozoic age comprising the Mt. Isa and Hilton deposits. Despite very intensive exploration around many of the *single* Sedex-type deposits, sometimes involving the drilling of deep holes aimed at conceptual targets based on models of ore deposits in third-order basins or associated for example with hinge lines (Large 1980, 1983; Morganti 1981 e.g.), a second exploitable Sedex-type deposit has rarely ever been found in the surrounding area.

MVT deposits, in contrast to Sedex-type deposits, are irregular in shape (for an example, see Fig. 1b). Their boundaries are usually not natural, but economic cutoff boundaries. MVT deposits normally occur in clusters (see examples in Table 1A). Owing to their discontinuous nature and irregular shape, MVT deposits are difficult to explore and thus the reserves are difficult to determine. Exploration is normally done with grid drilling. Koch and Schuenemeyer (1982) used the Tri-State and Upper Mississippi Valley districts as models to predict exploration success in the Middle Tennessee district in the USA using square drilling grids. The Austinville district in Virginia, USA, which consists of 17 irregulary shaped orebodies, was used by Miller (1991) as a basis for testing several theoretical drilling programs of 100 holes arranged in fences. He found that a grid with fences spaced 2000 ft apart with 600 ft between the holes and one cycle of followup drilling would have a 97% chance of hitting one or more large orebodies with at least one drill hole per orebody.

For the Pine Point district, Rhodes et al. (1984) noted that most of the property had been tested using 915 × 915 m or 915 × 1830 m grids, and some ore trends using 300 × 300 or 250 × 150 m grids. Individual orebodies were drilled at spacings varying between 20 and 35 m.

In other MVT deposits, e.g. Bleiberg, Austria (Polegeg 1975), the drill-hole spacing is even closer. This drill-hole spacing, which was used for detailed exploration, is quite comparable to that necessary to explore vein deposits but contrasts strongly with the wide spacing possible in the case of Sedex-type deposits as described above. The regularity or irregularity of these various deposit types, necessitating a wide or close drill-hole spacing, respectively, can be quantitatively described with the help of a variogram, a geostatistical tool, which will be discussed below.

Callahan (1977) described the discovery of the Central Tennessee district in the USA, a totally new MVT deposit district consisting of blind orebodies at depths of more than 400 m, using a random drilling pattern. The first hole was drilled in 1964. The first ore-grade intersection was achieved in 1967 in hole 79 at a depth of 421.8 m.

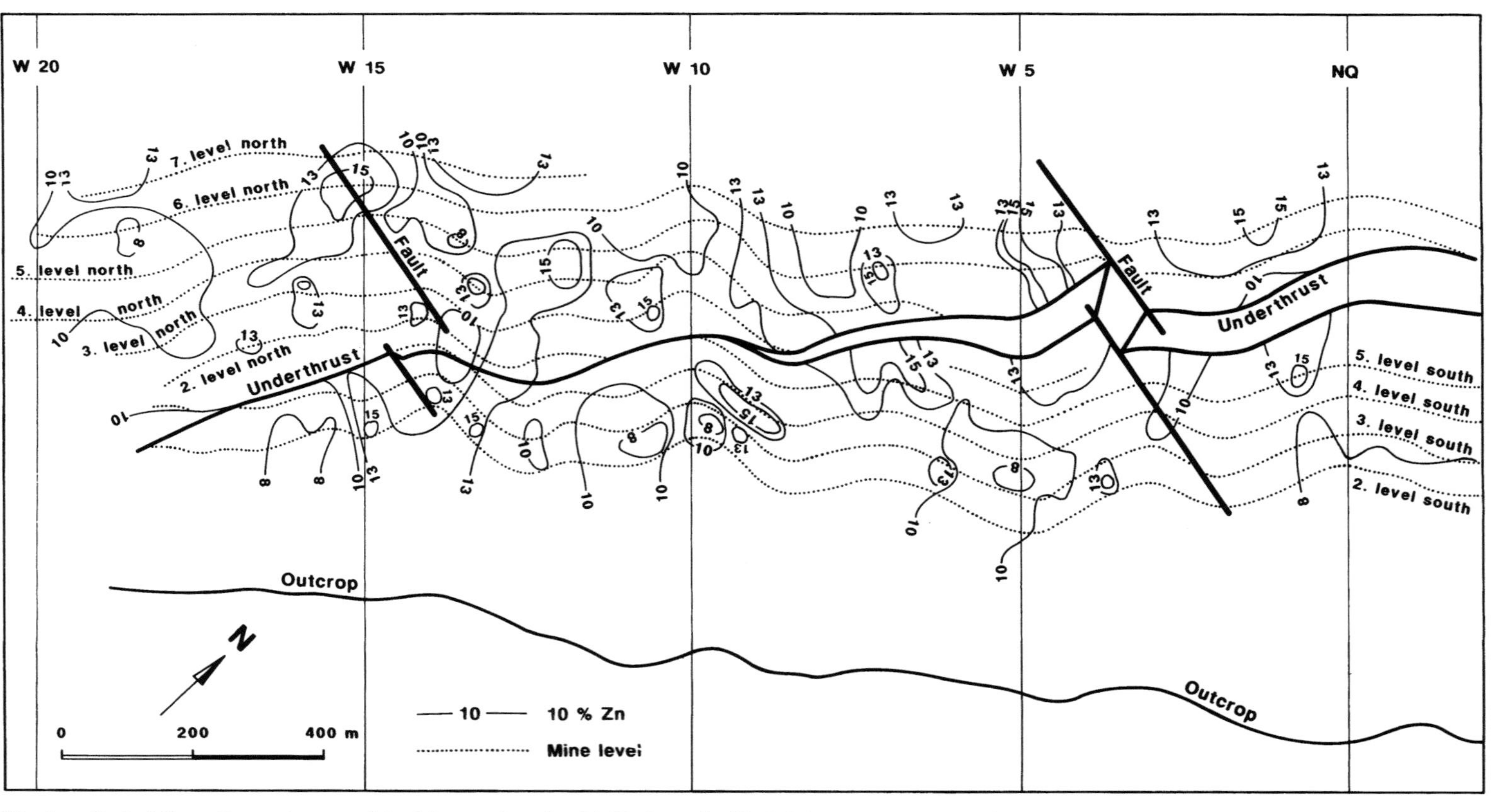

Fig. 1. **a** Part of the palinspastic map of the Meggen deposit with Zn isopachs (Sachtleben Bergbau GmbH). **b** Map of the central part of the Elmwood deposit, showing the distribution of orebodies. (After Gaylord and Briskey 1983)

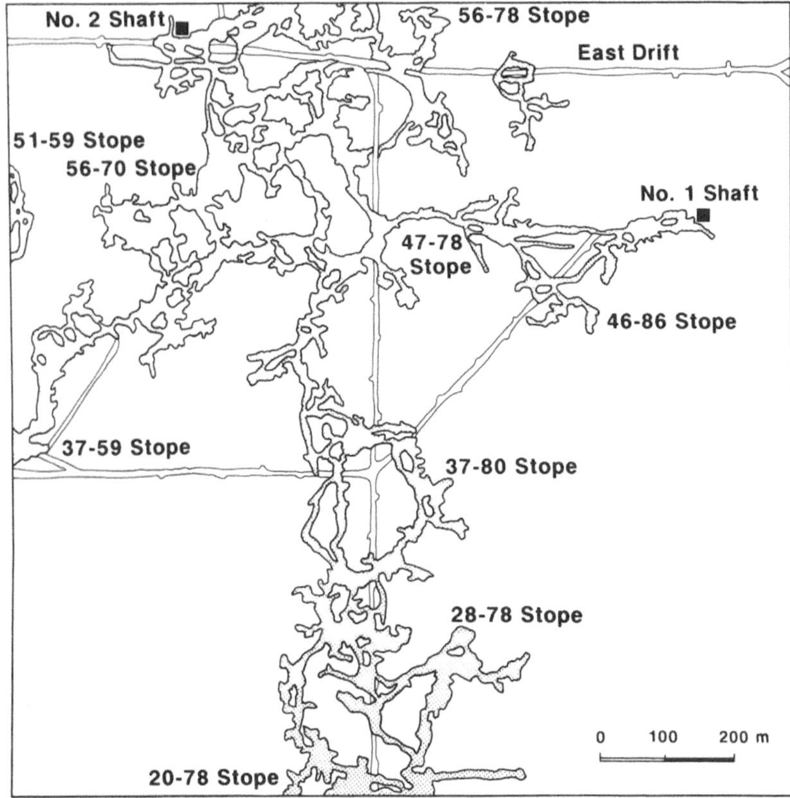

Fig. 1b.

Typical exploration costs in the 1970s for MVT deposits were US$0.60 to 0.80/t of delineated ore (Sames and Wellmer 1981). At today's prices this would be in the range of US$1.00 to 1.40/t.

When a Sedex-type deposit is drilled out and the reserves are outlined in a feasibility study, the deposit is normally quite well known. This is not the case for MVT deposits. Cranstone (1988) derived factors for the various types of deposits in Canada with which a final total production figure for the mined-out deposit can be estimated from the initial reserves figure at the commencement of production. According to Cranstone, the factor is 2 for Canadian non-porphyry, non-vein, base-metal sulphide deposits (excluding nickel-copper deposits). He did not consider MVT deposits as a separate category. Taking the Pine Point district in Canada as an example of MVT deposits and the start of production in 1966 as the starting point, the factor would be around 4.

The regularity of Sedex-type deposits or the irregularity of MVT deposits can be quantitatively described by a variogram (see Appendix I). In a variogram, regionalized variances are calculated as a function of the distances between sample pairs (called lag). Normally, deposits yield variograms of the transitive type (Appendix I).

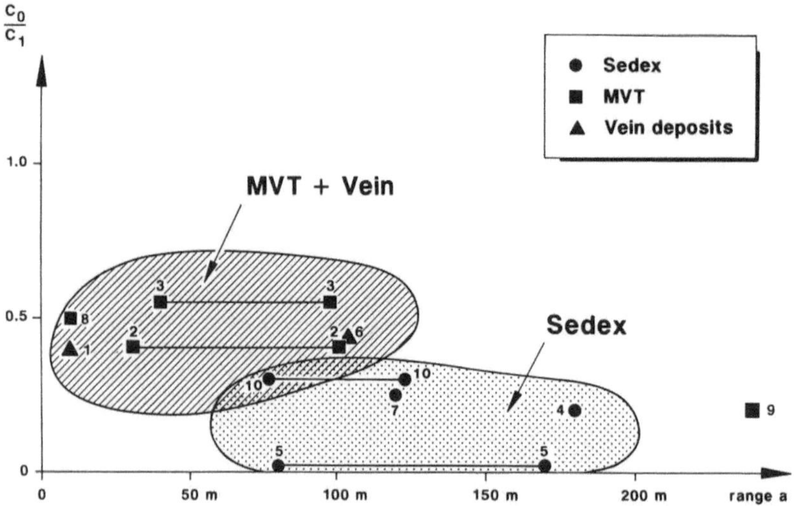

Fig. 2. Plot of the ratio nugget effect C_0/sill value C_1 against range a of variograms of Zn-Pb deposits for the product of grade x thickness. *1* Ramsbeck, Germany; *2* Nanisivik, Canada; *3* Pine Point, Canada; *4* Rampura-Agucha, India; *5* Mt. Isa, Australia; *6* Bad Grund, Germamy; *7* Meggen, Germany; *8* Bleiberg, Austria; *9* Song Toh, Thailand; *10* Tynagh, Ireland. (Bhatnagar and Haldar 1987; Diehl and Kern 1979; Dowd and Scott 1986; Polegeg 1975; Raymond 1984; Wellmer and Giroux 1980; Wellmer and Podufal 1977; Wilke 1981; pers. comm.; own calculations)

The larger the range a and the lower the nugget effect C_0, i.e. the lower the ratio of nugget effect C_0 to variance (called sill) C_1, the more uniform a deposit is. In a plot of C_0/C_1 versus a (Fig. 2), the MVT examples generally plot in the same field as vein deposits, which are characteristically very irregular, whereas Sedex-type deposits are much more uniform and continuous. An exception is the Song Toh deposit in Thailand (No. 9 in Fig. 2), classified by the mine geologists as an MVT deposit, but with some characteristics, e.g. fine-grained ore mineralogy to be described later, more akin to a Sedex-type deposit.

Concerning relationships between ore grade and the shape of an orebody, Sangster (1990) examined differently shaped orebodies in the Pine Point district, Canada, with regard to grade and tonnage. Rhodes et al. (1984) classified the individual Pine Point orebodies as either prismatic (i.e. essentially discordant) or tabular (i.e. essentially concordant) types. Grade-tonnage plots of Sangster (1990) show that the prismatic bodies tend to be larger and of slightly higher grade than the tabular bodies.

4.2 Host Rock and Mineralogy Aspects

MVT deposits are often hosted by highly brecciated dolomite. The breccia zones are healed with carbonates, quartz and ore minerals, mainly sphalerite and galena. From the point of view of rock mechanics, the host rock is very competent and stable, and allows frequently the application of cost-efficient room-and-pillar mining (see Table 5).

We have seen above that Sedex-type deposits are not restricted to clastic sedimentary rocks, although they commonly occur in shales, siltstones and fine- to coarse-grained turbidites. The minority occurs in calcareous shales and carbonates. As far as mining is concerned, the host rock is normally stable, commonly requiring minimal roof support, i.e. roof bolting. If the host rocks are shales, more extensive and therefore more costly roof-support systems frequently have to be used.

In MVT deposits the mineralogy is normally simple; usually sphalerite and galena are the only minerals of economic interest. The ores are also metallurgically simple, i.e. coarse-grained and devoid of intricate intergrowths, thus permitting the flotation of excellent, clean concentrates with good metal grades (Fig. 3). Since MVT deposits were formed at relatively low temperatures, the Fe content in sphalerites is frequently very low (Hutchinson and Scott 1981; Sangster 1990). This fact also contributes to the high Zn grade of sphalerite

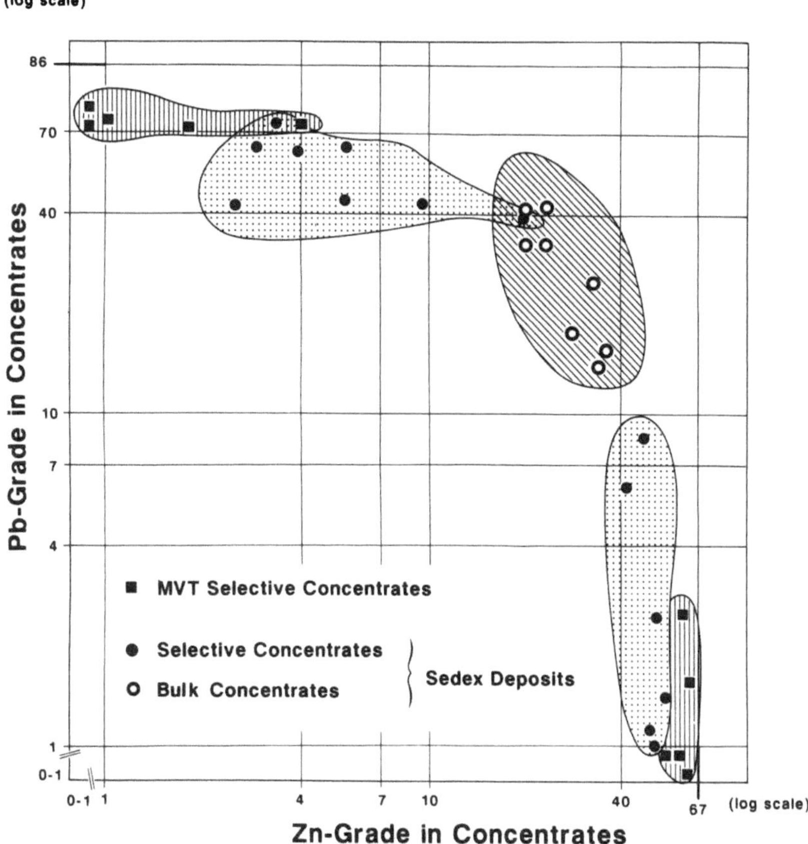

Fig. 3. Pb and Zn grades in concentrates of MVT and Sedex-type deposits (each *point*, *square* or: *circle* represents the quality of one concentrate

Table 4. Minor metals content of sphalerites from MVT deposits

Name	Country	Ge (ppm)	Ga (ppm)	In (ppm)	Cd (ppm)
Flat Gap mine (East Tennessee)	USA	20	7	?	7000
Jefferson City mine (East Tennessee)	USA	20	7	?	10000
Raibl	Italy	350–450	?	?	1000–2000
Bleiberg	Austria	240 (in one orebody 1200–1300)	25	?	1500
Mezica	Former Yugoslavia	60	?	?	3500

Sources: Feiser (1966), Gaylord and Briskey (1983), Cerny (1989), Wellmer et al. (1990)

concentrates. The silver content is normally low, but the cadmium, germanium, gallium and indium contents may be high (Table 4). In the case of cadmium and germanium, zinc concentrates from MVT deposits are a major source of these minor metals. Up to World War I the nearly exclusive source for cadmium was the MVT deposits of Upper Silesia (Meisner 1929).

In Sedex-type deposits the mineralogy is variable, the ore frequently fine-grained, and the metallurgy often complex, thus requiring a more complicated flotation circuit than is normal for ores from MVT deposits. The concentrates are normally of poorer quality than those from MVT deposits (Fig. 3). The higher Fe content of the sphalerite adversely influences the Zn content of the zinc concentrates. In many cases it is impossible to produce selective concentrates at an acceptable rate of recovery so that mixed (bulk) concentrates have to be produced. Poor quality selective concentrates, or bulk concentrates instead of selective concentrates, mean lower financial returns for the mine (the so-called net smelter return, see Sect. 6.1).

5 Mining and Milling Practices

Various statistical sources were used for the two following sections. The sources are not referenced individually, but are summarized in the list of references for Section 5.

5.1 Mining Practices

Lead and zinc ores are primarily exploited in underground mines. Out of the 75 producing mines investigated, 5% are open pits (e.g. Faro, Canada; Red Dog, USA; Rampura-Agucha, India). All these open pits mine Sedex-type deposits.

The selection of a certain mining method for the exploitation of an orebody depends on a number of factors such as dip and thickness and the rock-mechanical strength of the orebody and of the surrounding wall rock. Additional factors are the value of the ore and the local safety regulations.

Table 5. Underground mining methods used in various types of lead-zinc deposits

1) MVT deposits	
- Room-and-pillar	60%
- Room-and-pillar and cut-and-fill	5%
- Sublevel stoping	12%
- Sublevel stoping and cut-and-fill	13%
- Cut-and-fill	10%
2) Sedex deposits	
- Room-and-pillar	11%
- Sublevel stoping	22%
- Cut-and-fill	33%
- Open stoping and cut-and-fill	22%
- Cut-and-fill and square set	6%
- Sublevel stoping shrinkage	6%
3) Vein deposits	
- Cut-and-fill	64%
- Shrinkage	16%
- Cut-and-fill and shrinkage	12%
- Cut-and-fill-and sublevel stoping	8%

Sources: See references for Section 5, pers. comm.

Table 5 illustrates the underground mining methods which are used in MVT, Sedex- and vein-type deposits. In about 60% of the 40 investigated MVT deposits the room-and-pillar method is used. All MVT mines in Missouri and Tennessee apply room-and-pillar mining which is a very economic method as it allows a high degree of mechanization.

The important mining methods for Sedex deposits are sublevel stoping, open stoping, and cut-and-fill.

As Table 5 shows 64% of the vein deposits are mined by the cut-and-fill method.

The overall average ore production from MVT deposits amounts to 738 000 t/a and from Sedex-type deposits to 1 061 000 t/a.

5.2 Milling Practices

In MVT as well as in Sedex-type deposits, mainly galena and sphalerite are the sulphides of economic interest. In addition to these, some Sedex-type surface mines contain oxidized lead minerals (e.g. Red Dog, USA). MVT deposits commonly contain dolomite, calcite, and ankerite as gangue minerals. Common accompanying minerals in Sedex-type deposits are pyrite, pyrrhotite and chalcopyrite. Quartz, phyllosilicates and carbonates appear as gangue.

As outlined above, lead and zinc minerals are generally coarse-grained in MVT deposits and fine-grained in Sedex-type deposits. Therefore for beneficiation the lead and zinc ore from MVT deposits requires less fine grinding than the ore of Sedex deposits. The lead and zinc minerals are concentrated in the following processing stages: crushing, heavy-media separation (if the ore is coarse-grained), grinding, flotation and filtration.

The run-of-mine ore, crushed underground with a crusher to below 150 mm, is hoisted to the surface. From the shaft the ore is conveyed to bins in the concentrator or to the stockpile (e.g. Elura, Australia and Red Dog, USA, have crude ore stockpiles). The ore is comminuted in the secondary and tertiary crushers to around 15 mm in the concentrator. In some plants enough host rock has been liberated from the ore minerals, so that at this stage the crushed ore can undergo heavy-media separation (Bad Grund, Germany, Meggen, Germany, Bleiberg, Austria, Sullivan, Canada, and Young Mine, USA).

The heavy-media plant operates with a slurry of specific gravity 2.9 to 3.1 g/cm^3. Ferrosilicon or magnetite is used as a heavy medium. At Sullivan, Canada, galena containing 60% Pb is used as a heavy medium. The cone separator is widely applied for ore treatment in heavy-media plants. The feed is generally in the range of 30 to 5 mm. Float is removed as tailings. Tailings which contain less than 1% Pb and Zn constitute up to 30% of the feed. The sink is floated after grinding to about 80% under 75 μm.

In Sedex-type deposits the flotation feed is usually ground finer than in MVT deposits (e.g. Hilton, Australia, 80% – 60 μm, Elura, Australia, 80% – 45 μm).

In flotation plants selective concentrates are generally produced. In the first flotation stage lead is carried off at a natural pH and zinc is depressed. The flotation reagents used are ethyl xanthate as collector, methylisobutyl carbinol (MIBC) as frother, zinc sulphate as depressant for zinc sulphide and sodium cyanide as depressant for pyrite (if present).

Zinc flotation uses copper sulphate as a zinc activator, amyl xanthate as collector and MIBC as frother. Zinc flotation is also carried out at a natural pH. In some flotation plants the concentrates from the zinc rougher and scavenger are reground to about 40 μm (e.g. Faro, Canada 44 μm, or Red Dog, USA, 85% – 30 μm, deposits which belong to the Sedex-type group). After regrinding, the zinc rougher and scavenger concentrates are cleaned.

Lead and zinc concentrates are filtered and stockpiled for transport to metallurgical processing. The moisture of the concentrates is around 8–10%. Metal recovery is between 85 and 95%. For MVT deposits lead concentrates often contain 70–75% Pb and zinc concentrates 50–60% Zn. For Sedex-type deposits these figures are 50–70% Pb and 45–55% Zn (Fig. 3). In some plants, due to poor recoveries for selective concentrates bulk concentrates are also produced; these are suitable as feed for an Imperial Smelting furnace.

The flotation tailings are often used as backfill in the mines. Tailings, especially those of MVT deposits, are also used for building railroads and roads.

6 Economic Appraisal

In an economic appraisal of any mineral deposit, the cost and revenue aspects have to be considered before economic parameters can be calculated. The methods of preparation and interpretation of data and subsequent economic appraisal are described in detail in Wellmer (1989). They are only briefly mentioned in the present text and described in the appendices.

6.1 Revenue Aspects

The revenue side is determined by the metal prices, which show a high degree of fluctuation over the years. In Fig. 4a,b the prices of the revenue earners of MVT and Sedex-type deposits, i.e. Zn, Pb and Ag, are plotted for a 20-year period in (a) nominal and (b) real terms. One can see, firstly, that the price peaks for Zn and Pb are offset, but the price trends for Pb and Ag run roughly parallel to each other and, secondly, that the overall price trend is more or less horizontal, meaning that in the long run there is no indication of a price increase in real terms.

Since mines produce concentrates and not metal bars, which form the basis of price quotations on metal exchanges such as the London Metal Exchange (LME), it is the so-called net smelter return (NSR) of the mine that has to be calculated for the concentrates. Moreover, certain percentage of the metal in the concentrate is deducted as smelting loss. The smelter's treatment charge (TC) is deducted from the value of the metal in the concentrates (see Appendix II). Since the smelter's treatment charges also fluctuate, as they are influenced by the concentrate market (buyers' and sellers' markets controlled by supply and demand between mines and smelters), the net smelter return (NSR) of a mine includes two fluctuating components: the metal price and the treatment charge.

The revenues for a mine for a 20-year period producing a run-of-mine ore containing 10% combined Pb + Zn with varying Zn/(Zn + Pb) ratios and varying Pb:Ag ratios were calculated. It was assumed that Ag is always contained in the Pb concentrates, although some MVT deposits (e.g. the Nanisivik mine, Canada) are known in which a significant portion of the Ag occurs in the Zn concentrates. Examples of the fluctuating annual revenues (NSR) of a mine are given for different Zn/(Zn + Pb) ratios in Fig. 5a,b. One can see from the diagrams that the NSR fluctuates concomitantly to the price curves in Fig. 4a,b and that, in some years, Zn-rich mines fared better but in other years Pb-rich mines did better. If the mine exploits very Ag-rich ore, the NSR trend is largely dictated by the silver price.

These price fluctuations pose a major risk for metal mines. It was therefore investigated at what metal ratios the price risk will be at a minimum or a maximum. A price hausse (i.e. a high price period) can be considered a bonus for a mine and is not a risk situation. However, price baisses (i.e. low price periods) bring with them the highest risks. The highest risks are associated with price drops and extended periods of depressed prices. These two situations were considered in a mathematical model (see Appendix III) to derive metal ratios with minimum and maximum price risk. The pertinent result is that mines with high Zn/(Zn + Pb) ratios are subject to a substantially lower price risk than mines with a low Zn/Zn + Pb ratio, i.e. mines which are Pb-dominated (Fig. 6a,b). The interval of the relatively low price risk coincides with the main

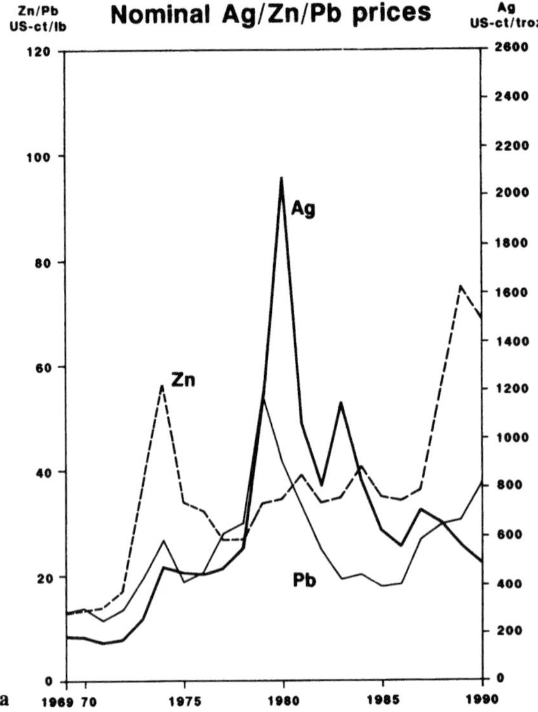

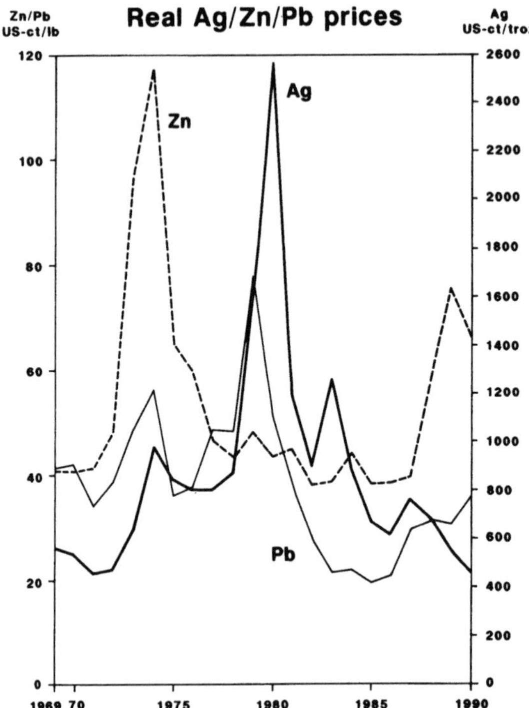

Fig. 4. **a** Nominal prices of Pb, Zn and Ag for the period 1969 to 1990 (annual averages) (Metallgesellschaft AG 1991 and previous years). **b** Real prices of Pb, Zn and Ag for the period 1969 to 1990 (annual averages). (Metallgesellschaft AG 1991 and previous years)

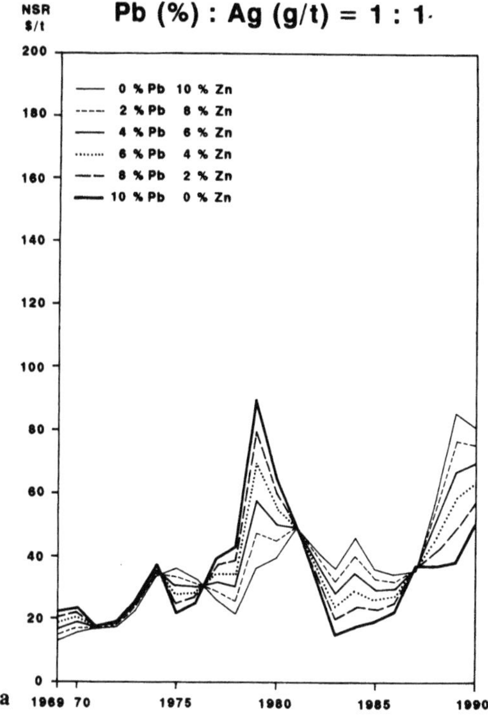

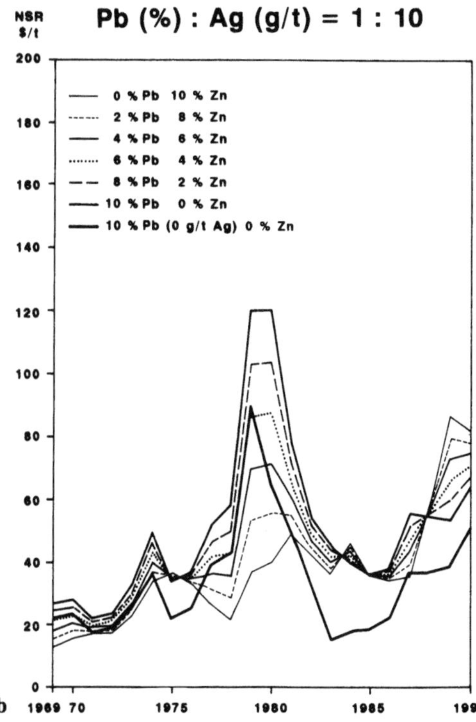

Fig. 5. **a** Net smelter return (NSR/t) for an ore with 10% combined Pb+Zn and various Pb/(Pb+Zn) ratios. **a** Pb (%)/Ag (g/t) = 1:1; **b** Pb (%)/Ag (g/t) = 1:10. NSR calculations with annual average of London Metal Exchange metal prices and annual averages of smelting charges (T/C)

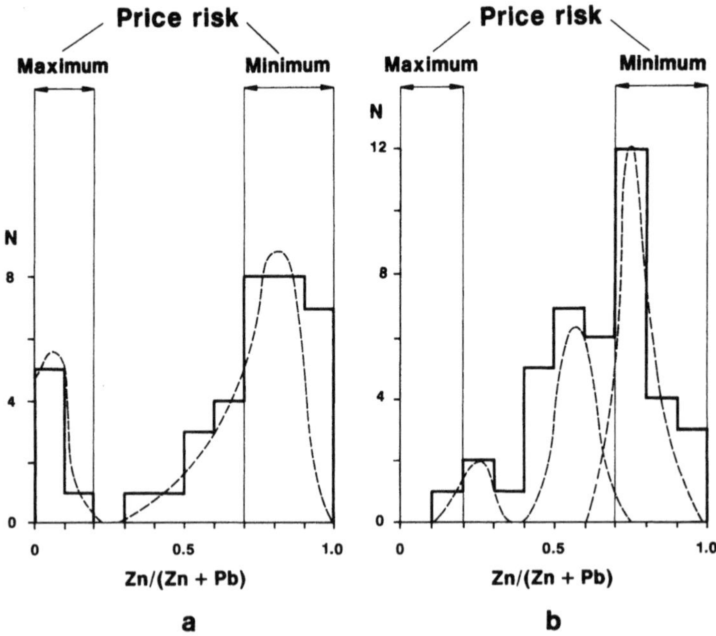

Fig. 6. Zn/(Zn+Pb) ratios of **a** MVT deposits and **b** Sedex-type deposits (Sangster 1990), with zones of minimum and maximum price risk; *N* number of deposits

maximum of the metal ratios of MVT and Sedex-type deposits of Sangster (1990).

6.2 Cost Aspects

Data on total operating (mining and milling, including services) and capital costs were modelled for mines exploiting MVT and Sedex-type deposits; the operating costs of mines based on vein-type deposits were also included. The costs per tonne of ore produced decrease with increasing capacity (economics of scale). For specific costs like operating costs (Fig. 7a), the data from a number of mines can normally be modelled with a power curve $y = a \cdot x^{-b}$, where a and b are constants ($b < 1$), y the specific cost per tonne ore and x the capacity. For absolute costs like capital costs (Fig. 7b) the power curve is $y = c \cdot x^{d}$, where c and d again are constants ($d < 1$), (US Bureau of Mines 1987). It is obvious from Fig. 7a that vein mines have the highest operating costs on average. One consequence of this is the declining number of vein mines in high-wage countries such as Germany, although some mine closures have also been due to depletion of ore reserves. Of the eight Zn-Pb vein mines operating in Germany in 1945, only three remained open in 1965, one in 1985 and all were finally closed down in 1992.

On average, mines exploiting MVT deposits have the lowest costs; this reflects the stable rock conditions allowing cost-efficient room-and-pillar mining

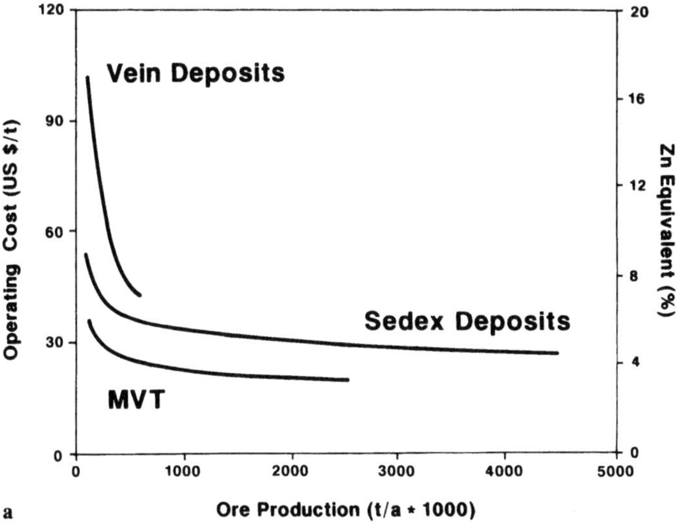

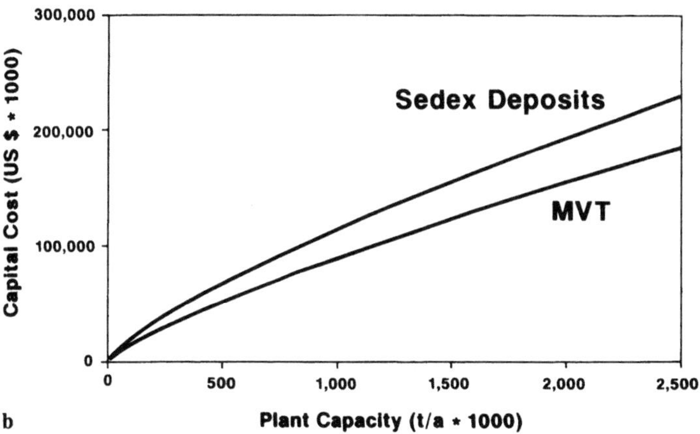

Fig. 7. a Plot of operating cost versus annual ore production. **b** Plot of capital cost versus plant capacity

methods to be used, and on the milling side it reflects the uncomplicated metallurgical properties mentioned above.

In the future, increased environmental costs might even further increase the relative cost advantage of MVT deposits. The waste rocks of MVT deposits, normally clean dolomites, which can be used as construction material, are environmentally far less problematic than clastic, especially shaley waste rocks of Sedex-type deposits, which frequently contain various amounts of sulphides like pyrite, which easily oxidizes to produce sulphate, and therefore have to be retained. On the other hand, credits for Cd, formerly available for some MVT concentrates, are no longer paid. Quite on the contrary, due to environmental restrictions and depressed Cd prices, high Cd values in concentrates might well attract penalties in the future.

6.3 Economic Comparisons

A simple way of comparing various deposits economically is to construct a threshold of viability, the so-called borderline of viability (Wellmer 1989). The borderline of viability, in general, separates the viable mines from uneconomic deposits. This is a somewhat crude separation, for in low-cost countries or countries with protected internal markets with higher metal prices, e.g. India, or in very favourable infrastructural locations, deposits that could not otherwise be mined at a profit might be economic. In the diagrams (Fig. 8a,b) of MVT and Sedex-type deposits, therefore, some deposits which are exploited and which plot to the left of the borderline of viability are exceptions.

In Fig. 8, comparing a with b, one can see that the field of viability of MVT deposits includes lower grade and lower tonnage deposits than the field of viability of Sedex-type deposits. This reflects the commonly better cost structure of mines exploiting MVT deposits as discussed above, but also frequently the better infrastructural situation of MVT deposits, which are concentrated in North America.

As economic parameters of dynamic appraisal, the net present value (NPV) and the internal rate of return (IRR) were calculated (see Appendix IV) for average grades and tonnages of MVT and Sedex-type deposits. For comparison the economic parameters of volcanogenic massive sulphide deposits were also determined. The results are shown in Table 6. The capital and operating costs used are those of Fig. 7a,b. Because the cost structure of Sedex-type deposits and of volcanogenic massive sulphide is similar, for the latter type the cost data for Sedex-type deposits were chosen. In order to use the cost diagrams, the average tonnages of the various deposit types had to be converted into theoretical mine capacities. The Taylor formula for the optimum lifetime of a mine was used for the conversion (Taylor 1977). As shown by Wellmer (1981) and McSpadden and Schaap (1984), the average lifetime of real mines as a function of reserves closely follows this theoretical formula. Table 6 shows that, on average, Sedex-type deposits yield better economic parameters than MVT deposits due to, on average, better grades despite the lower costs of the MVT deposits. Sedex-type deposits are also on average economically more attractive in comparison with the average volcanogenic massive sulphide deposit.

It has to be emphasized that these are comparisons of average tonnage and grade data. In individual cases the economics are determined not only by tonnage and grades of the individual deposit but by many other, very specific parameters such as infrastructure and the geological and metallurgical parameters. Giegerich (1986) published a diagram of the relative cost positions of zinc mines in the western world. Besides the cost structure of a mine, the relative position is also influenced by the prices and, consequently, a certain bonus due to the presence of the additional metals Pb and Ag, as well as Cu and Au in the case of volcanogenic deposits. This diagram is shown in Fig. 9 in which the type of deposit has been added. It clearly shows that all types of

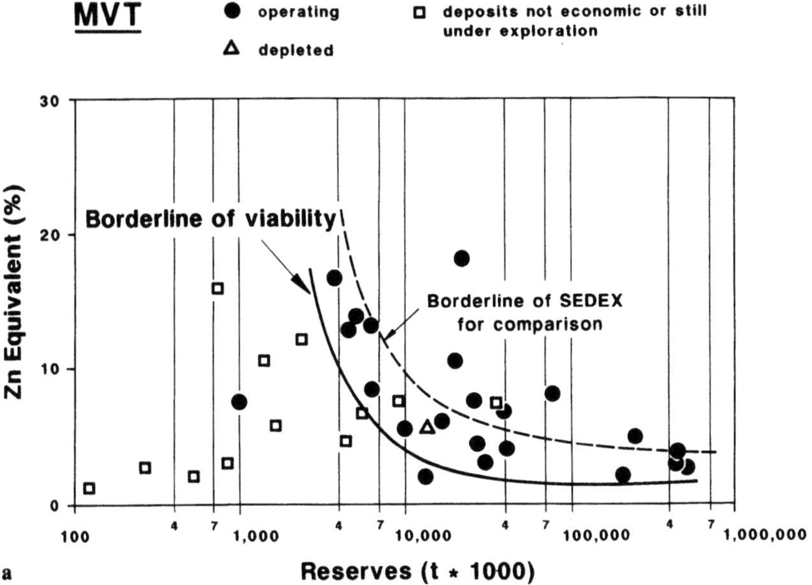

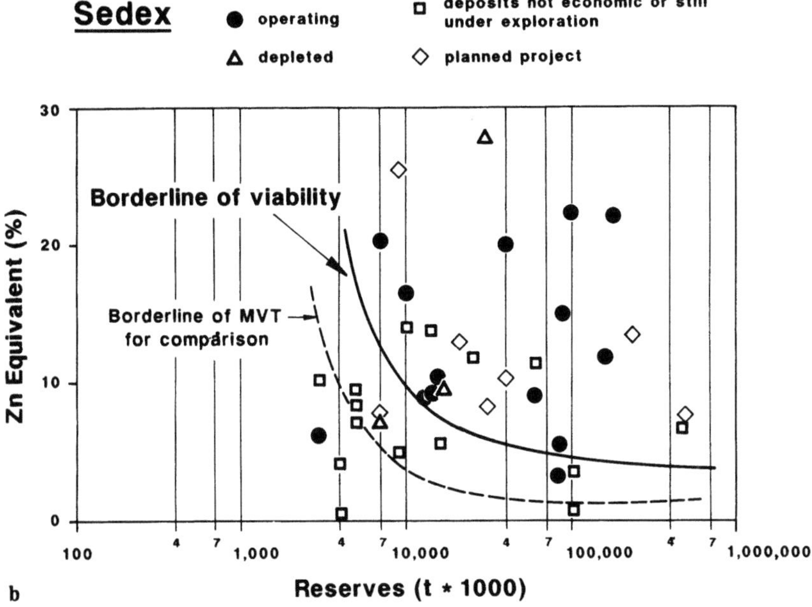

Fig. 8. Zn-equivalent grades versus reserves for **a** MVT deposits and **b** Sedex-type deposits showing borderline of viability. (Grade tonnage diagram modified after Sangster 1990; for derivation of Zn-equivalent grades, see Appendix V)

Table 6. Grade, tonnage and economic value of the various deposits

Calculations of average deposit data with sources[a]	Reserves ($\times 10^3$ t)	Lifetime (years)[d]	Mill capacity ($\times 10^3$ t/a)[f]	Ore grade Zn (%)	Pb (%)	Cu (%)	Ag (g/t)	Au (g/t)	Net present value[h] ($\times 10^6$ US-$)	Internal rate of return (%)
Mississippi Valley type:										
Average of all deposits (I)	65 459	18	3637	3.4	1.6		1		27.2	12
Average of deposits which are or have been mined (I)	100 647[b]	20	5032	2.9	1.4		0		−53.0	8
Average of producing mines (II)	12 400[c]	17[e]	738[g]	4.8	3.0		17		48.2	20
Southeast Missouri and Appalachian (III)	35 000	15	2330	4.0	0.9		0.5		−40.3	6
Sedex deposits:										
Average of all deposits (I)	65 049	18	3614	7.0	3.6		46		795.1	45
Average of deposits which are or have been mined (I)	61 467[b]	18	3415	7.7	5.9		88		1279.4	67
Average of producing mines (II)	27 500	26[e]	1061[g]	7.7	4.5		88		314.2	40
Phanerozoic-clastic (IV)	67 000	18	3720	7.2	2.9		26		712.3	41
Phanerozoic-carbonate (IV)	14 500	12	1208	8.1	2.5		8		103.8	24
Proterozoic-clastic (IV)	90 000	19	4737	7.1	5.1		71		1556.3	63
Proterozoic-carbonate (IV)	12 200	11	1109	11.5	1.8		9		199.3	39
Average of all Sedex-type deposits (IV, V)	60 000	17	3529	7.3	4.0		48		860.3	49
Sediment-hosted massive Pb-Zn (III)	15 000	12	1250	5.6	2.8		30		17.6	13
Volcanic-hosted massive sulphide deposits:										
With supergiants (VI)	9 000	11	818	4.1	0.8	1.2	42	0.69	23.1	15
Without supergiants (VI)	4 600	9	511	3.6	0.4	1.5	24	0.62	−18.3	3
Kuroko massive sulphide (III)	1 500	7	214	2.0	1.9	1.3	13	0.16	−34.0	0 (−41)

[a] Sources: (I): Grade-tonnage diagram of Sangster (1990), Fig. 5 (MVT districts counted as one deposit, see sect. 3). (II): Authors' compilation from the literature, company reports, pers. comm. (III): Cox and Singer (1986). (IV): Lydon (1983). (V): Eckstrand (1984). (VI): Eckstrand (1984) Average of volcanic-hosted deposits in the Abitibi Belt (Canada), Norwegian Caledonides, the Bathurst Camp (Canada) and the Green Tuff Belt (Japan).

[b] Average of deposits which are active. [c] Producing mines which were started in the last 10 years. [d] Calculated from $0.2\sqrt[4]{\text{reserves}}$ by Taylor, except Footnote. [e] Ratio of reserves/ore production. [f] Ratio of reserves/lifetime, except Footnote g. [g] Average ore production 1989. [h] Interest rate assumed 10%.

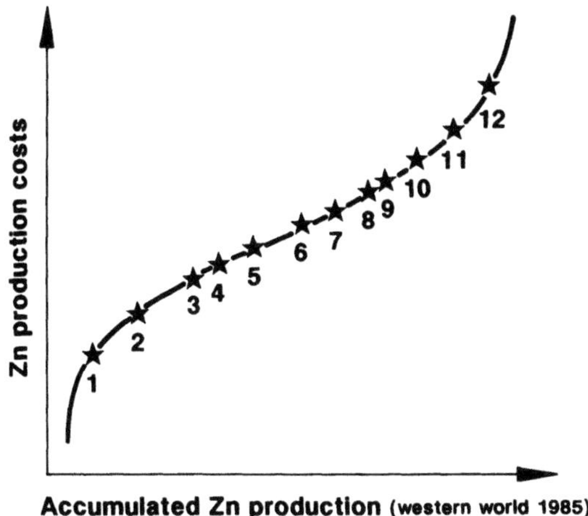

Deposit number	Mine	Type		
		MVT	Sedex	Volcanogenic
1	Kidd Creek, Canada			X
2	Red Dog, USA (Alaska)		X	
3	Polaris, Canada	X		
4	Rubiales, Spain		X	
5	Tara, Ireland	?	?	
6	Mt. Isa, Australia		X	
7	Pine Point, Canada	X		
8	Sullivan, Canada		X	
9	Boliden Group			X
10	Anvil, Canada		X	
11	Brunswick, Canada			X
12	Broken Hill, Australia		X	

Fig. 9 Zn production cost of various mines versus accumulated Zn production of western world 1985. (After Giegerich 1986)

deposits range over the whole cost spectrum, as might be expected, because in an investment decision each deposit has to be treated as an individual case.

The relative profitability of a given mine can also be illustrated by calculating the break-even grade, i.e. the Zn grade which is necessary to cover all costs, and comparing it with the real grade of a mine (Fig. 10). For this comparison Pb and Ag grades have to be converted into Zn-equivalent grades which is explained in Appendix V. If in the break-even diagram (Fig. 10) a mine plots on the 45° line, the grade is just high enough to cover all costs, but no profit is made. The further a mine plots to the right of the 45° line, the more profitable it is likely to be. It is obvious from Fig. 9 that all vein mines cluster around the 45° line, meaning they are break-even, marginal mines. According to this diagram, the most profitable mines are some of those exploiting Sedex-type deposits; this is in full agreement with the conclusions drawn from Table 6.

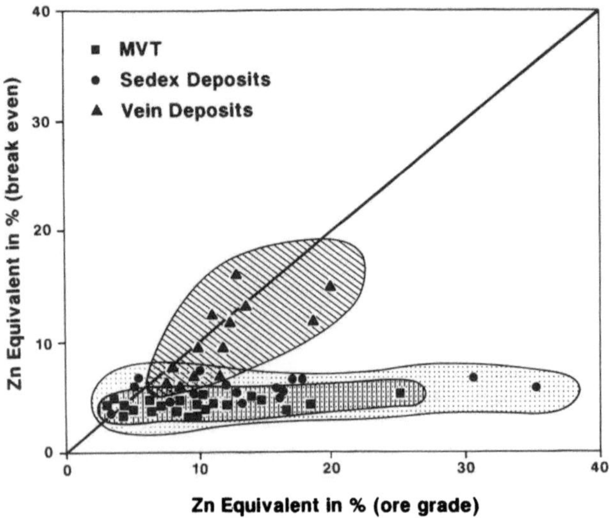

Fig. 10. Break-even diagram for the various types of deposits

Acknowledgements. The authors are grateful to Dr. D.F. Sangster, Geological Survey of Canada, and Dr. D.L. Leach, U.S. Geological Survey, for making available the original data used in Sangster's (1990) paper and a manuscript on MVT deposits. The authors thank Dr. Fuchs and Mr. Ehrhardt of Sachtleben Bergbau GmbH, Dr. Cerny of Bleiberger Bergwerks Union and Dr. Diehl of Preussag AG for unpublished data on the Bleiberg, Bad Grund and Meggen mines, Mr. H. Toms for constructive comments, Mrs. B. Piesker for preparing the figures, and Mrs. M. Simon for typing the manuscript.

General References

Bhatnagar SN, Haldar SK (1987) Sequential evaluation model of phased exploration data leading to sample optimization – a case study of Rampura-Agucha zinc-lead deposit, India. Can Inst Min Metall Bull 80(906):56–60

Bjørlykke A, Sangster PF (1981) An overview of sandstone lead deposits and their relation to red-bed copper and carbonate-hosted lead-zinc deposits. Econ Geol 75th Anniversary vol: 179–213

Callahan WHC (1977) The history of the discovery of the zinc deposit at Elmwood, Tennessee–concept and consequence. Econ Geol 72:1382–1392

Cerny I (1989) Die karbonatgebundenen Blei-Zink-Lagerstätten des alpinen und außeralpinen Mesozoikums. Die Bedeutung ihrer Geologie, Stratigraphie und Faziesgebundenheit für Prospektion und Bewertung. Arch Lagerstätten Forsch Geol Bundesanst Wien 11:5–125

Cox DP, Singer DA (eds) (1986) Mineral deposit models. US Geol Surv Bull 1693, 379 pp

Cranstone DA (1988) The Canadian mineral discovery experience since World War II. In: Tilton JE, Eggert RE, Landsberg HH (eds) World mineral exploration. Trends and economic issues. Resources for the Future, Washington DC, pp 283–328

Diehl P, Kern H (1979) Geostatistische Vorratsberechnung einer schichtgebundenen Pb-Zn-Lagerstätte in Karbonaten. Erzmetall 32(9):366–374

Dowd PA, Scott TR (1986) Geostatistics in the stratigraphic orebodies at Mount Isa. In: Woodcock JT (ed) 13th Congr Council Min Metall Institutes Singapore 11.–16.5.1986, vol 2. Geology and exploration. Aust Inst Min Metall, Parkville, Australia, pp 27–36

E&MJ Staff (1981) Hudson's Bay Oil & Gas buys Cyprus Anvil's long-term production potential. Eng Min J (Sept):35–38

Eckstrand OR (1984) Canadian mineral deposit-types: a geological synopsis. Geol Surv Can Econ Geol Rep 36, 86 pp

Einaudi MT, Meinert LD, Newberry RJ (1981) Skarn deposits. Econ Geol 75th Anniversary vol: 317–391

Feiser J (1966) Nebenmetalle. Series: Die Metallischen Rohstoffe, vol 17. Enke, Stuttgart, 247 pp

Fontboté L (1990) Stratabound ore deposits in the Pucaru Basin – an overview. In: Fontboté L, Amstutz GC, Cardozo M, Cedillo E, Frutos J (eds) Stratabound ore deposits of the Andes. Springer, Berlin Heidelberg New York, pp 253–266

Gaylord WB, Briskey JA (1983) Summary of the geology of the Elmwood-Gordonsville mining complex, Central Tennessee zinc district. In: Tennessee zinc deposits field trip guide book. Virginia Tech Dept of Geological Sciences Guidebook 9, Blacksburg, VA, pp 116–151

Giegerich HM (1986) Progress report on Cominco's Red Dog project in Alaska; second largest zinc deposit ever discovered. Min Eng (Dec): 1097–1101

Goodfellow WD (1984) Geochemistry of rocks hosting the Howards Pass (XY) strata-bound Zn-Pb deposit, Selwyn Basin, Yukon Territory, Canada. In: Janelidze, TV, Tralchrelidze AG (eds) Proc 6th Quadrennial IAGOD Symp. Schweizerbart, Stuttgart, 6(1):91–112

Gustafson LB, Williams N (1981) Sediment-hosted stratiform deposits of copper, lead, and zinc. Econ Geol 75th Anniversary vol: 139–178

Hewton RS (1982) Gayna River: a Proterozoic Mississippi Valley-type zinc-lead deposit. Geol Assoc Can Spec Pap 25:667–700

Hutchinson MN, Scott SD (1981) Sphalerite geobarometry in the Cu-Fe-Zn system. Econ Geol 76:8–25

Koch GS Jr, Schuenemeyer JH (1982) Exploration for zinc in middle Tennessee by drilling: a statistical analysis. Econ Geol 77:653–663

Large DE (1980) Geological parameters associated with sediment-hosted, submarine, exhalative Pb-Zn deposits: an empirical model for mineral exploration. Geol Jahrb D 40, Hannover: 59–129

Large DE (1983) Sediment-hosted massive sulphide lead-zinc deposits: an empiral model. In: Sangster DF (ed) Sediment-hosted stratiform lead-zinc deposits. Short Course Handbook 8. Min Assoc Can Toronto, pp 1–29

Lydon JW (1983) Chemical parameters controlling the origin and deposition of sediment-hosted stratiform lead-zinc deposits. In: Sangster DF (ed) Sediment-hosted stratiform lead-zinc deposits, Short Course Handbook 8. Min Assoc Can Toronto, pp 175–250

McSpadden G, Schaap W (1984) Technical note: Taylor's rule of mine life. Proc Aust Inst Min Metall 289(6):217–220

Meisner M (1929) Die Versorgung der Weltwirtschaft mit Bergwerkserzeugnissen I. 1860–1926, 2. Teil Erze und Nichterze. Enke, Stuttgart, 394 pp

Metallgesellschaft AG (1991) Metal statistic 1980–1990, and previous years. Metallgesellschaft AG Frankfurt am Main, pp 457–487

Miller JW Jr (1991) Optimization of grid-drilling using computer simulation. Math Geol 23:201–218

Morganti JM (1981) Sedimentary-type stratiform ore deposits: some models and a new classification. Geosci Can 8(2):65–75

Polegeg S (1975) Anwendung mathematischer Methoden für Such- und Erkundungsarbeiten im Raume Bleiberg-Kreuth. Berg- Hüttenm Monatsh 120(10):476–480

Raymond GF (1984) Geostatistical applications in tabular style lead-zinc ore at Pine Point, Canada. In: Verly G, David M, Journel AG, Marechal A (eds) Geostatistics for natural resources characterization, Part I. NATO ASI Ser C, vol 122. Reidel, Dordrecht, pp 469–483

Rhodes D, Lantos EA, Lantos JA, Webb RJ, Owens DC (1984) Pine Point orebodies and their relationship to stratigraphy, structure, dolomitization, and karstification of the Middle Devonian barrier complex. Econ Geol 79:991–1055

Sames W, Wellmer FW (1981) Exploration. Part I: Nothing ventured, nothing gained. Risks, strategies, costs, achievements. Glückauf + Translation 117(10):267–272

Sangster DF (1990) Mississippi Valley-type and Sedex lead-zinc deposits: a comparative examination. Trans Inst Min Metall Sect B Appl Earth Sci 99 (Jan–April):B21–B42

Taylor HK (1977) Mine valuation and feasibility studies. In: Hoskins JR, Green WR (eds) Mineral industry costs. Northwest Mining Assoc Spokane, pp 1–17

Tikkanen GD (1986) World resources and supply of lead and zinc. In: Bush WR (ed) Economics of internationally traded minerals Society of Mining Engineers Littleton, CO, pp 242–50

US Bureau of Mines (1987) Bureau of Mines cost estimating system handbook. Part 1: Inf Circ 9142, 631 pp; Part 2: Inf Circ 9143, 566 pp
Wellmer FW (1981) Reserve/consumption ratios – how can they be interpreted? Can Inst Min Metall Bull 74(831):59–62
Wellmer FW (1989) Economic evaluations in exploration. Springer, Berlin Heidelberg New York, 163 pp
Wellmer FW, Giroux GC (1980) Statistical and geostatistical methods applied to the exploration work of the Nanisivik Zn-Pb mine, Baffin Island, Canada. Math Geol 12(4):321–337
Wellmer FW, Podufal P (1977) A statistical model for exploration of the Ramsbeck Pb/Zn mine (FRG). In: Ramani, RV(ed) Application of computer methods in the mineral industry. Proc 14th Symp Oct 4.–8.1976. Soc Min Eng of AIME, New York, pp 431–440
Wellmer FW, Hannak W, Krauss U, Thormann A (1990) Deposits of rare metals. In: Kürsten M (ed) Raw materials for new technologies. 5th Int Symp Fed Inst Geosci & Natural Res Hannover, 19-21.10.1988. Schweizerbart, Stuttgart, pp 71–121
Wilke A (1983) Zur Ermittlung und Einteilung von Lagerstättenvorräten. In: Borchert H (ed) Monogragh Series on Mineral Deposits, Nr 22. Gebrüder Bomtraeger, Berlin, Stuttgart, pp 17–27
Williams NC (1978) Studies of the base metal sulfide deposits at McArthur River, Northern

References for Table 1A,B

Akande SO, Zentilli M (1984) Geologic, fluid inclusion, and stable isotope studies of the Gays River lead-zinc deposit, Nova Scotia, Canada. Econ Geol 79:1187–1211
Anger G (1991) Deutscher Auslandsbergbau – unternehmerische Aktivitäten und verbandliche Gemeinschaftsaufgaben. Jahrbuch 1991. Bergbau, Öl und Gas, Elektrizität, Chemie. Glückauf, Essen, pp 1–36
Briskey JA, Dingess PA, Smith F, Gilbert RC, Armstrong AK, Cole GP (1986) Localisation and source of Mississippi Valley-type zinc deposits in Tennessee, USA, and comparisons with Lower Carboniferous rock of Ireland. In: Andrew CJ, Crowe RWA, Finlay S, Pennell WM, Pyne JF (eds) Geology and genesis of minerals deposits in Ireland. Irish Association of Economic Geology, Dublin, pp 635–661
Gavin J (1975) Cyprus Anvil hammers out big expansion in the Yukon. Can Min J (Aug):25–29
Hermann J (1989) Entwicklung und Erprobung eines EDV-Modells für die kurz- und mittelfristige Produktionsplanung und -überwachung von Metallerzgruben mit komplexer Vererzung und komplizierter geologischer Struktur, ausgeführt am Beispiel des Betriebes Song Toh Thailand. PhD Thesis, Technical University Berlin, D 83, 135 pp
Jefferson CW, Kilby DB, Pigage LC, Roberts WJ (1983) The Cirque barite-zinc-lead deposits, northeastern British Columbia. In: Sangster DF (ed) Sediment-hosted stratiform lead-zinc deposits. Short Course Handbook 8 Min Ass Can Toronto, pp 121–140
Jorgensen GC, Dendle PK, Rowley M, Lee RJ (1990) Sorby lead-zinc-silver deposit. In: Hughes FE (ed) Geology of the mineral deposits of Australia and Papua New Guinea. Aust Inst Min Metall, Parkville, Australia, pp 1097–1101
Klau W, Large DE (1980) Submarine exhalative Cu-Pb-Zn deposits – a discussion of their classification and metallogenesis. Geol Jahrb D 40:13–58
Klau W, Mostler H (1986) On the formation of Alpine Middle and Upper Triassic Pb-Zn deposits, with some remarks on Irish carbonate-hosted base metal deposits. In: Andrew CJ, Crowe RWA, Finlay S, Pennell WM, Pyne JF (eds) Geology and genesis of mineral deposits in Ireland. Irish Association of Economic Geology, Dublin, pp 663–675
Murphy GC (1990) Lennard shelf lead-zinc deposits. In: Hughes FE (ed) Geology of the mineral deposits of Australia and Papua New Guinea. Aust Inst Min Metall, Parkville, Australia, pp 1103–1109
Nelson J, Macintyre D (1988) Metallogeny of northeastern British Columbia. Geosci Can 15(2):113–116
Nokleber WJ, Bundtzen TK, Berg HC, Brew DA, Grybeck D, Robinson MS, Smith TE, Yeend W (1987) Significant metalliferous lode deposits and placer district in Alaska. US Geol Surv Bull 1786, 104 pp
Randell RN (1989) The geology of the Polaris carbonate-hosted zinc-lead mine, Canadian Arctic Archipelago. Paper Geol Soc America, Annu Meet Nov 89, St Louis

Rosendaal A (1986) The Gamsberg zinc deposit. In: Anhaeuser CR, Maske S (eds) Mineral deposits of southern Africa, vol 2. Geol Soc South Africa, Johannesburg, pp 1477–1488
Sangster DF (1986) Classification, distribution and grade-tonnage summaries of Canadian lead-zinc deposits. Geol Surv Can Econ Geol Rep 37, 68 pp
Sureda RJ, Martin JL (1990) El Aguilar mine: an Ordovician sediment-hosted stratiform lead-zinc deposit in the central Andes. In: Fontboté L, Amstutz GC, Cardozo M, Cedillo E, Frutos J (eds) Stratabound ore deposits of the Andes. Springer, Berlin Heidelberg New York, pp 161–174
Thomassen B (1991) The Black Angel lead-zinc mine 1973–90. Gronlands Geol Unders Rapp 152 (current research including report of activities 1990): 46–50

References for Section 5

Canadian Mining Journal (1990, 1991) Mining Sourcebook. Can Min J, Don Mills, Ontario
Darling P (1989) Polymetallics in the heartland of the Union. Int Min 10:13–19
E & MJ (1989) 1989/1990 E & MJ International directory of mining Eng Min J, Chicago, 599 pp
Gardiner CD (1989/1990) American mines handbook 1989, 1990. Northern Miner Press, Toronto, Ontario, Canada
Kennedy A (1990) Red Dog zinc-lead mine – Alaskan success for Cominco. Min Mag (Dec): 418–425
Kennedy A (1991) Recent developments at Tara. Min Mag 6:352–358
Kilgore CC, Arbelbide SJ, Soja AA (1983) Lead and zinc availability – domestic. USBM Inf Circ 8962, 30 pp
Mining Magazine (1983) The Elura mine, New South Wales. Dec:436–443
Mining Magazine (1984) St Joe's Balmat mines. March:230–235
Peterson GR, Porter KE, Soja AA (1985) Primary lead and zinc availability – market economy countries. USBM Inf Circ 902, 44 pp
Scales M (1989) Sullivan – the grand old dame. Can Min J 6:64–76
Suttill KR (1990) Hilton inaugurated – a new step in Mt. Isa's future. Eng Min J 10:32–35
Weiss NL (1985) SME mineral processing handbook, vol 2. American Institute of Mining, Metallurgical, and Petroleum Engineers, New York, pp 15–1 to 15–49

Appendix I: The Variogram

The variogram, or more correctly semivariogram, represents variances between sample pairs as a function of distance (lag) between samples. The semivariogram function is:

$$\gamma(h) = \frac{1}{2n} \cdot \sum_{i=1}^{n} (\chi_i - \chi_{(i+h)})^2,$$

where χ_i and $\chi_{(i+h)}$ are the values at locations i and i + h, and n is the total number of sample pairs separated by the distance h. As an example, in the first step, the differences between samples which are 10 m apart are calculated, in a second step the differences of samples which are 20 m apart, etc.

In the variogram γ is plotted as a function of lag h (Fig. 11). Most ore deposits have variograms of the so-called transitive type: as long as the γ-value increases (within the range a), a regional dependence exists. Beyond the range a, the sample values are statistically independent. The variogram can be quantitatively described by the parameter range a, nugget effect C_0 and sill value C_1. (Often, the notation in geostatistical literature is only C instead of C_1). The sill value C_1 equals the variance s^2,

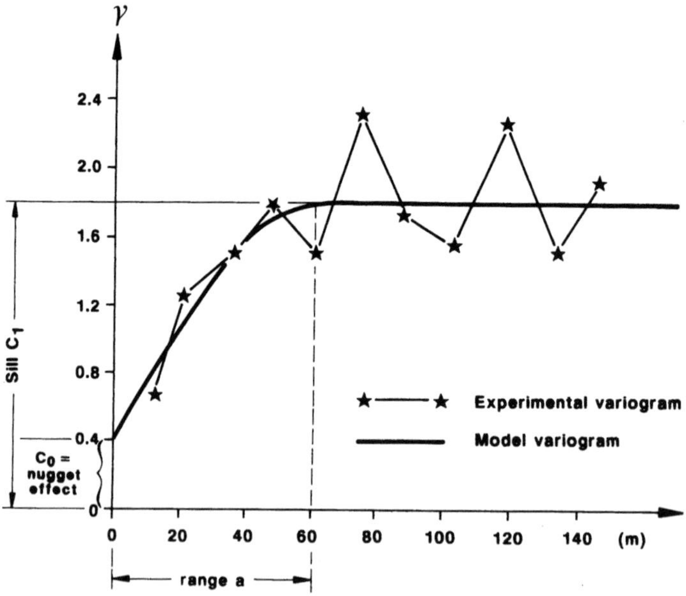

Fig. 11. Model of a variogram (see Appendix I)

$$s^2 = \sum_{i=1}^{n-1} (\chi_1 - \bar{\chi})^2.$$

These statistical values are dependent on the support of the data, meaning the mass of the samples. For example, samples with low mass, e.g. drill cores, have larger variances and therefore sills than bulk samples with a much larger mass. The supports of the geostatistical data in Fig. 2 are mainly drill holes, meaning they are roughly comparable.

With a variogram the anisotropy of deposits can be quantitatively described. The range of a variogram in anisotropic deposits calculated along the direction of the rake of a deposit is different from the range of a variogram in the direction perpendicular to the rake. In Fig. 2 horizontal bars display the range of anisotropies.

Appendix II: Calculation of the Net Smelter Returns (NSR)

The smelting formulas for zinc and lead concentrates are given in Wellmer (1989, Appendix, Table 11). The following example uses price and smelting charge data for 1985.

1. The following assumptions are made:

   a) Recovery in the mill 90% for Pb, Zn and Ag;
   b) Zn concentrates have a grade of 50% Zn;

c) Pb concentrates have a grade of 65% Pb;
d) The transport costs of the concentrates are neglected in this calculation.

2. The following rules for calculating the NSR apply:

a) For Zn concentrates the smelting loss is 8 units (percentage points), i.e. only 50 − 8 = 42% Zn is paid for;
b) For 65% Pb concentrates the smelting loss is 3 units, i.e. 62% Pb is paid for;
c) From the Ag content in the concentrate, 50 g is deducted;
d) The conversion factor from lb to % is 22.046, i.e. 22.046 lb/t is 1% in a metric tonne.

3. As an example for the calculation of the NSR/t of ore the grade of 4% Zn, 6% Pb and a Pb:Ag rate of 1:10 and the price data of 1985 are taken:

The metal prices are: 38 US-cts/lb Zn
18 US-cts/lb Pb
616 US-cts/troy oz Ag.

Treatment charge (T/C) for Zn concentrates: US-$ 157/t
T/C for Pb concentrates: US-$ 104/t

a) Calculation of the NSR for the Zn component.

i) At 90% recovery in the mill, 0.9 × 4 = 3.6% of the 4% Zn in the ore is recovered. To produce a 50% Zn concentrate:

$$\frac{50}{3.6} = 13.89\,\text{t}$$ of ore is needed to produce 1 t of concentrate (13.89 is the concentration factor KF)

ii) 42% of the 50% Zn in the concentrate is paid for (see 2a). The gross return of the concentrate, therefore, is:

$$42 \times 22.046 \times 0.38 = 351.85\,\$/\text{t}$$

From this amount the T/C has to be ducted −157.00 $/t

net revenue =194.85 $/t

iii) Using the concentration factor KF of 13.89, the net revenue per tonne of ore for the Zn component is 14.03 $/t.

b) Calculation of the NSR for the Pb component.

i) At 90% recovery in the mill, 5.4% of the 6% Pb in the ore is recovered to produce a 65% Pb concentrate:

$$\frac{65}{5.4} = 12.04\,\text{t}$$ of ore is needed to produce 1 t of concentrate (12.04 is the concentration factor KF)

ii) The Pb:Ag ratio was 1:10, meaning 60 g/t Ag in the ore. At 90% recovery 54 g/t is recovered.

All Ag is contained in the Pb concentrate. The KF factor for Pb, therefore, has to be used to calculate the Ag grade in the Pb concentrate.

The Ag grade in the Pb concentrate is $54 \times 12.04 = 650.16$ g/t Ag. 50 g has to be deducted from this, meaning that 600.16 g/t is paid for.

At 616 cts/oz (conversion factor from g to oz is 31.1035) the Ag content has a value of 118.86 $/t.

iii) 62% of the 65% Pb in the concentrate is paid for (see 2b). The gross return of the concentrate is therefore:

$62 \times 22.046 \times 0.18$	=	246.03 $/t
The Ag credit is added	=	+118.86 $/t
The treatment charge has to be deducted	=	−104.00 $/t
net revenue	=	260.89 $/t

iv) Using the concentration factor KF of 12.04 the net revenue per tonne of the Pb component is 21.67% $/t.

c) The sum of the Zn component and the Pb component is the net smelter return of the ore (NSR/t):

Zn component	14.03 $/t
Pb + Ag component	21.67 $/t
Sum = NSR	35.70 $/t

Appendix III: The Evaluation of Price Risk

The price risk was evaluated and compared for mines delivering run-of-mine ore of 10% combined Pb + Zn to the mill for different Pb/(Zn + Pb) and different Pb:Ag ratios. To evaluate the price risk, two methods were applied (Fig. 12):

1. The line of regression for the average yearly net smelter return (NSR) values was calculated. Because price cycles have a time span of about 4 to 5 years and by taking the 21-year period since 1969 as a base case, the four most *negative* residuals (i.e. the four most negative differences between the real values and the equivalent values of the line of regression) were added and the sums compared. The larger the sum of the four most negative price residuals, the higher the price risk.
2. Since price changes to lower prices are a major risk factor, the four largest negative price changes (i.e. changes from one year to the next year from a higher price to a lower price, but *not* vice versa) were added and the sums compared. The larger the sum, the higher the price risk.

Fig. 13. Sum of largest price changes from higher to lower prices within the periods 1969–1990, 1973–1990, 1977–1990

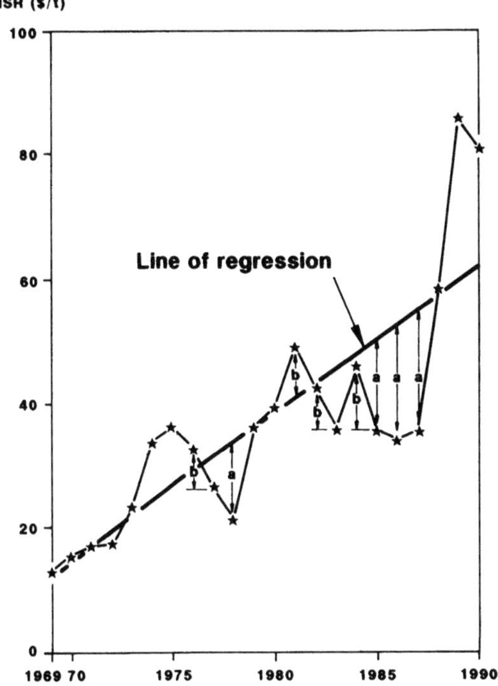

Fig. 12. Model for calculation of sum of the four most negative residuals (*a*) and the four largest negative price changes (*b*)

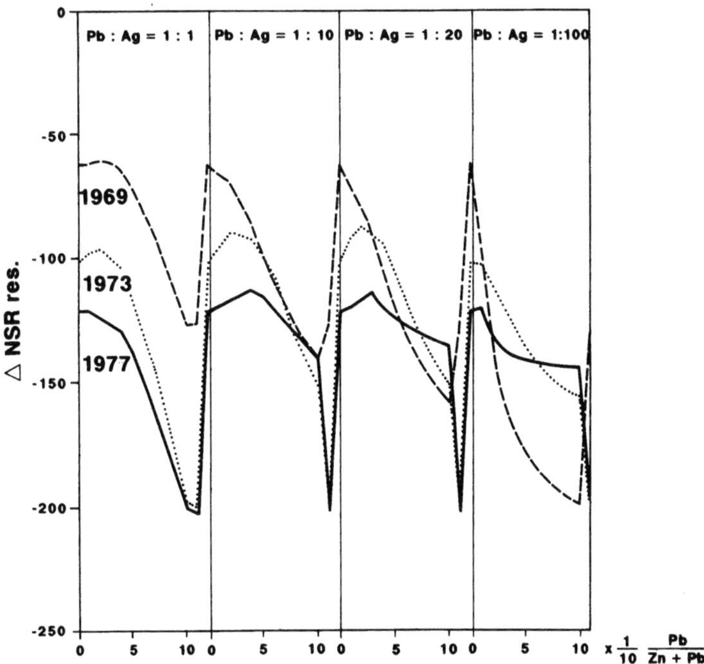

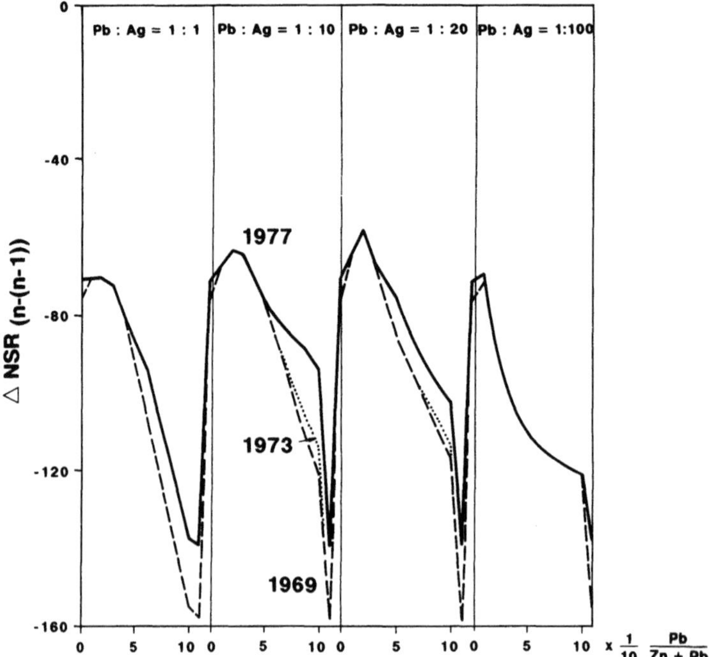

Fig. 14. Sum of the largest negative residuals against line of regression within the periods 1969–1990, 1973–1990, 1977–1990

To eliminate the effect of the time period, shorter periods were also considered (1973 to 1990 and 1977 to 1990). The results of the analyses are shown in Figs. 13 and 14. By using the criteria of residuals and price changes, in both cases regardless of the time periods considered, mines mining ore with high Pb/(Zn + Pb) ratios, meaning Pb-rich ores, have the highest price risk. The lowest price risk is associated with Zn-rich ore [i.e. low Pb/(Zn + Pb) or high Zn/(Zn + Pb) ratios].

Appendix IV: Calculation of the Economic Parameters Net Present Value (NPV) and Internal Rate of Return (IRR)

By calculating the economic parameters net present value (NPV) and internal rate of return (IRR), the annual cash flow will be compared with the investment, i.e. the capital cost. The simplified cash flows is the difference between revenues and operating costs. Therefore, first the revenues have to be calculated.

1. Whereas in Appendix II the net smelter return with yearly average prices and treatment charges was calculated, we now have to consider the long-term outlook of a mine. Reasonable price plateaus have to be assumed. Since capital and operating costs in Fig. 7a,b are adjusted to 1989 US-$, the price assumptions are also in 1989 US-$. The price assumptions are:

Zn 60 US-cts/lb Pb 35 US-cts/lb Ag 5 US-$/oz	all deposit types evaluated
Cu 1 US-$/lb Au 350 US-$/oz	only for volcanogenic massive sulphide deposits

Because the concentrate market changes over the years from a buyer's market to a seller's market and vice versa, we are not working with treatment charges, but with average percentage factors, the so-called net smelter return factor (NF), which is the long-term average percentage of the metal price that the mine receives as net smelter return (Wellmer 1989, Table VII). These factors are for:

Zn	50%
Pb	65%
Ag	95%
Cu	70%
Au	95%

Examples of calculations using the NF factors are given in Appendix V.

2. The operating costs are taken from the graph in Fig. 7a.

3. The capital costs, i.e. the investment (I), are taken from the graph in Fig. 7b.

4. The cash flow (CF) is the difference between the net smelter return (NSR) as calculated above under (1) and the operating costs.

5. We are evaluating a simple economic model:

   a) Each parameter NPV and IRR is determined without taxes, because tax rates differ in different countries and it would go beyond the scope of this chapter to consider different tax regimes.
   b) The investment I is equity financed by the mine owner without bank loans. This means that no interest has to be paid, which would influence our cash flow (CF).
   c) We assume constant grades over the lifetime of the mine. With constant prices we therefore have an economic case with constant annual cash flows (CF).

6. For constant annual cash flows (CF) and the investment (I), the net present value (NPV) is defined as $NPV = CF \cdot b_n - I$, where b_n is the annuity present value factor, which depends on the lifetime of the mine (n years) and the interest rate assumed. Our assumption is 10%.

7. The internal rate of return (IRR) for constant cash flows is defined:

$$I = CF \cdot b_n,$$

i.e. in this case the interest rate for the annuity present value factor is determined for the case: NPV = 0.

Appendix V: Calculation of Zn-Equivalent Grades

To calculate break-even grades (i.e. those grades which are necessary so that the mine can cover all its costs from its revenues, but do not generate a profit), the grades of multi-element deposits have to be converted into a common grade. In our case we take Zn as the main element and we therefore have to convert the Pb and Ag grades into Zn-equivalent grades. Again, we have to assume prices and recoveries in the mill. The prices are those in Appendix IV, the recoveries in the mill are listed below. To calculate the net smelter return, we use the NF factor (see Appendix IV).

	Price	Recovery	NF factor
Zn	0.60 US-$/lb	90%	0.5 (50%)
Pb	0.35 US-$/lb	90%	0.65 (65%)
Ag	5.00 US-$/oz	85%	0.95 (95%)

We now calculate the value of 1% Zn, 1% Pb and 1 oz Ag/t for the mine, i.e. the net smelter return (NSR). The conversion factor from pounds (lbs) into percentage metal is 22.046 and from ounces (oz) into g is 31.103. 1% Zn therefore has the NSR:

$$1\%\ \text{Zn} = 22.046 \times 0.60 \times 0.5 \times 0.9 = 5.95\,\$/\text{t}$$

$$1\%\ \text{Pb} = 22.046 \times 0.35 \times 0.65 \times 0.9 = 4.51\ \$/\text{t}$$

$$1\text{oz/t Ag} = 5 \times 0.95 \times 0.85 = 4.04\,\$/\text{t}$$

$$\text{or } 1\,\text{g/t Ag} = \frac{5 \times 0.95 \times 0.85}{31.103} = 0.130\,\$/\text{t}$$

We can see now that 1% Pb cannot just be added to 1% Zn, because it generates less revenue for the mine. So 1% Pb expressed as a Zn value is 4.51/5.95 = 0.76 and Ag expressed as a Zn value is 0.130/5.95 = 0.022.

The Zn-equivalent equation is therefore:

$$1\%\ \text{Zn equivalent} = 1\%\ \text{Zn} + 0.76 \times \%\text{Pb} + 0.022 \times \text{g/t Ag}$$

Subject Index

MIX
Papier aus verantwortungsvollen Quellen
Paper from responsible sources
FSC® C105338

If you have any concerns about our products,
you can contact us on
ProductSafety@springernature.com

In case Publisher is established outside the EU,
the EU authorized representative is:
Springer Nature Customer Service Center GmbH
Europaplatz 3, 69115 Heidelberg, Germany

Printed by Libri Plureos GmbH
in Hamburg, Germany